*INTERNATIONAL SERIES OF MONOGRAPHS ON PURE AND APPLIED BIOLOGY*

Division: **MODERN TRENDS IN PHYSIOLOGICAL SCIENCES**

GENERAL EDITORS: P. ALEXANDER and Z. M. BACQ

VOLUME 13

# ORGANOPHOSPHORUS POISONS

*OTHER TITLES IN THE SERIES*
*ON PURE AND APPLIED BIOLOGY*

MODERN TRENDS IN PHYSIOLOGICAL SCIENCES DIVISION

Vol. 1. FLORKIN—Unity and Diversity in Biochemistry
Vol. 2. BRACHET—The Biochemistry of Development
Vol. 3. GEREBTZOFF—Cholinesterases
Vol. 4. BROUHA—Physiology in Industry
Vol. 5. BACQ and ALEXANDER—Fundamentals of Radiobiology
Vol. 6. FLORKIN (Ed.)—Aspects of the Origin of Life
Vol. 7. HOLLAENDER (Ed.)—Radiation Protection and Recovery
Vol. 8. KAYSER—The Physiology of Natural Hibernation
Vol. 9. FRANÇON—Progress in Microscopy
Vol. 10. CHARLIER—Coronary Vasodilators
Vol. 11. GROSS—Oncogenic Viruses
Vol. 12. MERCER—Keratin and Keratinization

BOTANY DIVISION

Vol. 1. BOR—Grasses of Burma, Ceylon, India and Pakistan
Vol. 2. TURRILL (Ed.)—Vistas in Botany
Vol. 3. SCHULTES—Native Orchids of Trinidad and Tobago
Vol. 4. COOKE—Cork and the Cork Tree

BIOCHEMISTRY DIVISION

Vol. 1. PITT-RIVERS and TATA—The Thyroid Hormones
Vol. 2. BUSH—The Chromatography of Steroids

ZOOLOGY DIVISION

Vol. 1. RAVEN—An Outline to Developmental Physiology
Vol. 2. RAVEN—Morphogenesis: The Analysis of Molluscan Development
Vol. 3. SAVORY—Instinctive Living
Vol. 4. KERKUT—Implications of Evolution
Vol. 5. TARTAR—Biology of Stentor
Vol. 6. JENKIN—Animal Hormones
Vol. 7. CORLISS—The Ciliated Protozoa

# ORGANOPHOSPHORUS POISONS

## ANTICHOLINESTERASES and RELATED COMPOUNDS

D. F. HEATH

Toxicology Research Unit

Medical Research Council Laboratories

Carshalton, Surrey

PERGAMON PRESS

OXFORD · LONDON · NEW YORK · PARIS

1961

PERGAMON PRESS LTD.
*Headington Hill Hall, Oxford*
*4 & 5 Fitzroy Square, London W.1*

PERGAMON PRESS INC.
*122 East 55th Street, New York 22, N.Y.*
*1404 New York Avenue N.W.,*
*Washington 5 D.C.*

PERGAMON PRESS S.A.R.L.
*24 Rue des Écoles, Paris V^e^*

PERGAMON PRESS G.m.b.H.
*Kaiserstrasse 75, Frankfurt am Main*



Library of Congress Card Number 61-10655

*Set in Baskerville 10 on 12 point and printed in Great Britain by*
THE PITMAN PRESS, BATH

# Foreword

THIRTY years ago not one organophosphorus compound with anticholinesterase action had been characterized with certainty. Now several thousands have been prepared. Both their practical importance and their theoretical interest have led to this rapid advance. Some of these compounds have been studied as potential war gases, and others have found widespread use as insecticides. Fundamental work has been carried out both for its own sake and in attempts to discover compounds of greater or more selective toxicity.

A marked feature of fundamental studies has been the way in which they have cut across traditional disciplines. The pharmacologist, biochemist and physical chemist has each had need of the concepts developed by the others to further his own researches. In this book I have tried to meet this need, by presenting in a fairly simple way the main discoveries in each field. As well as trying to meet the needs of the expert in one subject who wishes to find out something of contingent ones, I have also tried to meet those of the beginner. There is, therefore, a good deal of explanatory matter that the expert will find redundant, and a fairly large number of references. On the matter of references, however, some compromise had to be reached. Neither space nor my capacity allowed me to write a series of authoritative reviews, and referencing has been selective. Controversial matters are heavily referenced, but on the more straightforward topics I have preferred to refer to the more recent publications, on the grounds that they themselves contain references to earlier work. In consequence the originators of important concepts are sometimes mentioned far less frequently than they deserve. This is the case, for example, with Dr. G. Schrader, who first synthesized many compounds of great interest, and whose monograph: *Die Entwicklung neuer Insektizide auf Grundlage organischer Fluor- und Phosphor-Verbindungen* contributed greatly to the rapid spread of knowledge about them. Many researchers, myself included, have to thank his kindness and skill, and those of his colleagues, for providing them with much-wanted specimens of pure compounds.

This book is on the fundamental aspects of the subject, and the results of applied research are only described when they are of fundamental importance. Applied research is space-consuming to describe, as it does not lend itself to systematization. Its omission has enabled me to deal with fundamental points in much more detail.

Even with the omission of technical work, the variety of subjects covered is still considerable, and I am very aware of the gaps in my knowledge of many of them. If these gaps are not obvious it is largely due to the helpful and ruthless criticism of Dr. W. N. Aldridge, Dr. J. M. Barnes, Dr. G. S. Hartley, Dr. H. S. Hopf, Mr. M. K. Johnson and Dr. H. B. Stoner, all of whom have read some part of this book, and have made many detailed comments. I am glad to express my thanks to them here.

Thanks to critics sometimes savours of an 'escape clause'. Many topics are still subjects of controversy, and to give a coherent account of them without making some judgements seemed to me unfair to the readers for whom this book is intended. I accept full responsibility for the views expressed.

My thanks are also due to the editors of Journals and authors who have given me permission to reproduce figures; and to the clerical staff of the Toxicology Research Unit, Mrs. A. Horne, and my assistant Mr. J. Rose, for their help in preparing the typescript.

Throughout the book brief names have been used for many compounds. These are often trade names. They are used purely for convenience, and their use does not imply recommendation of particular commercial products. The beginner is also warned that some of the compounds mentioned are exceedingly toxic.

D. F. HEATH

# Contents

# Introductory Chapter

THE COMPOUNDS which this book is about have not been much studied below the specialist level, so that a worker in one field may be unfamiliar with the principles on which work in another field is based. In this Chapter a simplified outline sketch is given of the main topics, by considering a few of the simpler compounds.

## COMPOUNDS

Three typical compounds, used as examples in the following pages, are Paraoxon, Parathion and DFP. Their structural formulae are:

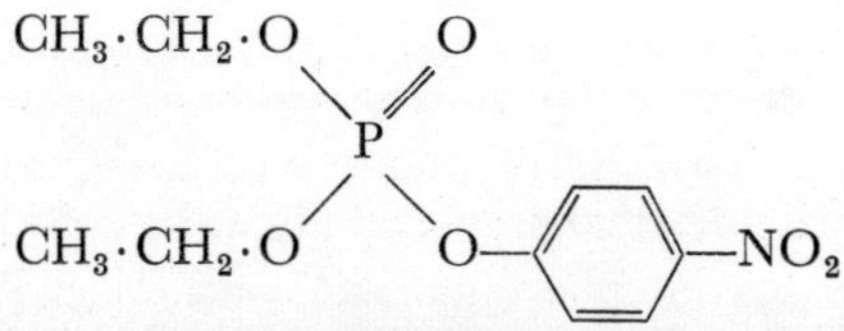


Paraoxon

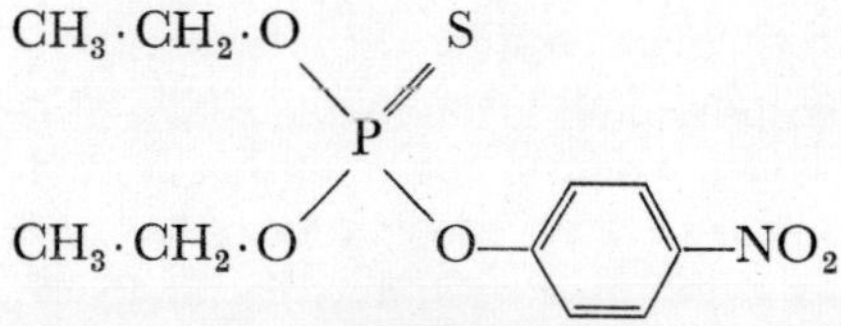


Parathion

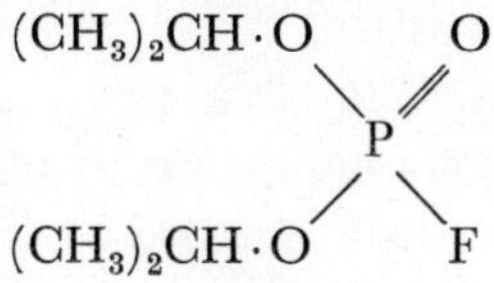


DFP

These and most other toxic phosphorus compounds fit the general formula:

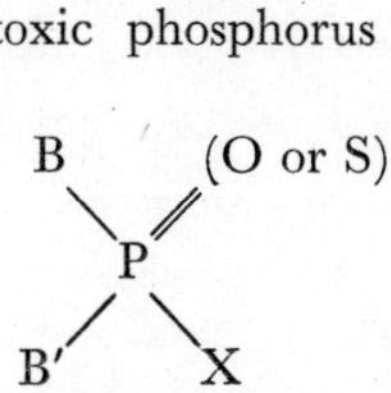

B and B′ are basic groups, such as alkoxy or alkylamino, and X is an acidic group. This use of the terms 'basic' and 'acidic' is explained later.

*Nomenclature*

Most of the compounds can be regarded as esters of phosphoric acid or its derivatives, and are described generically as phosphate esters. Names are built up in the same way as the names of organic esters such as ethyl acetate. Thus ethyl acetate is the ester formed by condensing ethanol with acetic acid:

$$\mathrm{EtOH + HO{\cdot}CO{\cdot}CH_3 \rightarrow EtO{\cdot}CO{\cdot}CH_3 + H_2O},$$

and diethyl 4-nitrophenyl phosphate (Paraoxon) is the ester formed (schematically only) by condensing ethanol and 4-nitrophenol with orthophosphoric acid according to the equation:

$$\mathrm{2EtOH + HO{\cdot}C_6H_4\text{-}4\text{-}NO_2 + (HO)_3PO \rightarrow (EtO)_2PO{\cdot}O{\cdot}C_6H_4\text{-}4\text{-}NO_2 + 3H_2O}.$$

Acids closely related to phosphoric acid have their own names, and these are used in naming all esters but those of orthophosphoric acid itself. Thus Parathion is regarded as an ester of phosphorothionic acid, $(HO)_3PS$, and is called diethyl 4-nitrophenyl phosphorothionate; and DFP is regarded as the di-isopropyl ester of phosphorofluoridic acid, $(HO)_2PO{\cdot}F$, and is called di-isopropyl phosphorofluoridate. Most names are formed in this way. In practice abbreviations are often used. Esters containing P=S bonds are called thionates, and those containing P—F bonds are called fluoridates, for example. The name 'phosphate' is reserved solely for compounds in which the only atoms directly bonded to the phosphorus atom are oxygen atoms.

Some abbreviations are always used in formulae: Me for methyl, Et for ethyl, $Pr^i$ for isopropyl, PO for P=O, and PS for P=S.

## REACTIONS

The compounds are all structurally related, and most undergo the same reactions. Of particular importance are those with hydroxyl ions, with water, and with esterases such as acetylcholinesterase (AChE). These reactions are all similar in form. Those with DFP can be written:

$$\mathrm{(Pr^iO)_2PO{\cdot}F + OH^{\ominus} \rightarrow (Pr^iO)_2PO{\cdot}OH + F^{\ominus}},$$
$$\mathrm{(Pr^iO)_2PO{\cdot}F + H{\cdot}OH \rightarrow (Pr^iO)_2PO{\cdot}OH + HF},$$
$$\mathrm{(Pr^iO)_2PO{\cdot}F + H{\cdot}E \rightarrow (Pr^iO)_2PO{\cdot}E + HF}.$$

(In the first reaction the solution must, of course, contain positive ions to keep the total electric charge zero, but they play no part in the reaction, which is the same whether sodium, potassium or calcium hydroxide is used. The positive ions are therefore omitted from such equations.) As explained later, these reactions do not only resemble each other formally. They can all be classified as nucleophilic displacement reactions, and the reagents are all of the same type. In these and some other reactions an enzyme can often be regarded solely as a chemical reagent, and not as a biochemical catalyst. The concept of an enzyme as a base proves particularly useful.

## RATES OF REACTION

Some of the most important differences between the compounds are the differences in the rates at which they undergo reactions, especially those of the nucleophilic displacement type. For example, one compound may be highly toxic and another of very low toxicity entirely because the first reacts rapidly with AChE and the second reacts only slowly. Again, it may be possible to store an aqueous solution of a compound hydrolysed only slowly by water, while it may be essential to use freshly prepared solutions of another which is hydrolysed rapidly. In many instances toxicity, biochemical activity, and storage and other properties can be related in a simple way to reaction rates. Reaction rates are measured in terms of rate constants, whose derivation is summarized in the Appendix.

Reaction rates are related to chemical structure. This relationship can often be explained by the electronic theory of reaction rates, which also has the advantages of providing a unified treatment of different reactions and a convenient terminology in which to discuss them. The theory is developed in Chapter I, but a brief summary of the main principles is given here in order to introduce the terminology. The terms used frequently in the book arc italicized.

Atoms of different elements have different affinities for electrons, i.e. they are of different *electronegativity*. When two such atoms are bonded the more electronegative tends to withdraw electrons from the less electronegative, so that the bond is *polarized*. Oxygen, for example, is more electronegative than phosphorus, so that a phosphorus-oxygen bond is polarized in the sense: $\overset{\delta\oplus}{\text{P}}\text{—}\overset{\delta\ominus}{\text{O}}$. ($\delta\ominus$ stands for 'delta minus', and represents a small negative charge of less than one electron.) As opposite charges attract each other the phosphorus atom in a molecule containing this type of bond tends to seek negatively charged reagents. For this reason the phosphorus is described as *electrophilic*. Conversely a negatively charged reagent is attracted to positively charged atoms or groups, and is said to be *nucleophilic*. Consider now the reaction of DFP with a hydroxyl ion:

$$(\text{Pr}^{\text{i}}\text{O})_2\text{PO}\cdot\text{F} + \text{OH}^{\ominus} \rightarrow (\text{Pr}^{\text{i}}\text{O})_2\text{PO}\cdot\text{OH} + \text{F}^{\ominus},$$

In DFP the phosphorus atom is surrounded by atoms more electronegative than itself, and is consequently highly electrophilic. It therefore attracts the nucleophilic hydroxyl ion. The reaction consists in the displacement of the most electronegative substituent (in this case the fluorine atom) by the nucleophilic hydroxyl ion, and is therefore classified as a *nucleophilic displacement reaction.*

The group displaced in the nucleophilic displacement reaction is always the group which is most stable as an anion. It is obvious that this should be so when the reaction is visualized in stages:

$$HO^{\ominus} + \overset{\delta\oplus}{P}—\overset{\delta\ominus}{F} \rightarrow \overset{\delta\ominus}{HO}\cdots\cdots\overset{\delta\oplus}{P}\cdots\cdots\overset{\delta\ominus}{F} \rightarrow HO—P + F^{\ominus}.$$

The group displaced accepts electrons, and finally separates as an anion, though it may, of course, only remain one for a very short time, as it may react to form a neutral molecule with, for example, a hydrogen ion.

The tendency of a group to separate as an anion can be measured in most instances by the stability of the corresponding hydrogen acid in water. Thus in water an acid dissociates according to the equation:

$$HA \rightleftharpoons \overset{\oplus}{H} + \overset{\ominus}{A}.$$

The more stable the anion, $A^{\ominus}$, the more the acid, HA, dissociates, i.e. the stronger the acid. It follows that in a nucleophilic displacement reaction the group most likely to be displaced is the one which forms the strongest acid with hydrogen.

We can now discuss the generalized structure, (B)(B′)P(O or S)·X, already given as representing the compounds under consideration. The group which forms the strongest acid is represented by X, and is called the *acidic group.* The acids corresponding to the groups B and B′, i.e. HB and HB′, are very much weaker acids, and the anions formed when they dissociate ($B^{\ominus}$ and $B'^{\ominus}$) are consequently very much stronger bases than $X^{\ominus}$ (see Appendix for proof). It is therefore convenient to regard B and B′ as *basic groups.* In DFP, for example, the acidic group is the fluorine atom, and the basic groups are the isopropoxy groups because hydrofluoric acid, HF, is a stronger acid than isopropanol, $Pr^iOH$. In practice the acid HX is nearly always a fairly strong acid in the usual sense—a solution of it in water gives an acid reaction with indicators, and it forms stable salts with bases. In some important compounds one of the 'B' groups is strictly neutral. Thus in Sarin, ($Pr^iO$)(Me)PO·F, one 'B' group is the methyl group. Such groups are very firmly bonded to the phosphorus atom, and are not displaced.

The P—X bond is of particular importance as it is the bond broken in nucleophilic displacement reactions. It can be regarded as formed by the combination of two acids with the loss of water:

$$>PO\cdot OH + HX \rightarrow >PO\cdot X + H_2O,$$

and is therefore called the *acid anhydride bond.*

Nucleophilic reagents are of two types: negative ions, such as hydroxyl ions, sometimes called *anionic reagents*; and uncharged molecules, one part of which carries a partial negative charge, just as the phosphorus atom carries a partial positive one. Examples of the latter are water: $\overset{\delta\oplus}{H}—\overset{\delta\ominus}{O}—\overset{\delta\oplus}{H}$, and many esterases: $\overset{\delta\ominus}{E}—\overset{\delta\oplus}{H}$. There is much evidence that the reactions are in many ways the same with the two types, and are influenced by the same factors, so that an esterase can be regarded as a particular nucleophilic reagent. In accordance with this, the reaction between a phosphorus compound and an enzyme takes place at a definite rate, as shown in Chapter VI. The reaction continues until all of one of the reagents is used up. Usually the enzyme is in the lower concentration, so this is used up first. As the *phosphorylated enzyme*, (B)(B′)PO·E, cannot hydrolyse substrates, the enzyme is *progressively* and *irreversibly inhibited.*

It can be shown theoretically how the rates of nucleophilic displacement reactions are likely to vary with the acidities of the various groups. The argument is given in Chapter I. The most important results are:

(1) The rate is faster the more acidic X is. Thus DFP reacts faster with hydroxyl ions and enzymes than $(Pr^iO)_2PO \cdot O \cdot C_6H_4$-4-$NO_2$, the di-isopropyl homologue of Paraoxon, because hydrofluoric acid, HF, is a stronger acid than 4-nitrophenol, $HO \cdot C_6H_4$-4-$NO_2$.

(2) When the basic groups are amino groups the compound is more stable than when they are alkoxy groups. Thus $(Me_2N)_2PO \cdot F$, Dimefox, is much less reactive than, e.g. $(Pr^iO)_2PO \cdot F$, DFP.

(3) In molecules in which only the alkyl groups in the basic radicals are changed, the rates increase in the order:

$$(Pr^iO)_2PO \cdot X < (PrO)_2PO \cdot X < (EtO)_2PO \cdot X < (MeO)_2PO \cdot X,$$

whatever X is, providing it is the same throughout the series.

The application of electronic theory to hydrolysis rates is given in Chapter IV, and to enzyme reactions in Chapter VI, and at several points in later Chapters.

## THE CHEMICAL BASIS OF TOXIC ACTION

Phosphorus compounds are toxic primarily because they react with acetylcholinesterase (AChE) in the way already described. This theory is discussed at length in Chapters VIII and XII–XIV and is returned to later in this Chapter.

All biochemical activity is destroyed by hydrolysis of any one bond between the phosphorus atom and either the acidic or a basic group, i.e. none of the products:

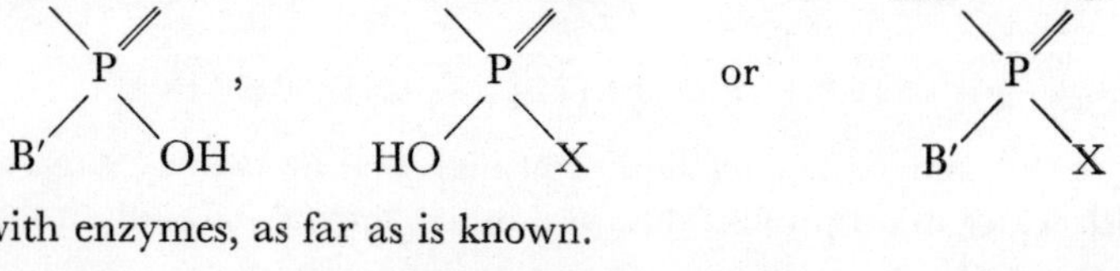


reacts with enzymes, as far as is known.

## ACTIVATION

Some compounds react only slowly with nucleophilic reagents, including AChE. *In vivo*, however, they are metabolized to compounds which are readily attacked by such reagents, i.e. they are *activated.* Activation is most commonly brought about by oxidation. For example, Parathion is oxidized enzymically to Paraoxon:

$$\underset{\text{Parathion}}{(EtO)_2PS\cdot O\cdot C_6H_4\text{-}4\text{-}NO_2} \xrightarrow{O} \underset{\text{Paraoxon}}{(EtO)_2PO\cdot O\cdot C_6H_4\text{-}4\text{-}NO_2}.$$

Paraoxon is more rapidly hydrolysed than Parathion and reacts about 10,000 times faster with AChE. The toxicity of Parathion, and of nearly all thionates, can be explained by their oxidation *in vivo* (Chapter IX).

Oxidation can also be brought about by the same enzyme systems *in vitro* (Chapter VII), and often very rapidly by ultra-violet light in the air, or by other common oxidizing systems (Chapter III).

Another important activating reaction is isomerization of thionates to thiolates. Parathion for example undergoes the reaction:

$$\underset{\substack{\text{diethyl 4-nitrophenyl}\\ \text{phosphorothionate, Parathion}}}{(EtO)_2PS\cdot O\cdot C_6H_4\text{-}4\text{-}NO_2} \rightarrow \underset{\substack{\text{O,S-diethyl 4-nitrophenyl}\\ \text{phosphorothiolate}}}{(EtO)(EtS)PO\cdot O\cdot C_6H_4\text{-}4\text{-}NO_2}.$$

The product reacts as rapidly as Paraoxon with AChE. This type of reaction does not take place *in vivo*, but it does take place on storage (Chapter III) and has made valueless many publications on the biochemical and toxic properties of Parathion and other thionates.

## METHODS OF PREPARATION

From the preceding discussion it can be seen that the compounds of interest are either readily attacked by nucleophilic reagents, or are oxidized or isomerized. They are therefore often hard to prepare, which accounts for the bewildering variety of methods described in Chapter II. No attempt to summarize them is made here. Chapter II stands on its own in most ways, and is rarely referred to later.

## PHYSICAL PROPERTIES AND PURITY

Most of the compounds are liquids of high boiling point. Some are miscible with water in all proportions, and others are substantially insoluble in water, but nearly all can be extracted from water with chloroform, and many with other solvents.

The extent to which a compound can be extracted from one solvent by another, immiscible solvent, is of great importance. This extractibility is measured by a partition coefficient, defined as follows. When a dilute solution of the compound in one solvent, A, is shaken up with another, immiscible solvent, B, the concentrations of the compound in the two solvents are in a constant ratio to each other, typical of the solvent system and the compound, i.e. $[C]_A/[C]_B = K$, where $[C]_A$ and $[C]_B$ are the concentrations in A and B, and $K$ is the partition coefficient. Differences in $K$ are made use of when compounds are separated by counter-current extraction or partition chromatography, and when they are characterized by values of $K$. Many compounds can only be purified or characterized adequately in these ways. Their melting points are too low for recrystallization to be practicable, and their boiling points too high for fractional distillation to be carried out even *in vacuo* without some decomposition.

While most compounds have been purified enough for chemical and toxicological work by distillation *in vacuo*, by counter-current extraction, or by other well-known methods, purification for biochemical work sets special problems. Some compounds cannot be purified enough for this by any method yet devised, and for others special methods must be used. This is generally the case with compounds which isomerize or which are oxidized readily, as the products are often more than 1000 times as active biochemically, and may be formed during attempted purification, especially if heat is applied. The problems of purification are discussed in Chapter III. A great many papers read in preparing this book contained results which could not be used because the authors did not pay enough attention to the purity of their compounds, or, if they did, had not described their methods of purification in sufficient detail.

## REACTIONS WITH ESTERASES *IN VITRO*

Only the reactions of phosphate esters with esterases are dealt with. There is evidence that some of the compounds react with enzymes which have no esteratic properties, but these reactions throw no light on their toxic effects.

Esterases can be divided into three types, according to their reactions with phosphate esters. These types are designated A, B and C. A-esterases hydrolyse some phosphate esters, B-esterases are inhibited by phosphate esters, and C-esterases do not react with phosphate esters in any way. We are concerned only with A- and B-esterases.

The mechanism of the reactions with A- and B-esterases is best approached by a consideration of the hydrolysis of acetylcholine (ACh) by AChE, and of the nature of the active centre of AChE. The reaction takes place in three stages:

$$\underset{\text{acetylcholine}}{MeCO\cdot O\cdot CH_2\cdot CH_2\cdot \overset{\oplus}{N}Me_3} + HE \rightleftharpoons \underset{\text{ACh adsorbed on AChE}}{MeCO\cdot O\cdot CH_2\cdot CH_2\cdot \overset{\oplus}{N}Me_3/HE} \rightarrow$$

$$\underset{\text{choline}}{HO\cdot CH_2\cdot CH_2\cdot \overset{\oplus}{N}Me_3} + \underset{\text{acetylated AChE}}{MeCO\cdot E} \xrightarrow{+H_2O} \underset{\text{acetic acid}}{MeCO\cdot OH} + HE.$$

The first stage consists of an equilibrium between the enzyme and its substrate on the one hand and an adsorption complex of the two on the other. In the second stage the complex reacts to give choline and acetylated AChE; and in the third stage the latter is hydrolysed to give acetic acid and the original enzyme.

The equilibrium stage has been much studied. The results are given in Chapter V, and the kinetic equations for this and other enzyme reactions are derived in the Appendix. It is found that ACh is strongly adsorbed on to the active centre of AChE, and that this adsorption is due to the operation of attractive forces at two sites, illustrated below:

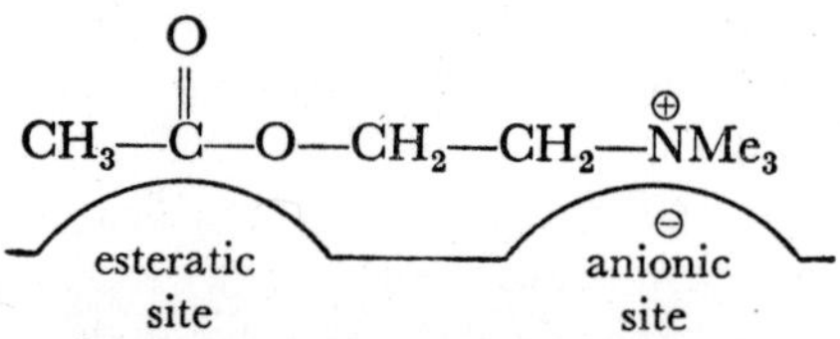


The ester bonding in ACh is attracted to the esteratic site, at which reaction takes place in the second stage, and the positively charged nitrogen atom is attracted to a negatively charged site a few Å away, called the anionic site. The anionic site is also stereospecific, and attracts the $—\overset{\oplus}{N}Me_3$ group by van der Waal's forces. In another enzyme, pseudocholinesterase (ΨChE, from the Greek 'Psi'), there is a similar stereospecific site, but it is not charged. Most esterases have an esteratic site only. These pictures of the enzyme surface are developed in Chapter V, and are used frequently in later Chapters.

The reactions of phosphate esters with A- and B-esterases apparently take place by a similar sequence of reactions. A complex is formed reversibly, which reacts to give a phosphorylated enzyme, and this in turn is hydrolysed to give a phosphorus acid and the original enzyme. The reactions with DFP can be written:

$$(Pr^iO)_2PO\cdot F + HE \rightleftharpoons (Pr^iO)_2PO\cdot F/HE \rightarrow HF + (Pr^iO)_2PO\cdot E$$
$$\xrightarrow{+H_2O} (Pr^iO)_2PO\cdot OH + HE.$$

These equations are exactly analogous to those which summarize the hydrolysis of ACh by AChE. A-esterases and B-esterases are distinguished by the rate at which the third stage, the hydrolysis of the phosphorylated enzyme, takes place. The phosphorylated A-esterase is hydrolysed rapidly, like the acetylated AChE in the ACh-AChE reaction. The phosphate ester therefore acts as a substrate for these enzymes. The phosphorylated B-esterase is hydrolysed very slowly, in some instances too slowly to measure. The site phosphorylated is the esteratic site in the active centre. This is thereby blocked, so that the enzyme is unable to hydrolyse its normal substrate, i.e. it is inhibited. There is no theoretical reason why a clear distinction between A- and B-esterases should be possible. If a phosphorylated enzyme was hydrolysed at an intermediate rate it would be hard to decide which group the enzyme belonged to. In practice the rates of hydrolysis fall into two groups, fast and slow, and the distinction is clear. The hydrolysis of phosphate esters by A-esterases *in vitro* is described in Chapter VII.

The reactions of phosphate esters with AChE, the most important B-esterase in this context, are now considered in more detail.

(1) The formation of an adsorption complex has only been demonstrated in a few instances, but there are good reasons for thinking that one is always formed (Chapter VI). In some cases the formation of a complex is the only simple way to explain results. For example, certain compounds which resemble ACh in having a cationic charge in the side chain react with AChE much faster than analogues which do not—$(EtO)_2PO\cdot S\cdot CH_2\cdot CH_2\cdot \overset{\oplus}{S}Et_2$, for instance, reacts 1000 times faster than $(EtO)_2PO\cdot S\cdot CH_2\cdot CH_2\cdot SEt$.

(2) The second stage is a nucleophilic displacement reaction in which the acidic group is displaced by the enzyme. Its rate should therefore vary with the structure of the phosphate ester according to the electronic theory of reaction rates. In practice one can only measure the overall rate of the first and second stages. As the stability of the adsorption complex influences the over-all rate, and the stability varies from compound to compound, an exact agreement with theory cannot be expected. There is, however, rough agreement, as shown in Chapter VI.

(3) A phosphorylated AChE is a phosphate ester, and it should therefore be possible to hydrolyse it, i.e. to re-form the enzyme by a stage 3 reaction. This is so. Hydrolysis by water is slow (it varies with structure as expected on electronic grounds), but more rapid hydrolysis can be brought about by various anions. The most important historically is P2-AM, pyridine-2-aldoxime methiodide (I):

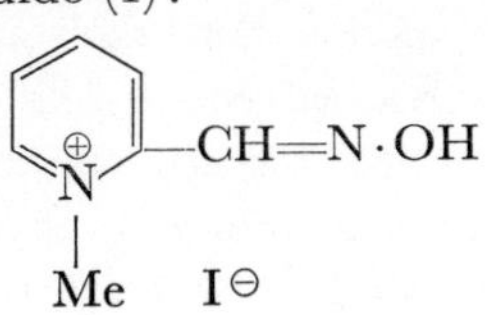


I

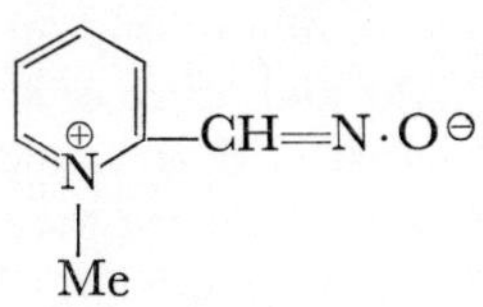


II

This reacts in the anionic form, II. The $\overset{\oplus}{N}$ atom is adsorbed on the anionic site of AChE and holds the molecule in position to react according to the equations:

$$=N\cdot O^{\ominus} + >PO\cdot AChE \rightarrow >PO\cdot O\cdot N= + {}^{\ominus}AChE,$$
$$>P\cdot O\cdot N= + H_2O \rightarrow >P\cdot OH + =N\cdot OH.$$

These are two successive nucleophilic displacement reactions. Thus P2-AM *re-activates* AChE. Some phosphorylated AChE's are substantially completely re-activated in a few seconds by $10^{-4}$ M P2-AM. Several more powerful re-activators have been synthesized. All the useful compounds are characterized by the group: —CH=N·OH, the hydroxyiminomethyl group. The reactions are described in Chapter VI.

## REACTIONS *IN VIVO*

Phosphorus compounds react with enzymes *in vivo* in the same three ways as *in vitro*.

(1) They may react with B-esterases, and phosphorylate them, thereby inhibiting them.

(2) They may be hydrolysed by A-esterases. In mammals most compounds reactive enough to inhibit B-esterases are attacked. A-esterases thus usually decompose direct inhibitors nearly completely within an hour of intravenous injection, so that inhibition of B-esterases then stops.

(3) They may be oxidized to compounds which react with A- and B-esterases.

The balance between the three types of reaction varies with species and compound, as described in Chapters VIII and IX.

The reactions which take place *in vivo* provide strong evidence that phosphorus compounds poison mammals by inhibiting AChE.

(1) Those compounds which are toxic either inhibit AChE directly, or are converted *in vivo* into compounds which do.

(2) AChE is the only enzyme known which is always greatly inhibited by toxic doses.

(3) The duration of signs of poisoning correlates well with the stability *in vivo* of the phosphorylated AChE. Thus AChE inhibited by a dimethyl phosphate ester, $(MeO)_2PO\cdot AChE$, is re-activated fairly rapidly by water, so that signs are of short duration. $(Pr^iO)_2PO\cdot AChE$ is stable, so that after administration of a di-isopropyl ester such as DFP signs are long-lasting.

(4) P2-AM and other AChE re-activators reduce toxic effects.

(5) There is a fairly good correlation for compounds which are not activated *in vivo* between the rates at which they inhibit AChE *in vitro*, and their toxicities to mammals. Perfect correlation cannot be expected, as the balance of reactions with A- and B-esterases varies.

In insects only the first finding can be repeated, so there is more doubt about the validity of the theory.

AChE is widely distributed within the body. It appears in erythrocytes, many nerve cells, ganglia and brain, and at myo-neural (muscle-nerve) junctions. The distribution of inhibition varies with the compound and route of application. The most important property of a compound in this respect is its fat–water partition coefficient. An indication of this can be obtained by determining oil–water or benzene–water partition coefficients. Only those compounds which can be extracted from water by lipids can penetrate the central nervous system to an appreciable extent. Ionic compounds, such as the compound $(EtO)_2PO \cdot S \cdot CH_2 \cdot CH_2 \cdot \overset{\oplus}{S}Et_2$ already mentioned, cannot, and nor do certain non-ionic compounds which are of low extractability. This does not imply that such compounds are non-toxic. Inhibition of brain AChE is not essential for toxic action, and in fact the most toxic compounds are ionic.

It is sometimes stated that only fat-soluble compounds can penetrate the central nervous system. This is correct, but misleading. Several compounds of appreciable fat-solubility cannot be extracted from water by fats, and these do not penetrate. Penetration involves the extraction of the compound from an aqueous medium, blood.

In insects all cholinergic synapses appear to be surrounded by a lipid sheath, so that ionic compounds are of relatively low toxicity to insects.

The balance between the various types of reaction, including some not summarized here, is often different in insects and mammals. Usually mammals have a wider range of esterases capable of hydrolysing phosphorus compounds, so that a compound may be able to kill insects at very low doses but can only kill mammals at high ones. Several compounds, e.g. Dimethoate, $(MeO)_2PO \cdot S \cdot CH_2 \cdot CO \cdot NHMe$, can be fed safely to cattle and kill the warble fly larvae in them. The balance of reactions in mammals and insects is described in Chapter IX.

Plants absorb some phosphorus compounds, so that they enter the sap. These are systemic insecticides. The plant system both oxidizes compounds, in many cases activating them, and hydrolyses them, perhaps, in part, by A-type esterases. B-type esterases do not seem to be important in plants. Reactions in plants are described in Chapter X.

## PHARMACOLOGY

In mammals and insects the signs of poisoning are all consistent with the theory that the transmission of nerve impulses is affected. In the insect system the precise mechanism is still an open question (discussed in Chapter XVII), and is not dealt with here. In mammals the interference results mainly from the inhibition of AChE. (In the following discussion it is assumed that the reader has some knowledge of the anatomy of the nervous system.)

Consider a myoneural junction. The inside of both the nerve fibre and the muscle fibre are at a negative electrical potential with respect to the outside. If anything is done to 'short' the inside and outside, so that the inside (axoplasm and sarcoplasm) becomes momentarily less negative, a wave of potential moves rapidly down the nerve or muscle fibre. This is the action potential. In the muscle fibre the action potential sets off a wave of contraction, i.e. the fibre twitches. When the action potentials in the muscle fibre follow each other rapidly the twitches fuse, and the fibre stays contracted until the action potentials cease, i.e. it holds a tetanus. The transmission of an impulse in a nerve fibre to a muscle fibre takes place in the following way. An action potential in the nerve reaches the nerve-ending at the myoneural junction, and releases ACh. The nerve-ending is separated from the surface of the muscle fibre by a small gap, about 100 Å wide. The ACh diffuses across the gap, and adsorbs on the receptor surface on the opposing muscle, the end-plate. It then depolarizes ('shorts') the fibre in this region, so that an action potential is set up in the muscle fibre, which contracts. The ACh is then hydrolysed by the AChE present, and the muscle repolarizes so that it can be stimulated again. An anti-AChE causes the ACh to remain on the receptor surface longer than usual—not indefinitely long as it can still diffuse away. The increased persistence of ACh has three effects, depending on the circumstances.

(1) After a single nerve stimulus the muscle remains depolarized longer, and in this condition several action potentials are propagated in it, so that the twitch becomes a tetanus, and the response is increased.

(2) If a second nerve stimulus arrives while the muscle fibre is still depolarized, the depolarization is increased. After a few such stimuli the muscle is depolarized too much to respond at all until it is rested for a second or so. Thus a tetanus can only be held for a very short time, i.e. the muscle becomes paralysed.

(3) If the ACh concentration is maintained for a fairly long time (a few seconds to a few minutes) the muscle becomes repolarized, but also refractory, i.e. it requires very much more ACh to set up a response at all. The muscle becomes partially adapted, at the expense of lowering its capacity to respond. This action also therefore tends to paralyse the musculature. Paralysis of the muscles involved in respiration is often a critical factor in bringing about asphyxia.

In muscles and glands innervated by the parasympathetic system the second, blocked stage is not normally attained, because these organs tolerate ACh better. The result of increasing the persistence of ACh is to keep the organ or gland in a perpetual state of stimulation. Thus the gastro-intestinal tract and bronchi contract, the lacrimatory and salivary glands excrete continuously, and the heart slows.

In ganglia AChE probably plays only a subsidiary role, and much of the ACh is lost by diffusion. It is not likely, therefore, that synaptic transmission in ganglia will be blocked by anti-AChE's, and it does not seem to be in most experiments. The stimulation of postganglionic fibres by preganglionic fibres is, of course, different from the stimulation of a muscle fibre by its nerve, in that each postganglionic fibre is innervated by a large number of preganglionic ones, and stimulation requires the reception of impulses nearly simultaneously from several. An anti-AChE increases the effect of each impulse, so that fewer are required to set off an action potential. It therefore facilitates synaptic transmission, as more postganglionic fibres are stimulated. This explains in particular much of the vasoconstriction which takes place in poisoned animals, so that the blood pressure is maintained fairly constant, although the heart is slowed, and the blood flow is reduced.

In the central nervous system only some synapses are mediated by ACh, and the effects of anti-AChE's are still obscure. Reflexes are first potentiated and then inhibited; convulsions take place; and the respiratory centre may be inhibited. Although these effects are undoubtedly increased in some instances by the action of the compounds on the central nervous system, they may in many cases be the results of peripheral action. Ionic inhibitors, for example, cannot penetrate the CNS, but they can cause convulsions. They paralyse the respiratory musculature or constrict the bronchi, so that the animal is asphyxiated, as a result of which convulsions take place. Asphyxia may also help to bring about paralysis of the respiratory centre.

The main signs of poisoning are thus:

(1) Lacrimation, salivation, violent contractions of the gastro-intestinal tract and bladder, bronchoconstriction and bradycardia, all caused by the accumulation of ACh at the junctions of the organs with the parasympathetic nerve endings.

(2) Muscular twitchings and paralysis, caused by ACh accumulation at the myoneural junctions of striated muscles.

(3) Convulsions, hyperexcitability and central paralysis of the respiratory system, effects complex in origin, but often partly due to accumulation of ACh in the CNS.

(4) Vasoconstriction, caused by ACh accumulation in sympathetic ganglia and the adrenal glands.

The balance of signs varies with the compound and animal, even within one species. The immediate cause of death is usually asphyxia.

The evidence that these effects are caused by accumulation of ACh is of three kinds.

(1) Reversible anti-AChE's which do not contain phosphorus cause similar effects to phosphorus anti-AChE's. The differences can be accounted

for by minor differences in distribution, and by the more progressive and persistent nature of AChE inhibition by phosphorus compounds.

(2) Signs of poisoning are reduced by compounds which cover the receptor surfaces, and so deny ACh access to them. Thus D-tubocurarine antagonizes anti-AChE's at myoneural junctions, atropine at parasympathetic nerve endings, and penta- and hexa-methonium at ganglia.

(3) Signs of poisoning are relieved by compounds which re-activate AChE, such as P2-AM.

The ACh theory of synaptic transmission and how transmission is affected by drugs, are described in Chapter XI. The effects of phosphorus compounds on isolated organs are described in Chapters XII and XIII, and the validity of the theory that toxic effects are caused by inhibition of AChE is discussed. The signs of poisoning are related to their causes in Chapter XIV, where the way signs vary with the compound and animal are also briefly discussed. Therapy in general, and symptoms in man are described in Chapter XV.

A few compounds poison by mechanisms which do not involve inhibition of AChE. These effects are described in Chapter XVI.

# Part I

# CHEMISTRY

## CHAPTER I

# Types of Compound, Nomenclature and Electronic Theory

THE SUBJECT of this book is a large group of pentacovalent phosphorus compounds which, with very few exceptions, can be represented by the general structural formula:

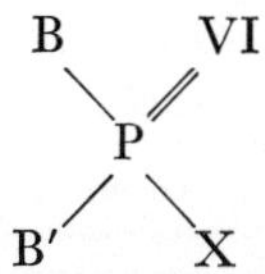


B and B′ represent alkyl, alkoxy, or amino groups, or, in a few compounds, aryl or aryloxy groups. They will be referred to as the basic groups or substituents, to distinguish them from groups represented by X. X is the most acidic group in the molecule, in the straightforward sense that the acid, HX, is stronger than BH or B′H. It will be called the acidic group and the P—X bond will be called the acid anhydride bond. In most compounds this corresponds to normal nomenclature; in others, such as diethyl phenyl phosphate, it would normally be described as an ester bond. As in both cases it is the acidity of X which confers anti-esterase properties on the compound, it is less confusing to hold to the nomenclature put forward. VI represents an atom of the sixth period of the periodic table, usually oxygen or sulphur, but sometimes selenium.*

Six compounds covered by the general formula are shown below:

I. 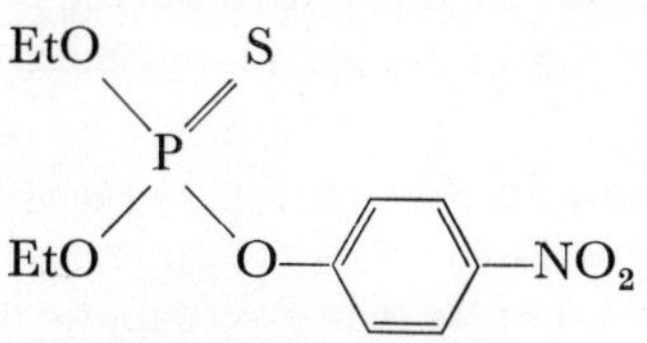


Diethyl 4-nitrophenyl phosphorothionate, Parathion

II. 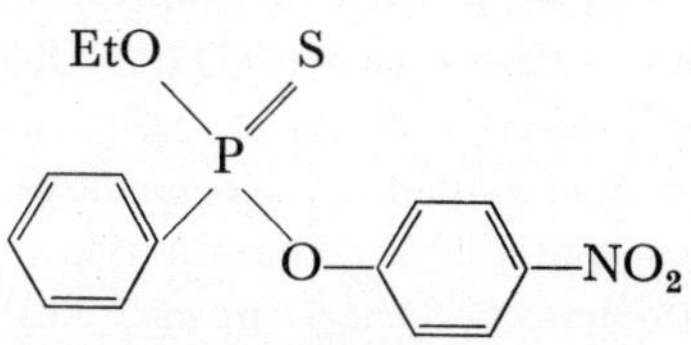


Ethyl 4-nitrophenyl phenylphosphonothionate, EPN

* The P=VI bond is represented as a double bond. The alternative is to represent it as dative bond, P→VI. The true nature of the bond is certainly very complex, by no means worked out as yet, and, when it is, will doubtless still have to be represented by a formal symbolism which does not indicate the real structure.

III. 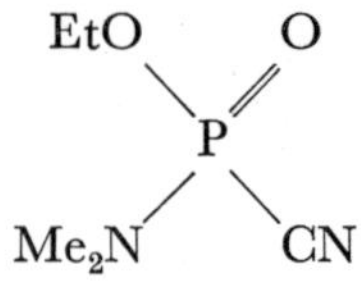

O-Ethyl NN-dimethylphosphoramidocyanidate, Tabun

IV. 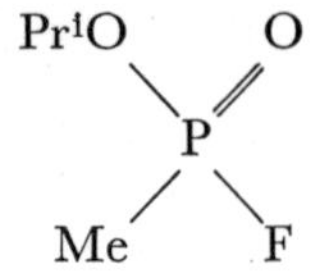

Isopropyl methylphosphonofluoridate, Sarin

V. 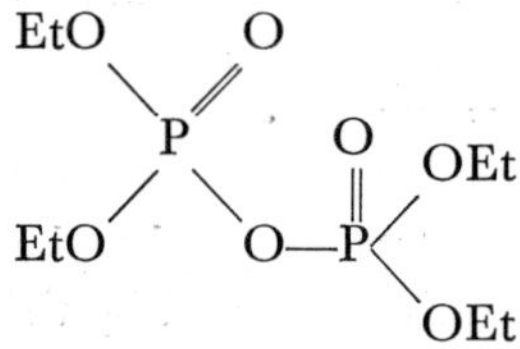

Tetraethyl pyrophosphate TEPP

VI. 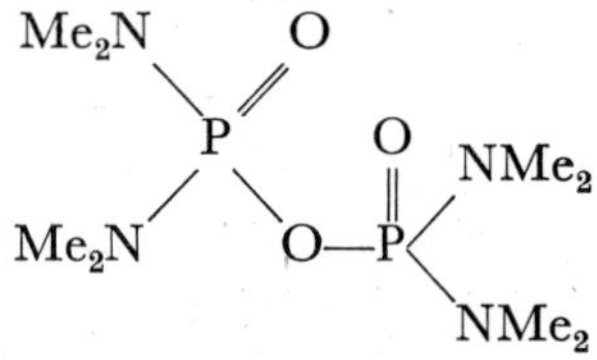

bis(Tetramethylphosphorodiamidic) anhydride, Schradan

VII and VIII below are examples of compounds which the general formula does not cover:

VII. 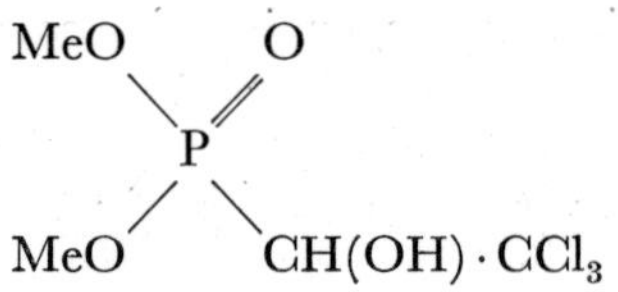

Dimethyl 1-hydroxy-2:2:2-trichloroethylphosphonate, Dipterex

VIII. 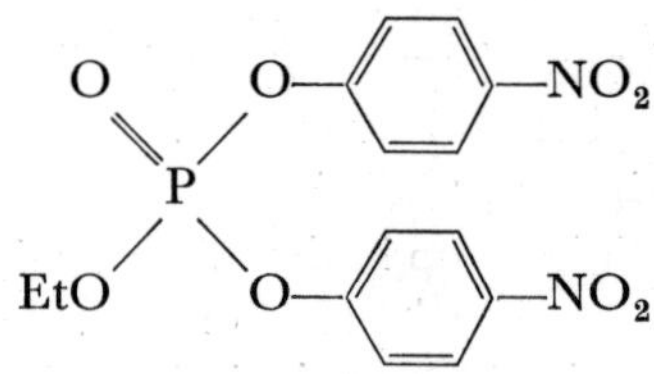

Ethyl di-4-nitrophenyl phosphate

Dipterex has no acid anhydride bond. It is rapidly converted by de-hydrochlorination and re-arrangement to dimethyl 2:2-dichlorovinyl phosphate: $(MeO)_2PO \cdot O \cdot CH:CCl_2$, to which its activity is attributed. This compound fits the general formula, where —X is $—O \cdot CH:CCl_2$. Compound VIII has two acid anhydride bonds. It fits the general formula: $BP:VI(X_2)$. Only a few such compounds have been prepared, and most are too unstable to be of interest.

*Nomenclature.* There are two main systems for naming organophosphorus compounds, the Scandinavian(1) and the Anglo-American.(2) The former is the less confusing, but most authors use the latter, so, when systematic names have to be used, the latter is adopted. I have, however, evaded the use of systematic names whenever possible, not mainly because of purist objections to a nomenclature which names certain derivatives of intransigently univalent fluorine fluoridic acids, but because it obscures close structural relationships. This is inevitable in any system. The compounds are complex in structure, and it is often convenient to use a different part of the molecule as the key in closely related molecules. It does, however, lead to situations in which, e.g. $(EtO)_2PO\cdot S\cdot CH_2\cdot CH_2\cdot SEt$ (Demeton-S), is named O,O-diethyl S-2-ethylthioethyl phosphorothiolate, and $(EtO)_2PO\cdot S\cdot CH_2\cdot CH_2\cdot \overset{\oplus}{S}Et_2$, its ethylsulphonium derivative, is named 2-(diethyoxyphosphinylthio)-ethyl-diethylsulphonium ion; and $(Pr^iO)_2PO\cdot F$ (DFP) is named di-isopropyl phosphorofluoridate, while $(Pr^iNH)_2PO\cdot F$ (Mipafox) is named N,N'-di-isopropylphosphorodiamidic fluoride. Much can, however, be done by the systematic use of trivial or shortened systematic names and in this book the following rules are adhered to.

(1) At the first mention of each compound the structural formula is given, and either its systematic name on the Anglo-American system, or, if it is referred to frequently, a brief name, which is used thereafter. Brief names have been chosen from those in common use, and, where possible, have been taken from the *Journal of Economic Entomology*[3] or British Standards.[4] Exceptions have been made when the recommended names include numerals, and other, more euphonious and memorable names exist—thus Dipterex, and not Bayer 13/59.

(2) Standard abbreviations are used in formulae: R = alkyl, Ar = aryl, Me = methyl, Et = ethyl, Pr = propyl, $Pr^i$ = isopropyl, Bu = butyl, Ph = phenyl, Pyr = pyridinyl. The double bond in P=O and P=S is not shown, PO and PS being regarded as hetero-atoms. Rings are always numbered; *o*-, *m*- and *p*- are not used.

(3) Groups of compounds are, when considered together, given group names, based on the Anglo-American system. They are listed in Table 1.1. The distinction between PS and P·S·C compounds is made by calling the former thionates and the latter thiolates. The alternative, in which both are called thioates, but the substituents are prefixed by O- or S-, is only unambiguous if these substituents are named, which is inconvenient. Thus O,O-diethyl S-ethyl phosphorothioate is unambiguous, but if for brevity the 'ethyls' are omitted, phosphorothioate includes both triethyl phosphorothiolate (the same compound), and triethyl phosphorothionate, $(EtO)_3PS$.

(4) Where a thionate has a trivial name ending in 'thion', the phosphate analogue is given the same prefix but the ending 'oxon'. Thus $(EtO)_2PS\cdot O\cdot C_6H_4\text{-}4\text{-}NO_2$ is Parathion, and $(EtO)_2PO\cdot O\cdot C_6H_4\text{-}4\text{-}NO_2$ is Paraoxon. Most of the names formed in this way are accepted names. When the oxygen analogue would have a particularly unpronounceable name, as Gu-oxon from Guthion, the 't' is retained, giving Gutoxon.

(5) When a trivial name exists for one member of a homologous series, other members in which *both* basic groups are changed are named by putting the formula of the basic group as a suffix to the trivial name. Thus the dimethyl homologue of Paraoxon, $(MeO)_2PO\cdot O\cdot C_6H_4\text{-}4\text{-}NO_2$, is Paraoxon-Me. This, again, only systematizes a fairly common practice. If only one

TABLE 1.1. GROUP NAMES

Chemical structures are given in the first column, and the name used for each type of compound in the second. These sub-groups are combined into groups in the third. In this table, R stands for any group, alkyl, aryl, or substituted aryl or alkyl, of which a carbon atom is bonded to the next atom in the formula. R in any one formula may stand for more than one group, e.g.

$(EtO)_3PO$ and $(EtO)(MeO)(PhO)PO$ are both represented by $(RO)_3PO$.

| Structure | Name of type | Group name |
|---|---|---|
| $(RO)_2PO \cdot R$ | phosphonates | phosphonates |
| $(RO)_3PO$ | phosphates | phosphates |
| $(RO)_2PS \cdot R$ | phosphonothionates | thionates |
| $(RO)_3PS$ | phosphorothionates | thionates |
| $(RO)(R)PO \cdot SR$ | phosphonothiolates | thiolates |
| $(RO)_2PO \cdot SR$ | phosphorothiolates | thiolates |
| $(RO)_2PS \cdot SR$ | phosphorodithioates | dithioates |
| $(RO)_2PO \cdot OH$ | hydrogen phosphates | hydrogen phosphates |
| $(RO)_2POSH$* | hydrogen phosphorothioates | hydrogen thioates |
| $(RO)_2PS \cdot SH$ | hydrogen phosphorodithioates | hydrogen dithioates |
| $R_2PO \cdot F$ | phosphinofluoridates | fluoridates |
| $(RO)(R)PO \cdot F$ | phosphonofluoridates | fluoridates |
| $(RO)_2PO \cdot F$ | phosphorofluoridates | fluoridates |
| $(RNH)_2PO \cdot F$, $(R_2N)_2PO \cdot F$ | phosphorodiamidic fluorides | fluoridates |
| $(RO)(R_2N)PO \cdot F$ | phosphoramidofluoridates | fluoridates |

Compounds containing P—Cl and P—CN bonds are named similarly to fluoridates, i.e. chloridates, cyanidates.

* This may be either $(RO)_2PO \cdot SH$ or $(RO)_2PS \cdot OH$, as the compounds are tautomeric.

group is changed, as in $(EtO)(MeO)PO \cdot O \cdot C_6H_4\text{-}4\text{-}NO_2$, the name or formula is given in full.

(6) When full systematic names are used in the text, the name is arranged so that the basic groups come first, regardless of alphabetical order. This is to avoid confusion when compounds such as $(MeO)_2PO \cdot O \cdot C_6H_4\text{-}4\text{-}Cl$ and $(MeO)_2PO \cdot O \cdot C_6H_4\text{-}4\text{-}NO_2$ are compared. On the alphabetical system they are called, respectively, 4-chlorophenyl dimethyl phosphate and dimethyl 4-nitrophenyl phosphate.

(7) A 'name' index is given at the back of the book, which includes all the trivial names used in this book, with their formulae and systematic names. It contains a cross-index giving most of the other trivial names in the literature. The remaining compounds are under their systematic names in the

ordinary index. In both indexes the names obey the alphabetical convention, to facilitate cross-referencing with the literature.

In addition to the trivial names given under the first seven compounds mentioned in this chapter, there are only a few which recur frequently, and these are listed in Table 1.2. The reader will find it helpful to memorize them.

*Optical isomerism.* In compounds of the type under consideration the four groups are arranged roughly tetrahedrally around the phosphorus atom, as in carbon compounds. Therefore, when the four groups are different, as in Tabun, $(Me_2N)(EtO)PO{\cdot}CN$, the compounds can exist in two optically active forms. Only very few optically active, toxic phosphorus compounds have been prepared, but there are several observations, described mainly in Chapters VI, VII and VIII, which can only readily be interpreted on the assumption that such forms exist, and have different biochemical and toxicological properties.

*Electronic theory.* Much of the first two parts of this book are concerned with relative rates of reaction. The electronic theory of reactions aims at providing an explanation of differences in rates in terms of the more general and fundamental concepts of charge distribution and electronegativity. Besides its intrinsic interest, it correlates many observations in a readily memorizable way, and provides a convenient nomenclature.

The section that follows is entirely theoretical. Electronic theory is successful if it predicts relative rates correctly; and it is shown, when particular reactions are discussed in later chapters, that it usually does so except when there are obvious additional factors which should be taken into consideration. The treatment given, in a somewhat simplified form, is confined to one type of reaction, the nucleophilic displacement reaction, which is the most important. The same principles can be applied to other types of reaction, and their application is given when these reactions are discussed. For the reader who wants more details, works on the electronic theory of organic reactions can be consulted, as it is assumed that the theory is basically the same for both carbon and phosphorus compounds.*

A typical nucleophilic displacement is the hydrolysis of DFP by hydroxyl ions:

$$HO^{\ominus} + (Pr^iO)_2PO{\cdot}F \rightarrow HO{\cdot}PO(OPr^i)_2 + F^{\ominus}.$$

The hydroxyl ion is negatively charged, and therefore seeks a positively charged atom, i.e. it is nucleophilic. The phosphorus atom carries a partial positive charge because it is surrounded by more electronegative atoms which withdraw electrons from it. It is therefore attracted to any negatively charged reagent, such as the hydroxyl ion. The phosphorus atom in such molecules is described as electrophilic. The overall reaction consists of the

* The exposition of theory which follows is very like that given by Dewar[5] for similar reactions in carbon chemistry. Certain details were taken from Watson.[6]

TABLE 1.2. TRIVIAL NAMES OF IMPORTANT COMPOUNDS

| Trivial name | Formula | Systematic name |
|---|---|---|
| DDVP | $(MeO)_2PO \cdot O \cdot CH{:}CCl_2$ | Dimethyl 2:2-dichlorovinyl phosphate |
| Demeton-O | $(EtO)_2PS \cdot O \cdot CH_2 \cdot CH_2 \cdot SEt$ | Diethyl 2-ethylthioethyl phosphorothionate |
| Demeton-S | $(EtO)_2PO \cdot S \cdot CH_2 \cdot CH_2 \cdot SEt$ | Diethyl S-2-ethylthioethyl phosphorothiolate |
| DFP | $(Pr^iO)_2PO \cdot F$ | Di-isopropyl phosphorofluoridate |
| Dimefox | $(Me_2N)_2PO \cdot F$ | Tetramethylphosphorodiamidic fluoride |
| Malathion | $(MeO)_2PS \cdot S \cdot CH(CO \cdot OEt) \cdot CH_2 \cdot CO \cdot OEt$ | Dimethyl S-(1:2-dicarbethoxyethyl) phosphorodithioate |
| Mipafox | $(Pr^iNH)_2PO \cdot F$ | N,N′-di-isopropylphosphorodiamidic fluoride |
| Paraoxon | $(EtO)_2PO \cdot O \cdot C_6H_4$-4-$NO_2$ | Diethyl 4-nitrophenyl phosphate |
| Phosdrin | $(MeO)_2PO \cdot O \cdot CMe{:}CH \cdot CO \cdot OMe$ | Dimethyl 1-carbomethoxy-1-propen-2-yl phosphate |
| Sulfotepp | $(EtO)_2PS \cdot O \cdot PS(OEt)_2$ | bis(Diethyl phosphorothionic) anhydride |
| TIPA | $(Pr^iNH)_2PO \cdot O \cdot PO(NHPr^i)_2$ | N,N′,N″,N‴-tetraisopropylpyrophosphoramide |

displacement from the phosphorus atom by the hydroxyl ion of the fluorine atom, which separates as a fluoride ion. At some point in the reaction both the hydroxyl group and the fluorine atom must be equally attached to the phosphorus, in which state it is equally possible for either to be lost, i.e. for the reaction to go to completion, or for the system to break up again into its original components. This is the transition state, and, confining attention to the atoms mainly affected, can be written: $H\cdot\overset{\delta\ominus}{O}\cdots\cdots\overset{\delta\oplus}{P}\cdots\cdots\overset{\delta\ominus}{F}$. The transition state only exists momentarily; so the rate of a reaction depends on the readiness with which the transition state is reached.

All the phosphorus compounds we are concerned with undergo the nucleophilic displacement reaction. A nucleophilic reagent is any compound or ion which is attracted to a partial positive charge, and which is reactive. The type includes the hydroxyl ion, water, several reagents used in preparing compounds, many esterases, and all the compounds or ions used successfully to re-activate inhibited esterases. Some are anions. Others are compounds containing labile hydrogen atoms attached to oxygen or nitrogen. In the latter the group attached to the hydrogen atom carries a partial negative charge. Thus water can be written: $\overset{\delta\oplus}{H}\text{—}\overset{\delta\ominus}{O}\text{—}\overset{\delta\oplus}{H}$, and an esterase: $\overset{\delta\ominus}{E}\text{—}\overset{\delta\oplus}{H}$. It proves very useful to regard many esterases as particular nucleophilic reagents, the reactivities of which are somewhat altered by the effects of the enzyme surface.

The reaction between a phosphorus compound and an anion is now considered in more detail.

The group displaced by the nucleophilic reagent must separate initially as an anion, although this may then react to give a neutral compound. It follows that the group displaced is the one which is most stable as an anion. The relative stabilities of groups as anions can be measured in most instances by the strengths of their hydrogen acids. Thus the group X gives a hydrogen acid HX, which dissociates in water according to the equation:*

$$HX \rightleftharpoons \overset{\oplus}{H} + \overset{\ominus}{X}.$$

The more stable the anion, $X^{\ominus}$, the more the acid dissociates, i.e. the stronger the acid. Thus the group displaced is the most acidic group. It is also the most electrophilic.

Consider the reactions between a given nucleophilic reagent and a group of phosphate esters in which X is the same but the other substituents are different. These substituents affect the rate by making the phosphorus atom more or less electrophilic. The more electrophilic the phosphorus atom, the more it attracts the reagent, and the faster the reaction.

Oxygen is much more electronegative than sulphur, so that the P=O

* Solvent molecules are, of course, involved in this reaction. For aqueous solutions it should be written: $HX + H_2O \rightleftharpoons H_3O^{\oplus} + X^{\ominus}$.

group is more polarized in the sense: $\overset{\delta\oplus}{P}=\overset{\delta\ominus}{O}$ in phosphates than the P=S bond in the corresponding thionates. The phosphorus is therefore more electrophilic in phosphates, and is more readily attacked, e.g. Paraoxon is more readily attacked by hydroxyl ions than Parathion.

Alkyl groups have a slight tendency to donate electrons. Consequently, if the direction of electron flow is shown by arrows, it is obvious that the tendency increases in the sense:

$$H < H_3C\rightarrow\ < CH_3\rightarrow CH_2\rightarrow\ < CH_3\rightarrow(CH_2)_n\rightarrow CH_2\rightarrow\ < \begin{array}{l} CH_3 \searrow \\ \qquad\quad CH\rightarrow. \\ CH_3 \nearrow \end{array}$$

These groups therefore lower the reactivity of the phosphorus atom by making it more negative, i.e. less electrophilic. Rates of alkaline hydrolysis therefore fall in the order:

$$Me_2PO\cdot F > (Et)(Me)PO\cdot F > Et_2PO\cdot F > Pr^i{}_2PO\cdot F.$$

The effects are also transmitted through intervening atoms, such as oxygen and nitrogen, so that rates also fall in the order:

$$(MeO)_2P > (EtO)_2P > (Pr^iO)_2P; \quad \text{and} \quad (Me_2N)_2P > (Et_2N)_2P,$$

e.g. Paraoxon-Me is more readily attacked than Paraoxon. These are the so-called inductive effects. Their direction can be reversed if negative substituents are introduced into the alkyl groups. Thus 2-chloroethyl is an electrophilic group: $Cl\leftarrow CH_2\leftarrow CH_2\leftarrow$, and compounds containing it hydrolyse faster than their analogues with ethyl groups.

Much more powerful *electromeric* effects are exerted by oxygen and nitrogen atoms singly bonded to the phosphorus atom. These atoms carry p electrons, which are not used to form bonds, and which they tend to donate to the phosphorus atom. A phosphorus–nitrogen single bond can be imagined as modified in three steps:

(1) The nitrogen donates an electron to the phosphorus:

$$>\ddot{N}-P \rightarrow\ >\underset{\oplus}{\dot{N}}-\underset{\ominus}{\dot{P}},$$

thereby acquiring a positive charge.

(2) The unshared p electron on the nitrogen forms a bond with the electron donated to the phosphorus, giving the structure:

$$>\overset{\oplus}{N}=\overset{\ominus}{P}.$$

(3) As nitrogen is more electronegative than phosphorus, it withdraws charge from the double bond until a balance is reached between the stability attained by forming a double bond and the affinity of nitrogen for electrons. The final state (which is, of course, the only state which corresponds to reality) is:

$$>\overset{\delta\oplus}{N}\overset{\cdots}{=}\overset{\delta\ominus}{P},$$

in which there is a partial double bond and partial charges on the nitrogen and phosphorus.

The overall effect is to make the phosphorus more negative, and less electrophilic and reactive. The more electronegative the attached group, the less the tendency to relinquish electrons, so reactivity increases in the order: $>N—P < —O—P$. Thus Dimefox, $(Me_2N)_2PO \cdot F$, is hydrolysed more slowly by alkalis, and reacts with enzymes more slowly, than DFP, $(Pr^iO)_2PO \cdot F$. Electromeric effects are confined to those atoms which readily form double bonds, so that sulphur groups show none. Aryl systems, however, can exert electromeric effects, while alkyl groups cannot, so that the order of reactivity can be extended to: $—O—P < Ph—P < R—P$.

There is a third type of effect which must be considered, the mesomeric effect. So far it has been assumed that the effects exerted by substituents are not changed as the nucleophilic reagent approaches, which is obviously incorrect. The approach of, e.g. a hydroxyl ion to the partial structure $\overset{\delta\ominus}{P}\cdots\overset{\delta\oplus}{N}<$, can hardly fail to make the phosphorus more positive:

$$HO^{\ominus} \curvearrowright \cdots P \overset{\curvearrowright}{\cdots} N,$$

denoting movement of electrons by arrows. Providing this takes place to an equal extent in different compounds, it is not observable, but it will be if an attached substituent is unusually polarizable. The assumption that P—S bonds are exceptionally polarizable explains the high reactivity of S-alkyl thiolates, such as $(EtO)(EtS)PO \cdot O \cdot C_6H_4\text{-}4\text{-}NO_2$, and also why thionates, although less reactive than the corresponding phosphates, are nowhere near as much less reactive as the great difference between the electronegativities of oxygen and sulphur leads one to expect.

The acidic group, X, can influence the rate in three ways: by its readiness to accept electrons, by the electronic effects it exerts on the phosphorus atom, and by the strength of its bond with the phosphorus atom. The last factor has not been discussed before. When the other factors are constant, the stronger the P—X bond the less readily it will be broken. Bond strengths in molecules as complex as these cannot be discussed theoretically. In practice, however, variations in bond strengths do not appear to be very important, and will usually be ignored. The readiness of X to accept an electron, is, as already pointed out, the greater the stronger the acid HX. There are exceptions when the acid strength of HX depends on factors absent from the group P—X. Thus in 2-chloro- or 2-nitro-phenol the phenolic hydrogen forms a hydrogen bond with the ortho-group:

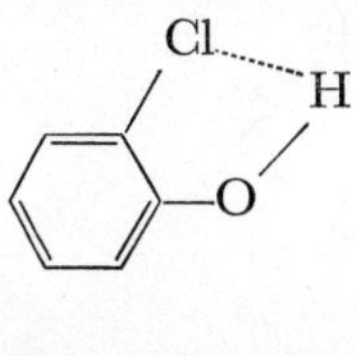

This facilitates the loss of a hydrogen atom as a proton, so that these compounds are markedly stronger acids than their para-isomers. This type of hydrogen bonding is absent from the phenyl phosphates. It is not, therefore, to be expected that the greater acid strengths of ortho-phenols will make 2-phenyl phosphates less stable than 4-phenyl phosphates to nucleophilic reagents. Other, similar, cases are dealt with as they arise. The effects of the acidic group on the polarity of the phosphorus is also dependent in simple cases on the acidity, as the group is more electrophilic the stronger an acid HX is. A partial exception is provided by some of the acids containing conjugated systems (alternative single and double bonds), such as some substituted phenols. In 4-nitrophenols, for example, the acids are strengthened by an electromeric effect which is absent in 3-nitrophenol, so that the former is the stronger acid:

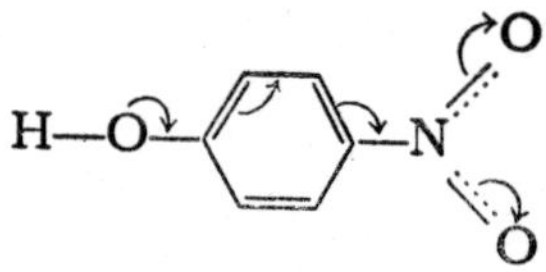


4-nitrophenol

In both, the nitro-group has a powerful inductive effect, so both are stronger acids than phenol itself. The electromeric effect cannot, however, be transmitted to the phosphorus atom though the intervening oxygen atom, except to the very limited extent made possible by the partial double bond formation between the oxygen and phosphorus already discussed. Consequently, if the rate of reaction mainly depended on the effect of the groups on the charge on the phosphorus atom, a 3-nitrophenyl phosphate would be attacked as fast as a 4-nitrophenyl phosphate, although 4-nitrophenol is the stronger acid; whereas, if it mainly depended on the capacity of the groups to separate as anions, the 4-nitrophenyl phosphate would be attacked faster.

The effects actually observed suggest that electromeric effects involving the phosphorus atom are very strong, and that the overall structure of Dimefox should be represented somewhat as below:

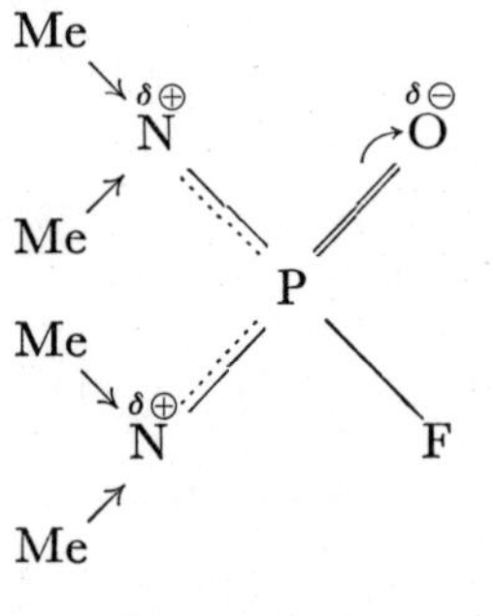

in which it is a moot point whether the phosphorus atom carries a partial negative or positive charge. As phosphorus can accommodate 12 electrons, it may well be negatively charged. All observations lead to the conclusion that the phosphorus carries a positive charge when under the influence of the approaching nucleophilic reagent, but this may be due to a mesomeric effect, the importance of which is obscured because it is similar in many different reactions.

The effects of varying the nucleophilic reagent are now considered. Both an anion, $A^{\ominus}$, and its acid, HA, are nucleophilic reagents. The two reactions can be written:

$$A^{\ominus} + P\text{—}X \rightarrow \overset{\delta\ominus}{A}\cdots\overset{\delta\oplus}{P}\cdots\overset{\delta\ominus}{X} \rightarrow A\text{—}P + X^{\ominus},$$

$$\overset{\delta\oplus}{H}\text{—}\overset{\delta\ominus}{A} + P\text{—}X \rightarrow \overset{\delta\oplus}{H}\cdots\overset{\delta\ominus}{A}\cdots\overset{\delta\oplus}{P}\cdots\overset{\delta\ominus}{X} \rightarrow A\text{—}P + H^{\oplus} + X^{\ominus}.$$

In the second case the partial charge on —A is less than in the anion, $A^{\ominus}$, and its bond with the hydrogen atom has to be broken for reaction to take place. As a result acids always react much more slowly than their conjugate bases. Water, for example, reacts more slowly than hydroxyl ions.

In the first type of reaction, the anion must lose charge to form the transition state, which should be easier the weaker the acid HA, i.e. the less the affinity of $A^{\ominus}$ for its electron. The order of reactivity should therefore fall in the order: $RO^{\ominus} > HO^{\ominus} > PhO^{\ominus}$, and carboxylates and phosphates should react still more slowly, both because the acids are much stronger than phenols and because the negative charge is shared between two oxygen atoms,

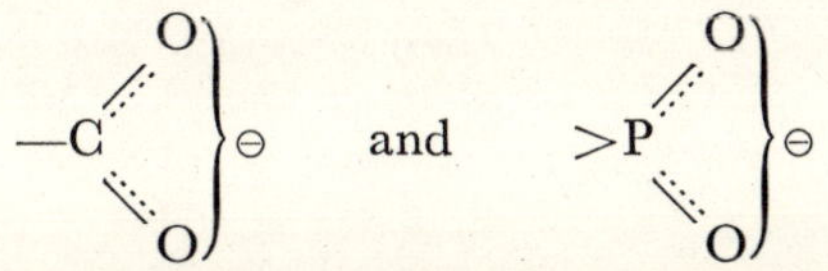

—only one of which can approach the phosphorus. In practice, $OH^{\ominus}$ is one of the most active anions, and carboxylates and phosphates have low activity. Fluoride ion is very much more active than the strength of hydrofluoric acid indicates, perhaps because it is so small. The order expected is, however, only adhered to very roughly, mainly because of solvent effects. Other findings are referred to at the end of Chapter IV.

In the second type of reaction, the stronger the acid, the faster the reaction, as the stronger acid loses a proton more readily in the transition state, and the group attached to the proton has a greater negative charge. This prediction has not been tested. As the first and second types of reaction depend in opposite ways on acid strength, it follows that the difference in reactivity between an acid and its anion should be least with the strongest acids. From the reactions favoured for preparing compounds this appears to be true.

Electronic theory does not indicate by what mechanism a reaction takes place, and the mechanism must be known before it can be applied. It is possible, for example, to assume that in aqueous solution, DFP is rapidly hydrated:

$$H_2O + (Pr^iO)_2PO \cdot F \rightarrow (Pr^iO)_2P(F)(OH)_2.$$

This primary product then breaks down slowly into hydrolysis products, bases catalysing the final stage by accepting hydrogen atoms:

$$(Pr^iO)_2P(F)(OH)_2 + OH^{\ominus} \rightarrow (Pr^iO)_2PO \cdot OH + H_2O + F^{\ominus}.$$

If this mechanism were correct, the reaction with hydroxyl ions would not be a nucleophilic displacement, and the theory would not hold. For reasons given in Chapter IV such mechanisms can probably be rejected, and the theory holds very well.

The theory also takes no account of steric hindrance. The groups in many compounds are crowded closely round the central phosphorus atom, and it appears reasonable to suppose that the larger nucleophilic reagents would reach the phosphorus atom with much more difficulty than the smaller ones. Steric hindrance seems to be important surprisingly rarely.

## REFERENCES

1. L. Larsson, B. Holmstedt and E. Tjus. *Acta chem. scand.* **8,** 1563 (1954)
2. The Editors. *J. chem. Soc.* 5122 (1952)
3. C. N. Smith. *J. econ. Ent.* **52,** 361 (1959). H. L. Haller, *J. econ. Ent.* **50,** 226 (1957); **48,** 112 (1955)
4. British Standards. 1831 (1957)
5. M. J. S. Dewar. *The Electronic Theory of Organic Chemistry*, Oxford, 1950
6. H. B. Watson. *Modern Theories of Organic Chemistry*, Oxford, 1941

CHAPTER II

# Methods of Preparation

A GENERAL survey is given in this Chapter of the methods which have been used to prepare organophosphorus poisons, with some examples to clarify the principles involved.

Many of the products are highly reactive, as biochemical and chemical reactivity are closely linked. Preparative reactions are therefore usually accompanied by side reactions, and good yields can only be got by paying careful attention to experimental details. Such details are omitted from this Chapter to save space, but the Name Index contains references to papers and patents in which detailed descriptions of the preparations of all the compounds named there may be found.

The phosphorus intermediates used are listed in Table 2.1. Their preparations are not given, as they are generally to be found in Kosolapoff's book,[81] or in the references to the methods for making the final products.

The conditions for carrying out reactions are represented as in:

$$A + B \xrightarrow[100°,\ 4\ \text{hr}]{\text{toluene, cat. Cu}} C + D,\ 70\%,$$

i.e. when A is allowed to react with B in toluene at 100°C for 4 hr in the presence of catalytic copper the required compound, D, is produced in 70% yield, estimated on the phosphorus intermediate. Solvents may be referred to by formulae in equations.

The reactions are grouped, each group being given a code letter and number, by which it will be referred to later.

### A. *Substitutions into Pentacovalent Phosphorus Compounds*

The substitutions may or may not take place at the phosphorus atoms.

A1. *Reactions between an acid or its salt and a halide.* This group comprises four general reactions, all of the nucleophilic displacement type:

$$\left.\begin{array}{l} >PO\cdot Cl + X^{\ominus} \rightarrow Cl^{\ominus} + >PO\cdot X \\ >PO_2^{\ominus} + Cl\cdot Y \rightarrow Cl^{\ominus} + >PO\cdot O\cdot Y \end{array}\right\} \text{halide + salt of acid,}$$

$$\left.\begin{array}{l} >PO\cdot Cl + HX \rightarrow HCl + >PO\cdot X \\ >PO\cdot OH + Cl\cdot Y \rightarrow HCl + >PO\cdot O\cdot Y \end{array}\right\} \text{halide + free acid.}$$

TABLE 2.1. PHOSPHORUS INTERMEDIATES USED IN PREPARING ORGANOPHOSPHORUS POISONS

| Formula | Name | Brief name |
| --- | --- | --- |
| $(RO)_2PO \cdot Cl$ | dialkyl phosphorochloridate | chloridate |
| $(RNH)_2PO \cdot Cl$ | N,N′-dialkylphosphorodiamidic chloride | chloridate |
| $(R_2N)_2PO \cdot Cl$ | tetra-alkylphosphorodiamidic chloride | chloridate |
| $(RO)_2PS \cdot Cl$ | dialkyl phosphorochloridothionate | chloridate |
| $(RO)_2PO \cdot OH$ | dialkyl hydrogen phosphate | hydrogen phosphate |
| $(RO)_2POSH$ | dialkyl hydrogen phosphorothioate | hydrogen thioate |
| $(RO)_2PS \cdot SH$ | dialkyl hydrogen phosphorodithioate | hydrogen dithioate |
| $(R_2N)_2PO \cdot OH$ | tetra-alkylphosphorodiamidic acid | diamidic acid |
| $POCl_2F$ | phosphoryl fluoride dichloride | fluorochloride |
| $PSCl_2F$ | thiophosphoryl fluoride dichloride | fluorochloride |
| $(RO)_3PO$ | trialkyl phosphate | same |
| $(RO)(R_2N)_2PO$ | O-alkyl N,N′-tetra-alkylphosphoro-diamidates | formula used |
| $(RO)_2PO \cdot S \cdot Cl$ | dialkyl S-chloro phosphorothiolate | formula used |
| $(RO)_2PO \cdot H$ | dialkyl phosphonate | same |
| $(RO)_3P$ | trialkyl phosphite | same |
| $(RO)_2PO^{\ominus}\ M^{\oplus}$ | metal dialkyl phosphite | metal phosphite |

Derivatives of phosphonic acid and phosphinic acid are known corresponding to many of the derivatives of phosphoric acid given above, and can be made to undergo most of the corresponding reactions.

Bromides can often be used instead of chlorides, and thioates or dithioates instead of phosphates. Fluorides can only be used for certain specific preparations, which are described.

A greater number of toxic compounds have been prepared by this group of reactions than by any other. A very wide variety of reaction conditions have been used, chosen not for theoretical reasons, but because they led to satisfactory yields. There are, however, three general principles:

(1) Salts react more readily than the hydrogen acids from which they are made. They are therefore generally preferred, as lower temperatures can be used, thereby minimizing side reactions.

(2) Salts are insoluble in non-polar solvents, so that polar solvents are used whenever possible. For most reactions the solvents must be dry, because water reacts preferentially, especially in reactions of the $PO \cdot Cl + X^{\ominus}$ type. Exceptions are given later. The product is often a good solvent; phosphate esters are highly polar, and resemble in structure both reagents. Therefore, for reactions of the $PO_2^{\ominus} + Cl \cdot Y$ type no solvent is added. The

reaction is initially slow, but accelerates as the product is formed and the reagents dissolve. A typical experimental instruction for this type of reaction is: 'Hold at 100° for 1 hr, allow to cool, and leave overnight'. These details are not given in the examples in this Chapter. In many other reactions only small quantities of solvents are used to start them, and to maintain temperature control by refluxing.

(3) In reactions of the $PO \cdot Cl + X^{\ominus}$ type the reactivity of the chlorides fall in the order:

$$(MeO)_2PO \cdot Cl > (EtO)_2PO \cdot Cl > (Pr^iO)_2PO \cdot Cl > (R_2N)_2PO \cdot Cl.$$

Unless there are special features, only one example will be given; with the help of the rule one can judge suitable conditions for the preparation of homologues and analogues. The preparations of dimethyl phosphate esters especially are accompanied by side reactions, and their yields are almost invariably lower than the yields of homologues.

Specific examples are now given with indications of the additional principles involved.

For reactions of the type: $>PO \cdot Cl + X^{\ominus}$, the anion is usually formed *in situ*, unless a salt is commercially available. When $H \cdot X$ is a weak acid, such as an alcohol or a thiol, the sodium salt is prepared from it by the action of metallic sodium in toluene, and is then allowed to react with the chloridate, as in the following preparation of Demeton-S:[1]

$$2Na + 2HS \cdot CH_2 \cdot CH_2 \cdot SEt \xrightarrow[40^\circ]{MePh} H_2 + 2NaS \cdot CH_2 \cdot CH_2 \cdot SEt.$$

$$(EtO)_2PO \cdot Cl + NaS \cdot CH_2 \cdot CH_2 \cdot SEt \xrightarrow[60^\circ,\ 1\ hr]{MePh}$$

$$NaCl + (EtO)_2PO \cdot S \cdot CH_2 \cdot CH_2 \cdot SEt, 50\%.$$

Demeton-S

In this instance the choice of solvent is controlled by the need to use metallic sodium in the first step. From stronger acids the salt is formed by the action of sodium carbonate in a polar solvent.* Thus Paraoxon may be prepared by the reaction:[2]

$$(EtO)_2PO \cdot Cl + HO \cdot C_6H_4\text{-}4\text{-}NO_2 + Na_2CO_3 \xrightarrow[reflux\ 2\ hr]{EtMeCO}$$

$$NaCl + NaHCO_3 + (EtO)_2PO \cdot O \cdot C_6H_4\text{-}4\text{-}NO_2, 93\%.$$

Sodium hydroxide,[3, 4] or strong tertiary amines,[5] such as triethylamine, may sometimes be used instead of sodium carbonate.

* Dimethylformamide, a non-inflammable, relatively non-toxic, and very polar solvent has, rather surprisingly, been rarely used,[114] although it would seem very promising for many of these reactions.

With strong acids, when the products are unstable in alkaline solution, sodium bicarbonate or pyridine may be used, as in preparing TEPP:[6]

$$2(EtO)_2PO\cdot Cl + 2NaHCO_3 \xrightarrow[30^\circ,\ \text{sev. hr}]{\text{none}} 2NaCl + CO_2 + (EtO)_2PO\cdot O\cdot PO(OEt)_2,\ 82\%.$$

$$2(EtO)_2PO\cdot Cl + H_2O + 2Pyr. \xrightarrow[<2^\circ,\ \text{sev. hr}]{\text{none}} 2Pyr\cdot HCl + TEPP,\ 89\%.$$

The first stage is the hydrolysis of some chloridate to dialkyl phosphate anion, which then reacts with more chloridate. Tetramethyl pyrophosphate is only obtained in very low yields by this reaction. The lower reactivity of tetramethylphosphorodiamidic compounds is illustrated by the much more drastic conditions required to prepare Schradan by an exactly analogous reaction:[7]

$$2(Me_2N)_2PO\cdot Cl + Na_2CO_3 \xrightarrow[100^\circ,\ 8\ \text{hr}]{\text{none}} 2NaCl + CO_2 + \underset{\text{Schradan}}{(Me_2N)_2PO\cdot O\cdot PO(NMe_2)_2},\ 70\%$$

Both the chloridate is less reactive than $(EtO)_2PO\cdot Cl$, and the diamidic acid, $(Me_2N)_2PO\cdot OH$, formed as an intermediate, is much weaker than $(EtO)_2PO\cdot OH$.[8]

The preparation of thionates by the analogous reactions with phosphorochloridothionates, $(RO)_2PS\cdot Cl$, is more difficult, because the chloridothionates are less reactive than the chloridates, and the products tend to isomerize on heating. Strong tertiary amines or copper may then be helpful as catalysts. With the chloridothionates they form unstable intermediates in which the phosphorus atom is more electrophilic. Triethylamine,[9] for example, is said to act by forming an ionic intermediate, $[(RO)_2PS\cdot NEt_3]^{\oplus} Cl^{\ominus}$, in which the positive charge is shared with the phosphorus atom; and copper,[10] by co-ordinating with the sulphur atom, increases the polarity of the P=S bond.

Water can be used as the solvent when the product is not readily hydrolysed, and a high concentration of the salt can be obtained in neutral solution.

$$(Me_2N)_2PO\cdot Cl \xrightarrow[20^\circ,\ 30\ \text{min}]{50\%\ \text{aq. KF}} KCl + \underset{\text{Dimefox}}{(Me_2N)_2PO\cdot F},\ 60\%.^{(11,12)}$$

$$(Me_2N)_2PO\cdot Cl \xrightarrow[20^\circ,\ 30\ \text{min}]{20\%\ \text{aq. NaN}_3} NaCl + \underset{\text{Mazidox}}{(Me_2N)_2PO\cdot N_3},\ 80\%.^{(13)}$$

Water may also be present when both the chloridate and the product are insoluble in it. Thus Parathion may be prepared in water:

$$(EtO)_2PS\cdot Cl + Na\cdot O\cdot C_6H_4\text{-4-}NO_2 \xrightarrow[70^\circ,\ 4\ \text{hr}]{\text{water}} NaCl + \underset{\text{Parathion}}{(EtO)_2PS\cdot O\cdot C_6H_4\text{-4-}NO_2};^{(14)}$$

and Trolene may be prepared in 20% NaOH:[59]

$$(MeO)_2PS\cdot Cl + HO\cdot C_6H_2\text{-}2{:}4{:}5\text{-}Cl_3 \xrightarrow[60^\circ,\ 1hr]{20\%\ aq.\ NaOH} HCl + \underset{\text{Trolene}}{(MeO)_2PS\cdot O\cdot C_6H_2\text{-}2{:}4{:}5\text{-}Cl_3},\ 91\%.$$

In these cases the reaction probably takes place in the non-aqueous phase, so that there is no direct competition between the phenate ion and the hydroxyl ion, as the latter is confined to the aqueous phase.

When more than one halogen atom is bonded to the phosphorus, the reactivity falls rapidly in the order: $PHal_3 \gg PHal_2 \gg PHal$. Chlorine atoms are also much more readily displaced than fluorine atoms. The polyhalidates are so much more reactive than the monohalidates that free acids are best used. When bases are added it is solely in order to neutralize the hydrogen halide evolved, which may otherwise catalyse side reactions. The following reactions exemplify these principles. Compounds of the type: $(R_2N)(R'O)PO\cdot F$ are prepared from the amidic difluorides and an alcohol:[11]

$$Me_2N\cdot PO\cdot F_2 + ROH + Base \rightarrow Base\cdot HF + (Me_2N)(RO)PO\cdot F,$$

and the corresponding cyanidates, isocyanitidates* and acetidates[11] from amidic dichlorides, as illustrated by the equation for the preparation of Tabun:[15]

$$Me_2N\cdot PO\cdot Cl_2 + 2NaCN + EtOH \xrightarrow[40^\circ,\ 30\ min]{PhCl} 2NaCl + HCN + \underset{\text{Tabun}}{(Me_2N)(EtO)PO\cdot CN},\ 75\%.$$

Tabun and its analogues can only be made by this type of reaction. The more obvious one: $(R_2N)(R'O)PO\cdot Cl + KCN$, does not take place, probably because the cyanide ion is not soluble enough except at high temperatures, which increase side reactions involving the very reactive P—CN system. By using the amidic dichloride, alcohol can be used as a solvent for the first stage of the reaction, the displacement of chlorine by cyanide ion, and then itself reacts in the second stage, displacing the remaining chlorine. Amidic dichlorides react differently with metal fluorides in alcohol, dialkyl fluoridates being produced for some unexplained reason:[11]

$$Me_2N\cdot PO\cdot Cl_2 + NaF + 2EtOH \xrightarrow[65^\circ,\ 1{\cdot}5\ hr]{benzene} Me_2NH_2Cl + NaCl + (EtO)_2PO\cdot F,\ 76\%.$$

The fluorochlorides of phosphorus may be used to prepare various P—F compounds in which the chlorine atoms are substituted by groups otherwise

* That metal cyanates give isocyanates is shown by the addition reactions described under D.

difficult to introduce. Thus several compounds have been prepared from $POCl_2F$[16] and $PSCl_2F$,[17] as, for example, $(EtS)_2PO \cdot F$ by the reaction:

$$POCl_2F + 2Et \cdot SH \xrightarrow[20°,\ 30\ min]{none} 2HCl + (EtS)_2PO \cdot F,\ 60\%,$$

and some extremely toxic analogues of acetylcholine from MePOCl(F).[18] Bases such as triethylamine have often been used in these reactions, but $POCl_2F$ gives satisfactory yields without them.

Except for these few reactions, in which the phosphorus atom is especially electrophilic, free acids have not often been used with success. Theoretically the advantage of using a salt would be least with a strong acid, and it has been found that O,O-dialkyl hydrogen phosphorodithioates ($pK \sim 1 \cdot 6$)[19] can displace chlorine from reactive C—Cl compounds under mild conditions. Thus Thimet may be prepared by the reaction:[20]

$$(EtO)_2PS \cdot SH + Cl \cdot CH_2 \cdot SEt \xrightarrow[35°,\ few\ min]{CCl_4} HCl + \underset{\text{Thimet}}{(EtO)_2PS \cdot S \cdot CH_2 \cdot SEt},\ 80\%.$$

The reaction is much slower with $Cl \cdot CH_2 \cdot CH_2 \cdot SEt$, 2-chloroethylthioethane, in which the C—Cl bond is less reactive than in the chloromethyl compound. It is also sometimes convenient to prepare mono-aryl phosphorodichloridates, converting these with alcohols to dialkyl aryl esters. Thus, if a phenol of p$K$ less than about 8 is refluxed with a large excess of phosphorus oxychloride in the presence of a catalyst, good yields (on the phenol) of the dichloridates may be prepared:[21]

$$POCl_3 + HO \cdot C_6H_4\text{-}4\text{-}NO_2 \xrightarrow[reflux\ 3\ hr]{POCl_3,\ cat.\ KCl} HCl + O_2N\text{-}4\text{-}C_6H_4 \cdot O \cdot POCl_2,\ 93\%.$$

Metal chlorides are used as catalysts, the order of effectiveness being: $FeCl_3 > CaCl_2 > RbCl > KCl > NaCl > BaCl_2$. Potassium chloride is recommended, as it is active enough, cheap, and easy to remove from the reaction mixture. Its effect is to lower the temperature at which the reactions take place to such an extent that a large excess of $POCl_3$ can be used as a solvent, thus favouring mono-substitution. Without a catalyst temperatures above the boiling point of $POCl_3$ are required.

Preparations from hydrogen phosphates, $(RO)_2PO \cdot OH$, phosphorothioates, $(RO)_2POSH$, and phosphorodithioates, $(RO)_2PS \cdot SH$, are usually carried out with the sodium or potassium salts,* previously prepared and

* Silver salts have been very little used, even in fundamental work, but one compound, linear pentaethyl triphosphate, has only been prepared by the reaction:[22]

$$Ag_5P_3O_5 + 5EtI \xrightarrow[reflux\ 4\ hr]{CHCl_3} 5AgI + (EtO)_2PO \cdot O \cdot PO(OEt) \cdot O \cdot PO(OEt)_2.$$

Alkaline hydrolysis gave $(EtO)_2PO_2^{\ominus}$ and $EtO \cdot PO_3^{\ominus}$ in the molecular ratio of 2:1, which proved the structure.

separated, which are allowed to react without a solvent with the chloride or bromide of the group to be introduced. Phosphates and dithioates can only give single products, but thioates can, in theory give two, a thiolate and a thionate:

$$(EtO)_2POS^{\ominus} + EtBr \begin{cases} \nearrow Br^{\ominus} + (EtO)_2PS\cdot OEt, \text{ triethyl phosphorothionate.} \\ \searrow Br^{\ominus} + (EtO)_2PO\cdot SEt, \text{ triethyl phosphorothiolate.} \end{cases}$$

In fact the reactions give almost pure thiolates, as in the preparation of Demeton-S and its analogues:[23]

$$(RO)_2POS\cdot Na + Cl\cdot CH_2\cdot CH_2\cdot SR' \xrightarrow[55°,\ 4\ hr]{none} NaCl + (RO)_2PO\cdot S\cdot CH_2\cdot CH_2\cdot SR',\ 80\%;$$

except when the products would contain P—S—P bonding. Thus diethyl phosphorothionate does not react with diethyl phosphorochloridate according to the equation:

$$(EtO)_2POS^{\ominus} + Cl\cdot PO(OEt)_2 \rightarrow Cl^{\ominus} + (EtO)_2PO\cdot S\cdot PO(OEt)_2,$$

but gives the pyrophosphoromonothionate, $(EtO)_2PS\cdot O\cdot PO(OEt)_2$ in 50% yield instead.[24] The structure is proved by the identity of the product with that from the reaction: $(EtO)_2PS\cdot Cl + NaO\cdot PO(OEt)_2$, and by infra-red spectroscopy, which shows that the P=O absorption band is only half as intense as in the pyrophosphate analogue, TEPP,[25] and that there is no band of the frequency expected for P—S—P bonding.[24] All attempts to make P—S—P compounds fail. If rearrangement to a thionate is possible, it takes place; and if it is not, as in the reaction $(EtO)_2PS_2^{\ominus} + Cl\cdot PS(OEt)_2$, the only products which can be separated are phosphorothioates, and neither P—O—P nor P—S—P bonding can be detected by infra-red spectroscopy.[24]

*A*2. The second basic reaction in this group is that in which a bipolar molecule, X·Y, displaces an alkyl radical, usually ethyl or methyl, from a phosphoric ester:

$$>P(O\text{ or }S)—\overset{\delta\ominus}{O}—\overset{\delta\oplus}{R} \rightarrow >P(O\text{ or }S)—O\cdot X + R\cdot Y.*$$
$$\overset{\delta\oplus}{X}—\overset{\delta\ominus}{Y}$$

* This is only a formal representation of the reaction, the precise mechanism of which is unknown. Thus the Y radical could attach itself to the R radical, to give the mechanism: P—O—R + X·Y → P—O⋯R⋯Y⋯X → P—O$^{\ominus}$ + R·Y + X$^{\oplus}$ → P—O·X; or the X radical might displace the R from the oxygen atom:

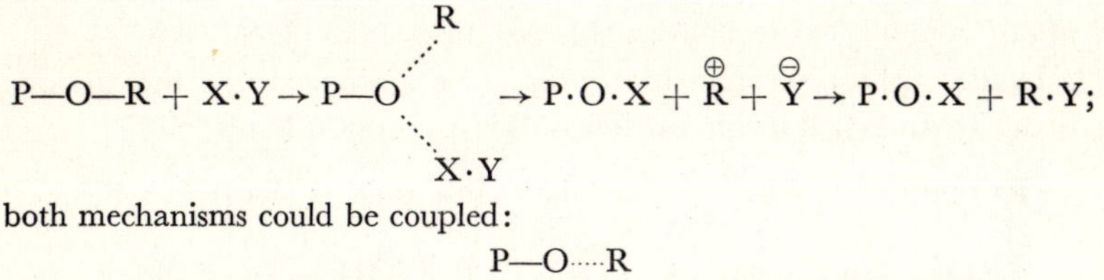

or both mechanisms could be coupled:

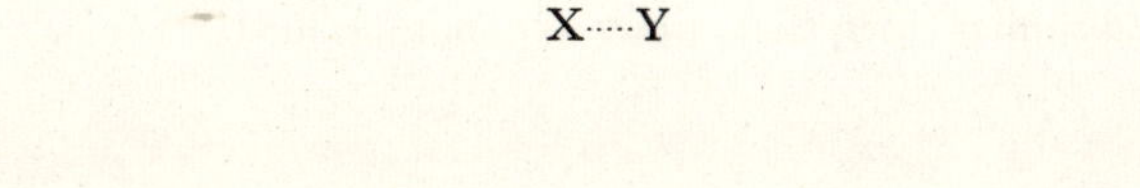

The polar compound, X·Y, can be an acid anhydride, acid chloride or halogen.

This type of reaction is especially useful for preparing asymetrical or unstable derivatives of pyro- or tri-phosphoric acids:

$$(R_2N)_2PO\cdot OEt + Cl\cdot PO(OR')_2 \xrightarrow[20°,\ 30\ min]{none} EtCl + (R_2N)_2PO\cdot O\cdot PO(OR')_2.$$ [26]

$$(MeO)_3PO + Cl\cdot PO(OMe)_2 \xrightarrow[113°,\ 2\cdot5\ hr]{none} MeCl + (MeO)_2PO\cdot O\cdot PO(OMe)_2,\ 85\%.$$ [27]

tetramethyl pyrophosphate

$$2(Me_2N)_2PO\cdot OEt + Cl_2\cdot PO\cdot NMe_2 \xrightarrow[150°,\ 1\ hr]{none} 2EtCl + (Me_2N)_2PO\cdot O\cdot PO(NMe_2)\cdot O\cdot PO(NMe_2)_2,\ 70\%.$$ [28]

decamethyltriphosphoramide

Other types of anhydride have also been prepared, as, for example, derivatives of thiocarbamic acid:[29]

$$(EtO)_3PO + Me_2N\cdot CS\cdot Cl \xrightarrow[150°,\ 1\ hr]{none} EtCl + (EtO)_2PO\cdot O\cdot CS\cdot NMe_2,\ 80\%.$$

O-(diethyl phosphoric) N-dimethylthiocarbamic anhydride

When the substituent introduced in the first stage of the reaction is very electrophilic, the initial product may react with more of the starting material. Several methods of preparing TEPP from triethyl phosphate depend on this. The reactions using phosgene may be written:[30]

$$(EtO)_2PO\cdot OEt + \underset{Y-X}{Cl-CO\cdot Cl} \rightarrow EtCl + (EtO)_2PO\cdot O\cdot CO\cdot Cl,$$

$$(EtO)_2PO\cdot OEt + \underset{X-Y}{(EtO)_2PO-O\cdot CO\cdot Cl} \rightarrow EtO\cdot CO\cdot Cl + TEPP,$$

$$EtO\cdot CO\cdot Cl \rightarrow EtCl + CO_2.$$

The overall reaction is thus:

$$2(EtO)_3PO + COCl_2 \xrightarrow[130°,\ 2\ hr]{none} CO_2 + 2EtCl + TEPP,\ 60\%.$$

Chlorine,[31] thionyl,[32] phosphoryl[35] and acetyl chlorides[33] and phosphorus pentoxide[34] have been used instead of phosgene. Yields are usually only about 50–60%, as numerous side reactions take place.[35]

*A*3. Dialkyl phosphorofluoridates may be prepared by the action of anhydrous hydrogen fluoride on tetra-alkyl pyrophosphates:[36, 37, 38]

$$(RO)_4P_2O_3 + HF \xrightarrow[0°,\ min]{none} (RO)_2PO\cdot OH + (RO)_2PO\cdot F,^*$$

$$[(RO)_2PS]_2O + HF \xrightarrow[50°,\ 1\ hr]{none} (RO)_2PSOH + (RO)_2PS\cdot F.$$

* In this and similar cases, exact times were not specified.

The reactions go to completion, and give pure products easy to separate, but special equipment is needed to handle anhydrous hydrogen fluoride. By a similar reaction tetra-alkyl pyrophosphorothionates may be prepared in excellent yields by the reaction between potassium cyanide and phosphinyl disulphides:[39]

$$[(RO)_2PO\cdot S]_2 + KCN \xrightarrow[20^\circ]{\text{acetone}} (RO)_2PO\cdot S\cdot CN + (RO)_2POS^\ominus + K^\oplus \rightarrow KSCN + (RO)_2PO\cdot O\cdot PS(OR)_2.$$

When R is Et, the yield is 82%. The second step is an A1 type displacement. The disulphide intermediates are readily prepared by mild oxidation of hydrogen thioates or dithioates, with iodine, for example:[40]

$$2(RO)_2POS^\ominus + I_2 \xrightarrow[\text{v. fast}]{} 2I^\ominus + [(RO)_2PO\cdot S]_2,\ 100\%.$$

*A*4. An ether and phosphorus pentoxide give a pyrophosphate:[41]

$$P_2O_5 + 2R_2O \xrightarrow[180^\circ,\ \text{high pressure}]{\text{none}} (RO)_4P_2O_3.$$

There are many side reactions, yielding phosphates, triphosphates, and metaphosphates.[42]

*B. Additions of Phosphates and Phosphorodithioates to Unsaturated Compounds*

With two exceptions, B4 and B5, these reactions involve the addition of dialkyl hydrogen dithioates or their anions to olefinic, aldehydic or ketonic double bonds.

*B*1. Hydrogen dithioates add to olefins without polar substituents.

When the double-bonded carbon atoms are attached to different nucleophilic groups, such as different alkyl groups, two products can be obtained. Thus 1-octene can give either the 1-octyl or the 2-octyl ester:[43]

$$(RO)_2PS\cdot SH + CH_2{:}CH\cdot C_6H_{13} \nearrow (RO)_2PS\cdot S\cdot CH_2\cdot CH_2\cdot C_6H_{13}.\ \text{anti-Markownikov addition}$$

$$(RO)_2PS\cdot SH + CH_2{:}CH\cdot C_6H_{13} \searrow (RO)_2PS\cdot S\cdot CH(Me)\cdot C_6H_1\ \text{Markownikov addition}$$

A similar situation is found in carbon chemistry. Hydrogen halides add to pure olefins in such a way that the halogen atom goes on to the more highly substituted carbon. This is Markownikov's rule. In the presence of traces of peroxides the halogen atom adds to the less substituted carbon by a much faster, free radical mechanism, i.e. the addition is 'anti-Markownikov'. The two mechanisms are discussed by Dewar.[53] In the addition reactions under consideration, most workers have found addition to take place according to the rule; some, however, have found the opposite. The discrepancy can be explained as follows.[43] When the dithioates are purified, most olefins contain enough peroxides to cause the addition to disobey the rule. When crude dithioates, prepared by the action of alcohols on phosphorus pentasulphide, are used, they contain enough $P_4S_3$ to reduce the peroxides, and the addition obeys Markownikov's rule. It is also obeyed if both the olefins and the dithioates are carefully purified, as then there are, of course, no peroxides. In this case apparently a little purification is worse than none. One can now cause the addition to go either way at will. If crude dithioates are used, or a little hydroquinone is added to destroy the peroxides, the rule is obeyed. If the dithioates are purified, and a little cumene peroxide is added, it is not.

As in carbon chemistry, the rule is only obeyed when the substituents tend to donate electrons, e.g. when they are aryl, alkyl or thio-alkyl groups. When one carbon atom is attached to a very electrophilic group, such as a carboxylic ester group, the reaction is of a different type, B2, and the rule does not apply. If the substituents are of an intermediate type, such as chlorine, it is not clear what will happen, as the polarity of the C—Cl bond may be reversed under the influence of the reagents when the carbon is double-bonded. This possibility has not so far been investigated under conditions which rule out interference by peroxides.

*B*2. Addition to olefinic and treble[113] bonds also takes place when the olefin is substituted by an electrophilic radical, especially one with which it can conjugate. The reaction has been studied with carbonyl,[44] sulphonyl[45] and cyanide[46] substituents. The sulphur always attaches itself to the carbon farther from the polar group, and the reaction may be represented by:

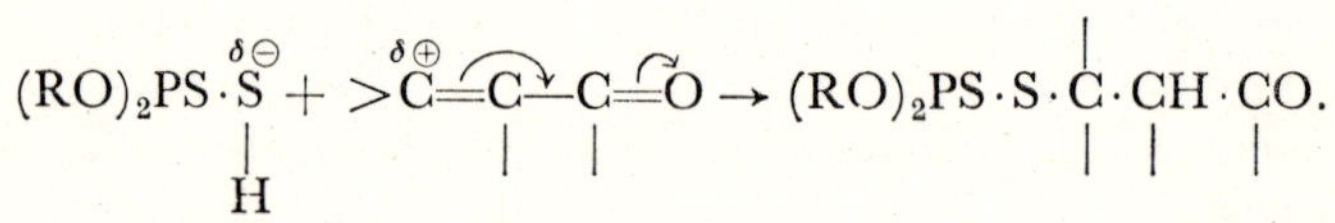

As is to be expected from the mechanism, the reaction is catalysed by bases, the dithioate ion reacting more rapidly than the acid, but the reaction is often so rapid that no catalyst is required.[46, 57, 58] Thus O,O-diethyl S-2-methylsulphonylethyl phosphorodithioate is prepared by the reaction:[45]

$$(EtO)_2PS\cdot SH + CH_2{:}CH\cdot SO_2\cdot Me \xrightarrow[100°,\ \text{few min}]{\text{none, cat. }Et_3N} (EtO)_2PS\cdot S\cdot CH_2\cdot CH_2\cdot SO_2\cdot Me,$$

and Malathion by the reaction:[47]

$$\underset{\displaystyle H}{(MeO)_2PS\cdot \overset{|}{S}} + \underset{\displaystyle CH\cdot CO\cdot OEt}{\overset{\|}{CH\cdot CO\cdot OEt}} \xrightarrow[65°,\ 24\ \text{hr}]{\text{none}} \underset{\displaystyle CH_2\cdot CO\cdot OEt,}{(MeO)_2PS\cdot S\cdot \overset{|}{CH}\cdot CO\cdot OEt}\ 90\%.$$

Malathion

*B*3. Hydrogen dithioates add to aldehydic and ketonic groups, the sulphur becoming attached to the carbonyl carbon, i.e. to the more positive atom in the group, as in B2 type reactions. The addition to aldehydes is particularly useful, as the products condense with thiols(48, 49) and amides(50, 51) to give useful insecticides:

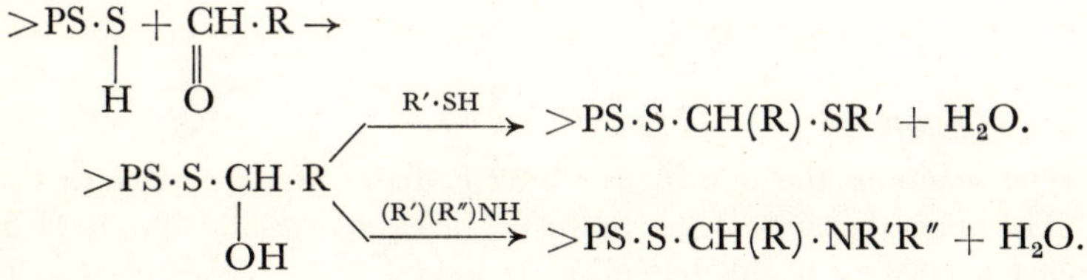

R and R′ = H or alk; R″ = H, alk, or CO·alk.

Both the addition and the condensation reactions are usually carried out simultaneously, as in the preparation of Thimet:(48)

$$(EtO)_2PS\cdot SH + CH_2{:}O + EtSH \xrightarrow[<50°,\ 4\ hr]{water} H_2O + \underset{\text{Thimet}}{(EtO)_2PS\cdot S\cdot CH_2\cdot SEt}, 90\%.$$

Theoretically, the condensation stage of this reaction could give numerous other products; e.g. the hydroxymethyl compound might condense with more dithioate to give $[(EtO)_2PS\cdot S]_2CH_2$; or $H_2S$ might be eliminated instead of $H_2O$ to give $(EtO)_2PS\cdot S\cdot CH_2\cdot OEt$. It has, however, been shown that such side reactions are not important.

*B*4. Diaryl and dialkyl hydrogen phosphates (but not phosphorothioates)(24) react at room temperature in dioxane with dicyclohexyl carbodiimide to give nearly quantitative yields of pyrophosphates.(52, 54) The yields are not affected by traces of water.

Probably there are three steps in the reaction:

(1) Addition:

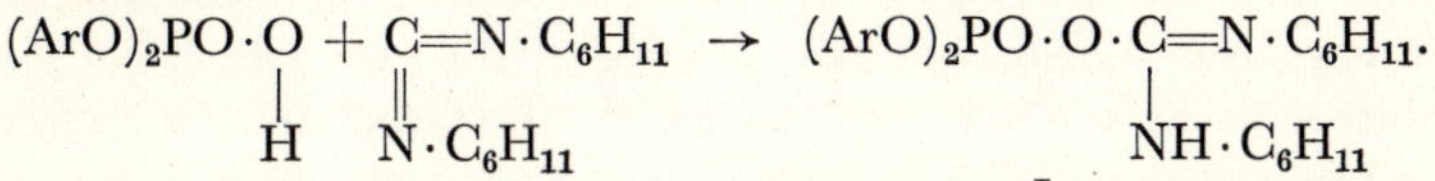

I.

(2) Ionization:

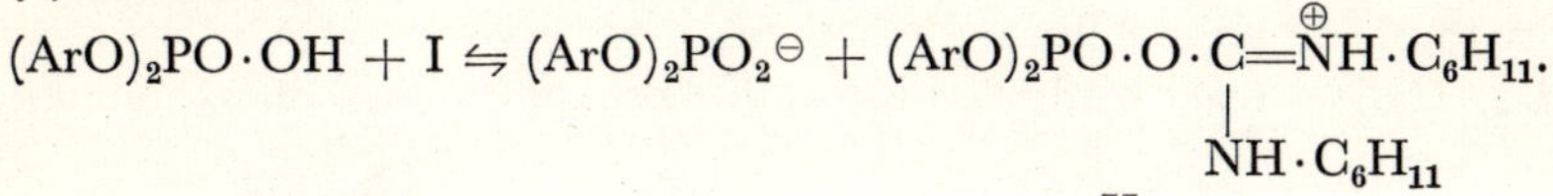

II.

(3) A1 type reaction:

$$(ArO)_2PO_2^{\ominus} + II \rightarrow (C_6H_{11}\cdot NH)_2CO + (ArO)_4P_2O_3.$$

*B*5. Addition to diazo compounds. Hydrogen dithioates add readily to many diazo compounds by the reaction:[55]

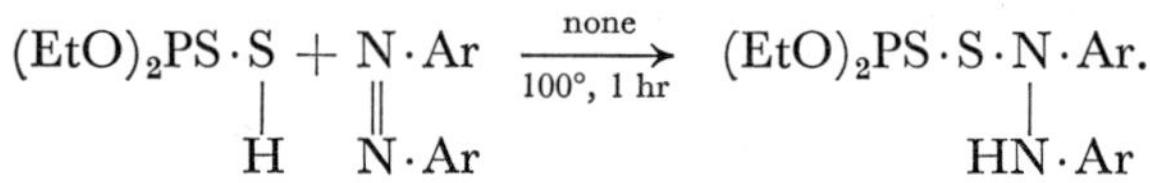


*C. Additions of Compounds of the Type* $(RO)_2P$(O *or* S)·S·X *to Olefins*

In B type reactions the phosphorus intermediates are free acids or their anions. The general reaction can, however, be extended, addition to olefins taking place according to the general equation:

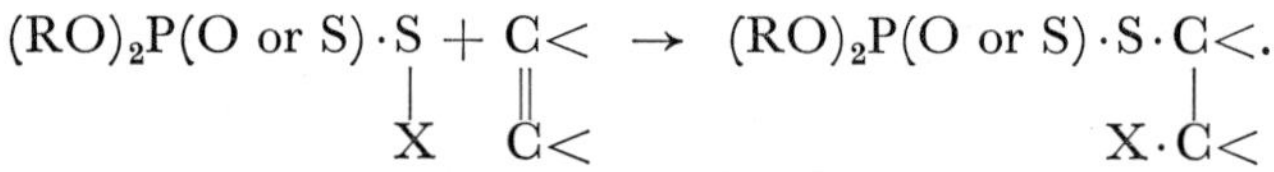


The reaction has been studied with X as Cl[56] and S·PS$(OR)_2$. Delnav, for example, can be prepared by the reaction:[60]

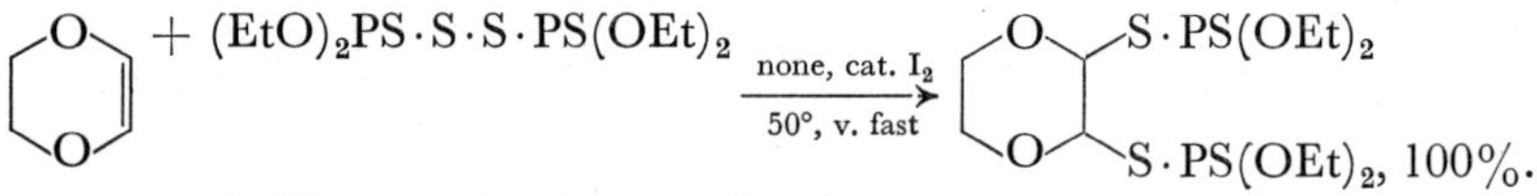


bis-(diethoxyphosphinothioyl) disulphide

Delnav

*D. Substitutions into Dialkyl Hydrogen Phosphonates*

Compounds of the general formula $(RO)_2POH$ exist in the phosphonate form, $(RO)_2PO·H$, and not to any appreciable extent in the phosphite form, $(RO)_2P·OH$.[61, 62, 63, 64] This corresponds to the nature of the products obtained from substitution reactions, which have the structure to be expected if the hydrogen atom in the phosphonate form is displaced:

$$(RO)_2PO\cdot H + X\cdot Y \rightarrow (RO)_2PO\cdot X + H\cdot Y.$$

In several of the reactions, however, the sodium salts are used, in which the phosphite structure must predominate, and in one where the acid is used, D2, it has been shown that the kinetics of the reactions greatly favour phosphite ions as the reactive species.[65] Probably they always are in substitution reactions. Compounds of tri-covalent phosphorus are unsaturated, and readily form covalent bonds with electrophilic groups and atoms; whereas it is difficult to see why a phosphonate should show the high reactivity implied by the following reactions.

*D*1. Halides react readily with metal phosphites to give phosphonates:

$$\overset{\delta\ominus}{Hal}—\overset{\delta\oplus}{X} + :PO(OR)_2^{\ominus} \rightarrow Hal^{\ominus} + X—PO(OR)_2.$$

As, to be of toxicological interest, the products must contain acid anhydride bonds, the reactions of interest here are those with acyl halides or pseudohalides. Thus sulphenyl chlorides, R·S·Cl, and alkyl thiocyanates,

R·S·CN give compounds containing C·S·P bonding, and selenocyanates, R·Se·CN, give R·Se·P compounds.[11, 66] For example, the cyanide group in thiocyanates acts as a pseudo-halogen in the preparation of Amiton:[67]

$$(EtO)_2P\cdot ONa + NC\cdot S\cdot CH_2\cdot CH_2\cdot NEt_2 \xrightarrow[45°,\ 1\cdot 5\ hr]{benzene} NaCN + \underset{\text{Amiton}}{(EtO)_2PO\cdot S\cdot CH_2\cdot CH_2\cdot NEt_2}.$$

Thiocyanogen, $(CNS)_2$, gives phosphoro-isothiocyanitidates, which add amines, alcohols, etc., to yield derivatives of thiourea, as in the reaction:[68]

$$(RO)_2PO\cdot NCS + R'NH_2 \xrightarrow[20°,\ min]{ether} (RO)_2PO\cdot NH\cdot CS\cdot NHR'.$$

N-alkyl N′-(dialkoxyphosphinyl) thiourea

*D*2. Carbon tetrachloride reacts with hydrogen phosphonates in the presence of catalytic base to give phosphorochloridates:[65]

$$(RO)_2PO\cdot H + CCl_4 \xrightarrow[20°,\ 3\ hr]{CCl_4,\ cat.\ Et_3N} CHCl_3 + (RO)_2PO\cdot Cl,\ 80\%.$$

The mechanism of this reaction, which is highly complex, has already been referred to. If the reaction is carried out in the presence of amines, alcohols or acids, A1 type reactions take place in the way expected with the chloridates formed.

Phosphorochloridates are also formed, without the addition of a base, by the action of unstable N—Cl compounds, e.g. chloramine-T[69] and N-chlorosuccinimide:[70]

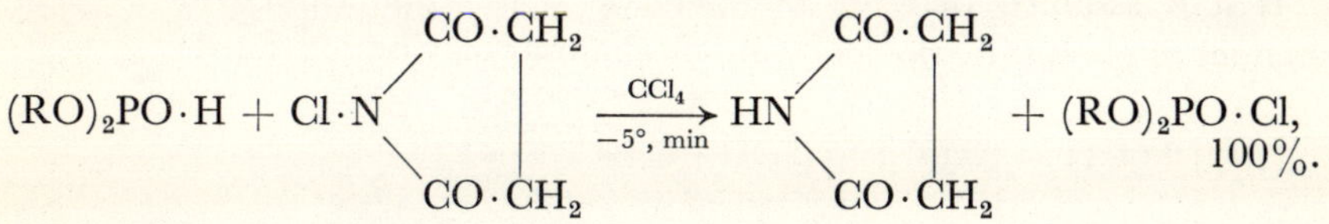

The mechanism has not been studied. As no free acid is liberated in the reaction, the chloridates are easily fractionated.

*D*3. Hydrogen phosphonates or their salts react with phosphorochloridites in the presence of sulphur. The best method of preparing Sulfotepp is by the reaction:[71, 72]

$$(EtO)_2P\cdot ONa + (EtO)_2P\cdot Cl + 2S \xrightarrow[110°,\ 2\ hr]{MePh} NaCl + \underset{\text{Sulfotepp}}{(EtO)_2PS\cdot O\cdot PS(OEt)_2},\ 75\%.$$

Until the sulphur is added there is no reaction. This suggests that the first step is the addition of sulphur to the chloridite to give the chloridothionate: $(EtO)_2PS\cdot Cl$. In the chloridite the phosphorus atom is trivalent, and therefore nucleophilic, but in the chloridothionate it is pentavalent and electrophilic, and is therefore open to attack by the phosphite ion (an A1 type of reaction). The mixed anhydride formed: $(EtO)_2P\cdot O\cdot PS\cdot(OEt)_2$, then adds more sulphur to give the final product.

*D4*. Hydrogen phosphonates form Grignard reagents, from which 1-hydroxyalkylphosphonates may be prepared:[73]

$$(EtO)_2PO\cdot H + R'\cdot Mg\cdot Br \xrightarrow[20°,\ 20\ \text{min}]{\text{ether}} R'\cdot H + (EtO)_2PO\cdot Mg\cdot Br,\ 97\%$$

$$\Big\downarrow \text{reflux, few hr};\ + R''\cdot CHO,\ \text{ether}$$

$$(EtO)_2PO\cdot CH(OH)\cdot R'' + MgClBr \xleftarrow{\text{HCl aq.}} (EtO)_2PO\cdot CH(R'')\cdot O\cdot Mg\cdot Br$$

*E. Addition to Dialkyl Hydrogen Phosphonates*

Ketones add to hydrogen phosphonates in the presence of catalytic amine, traces of benzoyl peroxide greatly increasing the rate. The products are 1-hydroxyalkylphonates.[74, 75] A more important reaction, which requires no catalyst, takes place with chloral, to give 2:2:2-trichloro-1-hydroxyethylphosphonates, of which Dipterex is the most important.[76]

$$(MeO)_2PO\cdot H + CCl_3\cdot CHO \xrightarrow[60°,\ \text{min}]{\text{none}} \underset{\text{Dipterex}}{(MeO)_2PO\cdot CH(OH)\cdot CCl_3},\ > 90\%.$$

The structure is proved by elementary analysis, and by the observations that no hydrochloric acid is liberated in the reaction and that the product can be acetylated, indicating the presence of a hydroxyl group.

If it is assumed that the phosphonate is in equilibrium with a small amount of phosphite, the following mechanism may be suggested:

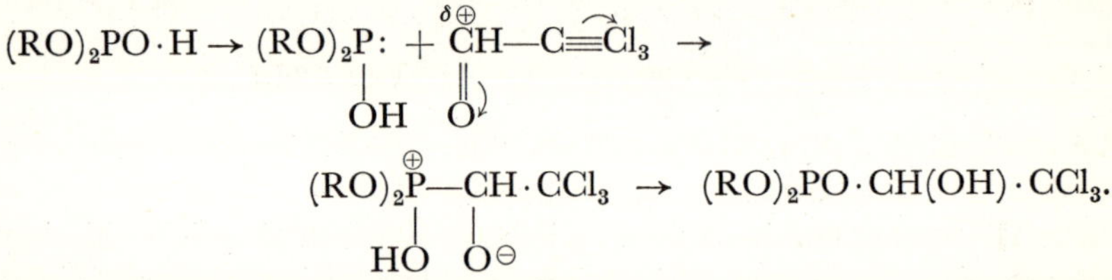

The carbonyl carbon is made unusually positive by the $—CCl_3$ group. In the first reaction, that between ketones and phosphonates, a base is required to raise the concentration of the phosphite to a level such that reaction with the much less positive carbon in the ketone can go forward at a reasonable rate.

*F. Substitutions into Trialkyl Phosphites*

One of the best known reactions in phosphorus chemistry is that discovered by Michaelis[78, 79] and developed and studied by Arbuzov.[80] An alkyl halide reacts with a trialkyl phosphite to give a dialkyl alkylphosphonate:

$$(RO)_3P + R'\cdot Hal \rightarrow (RO)_2R'\cdot PO + R\cdot Hal.$$

The intermediate probably has a phosphonium structure,[81]

$$[(RO)_3P\cdot R']^{\oplus}Hal^{\ominus}.$$

The reactions of interest here are of two types: those in which the reaction takes place in the normal way, such that the atom which was attached to the halogen becomes attached to the phosphorus; and those, abnormal, reactions in which an oxygen atom becomes attached to the phosphorus.

*F*1. *Normal reactions.* As the desired products are acid anhydrides and esters, acid halides or pseudo-halides are allowed to react with trialkyl phosphites[82], as in the preparation of the S-phenyl isomer of Parathion:[83]

$$(EtO)_3P + Cl\cdot S\cdot C_6H_4\text{-}4\text{-}NO_2 \xrightarrow[20°,\ min]{benzene} EtCl + (EtO)_2PO\cdot S\cdot C_6H_4\text{-}4\text{-}NO_2.$$

Thiocyanates,[84] selenocyanates[11] and cyanogen iodide react similarly, the last yielding cyanidates:

$$(EtO)_3P + CN\cdot I \xrightarrow[20°,\ 1{\cdot}5\ hr]{ether} EtI + (EtO)_2PO\cdot CN.$$ [85]

$$(Me_2N)_2P\cdot OEt + CN\cdot I \xrightarrow[20°,\ min]{ether} EtI + (Me_2N)_2PO\cdot CN.$$ [86]

*F*2. *Abnormal reactions.* Any trialkyl phosphite reacts rapidly with chloral, but the product is not the expected phosphonate, $(RO)_2PO\cdot CCl_2\cdot CHO$, but a 2:2-dichlorovinyl phosphate:

$$(RO)_3P + CCl_3\cdot CHO \rightarrow RCl + (RO)_2PO\cdot O\cdot CH{:}CCl_2.$$

The structure of the diethyl homologue has been proved as follows:[77]

(1) It gives no reactions for an aldehyde group, e.g. it will not add disulphite, reduce Fehling's solution, or give a 2:4-dinitrophenylhydrazone. The dimethyl homologue, DDVP, does give some of these reactions, but it has been shown that it is hydrolysed to dichloroacetaldehyde under the test conditions:

$$\underset{\text{DDVP}}{(MeO)_2PO\cdot OCH{:}CCl_2} + OH^{\ominus} \rightarrow (MeO)_2PO_2^{\ominus} + CHCl_2\cdot CHO.$$

(2) The compound takes up an equimolar quantity of chlorine without liberating HCl, which indicates the double bond.

(3) It cannot be acetylated, and gives no other reactions indicative of a hydroxyl group.

The mechanism of the reaction has been much discussed. A modification of a theory already put forward seems most plausible to me:

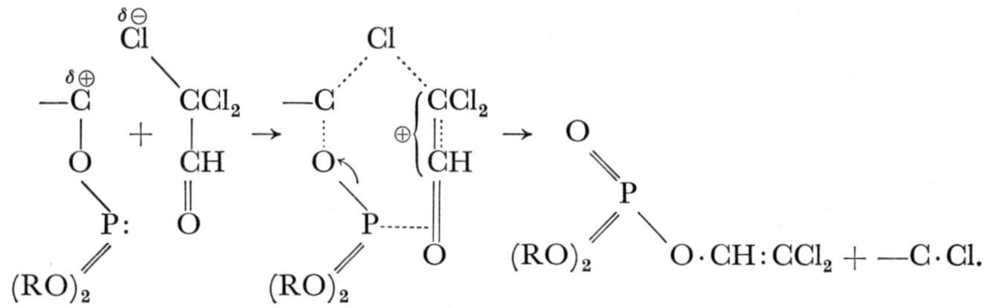

The first step is the formation of a $\pi$ complex between the C═O $\pi$ electrons and the un-bonded p electrons on the phosphorus. Such a bond is not localized, but will hold the molecules in the position shown, which is a very favourable one for the nucleophilic attack of a chlorine atom on the α-carbon of the alkyl group. This attack makes the $CCl_3$ carbon more positive, so that electrons are withdrawn from the carbonyl bond, giving a mesomeric state in which the positive charge is spread, but in which the $\pi$ bond is weakened. The removal of the alkyl group by the chlorine leads to the polarization of the P—O bond, so that the phosphorus atom becomes electrophilic. It therefore attracts the carbonyl oxygen atom, and repels the positively charged ┈CH┈$CCl_2$ system, and gives the final products.

This theory agrees with the following observations:

(a) No chloride ion is found in the intermediates which can be separated in some instances; nor does infra-red spectroscopy reveal vinyl groups in them.

(b) The reactions take place under mild conditions, which suggests that P—C bonds are not formed and then broken.

(c) The rate of reaction is not much influenced by the introduction of two relatively positive groups such as ($Me_2N$) on the phosphorus. Such groups should slow the last stage of the reaction more than alkoxy groups, but they should also stabilize the $\pi$ complex.

(d) The reaction is most rapid with chloro-ketones or -aldehydes in which the mesomeric cation postulated would be stabilized either by electronegative substituents or by an extended conjugated system over which the charge can be spread. Dichloroacetaldehyde reacts fairly readily, but monochloroacetaldehyde reacts very slowly. α-Chloro-β-diketones and α-chloro-β-keto-esters react very rapidly, as in the preparation of Phosdrin:[87]

$$(MeO)_3P + MeCO\cdot CH(Cl)\cdot CO\cdot OEt \xrightarrow[<85^\circ]{\text{none}} MeCl + (MeO)_2PO\cdot O\cdot CMe{:}CH\cdot CO\cdot OEt.$$

Phosdrin

In the transition state the positive charge could be spread over three carbon atoms:

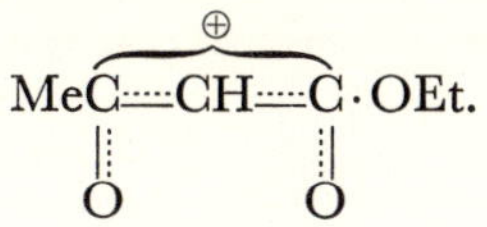

The reaction can also take place with the carbonyl group of trichloroacetic esters:[88]

$$(MeO)_3P + Cl_3C\cdot CO\cdot OR' \xrightarrow[60°,\ 2\ hr]{none} MeCl + (MeO)_2PO\cdot O\cdot C(OR'):CCl_2,$$

α-chloro-, bromo- and iodo-ketones may either react in this way or in the normal way to give phosphonates.[89, 90] Usually mixtures are produced, the proportion of phosphonate increasing in the order: —Cl, —Br, —I.[91] This is the order in which the tendency to form quaternary derivatives, as in the first stage of the Arbuzov reaction, increases, and in which the C—Hal polarity, and hence reactivity in the abnormal reaction, decreases.

α-Diketones react abnormally,[92] but α-chloroacyl chlorides undergo both types of reaction in turn:[93]

$$(EtO)_3P + CCl_3\cdot CO\cdot Cl \rightarrow EtCl + (EtO)_2PO\cdot CO\cdot CCl_3,\ (I)$$
(normal reaction)

$$I + (EtO)_3P \rightarrow EtCl + (EtO)_2PO\cdot C(:CCl_2)\cdot O\cdot PO(OEt)_2.$$
(abnormal reaction)

When the olefinic carbon atoms are asymetrically substituted, there are two stereo-isomers of each compound. Phosdrin, for example, is a mixture of the two isomers:

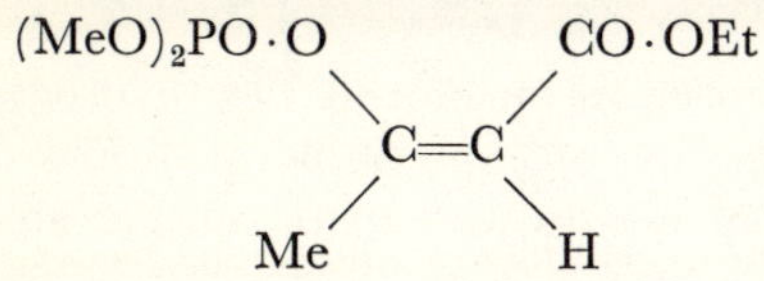

and

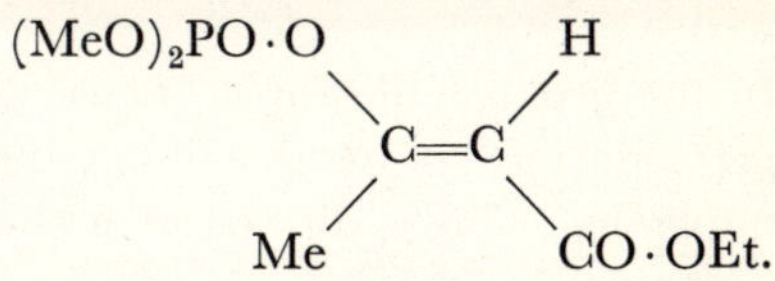

The stereo-isomers of most compounds of this type can be separated on a silica column.[94] In each case it is found that one isomer, called the α-isomer, is more polar, less stable to alkalis and to oxidizing agents, and much more toxic than the other, β-isomer, and is therefore probably the *cis*-isomer. As this is not established, the isomers will be referred to as α and β. F2 type reactions give the isomers in the proportions, α:β, of about 70:30. Some of the compounds may be prepared also from β-keto esters and dialkyl phosphorochloridates by A1 type reactions:

$$(MeO)_2PO\cdot Cl + MeC(ONa):CH\cdot CO\cdot OEt \rightarrow NaCl + Phosdrin.$$

The products obtained in this way contain only about 6% of the α-isomer, and are therefore much less active poisons. Ultra-violet light converts either isomer to a mixture, containing, in the case of Phosdrin, 30% of α-isomer.

### *G. Oxidation of Phosphate Esters*

The reactions are of two types; oxidation of thiono-sulphur atoms, and side chain oxidations. They are important mainly for preparing compounds which are also metabolites of the parent compounds, but they have been used commercially to give more water-soluble and less malodorous products.

*G*1. Oxidation of thiono sulphur is usually carried out with nitric acid (s.g. 1·4), and nearly quantitative yields are obtained. Thus Malathion can be converted to Malaoxon:[112]

$$(MeO)_2PS\cdot S\cdot CH(CH_2\cdot CO\cdot OEt)\cdot CO\cdot OEt \xrightarrow[25°,\ hr]{HNO_3} \underset{\text{Malaoxon}}{(MeO)_2PO\cdot S\cdot CH(CH_2\cdot CO\cdot OEt)\cdot CO\cdot OEt.}$$

Bromine water may be used in some cases.

*G*2. Sulphides may be oxidized to sulphoxides with hydrogen peroxide, chlorine[115] or bromine,[98] and to sulphones with potassium permanganate in weakly acid solution or with hydrogen peroxide, as in the following examples:

$$(EtO)_2PO\cdot S\cdot CH_2\cdot CH_2\cdot SEt + H_2O_2 \xrightarrow[80°,\ 30\ min]{none} H_2O + (EtO)_2PO\cdot S\cdot CH_2\cdot CH_2\cdot SO\cdot Et,\ 90\%.^{(96)}$$

$$(EtO)_2PO\cdot S\cdot CH_2\cdot CH_2\cdot SEt \xrightarrow[0°,\ 1\ hr]{8\%\ aq.\ KMnO_4 + H_2SO_4} (EtO)_2PO\cdot S\cdot CH_2\cdot CH_2\cdot SO_2\cdot Et,\ 80\%.^{(97)}$$

In the first reaction up to twice the theoretical amount of 15% hydrogen peroxide may be used without the formation of appreciable quantities of sulphone. This is formed at higher temperatures with more concentrated peroxide.

N-dimethylphosphoramidic compounds have been oxidized with many different agents. The structure of the products is not known with certainty. These reactions are discussed in Chapter IX.

### *H. Quaternization Reactions*

If a phosphate ester contains a tertiary amino group or a thio group in the side chain, ammonium or sulphonium derivatives may be prepared by the action of alkyl halides and sulphates. The products are structurally similar to acetylcholine, $MeCO\cdot O\cdot CH_2\cdot CH_2\cdot \overset{\oplus}{N}Me_3$, and the most toxic phosphorus compounds known are prepared in this way. Two examples are given below. More may be found in the next Chapter.

$$(EtO)_2PO\cdot O\cdot C_6H_4\text{-3-}NMe_2 + Me_2SO_4 \xrightarrow[20°,\ 29\ hr]{benzene} (EtO)_2PO\cdot O\cdot C_6H_4\text{-3-}\overset{\oplus}{N}Me_3\ MeHSO_4^{\ominus},\ 66\%.^{(99)}$$

$$(MeO)_2PO\cdot S\cdot CH_2\cdot CH_2\cdot SEt + MeI \xrightarrow[37°,\ 39\ hr]{EtOH} (MeO)_2PO\cdot S\cdot CH_2\cdot CH_2\cdot \overset{\oplus}{S}EtMe\ I^{\ominus},\ 9\%.^{(100)}$$

*I. Isomerization Reactions*

Phosphorothionates isomerize on heating to phosphorothiolates:(101) PS·OR → PO·SR. Thus Parathion isomerizes to its S-ethyl isomer:(102)

$$(EtO)_2PS\cdot O\cdot C_6H_4\text{-4-}NO_2 \xrightarrow[170°,\ 10\ hr]{none} (EtO)(EtS)PO\cdot O\cdot C_6H_4\text{-4-}NO_2,\ 80\%.$$

Side reactions take place, and in the case of dimethyl phosphorothionates may become explosively rapid.(101)

The reaction takes place more readily when a cyclic immonium or sulphonium intermediate can be formed. The mechanism in these cases is well known,(95, 103–106) and follows the course illustrated by the two examples given below:

$$(EtO)_2PS\cdot O\cdot CH_2\cdot CH_2\cdot NEt_2 \rightarrow (EtO)_2POS^{\ominus} + \left[\begin{array}{c} CH_2 \\ | \\ CH_2 \end{array}\right\rangle \overset{\oplus}{N}Et_2 \rightarrow (EtO)_2PO\cdot S\cdot CH_2\cdot CH_2\cdot NEt_2.$$

$$(MeO)_2PS\cdot O\cdot CH_2\cdot CH_2\cdot SEt \rightarrow (MeO)_2POS^{\ominus} + \left[\begin{array}{c} CH_2 \\ | \\ CH_2 \end{array}\right\rangle \overset{\oplus}{S}Et \rightarrow (MeO)_2PO\cdot S\cdot CH_2\cdot CH_2\cdot SEt.$$

As the intermediate is ionic, the reaction is most rapid in polar solvents, and can be carried out satisfactorily with no solvent at all. In polar, non-aqueous solvents yields are better than 80%, but side reactions are more important in water.(106) These are discussed in the next Chapter.

*J. Dehydrochlorination of 2-Chloro-1-hydroxyalkylphosphonates and Similar Reactions*

Dipterex and its analogues undergo a very rapid reaction in alkaline solution. From Dipterex itself the product is DDVP:(107)

$$(MeO)_2PO\cdot CH(OH)\cdot CCl_3 + OH^{\ominus} \xrightarrow[20°,\ min]{water} Cl^{\ominus} + H_2O + \underset{DDVP}{(MeO)_2PO\cdot O\cdot CH{:}CCl_2},\ 50\%.$$

The products are unstable in alkaline solution, so only theoretical quantities of base are used. The dichloro analogues give low yields of the mono-chlorovinyl compounds, and the mono-chloro analogues do not react. A similar reaction takes place when α-keto-phosphonate esters are treated in turn with sodium bisulphite and sodium cyanide:[108]

$$(RO)_2PO\cdot CO\cdot R' \xrightarrow[20^\circ,\ min]{water} (RO)_2PO\cdot O\cdot CH(CN)\cdot R',$$

and when compounds of the type: $(RO)_2PO\cdot CMe(OH)\cdot CO\cdot Me$ are warmed with sodium ethoxide in ethanol.[116] The following mechanism has been suggested:

$$>PO\cdot CO\cdot R' \xrightarrow{NaHSO_3} >PO\cdot C(OH)(SO_3Na)\cdot R' \xrightarrow{CN^\ominus}$$

$$>PO\cdot \overset{\displaystyle O^\ominus}{\overset{|}{\underset{\underset{\displaystyle CN}{|}}{C}}}\cdot R' \rightarrow >PO\cdot O\cdot \overset{\ominus}{C}(CN)R' \xrightarrow{+H^\oplus} \text{product.}$$

These reactions are the only cases known of the conversion of phosphonates to phosphates.

*K. Miscellaneous Reactions*

*K*1. If a trialkyl phosphite is heated with an involatile alcohol, trans-esterification leads to the formation of the mixed ester. This can add sulphur under mild conditions, and so provide a method for preparing unstable thionates, e.g. Demeton-O:[109]

$$(EtO)_3P + HO\cdot CH_2\cdot CH_2\cdot SEt \xrightarrow[205^\circ,\ 5\ hr]{none,\ cat.\ Na}$$

$$EtOH + (EtO)_2P\cdot O\cdot CH_2\cdot CH_2\cdot SEt\ (I).$$

$$I + S \xrightarrow[15^\circ,\ 16\ hr]{none} (EtO)_2PS\cdot O\cdot CH_2\cdot CH_2\cdot SEt.$$

*K*2. Side chains can generally be made to undergo organic reactions in the usual way. The main limitations are that basic reagents cannot be used, as they are liable to break the anhydride bond by A1 type nucleophilic attack, and strong oxidizing agents cannot be used with thionates. A few examples only are given.

Vinyl phosphates may take part in Diels–Alder reactions, for example with hexachlorocyclopentadiene:[110]

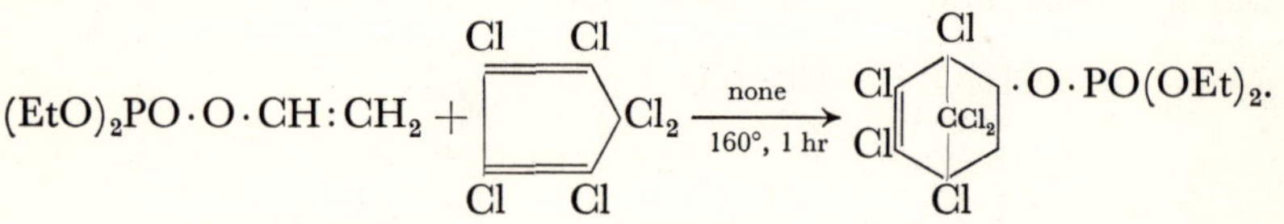

The product of the B1 type reaction between maleic anhydride and a dialkyl phosphorodithioate can undergo numerous reactions, as in the sequence which follows ($X = (RO)_2PS \cdot S—$):[111]

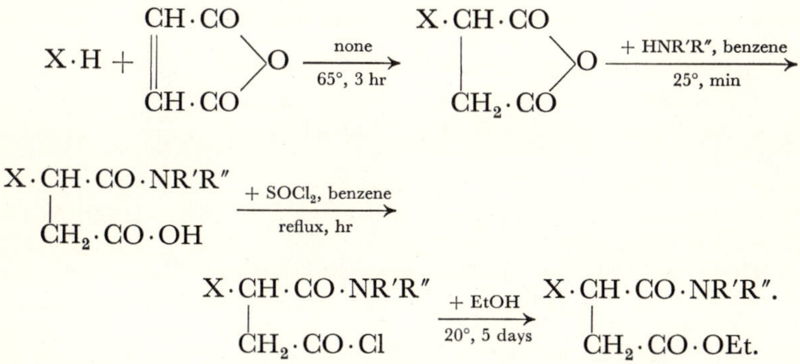


More examples of such reactions appear in later Chapters.

## REFERENCES

Abbreviation for Patent literature: *U.S.* = American; *Brit.* = British; *Ger.* = German; *Sw.* = Swiss. Authors are not given, as some patents give the names of industrial concerns as authors.


1. *Ger.* 830,508
2. *Brit.* 644,616
3. *Brit.* 703,160
4. *U.S.* 2,810,756
5. *Brit.* 690,835–6
6. *U.S.* 2,504,165
7. *U.S.* 2,810,757
8. D. F. Heath and P. Casapieri. *Trans. Faraday Soc.* **47,** 1093 (1951)
9. A. D. F. Toy and T. M. Beck. *J. Amer. chem. Soc.* **72,** 3191 (1950)
10. *U.S.* 2,784,207
11. G. Schrader. *Die Entwicklung neuer Insektizide u.s.w.*, Verlag Chemie, Weinheim, 1952
12. *Brit.* 688,760
13. *U.S.* 2,649,464
14. *Brit.* 668,536
15. B. Holmstedt. *Acta physiol. scand.* **25,** Suppl. 90 (1951)
16. *Brit.* 602,446
17. G. Olah and A. Oswald. *Liebigs Ann.* **602,** 118 (1957)
18. L-E. Tammelin. *Acta chem. scand.* **11,** 859 (1957)
19. M. I. Kabachnik, S. T. Joffe and T. A. Mastryukova. *J. gen. Chem., Moscow* **25,** 653 (1955)
20. *Brit.* 797,307
21. V. V. Katyshkina and M. Y. Kraft. *J. gen. Chem., Moscow* **26,** 3407 (1956)
22. R. Rätz and E. Thilo. *Z. anorg. Chem.* **272,** 333 (1953)
23. *Ger.* 830,509

24. R. A. McIvor, G. D. McCarthy and G. A. Grant. *Canad. J. Chem.* **34,** 1819 (1956)
25. D. G. Coe, B. J. Perry and R. K. Brown. *J. chem. Soc.* 3604 (1957)
26. *Brit.* 732,384
27. A. D. F. Toy. *J. Amer. chem. Soc.* **71,** 2268 (1949)
28. *U.S.* 2,673,876
29. *U.S.* 2,648,696
30. French patent 1,092,956
31. *U.S.* 2,502,400
32. J. R. Robinson. *Canad. J. Chem.* **33,** 722 (1958)
33. *U.S.* 2,748,153
34. *U.S.* 2,402,703
35. E. Cherbuliez, G. Cordahi and J. Rabinowitz. *Helv. chim. Acta* **42,** 590 (1959)
36. *U.S.* 2,614,116
37. A. Hood and W. Lange. *J. Amer. chem. Soc.* **72,** 4956 (1950)
38. W. E. White and A. Hood. *J. Amer. chem. Soc.* **74,** 853 (1952)
39. *Ger.* 947,366
40. O. Foss. *Acta chem. scand.* **1,** 307 (1947)
41. *Brit.* 649,584
42. *U.S.* 2,492,153
43. W. E. Bacon and W. M. LeSuer. *J. Amer. chem. Soc.* **76,** 670 (1954)
44. *U.S.* 2,632,020
45. *Ger.* 876,691
46. N. N. Melnikov and K. D. Shvetsova-Shiloskaya. *J. gen. Chem., Moscow* **23,** 1417 (1953)
47. *Ger.* 847,897
48. E. L. Clark, G. A. Johnson and E. L. Mattson. *J. Agric. Food Chem.* **3,** 834 (1955)
49. B. A. Arbuzov, K. V. Nikonorov and G. M. Vinokurova. *Bull. Acad. Sci.* (*Chem.*) 597 (1955)
50. D. Shvetsova-Shilovskaya, N. N. Melnikov and N. I. Martemyanova. *J. gen. Chem.* **26,** 523 (1956)
51. *U.S.* 2,566,129
52. J. G. Moffatt and H. G. Khorana. *J. Amer. chem. Soc.* **79,** 3741 (1957)
53. M. J. S. Dewar. *The Electronic Theory of Organic Chemistry*, Oxford Univ., 1949
54. H. G. Khorana and A. R. Todd. *J. chem. Soc.* 2257 (1953)
55. *U.S.* 2,865,949
56. J. Michalski and B. Lenard. *Roczniki Chem.* **30,** 655 (1956)
57. *U.S.* 2,578,652
58. G. R. Norman, W. M. LeSuer and T. W. Mastin. *J. Amer. chem. Soc.* **74,** 161 (1952)
59. *U.S.* 2,928,864
60. W. R. Drively, A. H. Haubein, A. D. Lohr and P. B. Moseley. *J. Amer. chem. Soc.* **81,** 139 (1959)
61. A. E. Arbuzov, M. I. Batnev and V. S. Vinogradova. *C.R. Acad. Sci. U.R.S.S.* **54,** 599 (1946)
62. M. I. Kabachnik and T. A. Mastryukova. *Chem. Abstr.* **48,** 3243e (1954)
63. C. F. Callis, J. R. van Wazer, J. N. Shoolery and W. A. Anderson. *J. Amer. chem. Soc.* **79,** 2719 (1957)
64. L. W. Daasch. *J. Amer. chem. Soc.* **80,** 5301 (1958)
65. G. M. Steinberg. *J. org. Chem.* **15,** 637 (1950)

66. *Ger.* 830,262
67. *Brit.* 763,516
68. J. Michalski and J. Wieczorkowski. *Roczniki Chem.* **31,** 585 (1957)
69. S. Chwalinski and W. Rypinska. *Roczniki Chem.* **31,** 539 (1957)
70. H. Goldwhite and B. C. Saunders. *J. chem. Soc.* 2040 (1955)
71. A. E. Arbuzov, B. A. Arbuzov, P. I. Alimov, K. V. Nikonorov, N. I. Rizpolozhenskii and O. M. Fedorova. *Chem. Abstr.* **54,** 10636 (1957)
72. *Ger.* 896,643
73. O. Gawron, C. Grelecki, W. Reilly and J. Sands. *J. Amer. chem. Soc.* **75,** 3591 (1953)
74. *U.S.* 2,579,810
75. *U.S.* 2,593,213
76. *U.S.* 2,701,225
77. W. Perkow. *Ber. dtsch. chem. Ges.* **87,** 755 (1954)
78. A. Michaelis and W. La Coste. *Ber. dtsch. chem. Ges.* **18,** 2109 (1885)
79. A. Michaelis and R. Kaehne. *Ber. dtsch. chem. Ges.* **31,** 1048 (1898)
80. A. E. Arbuzov. *Structure of Phosphorous Acid and its Derivatives,* St. Petersburg, 1905
81. G. N. Kosolapoff. *Organophosphorus Compounds,* p. 326. John Wiley, New York, 1950
82. D. C. Morrison. *J. Amer. chem. Soc.* **77,** 181 (1955)
83. *Brit.* 739,726
84. *Ger.* 926,488
85. B. C. Saunders, G. J. Stacey and J. G. E. Wilding. *J. chem. Soc.* 699 (1948)
86. L. Larsson. *Acta chem. scand.* **6,** 1470 (1952)
87. *U.S.* 2,685,552
88. *Brit.* 767,225
89. N. Kreuzkamp and H. Kayser. *Ber. dtsch. chem. Ges.* **89,** 1614 (1956)
90. B. A. Arbuzov, V. S. Vinogradova and N. A. Polezhaeva. *Chem. Abstr.* **53,** 1180i (1959)
91. H. I. Jacobson, M. J. Griffin, S. Preis and E. V. Jensen. *J. Amer. chem. Soc.* **79,** 2608 (1957)
92. V. A. Kukhtin. *Chem. Abstr.* **53,** 1105 (1959)
93. *Brit.* 783,697
94. P. E. Gatterdam, J. E. Casida and D. W. Stoutamire. *J. econ. Ent.* **52,** 270 (1959)
95. T. R. Fukuto and R. L. Metcalf. *J. Amer. chem. Soc.* **76,** 5103 (1954)
96. *U.S.* 2,791,599
97. *Ger.* 948,241
98. K. Groves. *J. Agric. Food Chem.* **6,** 30 (1958)
99. K. J. M. Andrews, F. R. Atherton, F. Bergel and A. L. Morrison. *J. chem. Soc.* 780 (1952)
100. D. F. Heath and M. Vandekar. *Biochem. J.* **67,** 187 (1957)
101. J. B. McPherson, Jr. and G. A. Johnson. *J. Agric. Food Chem.* **4,** 42 (1956)
102. *U.S.* 2,601,219
103. D. F. Heath. *J. chem. Soc.* 1643 (1958)
104. N. Muller and J. Goldenson. *J. Amer. chem. Soc.* **78,** 5182 (1956)
105. T. R. Fukuto and E. M. Stafford. *J. Amer. chem. Soc.* **79,** 6083 (1957)
106. A. Henglein and G. Schrader. *Naturforsch.* **10B,** 12 (1955)
107. W. Lorenz, A. Henglein and G. Schrader. *J. Amer. chem. Soc.* **77,** 2554 (1955)

108. L. A. R. Hall, C. W. Stephens and J. J. Drysdale. *J. Amer. chem. Soc.* **79,** 1768 (1957)
109. F. W. Hoffmann and T. R. Moore. *J. Amer. chem. Soc.* **80,** 1150 (1958)
110. *U.S.* 2,767,206
111. *U.S.* 2,630,451
112. *U.S.* 2,713,018
113. *U.S.* 2,912,450
114. *U.S.* 2,910,500
115. *U.S.* 2,898,341
116. V. A. Kukhtin, V. S. Abramov and K. M. Orekhova. *Chem. Abstr.* **54,** 7536a (1960)

## CHAPTER III

# Chemical Reactions and Methods of Purification

THE REACTIONS considered in this Chapter are those which take place on heating or storage, or with common reagents. The former particularly often have profound effects, illustrated in the tables, on the toxicities and biochemical activities of the compounds. Their study forms a natural prelude to a description of methods of purification, which have to be devised in such a way as to minimize the formation of toxic products during the purification procedures.

Hydrolytic reactions are described in the next Chapter; some reactions studied only in attempts to solve the structure of toxic metabolites are described in Chapter IX; and no attempt has been made to describe analytical reactions systematically, though analytical uses of the reactions described are indicated. Methods of analysis are reviewed annually in *Analytical Chemistry*.

## REACTIONS

The effects of a reaction on toxicity are measured by the difference between the $LD_{50}$'s of the parent compound and the product. The $LD_{50}$ is the dose which kills half the animals under the conditions of the test. The $LD_{50}$ found depends on the species, strain, sex and age of the test animal, the way the compound is given, and in which solvent, if any, and sometimes even on the time of day. Some of these variations are discussed more fully in Chapter VIII. Different workers using different strains of the same species usually agree on an $LD_{50}$ to within a factor of 2. Smaller differences have little significance, except in strictly comparable experiments carried out by the same people. In this Chapter and throughout the rest of the book such internally standardized comparisons are used whenever possible. More weight should therefore be given to differences between $LD_{50}$'s obtained within a particular experiment than to absolute values.

The effects of a reaction on biochemical activity are measured by the differences between the $I_{50}$'s to enzymes of the parent compound and the product. The $I_{50}$ is the molar concentration of an enzyme inhibitor which halves the rate at which the enzyme converts its substrate. The enzyme used

TABLE 3.1. THE EFFECTS OF ISOMERIZATION, DISPROPORTIONATION AND OXIDATION ON TOXICITY AND BIOCHEMICAL ACTIVITY

| Parent compound | Reaction | Product | $LD_{50}$ (mg/kg) | $I_{50}$ (molar) | References |
|---|---|---|---|---|---|
| α-Phosdrin<br>$(MeO)_2PO\cdot O\cdot CMe{:}CH\cdot CO\cdot OMe$ | — | — | 0·35, i-p., rat | $1{\cdot}7 \times 10^{-8}$ | (1) |
| | isomerization | β-isomer | 35, i-p., rat | $1{\cdot}2 \times 10^{-6}$ | |
| Parathion<br>$(EtO)_2PS\cdot O\cdot C_6H_4\text{-}4\text{-}NO_2$ | — | — | 3, i-v., rat | $2 \times 10^{-4}$ | (3, 4, 5) |
| | isomerization | $(EtO)(EtS)PO\cdot O\cdot C_6H_4\text{-}4\text{-}NO_2$<br>(S-ethyl isomer) | 1·2, i-v., rat | $1 \times 10^{-6}$ | |
| | oxidation and disproportionation | $(EtO)_2PO\cdot O\cdot C_6H_4\text{-}4\text{-}NO_2$<br>(Paraoxon) | 0·4, i-v., rat | $6{\cdot}6 \times 10^{-9}$ | |
| | disproportionation | $EtO\cdot PS(O\cdot C_6H_4\text{-}4\text{-}NO_2)_2$ | — | $>5 \times 10^{-3}$ | |
| | and transesterification | $EtS\cdot PO(O\cdot C_6H_4\text{-}4\text{-}NO_2)_2$ | — | $6 \times 10^{-9}$ | (6) |
| Demeton-O-Me<br>$(MeO)_2PS\cdot O\cdot CH_2\cdot CH_2\cdot SEt$ | — | — | 216, i-v., rat | — | (13) |
| | isomerization | Demeton-S-Me | 65, i-v., rat | $6{\cdot}5 \times 10^{-5}$ | |
| Demeton-S-Me | self-alkylation | $(MeO)_2PO\cdot S\cdot CH_2\cdot CH_2\cdot \overset{\oplus}{S}EtMe$ | 0·062, i-v., rat | $3{\cdot}9 \times 10^{-8}$ | |
| | | $(MeO)_2PO\cdot S\cdot CH_2\cdot CH_2\cdot \overset{\oplus}{S}Et\cdot CH_2\cdot CH_2\cdot SEt$ | 0·007, i-v., rat | $4 \times 10^{-9}$ | (13, 20) |
| | oxidation | $(MeO)_2PO\cdot S\cdot CH_2\cdot CH_2\cdot SO\cdot Et$ | 47, i-v., rat | $4{\cdot}1 \times 10^{-5}$ | (13) |
| | | $(MeO)_2PO\cdot S\cdot CH_2\cdot CH_2\cdot SO_2\cdot Et$ | 22, i-v., rat | $2{\cdot}3 \times 10^{-5}$ | |
| Diazinon<br>$(EtO)_2PS\cdot O\cdot$ 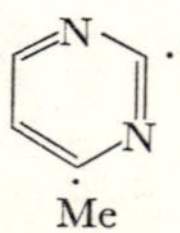 $\cdot CHMe_2$ | | | 120, oral, mouse | $3 \times 10^{-5}$ | (11) |
| | oxidation and disproportionation | $(EtO)_2PS\cdot O\cdot PO(OEt)_2$ | 4·2, oral, mouse | $3 \times 10^{-9}$ | |
| | | $[(EtO)_2PS]_2O$ | — | $4 \times 10^{-8}$* | |

Routes of administration of the compounds are indicated by: i-v. = intravenous; i-p. = intraperitoneal; oral = by mouth.
* The biochemical activity given here is surprisingly high for a compound of this structure, and is probably incorrect.

for comparisons is acetylcholinesterase (AChE), as results obtained on this enzyme are relevant to the study of toxic action. $I_{50}$'s have little absolute significance unless several stringent conditions are satisfied in their determination, as discussed in Chapter VI. Relative values obtained by the same method do, however, provide a valid basis for comparison.

A low $LD_{50}$ and $I_{50}$ correspond, of course, to a high toxicity and biochemical activity respectively.

*Reactions of Compounds with Themselves*

The first group of reactions to be considered are those in which a compound reacts with itself. Although these reactions only take place rapidly on heating, or, sometimes, on exposure to ultra-violet light, they also take place on storage at any temperature at which the compound remains liquid. A very rough guide to the effect of temperature is that these reactions take place about 1000 times more rapidly at 100° than at 20°.

(1) A compound containing substituents asymmetrically placed about a double bond goes to the equilibrium mixture of *cis* and *trans* isomers, slowly in the dark, and rapidly in ultra-violet light (Chapter II, F2). Thus, pure α-Phosdrin is converted to a mixture containing about 70% of the much less active and toxic β-isomer (see Table 3.1). Several compounds of this type have been studied, and all undergo this reaction.[1]

(2) Phosphorothionates with at least one alkyl substituent give numerous products on heating, of which the S-alkyl thiolate predominates unless heating is very prolonged. The first reaction to be studied was the isomerization of triethyl phosphorothionate to O,O,S-triethyl phosphorothiolate:[2]

$$(EtO)_3PS \rightarrow (EtO)_2PO \cdot SEt.$$

There is some evidence that $(MeO)_2PS \cdot S \cdot CH(CO \cdot OEt) \cdot CH_2 \cdot CO \cdot OEt$, Malathion, isomerizes in the same way to its S-methyl isomer.[3] More is known, however, about the reactions of dialkyl aryl thionates, and of a few compounds with nitrogen or sulphur atoms in aliphatic side chains. The basic reactions of these two types are different.

The dialkyl aryl thionates give S-alkyl isomers rapidly on heating above 150°. The S-aryl isomers are, apparently, not formed. As under mild conditions the S-alkyl isomers are the main products, and are highly reactive compounds, the numerous other products found are presumably formed by secondary transesterification reactions. Those found by different workers on heating Parathion are listed in Table 3.2, and their toxic properties in Table 3.1. A similar range of products has been obtained on heating a number of other thionates, e.g. Parathion-Me[3, 4] and -$Pr^i$, EPN[3], Chlorthion,[4] and Diazinon,[7, 9, 10] though in some instances only one or two have been positively identified. Some possible reactions which would yield the products found on heating Parathion are listed below, but there are many alternatives:

$$\mathrm{Ar} = \cdot \mathrm{C_6H_4\text{-}4\text{-}NO_2}.$$

$$\mathrm{(EtO)_2PS \cdot OAr} \rightarrow \mathrm{(EtO)(EtS)PO \cdot OAr}.$$

$$\mathrm{(EtO)(EtS)PO \cdot OAr} + \mathrm{(EtO)_2PS \cdot OAr} \begin{cases} \nearrow \mathrm{EtO \cdot PO(OAr)_2} + \mathrm{(EtO)_2PS \cdot SEt}. \\ \rightarrow \mathrm{(EtO)_2PO \cdot OAr} + \mathrm{(EtO)(EtS)PS \cdot OAr}. \\ \searrow \mathrm{EtO \cdot PO(OAr)_2} + \mathrm{(EtO)_2PO \cdot SEt}. \end{cases}$$

$$\mathrm{(EtO)_2PO \cdot SEt} + \mathrm{(EtO)_2PS \cdot OAr} \rightarrow \mathrm{EtSAr} + \mathrm{(EtO)_2PO \cdot O \cdot PS(OEt)_2}.$$

TABLE 3.2. THE PRODUCTS OBTAINED ON HEATING PARATHION

In the first experiment Parathion was heated at 150° for 24 hr,[3] and in the second at 140° for 40 hr.[4] The products were separated by reverse phase partition chromatography, in the same way in both experiments, and the 4-nitrophenol content of the spots was determined, quantitatively in the first experiment, semi-quantitatively in the second. In the second the capacity of the spots to inhibit cholinesterase was also determined. Ar = $\cdot \mathrm{C_6H_4\text{-}4\text{-}NO_2}$ in the tables.

| First experiment[3] | | | Second experiment[4] | | | |
|---|---|---|---|---|---|---|
| Product identified | $R_f$ | % | Product identified | $R_f$* | Density of spot | Inhibition of ChE |
| $\mathrm{EtO \cdot PS(OAr)_2}$ | 0·01 | 2·8 | | | | |
| $\mathrm{(EtO)_2PS \cdot OAr}$ (Parathion) | 0·04 | 4·6 | Parathion† | 0·03 | v. low | none |
| ? | 0·14 | 8·8 | ? | 0·12 | v. low | none |
| ? | 0·25 | 6·5 | ? | 0·17 | v. low | none |
| ? | 0·36 | 2·4 | $\mathrm{EtO \cdot PO(OAr)_2}$ | 0·24 | high | great |
| $\mathrm{(EtO)(EtS)PO \cdot OAr}$ (S-ethyl isomer) | 0·46 | 57·3 | S-ethyl isomer | 0·36 | high | great |
| $\mathrm{(EtO)_2PO \cdot OAr}$ (Paraoxon) | 0·71 | 10·9 | Paraoxon | 0·56 | high | great |
| HO·Ar | 0·86 | 9·0 | HO·Ar | 0·76 | high | none |
| | | | P·O·P compound** | 0·90 | v. low | moderate |

* The $R_f$'s in this column are from the figure, not the text, in the original paper.

† $\mathrm{EtO \cdot PS(OAr)_2}$ would not have been distinguished from Parathion in this experiment.

** The author did not suggest any structure for this compound. From the results obtained on Diazinon (see main text) it is likely to have been tetraethyl pyrophosphorothionate, Thionotepp.

The formation of Thionotepp, $(EtO)_2PO \cdot O \cdot PS(OEt)_2$, from Parathion has not been established, but it is formed from Diazinon in 14% yield along with 3% of Sulfotepp, $[(EtO)_2PS]_2O$[11] (see Table 3.1), and its solubility properties are consistent with those of an inhibitor found in heated Parathion (see Table 3.2).

Parathion and several other thionates have been shown to produce the same compounds on storage at room temperatures as on heating[3, 4] and the reactions are markedly catalysed by ultra-violet light.[3, 32, 56] The last observation suggests that free-radical mechanisms are involved.

The mixtures resulting from these reactions are much more biochemically active than the parent compounds, which are almost inert, whereas some of the products are amongst the most active known. The toxicity is often little affected, though the action of the poison is usually more rapid.

Reactions of the type already described are accompanied by others which give degradation products such as alkyl metaphosphates, sulphur and sulphur dioxide. Under more drastic conditions, especially in the presence of metal catalysts, these reactions predominate, and may reach explosive violence. The product is a charred mass.[12]

The few compounds studied with a sulphur or nitrogen atom in the side chain react somewhat differently on heating. O-dialkylaminoethyl and O-alkylthioethyl dialkyl phosphorothionates isomerize readily, as described in Chapter II, I. In the cases investigated in detail, the isomerizations of diethyl 2-diethylaminoethyl and dimethyl and diethyl 2-ethylthioethyl phosphorothionates, the reactions give good yields of the S-2-diethylamino-ethyl and S-2-ethylthioethyl isomers at moderate temperatures. The S-ethyl and S-methyl isomers do not seem to be formed. The reaction proceeds via a cyclic intermediate.[13, 14] Thus Demeton-O-Me gives the more toxic Demeton-S-Me (see Table 3.1) by the reaction:

$$(MeO)_2PS \cdot O \cdot CH_2 \cdot CH_2 \cdot SEt \rightarrow$$

Demeton-O-Me

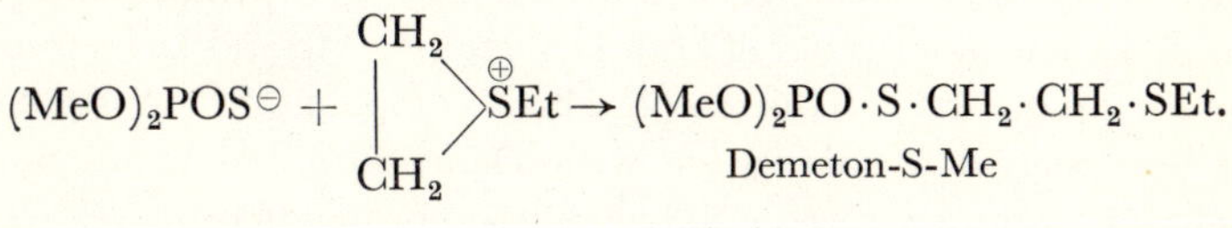


It follows from the mechanism of this reaction that its rate and the nature of the products should depend upon chain length, i.e. upon n in the formula: $(RO)_2PS \cdot O \cdot (CH_2)_n \cdot SEt$. When $n = 1$, a cyclic intermediate cannot be formed, and the reaction should go slowly or not at all. As n increases from 2 to 5 the formation of the ionic intermediate should take place faster, for the ring becomes more stable, but recombination of the ions to give the isomer should become more difficult. It is therefore possible that when n is greater than 2 the main product would be a non-toxic salt, and not the S-ethylthioalkyl isomer.* These possibilities have not been investigated.

In water the reactions are greatly accelerated, and the main products from Demeton-O-Me are ions, not Demeton-S-Me. The proportion of ionic products increases as the concentration of Demeton-O-Me is decreased.[15] These are formed because the sulphonium ion is unstable, and decomposes faster than it reacts with the phosphorothioate ion. The effect is probably one of concentration rather than of solvent—even in the absence of water or solvent of any kind some ionic products are formed alongside the isomer.

(3) Compounds containing secondary sulphur or ternary nitrogen atoms may alkylate themselves. Demeton-O-Me in water provides an unusual example, the cyclic sulphonium intermediate reacting with the thiolate already formed:

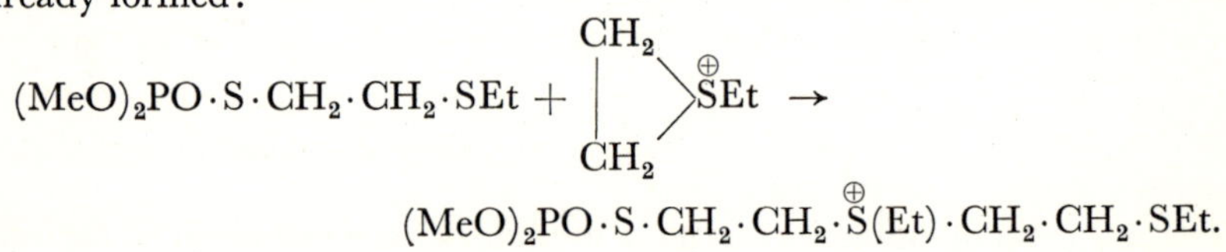

The yields are less than 1%, and the product is rather unstable, which is probably why it has not been found in old samples of Demeton-O-Me.[15] The compound formed is, however, of such startling toxicity (Table 3.1) that within a few minutes of preparing a 1% aqueous suspension of the parent compound the intravenous toxicity and biochemical activity are increased by one-hundred-fold.[16] Demeton-S-Me undergoes a more common reaction both in water and on storage, one molecule of the compound methylating another:†

$$2(MeO)_2PO \cdot S \cdot CH_2 \cdot CH_2 \cdot SEt \rightarrow$$
$$MeO \cdot PO_2^{\ominus} \cdot S \cdot CH_2 \cdot CH_2 \cdot SEt + (MeO)_2PO \cdot S \cdot CH_2 \cdot CH_2 \cdot \overset{\oplus}{S}EtMe.$$

The toxicity and biochemical activity are again greatly increased (Table 3.1). Samples a few months old of both Demeton-O-Me and Demeton-S-Me may contain up to 4% of this sulphonium compound.[13] A similar reaction has been shown to take place with Demeton-S.[13] As alkyl phosphates (including, contrary to general belief, tri-isopropyl phosphate) are alkylating agents,[18, 19] this type of reaction may take place frequently. The reactivity of the trialkyl phosphate varies with the alkyl group, falling in the order: $Me \gg Et > Pr, Bu \gg Pr^i$, and probably only dimethyl phosphates react

* For $n > 5$, the rings should be difficult to form, but it may be possible in this way to make rings of unusual size.

† In water Demeton-S-Me also gives the unstable inhibitor, showing that the last step of the isomerization reaction is to some extent reversible.

readily enough to form more than 1% of the -onium derivatives on storage or heating, as these are not very stable. In the cases of Demeton-S and Demeton-S-Me 1% is enough to increase the biochemical activity and intravenous toxicity 10–20 times. Any -onium derivative formed from a compound of the general structure:

$$(RO)_2PO\cdot(O \text{ or } S)\cdot(C)_{2 \text{ or } 3}\cdot(N< \text{ or } S—)$$

is expected to be much more biochemically active than its parent compound, for reasons given in Chapter VI. In other cases the effect of alkylation on the toxicity cannot be foretold.

Intramolecular alkylation can also take place, with the formation of a betaine,[17] e.g.

$$(MeO)_2PO\cdot O\cdot C_6H_4\text{-}3\text{-}NMe_2 \xrightarrow[160°,\ 30\ \text{min}]{\text{none}} MeO\cdot PO_2^{\ominus}\cdot O\cdot C_6H_4\text{-}3\text{-}\overset{\oplus}{N}Me_3,\ 50\%.$$

3-(dimethoxyphosphinyloxy)-N-dimethylaniline

3-(methyl phosphato)-trimethylanilinium betaine

The higher dialkyl homologues react more slowly, so they can be distilled under reduced pressure without extensive decomposition. Toxicity is destroyed by these reactions.

2-(methylfluorophosphinyloxy)-ethyldimethylamine is also destroyed by a quaternization reaction to a compound of low toxicity:[21]

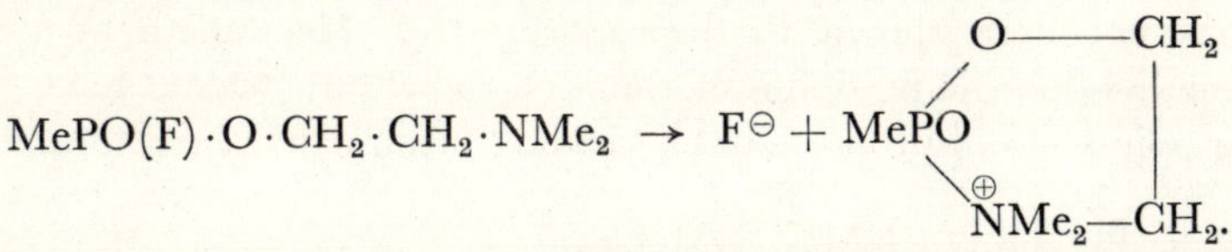


(4) Compounds containing pyrophosphate bonding, P—O—P, tend to disproportionate on heating. Thus pentaethyl triphosphate is said to go to triethyl phosphate, TEPP and ethyl metaphosphate on attempted distillation,[22] and TEPP itself disproportionates to triethyl phosphate and ethyl metaphosphates:[23, 24]

$$(EtO)_2PO\cdot O\cdot PO(OEt)_2 \xrightarrow[>100°]{\text{none}} (EtO)_3PO + \frac{1}{n}(EtO\cdot PO_2)_n.$$

The reactions are catalysed by acids, which are readily formed by hydrolysis if water is present. Consequently TEPP can only be distilled successfully if it is completely dry. Prior separation from metaphosphates is also necessary, as they raise the boiling point of the mixture, and hence the rate of the destructive reactions. Tetramethyl pyrophosphate is less stable than TEPP, and the higher homologues are more stable, especially those containing secondary alkyl groups.

Only TEPP itself is appreciably toxic in the mixtures of ethyl phosphates and polyphosphates obtained in the manufacture of TEPP. Crude preparations of other polyphosphates probably owe their toxicity entirely to their pyrophosphate content.[25] This is the case with the insecticide which used to be called hexaethyl tetraphosphate, HETP, and the name is now used to denote a mixture containing 20% of TEPP. The reactions taking place on heating TEPP obviously lower the toxicity, but those taking place when crude commercial products containing a high proportion of polyphosphates are heated may either raise or lower the toxicity, according as to whether the TEPP content is raised or lowered.

The dimethylamides of triphosphoric and higher acids also disproportionate, but the main product is Schradan, not hexamethylphosphoramide. Thus decamethyltriphosphoramide decomposes mainly as follows:

$$\underset{\text{Decamethyltriphosphoramide}}{(Me_2N)_2PO \cdot O \cdot PO(NMe_2) \cdot O \cdot PO(NMe_2)_2} \xrightarrow[160^\circ]{\text{none}} \underset{\text{Schradan}}{(Me_2N)_4P_2O_3} + \frac{1}{n}(Me_2N \cdot PO_2)_n.$$

It can, however, be distilled at lower temperatures, unlike pentaethyl triphosphate. At about 200°/1 mm a cyclic decamethylpentaphosphoramide distils from the mixture, and a very acidic residue is left behind.[26] Two isomeric dodecamethyltetraphosphoramides have been prepared, and also give Schradan on heating.[27] Schradan is much more stable than TEPP, and has not been shown to disproportionate. Heptamethyl-octyl- and heptamethyl-dodecyl-pyrophosphoramides, however, which have higher boiling points, do disproportionate, yielding hexamethylphosphoramide:[28]

$$(Me_2N)_2PO \cdot O \cdot PO(NMe_2)(NMeOct) \rightarrow (Me_2N)_3PO + \text{metaphosphoramides.}$$

Of the phosphoramides discussed, hexamethylphosphoramide is nearly non-toxic, Schradan and decamethyltriphosphoramide are highly toxic, and the higher members are believed to be of low toxicity, as they are readily hydrolysed by water.

(5) Two compounds containing sulphoxide groups in the side chain are reduced to their sulphide analogues on heating and on storage: the sulphoxide of Demeton-S-Me,[13] $(MeO)_2PO \cdot S \cdot CH_2 \cdot CH_2 \cdot SO \cdot Et$, and $(EtO)_2PS \cdot O \cdot C_6H_4\text{-}4\text{-}SO \cdot Me$.[29] This is not a disproportionation reaction, as the sulphones are not produced, and the mechanism is unknown. The reaction is accompanied in both cases by others which have a greater effect on the toxicity and activity.

(6) It has been found that many phosphate esters contain traces of tetra-alkyl pyrophosphates.[30] Their presence could be attributed in many

instances to traces of water in the reagents used in the preparations, especially when these were carried out by the A1 type of reaction:

$$(RO)_2PO\cdot Cl + {}^{\ominus}OR' \rightarrow Cl^{\ominus} + (RO)_2PO\cdot OR'.$$

0·1% w/w of water in either reagent, or 0·01% in the solvent, is enough, theoretically, to give above 1% of tetra-alkyl pyrophosphate in the product by the reaction:

$$2(RO)_2PO\cdot Cl + H_2O \rightarrow (RO)_4P_2O_3 + 2HCl.$$

The compounds were, however, purified by methods which seem likely to have separated them fairly efficiently from pyrophosphates, and it is possible that the latter were formed by transesterification reactions during distillation, especially if a little water was present to form some hydrogen phosphate by hydrolysis:[31]

$$(RO)_2PO\cdot OR' + (RO)_2PO_2^{\ominus} \rightarrow (RO)_4P_2O_3 + {}^{\ominus}OR'.$$

The effects on biochemical activity of small quantities of TEPP in various diethyl aryl phosphates are shown in Table 3.3.

TABLE 3.3. THE EFFECT OF TEPP IN SOME DIETHYL ARYL PHOSPHATES ON THEIR BIOCHEMICAL ACTIVITY[30]

$I_{50}$'s are given for some diethyl aryl phosphates, previously supposed to be pure, before and after purification by selective hydrolysis. The active impurity was shown to be TEPP, whose percentage in each specimen was calculated from the difference. The 4-nitrophenyl compound, Paraoxon, was the only one so active itself that a small proportion of TEPP would not have affected its activity.

| Compound $(EtO)_2PO\cdot O\cdot C_6H_4$—X | $I_{50}$'s | | |
|---|---|---|---|
| X | impure | purified | % TEPP |
| 4-chloro | $5{\cdot}75 \times 10^{-6}$ | $2{\cdot}10 \times 10^{-4}$ | 0·68 |
| 2-chloro | $2{\cdot}24 \times 10^{-5}$ | $1{\cdot}1 \times 10^{-4}$ | 0·053 |
| 4-nitro | $2{\cdot}0 \times 10^{-8}$ | $2{\cdot}01 \times 10^{-8}$ | — |
| 2-nitro | $4{\cdot}58 \times 10^{-7}$ | $1{\cdot}48 \times 10^{-6}$ | 0·33 |
| 3-nitro | $1{\cdot}2 \times 10^{-6}$ | $3{\cdot}0 \times 10^{-6}$ | 0·55 |
| H (phenyl) | $1{\cdot}35 \times 10^{-5}$ | $3{\cdot}77 \times 10^{-3}$ | 0·025 |

From: W. N. Aldridge and A. N. Davison, *Biochem. J.* **51,** 62 (1952), with permission. The table is slightly shortened.

*Reactions in the Presence of Air and Light*

Air does not attack most compounds rapidly in the absence of light. Many experiments have, however, been carried out in which thin films of thionates (Parathion and its analogues, EPN, etc.) and other phosphorus compounds containing sulphur were exposed to air and ultra-violet light at room temperature in order to simulate the condition of spray residues. Chromatography then revealed numerous products, similar to those obtained on heating the same compounds. Each thionate gave at least two inhibitors of acetylcholinesterase, one of which chromatographed like the corresponding phosphate.(32, 56) In addition some compounds gave less polar products, which were not powerful inhibitors.(7) Parathion was also reduced in ultra-violet light to $(EtO)_2PS \cdot O \cdot C_6H_4\text{-}4\text{-}NH_2$ (amino-Parathion).(33) There is no evidence that these reactions require oxygen, as it is known that ultra-violet light catalyses the conversion of thionates to thiolates, and the other products could be formed from this by transesterification, as already described. Perhaps amino-Parathion is formed by the reduction of Parathion by thiols released in other reactions. The formation of less polar and reduction products is easily explained in this way, whereas it is difficult to see how they could be formed by oxidation. Thio groups in side chains are oxidized. Thus, when Demeton-O and Demeton-S are exposed to light the thio sulphur atoms are oxidized to sulphoxide groups.(8) The thionate sulphur in Demeton-O is not attacked to any marked extent. Side-chain oxidation increases toxicity and biochemical activity.

The effect of these reactions is to increase rapidly the biochemical activity of thin films, perhaps at the same time reducing the toxicity.(34) The effects are not confined to spray deposits; they also take place in small samples left around in weighing bottles.

*Reactions with Halogens*

Chlorine and bromine can act either as oxidizing or as halogenating agents. They attack many organophosphorus compounds, degrading some, but oxidizing others to compounds of higher biochemical activity and greater solubility in polar solvents. Thus bromine reacts quantitatively in water with Demeton-O and Demeton-S according to the equations:(35)

$$(EtO)_2PS \cdot O \cdot CH_2 \cdot CH_2 \cdot SEt + 5Br_2 + 6H_2O \rightarrow (EtO)_2PO \cdot O \cdot CH_2 \cdot CH_2 \cdot SO \cdot Et + H_2SO_4 + 10HBr.$$

$$(EtO)_2PO \cdot S \cdot CH_2 \cdot CH_2 \cdot SEt + Br_2 + H_2O \rightarrow (EtO)_2PO \cdot S \cdot CH_2 \cdot CH_2 \cdot SO \cdot Et + 2HBr.$$

The thionate sulphur is replaced by oxygen, and the thio sulphur is oxidized to a sulphoxide group, but the thiolate sulphur is not attacked. The reactions have been used to analyse mixtures of the two Demetons. Parathion is similarly converted by bromine water to a compound which, from its solubility and biochemical properties, is concluded to be Paraoxon; and many other thionates, amongst which are EPN, Diazinon, Malathion,

Chlorthion, and homologues of Parathion, are converted to more active and polar compounds, which, by analogy, are considered to be the corresponding P:O compounds.[36] The increases in biochemical activity observed are summarized in Table 3.4. Oxidation of phosphorothionate and sulphide groups appears to be general.[36] N-bromosuccinimide reacts similarly to bromine water, and can be used in non-aqueous solutions.

TABLE 3.4. THE EFFECT OF OXIDATION WITH BROMINE WATER ON THE $I_{50}$'S OF SOME INSECTICIDES[36]

Several phosphorothionate insecticides were oxidized with bromine water. It was shown that Parathion gave Paraoxon by oxidation of the P:S group to a P:O group. Paraoxon itself was not affected by the treatment. Only a selection of the results are given here. The complete set demonstrates that different specimens of the parent compounds gave different $I_{50}$'s, but that the oxidation products of different specimens of the same compound always gave the same $I_{50}$. The $I_{50}$'s quoted here for the parent compounds may therefore be spurious, the inhibition observed being due to impurities.

| Compound | $I_{50}$'s | |
|---|---|---|
| | before | after |
| Parathion | $1{\cdot}1 \times 10^{-4}$ | $1 \times 10^{-8}$ |
| Diazinon | $1{\cdot}8 \times 10^{-5}$ | $5 \times 10^{-9}$ |
| Malathion | $1{\cdot}0 \times 10^{-4}$ | $7 \times 10^{-7}$ |
| Chlorthion | $1{\cdot}4 \times 10^{-4}$ | $4 \times 10^{-7}$ |
| Parathion-Me | $1{\cdot}5 \times 10^{-5}$ | $5 \times 10^{-7}$ |
| EPN | $2{\cdot}0 \times 10^{-5}$ | $6 \times 10^{-7}$ |
| Paraoxon | $1{\cdot}0 \times 10^{-8}$ | $1{\cdot}0 \times 10^{-8}$ |

These reactions have several analytical applications:

(a) In the colorimetric micro-determination of phosphorus in organic compounds by the molybdenum blue method, the compound in aqueous solution is concentrated, and then wet ashed with perchloric acid. Thionates tend to steam-distil during concentration, and are difficult to ash. By adding bromine before concentrating one can aid the oxidation, and also avoid losses by evaporation, as the phosphates produced are less volatile in steam.[37] For the first reason, paper chromatograms of thionates to be developed by molybdenum blue techniques are best first sprayed with N-bromosuccinimide in acetone.[38]

(b) Paper chromatograms may be sprayed with N-bromosuccinimide, and then with fluorescein. Thionates and sulphides use up the bromine, so that the fluorescein is unchanged where they are, and show up green in ultra-violet light; but elsewhere the fluorescein is brominated, and becomes pink.[32]

Spraying with 2:6-dibromo-N-chloroquinone-imine, followed by heating, also shows up sulphur compounds on paper chromatograms. Thionates appear as brown spots, and sulphides as yellow ones. The mechanism is not known, but the first stage is presumably oxidation.[39]

(c) It is often convenient to estimate residues on crops by their capacity to inhibit cholinesterase. Thionates are poor inhibitors, and are not retained as such by most crops for long. The only residues of importance are oxidation products, often the same as those obtained by brominating the parent compounds. The parent compound and its products are extracted together by a non-aqueous solvent, and the extract is treated with bromine water. The capacity of the extract to inhibit AChE is compared with that of a standard prepared by brominating a dilute solution of the parent compound.[36] The method is also used for compounds with side chains containing sulphur, such as the Demetons. The precision is usually low, but the sensitivity great.

Chlorine and bromine in anhydrous media attack S-alkyl phosphorothiolates to give phosphorohalidates and sulphenyl halides:[40]

$$(EtO)_2PO \cdot SR + Cl_2 \rightarrow (EtO)_2PO \cdot Cl + Cl \cdot SR.$$

In water the active species in the action of chlorine on triethyl phosphorothiolate is hypochlorous acid, which reacts very rapidly to give the hydrogen phosphate. The reactions follow the equations:[41]

$$(EtO)_2PO \cdot SEt + 3Cl_2 + H_2O \xrightarrow{k_1} (EtO)_2PO \cdot OH + EtSO_2Cl + 5HCl,$$

$$EtSO_2 \cdot Cl + H_2O \xrightarrow{\text{slow}} EtSO_2 \cdot OH + HCl.$$

At 25°, $k_1 = 2{\cdot}4$ l.mole$^{-1}$min$^{-1}$ with respect to hypochlorite concentration. The acids produced are practically without toxicity or biochemical activity. It is not known why these reactions do not interfere more than they do with the use of halogens for oxidation in residue analysis.

Phosphoramides also are attacked by halogens. The alkyl groups in Schradan and its octaethyl homologue are successively chlorinated in non-aqueous solvents, giving chlorinated products, for example:

$$(Me_2N)_2PO \cdot O \cdot PO{-}(NMe_2)_2 \xrightarrow{Cl_2} {-}(NMe_2)(NMe \cdot CH_2 \cdot Cl) \text{ and } {-}(NMe \cdot CH_2 \cdot Cl)_2.$$

Each stage of chlorination decreases the stability to hydrolysis, but the mono-chlorinated products are powerful inhibitors of cholinesterase.[42] In water an excess of chlorine rapidly leads to complete breakdown to non-toxic compounds, presumably via the hydrolysis of the chlorinated products. Hypochlorites have therefore been used to clean equipment contaminated with Schradan and Dimefox, $(Me_2N)_2PO \cdot F$.[20]

Iodine also reacts with phosphoric and pyrophosphoric dimethylamides, giving in water black oily precipitates containing a high proportion of iodine. Their structure is unknown.[43]

*Reactions with other Oxidizing Agents*

Powerful oxidizing agents, such as perchloric acid, degrade all the compounds to phosphoric acid, and are used in the estimation of phosphorus for this reason.

Potassium permanganate in weakly acid solution oxidizes thionates to phosphates, sulphides to sulphones, and dimethylamides to various products, discussed in Chapter IX, which hydrolyse readily. Hydrogen peroxide in glacial acetic acid has a similar, though milder, action. It has been used instead of bromine water to oxidize Trithion and Thimet to the sulphones of their oxygen analogues and Guthion to Gutoxon before estimating them with cholinesterase.[44] The sulphones are usually more biochemically active than the sulphoxides (see Table 3.1, where the sulphoxide and sulphone of Demeton-S-Me are given as examples). Nitric acid oxidizes thionates to phosphates quantitatively under suitable conditions, a reaction which is the basis of a method for estimating P:S groups as sulphate,[45, 46] and which has also been used to prepare certain compounds. (Chapter II, G.)

## METHODS OF PURIFICATION

It is obvious from the nature of the reactions which have just been described and their effects on biochemical activity that it may be very difficult to make some compounds pure enough for biochemical work. In fact many papers have later needed revision because their authors had not realized the importance of impurities, or, if they had, had been satisfied with methods of purification which were wholly inadequate. It is not my intention to criticize specific papers—this section is not pointed with many references—but to describe in a general way the problems which have been found to arise and how they have been overcome in certain instances; and to suggest criteria whereby the reader can judge whether the methods described in a particular paper were likely to have been adequate for the purposes for which the compounds were prepared and purified.

The problems encountered in obtaining pure specimens are acute for three reasons: the degree of purity which must be achieved is often exceptionally high; the compounds are often very reactive; and the physicochemical properties of many of the compounds rule out the methods of purification most commonly used by organic chemists. These three factors are considered in the next three paragraphs.

The degree of purity to be attained varies with the purpose for which the compound is to be used. The ordinary degree of purity attained by most methods, 98+ %, is adequate for analysing structure by chemical reactions and infra-red spectroscopy, for finding melting and boiling points, and for determining hydrolysis rates. This degree of purity is often wholly inadequate for biochemical work. Thionates, for example, must usually contain less than 0·005% of phosphates or thiolate isomers; many compounds must contain less than this percentage of TEPP; and compounds such as Demeton-S and Demeton-S-Me must contain less than this percentage of sulphonium compounds. A few per cent of solvent or hydrolysis products are, however, often harmless in specimens used for biochemical work, whereas they can interfere seriously with chemical and physico-chemical investigations. Toxicity testing does not often require such pure specimens as biochemical work, but this cannot be taken for granted. The two cases of phosphates in thionates and sulphonium compounds in a compound with a 'thio' side-chain, such as Demeton-S-Me, illustrate the point. The oxidation of a thionate to the analogous phosphate, e.g. Parathion to Paraoxon, increases the toxicity about 10 times, and the biochemical activity about 10 000 times: 1% of a phosphate in a thionate will not, therefore, affect the $LD_{50}$ significantly, but will increase the biochemical activity 100 times. The effect on toxicity is low because the thionate is toxic only by virtue of its conversion to the phosphate *in vivo*, so that the $LD_{50}$'s and $I_{50}$'s are not correlated. In the case of thio compounds and their sulphonium derivatives the $LD_{50}$'s and $I_{50}$'s are correlated. 1% of sulphonium derivative in a specimen of Demeton-S-Me increases both its biochemical activity and its toxicity about tenfold. In practice the presence of impurities has affected toxicity testing much less than biochemical work, and it is not uncommon to find papers in which the toxicities and chemical properties of a group of compounds are given correctly in the same table as $I_{50}$'s wrong by a factor of a thousand or more.

Some compounds are so reactive that they cannot be purified enough for biochemical work. Demeton-O-Me, for example, isomerizes so rapidly to Demeton-S-Me in water, forming sulphonium compounds as byproducts, that its own biochemical properties are obscured in a few seconds;[15] but by solvent extraction in non-aqueous solvents it can be purified sufficiently for a study of its chemical and toxic properties.[13] It is also probable that the biochemical activity shown by aqueous solutions of Dipterex, $(MeO)_2PO \cdot CH(OH) \cdot CCl_3$, should be attributed to DDVP,[47] which is formed fairly rapidly by dehydrochlorination and re-arrangement (Chapter II, J) even in neutral solution. Some other compounds can only be purified in dilute solution, and must then be used immediately. Thus Demeton-S-Me alkylates itself to sulphonium compounds in water, the rate depending on the square of the concentration.[13] When a fairly pure sample is dissolved in chloroform, the active sulphonium impurities can be removed by extracting with water. The chloroform solution can then be concentrated, water added, and a dilute aqueous solution prepared by evaporating the rest of the chloroform under reduced pressure. If, however, an attempt is made to remove the chloroform in the absence of water to obtain a pure compound suitable for the determination of physical constants, it is found that the last few per cent of chloroform cannot be removed except by heating, whereupon sulphonium compounds are again formed. It may therefore be difficult to obtain a compound in a state suitable for both chemical and biochemical work, but there is often no need to try.

Most phosphate esters are colourless or pale yellow oils, which crystallize below 0°, usually with difficulty. They are involatile, often with vapour pressures of $10^{-6}$–$10^{-4}$ mm at room temperature, and boiling points above 100° at 1 mm. Recrystallization and fractional distillation are, therefore, methods of limited application. Compounds containing P:O groups are much more soluble in water than carboxylic esters containing the same number of carbon atoms. Thus water dissolves 2 g/litre at room temperature of Paraoxon, which contains 10 carbon atoms, and phosphoramidic compounds containing up to 6 carbon atoms are usually miscible with water in all proportions. Many compounds therefore take up water from the atmosphere and may be hydrolysed. These high solubilities can be attributed to the unusual capacity of the P:O group to form hydrogen bonds, which the P:S group does not share. The corresponding thionates are accordingly much less soluble. Parathion, for example, is only one-hundredth as soluble as Paraoxon. Except for ionic compounds, nearly all phosphorus compounds can be extracted from water by chloroform, methylene chloride and tetrachloroethane. These are exceptionally good extracting solvents, especially for phosphoramidic compounds such as Schradan and Dimefox, although they are not outstanding for the extraction of most carbon compounds. Thus chloroform is usually not much better than ether for extracting carbon compounds from water, but is about 100 times better (measured by partition coefficients) at extracting phosphoramidic compounds. Chloroform and presumably the other partially chlorinated hydrocarbons are powerful hydrogen bonding agents,[53] which explains their superiority. Carbon tetrachloride is not, and is a poor extracting solvent. Chloroform is, therefore, much used in purification procedures. The compound is extracted from neutral, aqueous solution, when any hydrolysis products are left behind, and the chloroform evaporated under vacuum. As chloroform forms an azeotrope with water, the compound is dried at an early stage in the procedure.

It is obvious that each compound presents individual problems. Many have not been purified. Some examples of the methods used for specific compounds are now described, taking the easiest first.

A very few compounds are crystalline, and can be obtained pure by crystallization. TIPA [NN′N″N‴-tetraisopropylpyrophosphoramide, $(Pr^iNH)_4P_2O_3$], EPN and Parathion-Me and -$Pr^i$ are among these.(48) Some of the crystalline compounds are hygroscopic and readily hydrolysed, e.g. Mipafox, $(Pr^iNH)_2PO \cdot F$,(49) or tend to isomerize readily, but one can avoid the resulting difficulties by taking the obvious precautions.

Another small, but important, group of compounds is fairly volatile. It includes Tabun, $(Me_2N)(EtO)PO \cdot CN$, and many fluoridates and amidic fluorides. Compounds of this type are usually made by A1 type reactions in organic solvents with insoluble metal chlorides as by-products. The metal chlorides are removed by filtration, and the filtrates fractionated under reduced pressure. Several of the compounds react rapidly with water, and the reaction is auto-catalysed, so care must be taken to keep everything dry.

Dialkyl aryl phosphates, e.g. Paraoxon, are less volatile, but can be purified in the same way except for biochemical work. The fractionated products may contain tetra-alkyl pyrophosphates, which are usually more active (see Table 3.3) but they are also more readily hydrolysed, and so may be removed from the fractionated products by controlled hydrolysis.(30, 50) The purified compounds are then extracted from water, and the extracting solvent evaporated *in vacuo*. The removal of the last traces of solvent may be difficult except at high temperatures, exposure to which may lead to the formation of tetra-alkyl pyrophosphates again. Probably most compounds of this type could be obtained pure and free from solvent in a molecular still.

Tetra-alkyl pyrophosphates, e.g. TEPP, are obtained mixed with phosphates and metaphosphates. The metaphosphates are removed by controlled hydrolysis in aqueous solution after which the pyro- and ortho-phosphates are extracted and fractionated.(25, 51, 52)* The extraction is best carried out with solvents which form azeotropes with water, such as chloroform or benzene, as water is then removed completely at an early stage of the distillation, and hydrolysis and other reactions are avoided. The compounds are more toxic and biochemically active than any likely impurities, but the lower homologues pick up water readily from the air and then hydrolyse. Some of the early work on TEPP seems to have been carried out with preparations containing more inert hydrolysis products than TEPP itself.

Pyrophosphoramides also contain amides of ortho- and poly-phosphoric acids. Schradan, the octamethyl compound, is the only important member of this group, except for TIPA which has been discussed already. The polyphosphoramides are removed by selective hydrolysis, the Schradan

* *Added in proof.* E. Schwarzmann and J. R. van Wazer (*J. Amer. chem. Soc.* **83,** 365 (1961)) have shown that on storage and heating TEPP goes to an equilibrium mixture of triethyl phosphate, TEPP, and higher peralkylated polyphosphates. A specimen of TEPP is only pure shortly after distillation at room temperature or below. (Probably such a sample remains pure enough for biochemical and toxicological work for several months if stored at room temperature.)

and hexamethylphosphoramide extracted with chloroform, and the solution is fractionated.[26] Pure Schradan is a hygroscopic solid, m.p. about 20°.

Thiolates present varying degrees of difficulty. Some of the more volatile such as Demeton-S-Me and Demeton-S, may be treated like the dialkyl aryl phosphates. Others undergo transesterification reactions readily, and have not been obtained pure by any standards. Thionates can often be purified enough for chemical and toxicological work by rapid distillation through a column of low resistance at pressures under 1 mm, but they have never been obtained pure enough for biochemical work in this way. Falling film evaporation at $58°/10^{-4}$ mm gave Demeton-O, one of the less stable thionates, in a high state of chemical purity,[54] and might give more stable thionates pure enough for biochemical work.[12] There are, however, few exceptions to the generalization that thioates and dithioates have only been obtained free from more active impurities by chromatography or fairly complex solvent extraction procedures.

There is no doubt that the best way to purify many compounds would be by counter-current separation, as a few grams can be purified in each run, and the separations are highly reproducible. If the compound can be prepared labelled with radioactive phosphorus, $^{32}P$, the efficiency of the separation and the purity of the product are easily tested by determining the radioactivity of the solutions in the different tubes at the end of the separation. When the starting concentration is under 1%, the ratios of the concentrations in the different tubes usually agree precisely with those which can be calculated from the theory of the method, so that discrepancies between the theoretical and experimental ratios indicate the presence of impurities.[55] It is, of course, necessary to avoid using aqueous solvents with some compounds, and this limits the general utility of solvent extraction procedures.

When a compound has been purified, one must decide which physical properties should be recorded so that it can be identified by other workers. The custom in carbon chemistry is to record boiling points or melting points, and those of simple derivatives. These constants are not of much help in characterizing the compounds considered in this book; few are crystalline, and many can only be distilled at pressures below 1 mm, so that it is difficult to find accurate boiling points. The same problems arise in characterizing derivatives, even when the compound has a reactive side-chain from which they can be formed. Refractive indices have been used, especially in Russia, but the discrepancies in the literature between the values given by different workers are discouraging. The two most useful physical measurements are of infra-red absorption spectra and of partition coefficients. The use of the first is limited by the high cost of the spectroscope. A partition coefficient (the ratio of the concentrations of a solute in two immiscible solvents in equilibrium) is readily determined, either by chemical analysis, or, on

$^{32}$P-labelled compounds, by radiochemical analysis of the concentrations in the two solvent layers after separation. As there is a wide choice of solvent pairs, the characterization is unusually specific. Combined with infra-red spectroscopy a high degree of certainty can be attained.

A comparison between the use of partition coefficients and melting points to characterize compounds is interesting. The advantage of melting points is that a mixed melting point with an authentic sample can be determined. In every other way partition coefficients are more specific, and often easier to determine. The practicable range over which melting points can be determined is 20–300°. If the precision is 1°, then there are 280 distinctive melting points. Partition coefficients can be determined to an accuracy of about 3% over the range 0·1–10, giving 140 distinctive values with one pair of solvents. There are about 4 solvent pairs such that a solute partitioning 10:1 in favour of the less polar solvent in one pair will only partition about 1:10 in favour of the less polar solvent of the next pair. The number of distinctive values is therefore about 4 × 140, i.e. 560. Four solvent pairs which nearly meet the requirements of this calculation are: water–chloroform, water–benzene, water–hexane and methanol–iso-octane. Much greater specificity can be attained by determining partition coefficients in more than one solvent pair—two compounds may, by chance, partition 7·0:1 in favour of chloroform from water, but it is improbable that they will both, for example, partition 2·3:1 in favour of trichloroethylene from water. Many workers carry out all preparations with $^{32}$P-labelled compounds. This makes determinations of partition coefficients very simple, and they can be carried out on specimens containing traces of solvent, and on specimens of biological origin, such as plant and animal metabolites. The fact that impurities, providing they are not radioactive, do not interfere, is a great advantage, in view of the difficulty of obtaining many compounds free of solvent without partially decomposing them. The very great sensitivity of radiotracer techniques is, of course, very helpful in characterizing metabolites. At the fairly low specific activity of 1 mc/mmole accurate partition coefficients can be determined on under 10 $\mu$g of material.

## REFERENCES


1. J. E. Casida. *Science* **122,** 597 (1955)
2. W. A. Emmett and H. O. Jones. *J. chem. Soc.* **99,** 713 (1911)
3. R. L. Metcalf and R. B. March. *J. econ. Ent.* **46,** 288 (1953)
4. K-B. Augustinsson and G. Jonsson. *Acta chem. scand.* **11,** 275 (1957)
5. W. N. Aldridge and A. N. Davison. *Biochem. J.* **52,** 663 (1952)
6. Z. M. Bakanova, Y. A. Mandel'baum and N. N. Mel'nikov. *Chem. Abstr.* **51,** 1824 (1957)
7. J. W. Cook and R. Ottes. *J. Ass. off. Agric. Chem., Wash.* **42,** 211 (1959)
8. J. W. Cook. *J. Ass. off. Agric. Chem., Wash.* **37,** 989 (1954)
9. J. Chouteau, J. P. Vigne, A. Karamanian and R. L. Tabau. *C.R. Soc. Biol. Paris* **150,** 1773 (1956)

10. K-B. Augustinsson and G. Jonsson. *Experientia* **13,** 438 (1957)
11. A. Margot and H. Gysin. *Helv. chim. Acta* **40,** 1562 (1957)
12. J. B. McPherson, Jr., and G. A. Johnson. *J. Agric. Food Chem.* **4,** 42 (1956)
13. D. F. Heath and M. Vandekar. *Biochem. J.* **67,** 187 (1957)
14. T. R. Fukuto and E. M. Stafford. *J. Amer. chem. Soc.* **79,** 6083 (1957)
15. D. F. Heath. *J. chem. Soc.* 1643 (1958)
16. M. Vandekar. *Brit. J. indust. Med.* **15,** 158 (1958)
17. K. J. M. Andrews, F. R. Atherton, F. Bergel and A. L. Morrison. *J. chem. Soc.* 780 (1952)
18. J. H. Billman, A. Radike and B. W. Mundy. *J. Amer. chem. Soc.* **64,** 2977 (1942)
19. D. G. Thomas, J. H. Billman and C. E. Davis. *J. Amer. chem. Soc.* **68,** 895 (1946)
20. D. F. Heath, unpublished.
21. L. Larsson. *Acta chem. scand.* **12,** 587 (1958)
22. A. Simon and C. Stölzer. *Ber. Dtsch. Chem. Ges.* **89,** 2253 (1956)
23. S. A. Hall and M. Jacobson. *Ind. eng. Chem.* **40,** 694 (1948)
24. *U.S. Pat.* 2,523,243
25. M. N. Dvornikoff and H. L. Morrill. *Analyt. Chem.* **20,** 939 (1948)
26. G. S. Hartley, D. F. Heath, J. M. Hulme, D. W. Pound and M. Whittaker. *J. sci. Food and Agric.* **2,** 303 (1951)
27. H. Tolkmith. *J. Amer. chem. Soc.* **75,** 5270 (1953)
28. J. Matt and H. J. Harwood. *J. Amer. chem. Soc.* **77,** 6068 (1955)
29. E. Benjamini, R. L. Metcalf and T. R. Fukuto. *J. econ. Ent.* **52,** 94, 99 (1959)
30. W. N. Aldridge and A. N. Davison. *Biochem. J.* **51,** 62 (1952)
31. *U.S. Pat.* 2,594,455
32. J. W. Cook. *J. Ass. off. Agric. Chem., Wash.* **38,** 826 (1955)
33. E. Sandi. *Nature, Lond.* **181,** 499 (1958)
34. J. P. Frawley, J. W. Cook, J. R. Blake and O. G. Fitzhugh. *J. Agric. Food Chem.* **6,** 28 (1958)
35. K. Groves. *J. Agric. Food Chem.* **6,** 30 (1958)
36. H. O. Fallscheer and J. W. Cook. *J. Ass. off. Agric. Chem., Wash.* **39,** 691 (1956)
37. K. Gardner and D. F. Heath. *Analyt. Chem.* **25,** 1849 (1953)
38. I. K. H. Otter. *Nature, Lond.* **176,** 1078 (1955)
39. J. J. Menn, W. R. Erwin and H. T. Gordon. *J. Agric. Food Chem.* **5,** 601 (1957)
40. C. J. M. Stirling. *J. chem. Soc.* 3597 (1957)
41. N. G. Lordi and J. Epstein. *J. Amer. chem. Soc.* **80,** 509 (1958)
42. E. Y. Spencer and R. D. O'Brien. *J. Agric. Food Chem.* **1,** 716 (1953)
43. G. S. Hartley. *First Steps in the Biochemistry of Systemic Insecticides.* W. Heffer, Cambridge, 1951. (Address to XV Internat. chem. Congress, N.Y., 1951)
44. T. E. Archer and G. Zweig. *J. Agric. Food Chem.* **7,** 178 (1959)
45. G. Schrader. *Die Entwicklung neuer Insektizide u.s.w.*, Verlag Chemie, Weinheim, 1952
46. P. A. Giang and S. A. Hall. *Analyt. Chem.* **23,** 1830 (1951)
47. J. van Die. *Proc. Koninkle Ned. Akad. Wetenschap.* **60B,** 227 (1957)
48. R. L. Metcalf and R. B. March. *Ann. ent. Soc. Amer.* **46,** 63 (1953)
49. *Brit. Pat.* 688,787
50. T. R. Fukuto and R. L. Metcalf. *J. Agric. Food Chem.* **4,** 930 (1956)
51. S. A. Hall. *J. Ass. off. Agric. Chem., Wash.* **32,** 377 (1949)
52. J. A. A. Ketelaar and H. A. Bloksma. *Rec. Trav. Chim.* **67,** 665 (1948)
53. L. F. Audrieth and A. D. F. Toy. *J. Amer. chem. Soc.* **64,** 1553 (1942)
54. N. Muller and J. Goldenson. *J. Amer. chem. Soc.* **78,** 5182 (1956)
55. B. Baggett and L. L. Engel. *J. biol. Chem.* **229,** 443 (1957)
56. J. W. Cook. *J. Ass. off. Agric. Chem., Wash.* **38,** 664 (1955)

CHAPTER IV

# Hydrolytic and Related Reactions

MOST of this Chapter is on hydrolysis. There are also two short sections on reactions which either give similar products, or which take place by similar mechanisms.

Hydrolysis is a very general reaction. Every acid anhydride, amide or ester bond in every pentacovalent phosphorus compound so far studied, whether it is biochemically active or not, can be hydrolysed. Tabun, for example, undergoes all the following reactions:

$$(Me_2N)(EtO)PO \cdot CN + H_2O \rightarrow HCN + (Me_2N)(EtO)PO \cdot OH,$$
$$(Me_2N)(EtO)PO \cdot OH + H_2O \rightarrow Me_2NH + EtO \cdot PO(OH)_2,$$
$$EtO \cdot PO(OH)_2 + H_2O \rightarrow EtOH + H_3PO_4,$$
$$(Me_2N)(EtO)PO \cdot CN + H_2O \rightarrow Me_2NH + (HO)(EtO)PO \cdot CN.$$

(In the hydrolysis of Tabun, as in the hydrolysis of many other compounds, several reactions take place at the same time, and the relative rates depend on pH.)

It must be emphasized that a knowledge of hydrolytic reactions is essential for an understanding of phosphorus chemistry and biochemistry. The main reasons are summarized below.

In biochemically active compounds, or in those which are metabolized *in vivo* to biochemically active compounds, the breaking of any one phosphorus bond by hydrolysis destroys the actual or potential* biochemical activity of the phosphorus moiety. Hydrolysis is indeed closely related to biochemical de-toxication reactions (Chapters VII, IX and X). As the first stage of hydrolysis destroys the toxic properties of the compounds, later stages are only touched upon, as they are outside the limits of this book.

Most toxic compounds can be extracted from water by suitable solvents, especially by chloroform. The products of hydrolysis contain P·OH groups, i.e. they are acids, and their salts, formed in alkaline solution, cannot be extracted. This provides the basis for many methods of analysis and purification, and for some methods of determining hydrolysis rates.

As all compounds are hydrolysed, rates of hydrolysis provide measures of

* Only some compounds give biochemically active metabolites. Hydrolysis products do not, so hydrolysis destroys the potential activity of the parent compounds.

the general reactivity of the compounds. Biochemical activity and ease of hydrolysis are closely related.

Finally, hydrolytic reactions are fairly simple to study, and most of our knowledge of reaction mechanisms in phosphorus chemistry comes from them. In the discussion of mechanism several examples are taken from work on non-toxic compounds, as there are none of comparable detail on toxic ones. The physico-chemical differences, however, between non-toxic trimethyl phosphate, $(MeO)_2PO \cdot OMe$, and highly toxic Paraoxon-Me, $(MeO)_2PO \cdot O \cdot C_6H_4\text{-}4\text{-}NO_2$, are likely to be only ones of degree, and it has been assumed that this is so.

Some non-hydrolytic reactions also give acids of phosphorus, and are described in this Chapter. They have analytical applications and some biochemical counterparts.

The rest of this Chapter is devoted to the anionic displacement reaction in non-aqueous solvents, i.e. to the A1 reaction of Chapter II. The preparation of DFP by the reaction:

$$(Pr^iO)_2PO \cdot Cl + F^{\ominus} \rightarrow (Pr^iO)_2PO \cdot F + Cl^{\ominus},$$

is an example. Rates and mechanism are described. The products of these reactions are not, of course, necessarily non-toxic.

The treatment given in this Chapter assumes that the reader is acquainted with the elementary part of the theory of reaction rates and acid-base catalysis. A simple treatment of the essentials is given in the Appendix.

## HYDROLYSIS BY ACIDS, BASES AND WATER

The results considered in this section were mainly obtained in dilute aqueous solution with hydrogen ions (protons*) as acid catalysts, or hydroxyl ions as basic catalysts, but other acids and bases also catalyse hydrolysis. Attention here is confined to those acids and bases which are commonly used in buffer solutions, and some closely related to them, such as simple amines. Anions possessing unusual activity, or reacting in unusual ways, are described in later sections.

### *Products of Hydrolysis*

Hydroxyl ions in water invariably displace the most acidic group, i.e. they break the acid anhydride bond.

Attack by hydrogen ions usually leads to the displacement of the most basic group. Phosphoramidic compounds give amines readily, and dialkyl phosphate esters give alkanols more slowly. P—F bonds are, however,

* The active species is the oxonium ion, $\overset{\oplus}{H_3O}$. $\overset{\oplus}{H}$ is to be understood as standing for this in equations.

attacked by both acids and bases, so that fluoridates may give either the same or different products in acids and bases. Thus DFP gives the same products:[1, 2]

$$(\mathrm{Pr^iO})_2\mathrm{PO}\cdot\mathrm{F} + \mathrm{H_2O} \rightarrow (\mathrm{Pr^iO})_2\mathrm{PO}\cdot\mathrm{OH} + \mathrm{HF};$$

Dimefox gives different products:[3]

$$(\mathrm{Me_2N})_2\mathrm{PO}\cdot\mathrm{F} + \mathrm{H_2O} \begin{cases} \xrightarrow{\overset{\oplus}{\mathrm{H}}} (\mathrm{Me_2N})(\mathrm{HO})\mathrm{PO}\cdot\mathrm{F} + \mathrm{Me_2NH}, \\ \xrightarrow{\mathrm{OH}^{\ominus}} (\mathrm{Me_2N})_2\mathrm{PO}\cdot\mathrm{OH} + \mathrm{HF}; \end{cases}$$

and Mipafox, $(\mathrm{Pr^iNH})_2\mathrm{PO}\cdot\mathrm{F}$, and some other compounds containing primary alkylamino groups undergo both types of reaction in acids.[4] In Tabun also acids break both P—N and P—CN bonds.[5]

Water may displace either the most acidic or the most basic group, according to the structure of the compound.

The products of the first stage of hydrolysis are still esters or amides of acids of phosphorus, and can be hydrolysed. They are acids, the anions of which are stable, whereas the free acids are not. In alkaline solution, therefore, further hydrolysis takes place very slowly, but in acid solution it takes place rapidly. In neutral solution dialkyl hydrogen phosphates are completely ionized, and are stable; but phosphorodiamidic acids are weak, exist mainly as the free, unionized acids, and consequently they are hydrolysed. The hydrolysis of most products is catalysed by acids, so that the products from several stages of hydrolysis may be present simultaneously. This is the case during the acid hydrolysis of Schradan, which must involve at least the following reactions:[3]

$$(\mathrm{Me_2N})_2\mathrm{PO}\cdot\mathrm{O}\cdot\mathrm{PO}(\mathrm{NMe_2})_2 + \mathrm{H_2O} + \overset{\oplus}{\mathrm{H}} \rightarrow \mathrm{Me_2}\overset{\oplus}{\mathrm{N}}\mathrm{H_2} + \underset{\mathrm{I}}{(\mathrm{Me_2N})(\mathrm{HO})\mathrm{PO}\cdot\mathrm{O}\cdot\mathrm{PO}(\mathrm{NMe_2})_2} \quad (1)$$

$$\mathrm{I} + \mathrm{H_2O} + \overset{\oplus}{\mathrm{H}} \rightarrow \mathrm{Me_2}\overset{\oplus}{\mathrm{N}}\mathrm{H_2} + \underset{\mathrm{II}}{(\mathrm{HO})_2\mathrm{PO}\cdot\mathrm{O}\cdot\mathrm{PO}(\mathrm{NMe_2})_2} \quad (2)$$

$$\mathrm{II} + \mathrm{H_2O} \rightarrow \mathrm{H_3PO_4} + (\mathrm{Me_2N})_2\mathrm{PO}\cdot\mathrm{OH} \quad (3)$$

or:

$$\mathrm{I} + \mathrm{H_2O} \rightarrow \mathrm{Me_2N}\cdot\mathrm{PO}(\mathrm{OH})_2 + (\mathrm{Me_2N})_2\mathrm{PO}\cdot\mathrm{OH} \quad (2a)$$

$$(\mathrm{Me_2N})_2\mathrm{PO}\cdot\mathrm{OH} + \mathrm{H_2O} + \overset{\oplus}{\mathrm{H}} \rightarrow \mathrm{Me_2N}\cdot\mathrm{PO}(\mathrm{OH})_2 + \mathrm{Me_2}\overset{\oplus}{\mathrm{N}}\mathrm{H_2} \quad (4)$$

$$\mathrm{Me_2N}\cdot\mathrm{PO}(\mathrm{OH})_2 + \mathrm{H_2O} + \overset{\oplus}{\mathrm{H}} \rightarrow \mathrm{H_3PO_4} + \mathrm{Me_2}\overset{\oplus}{\mathrm{N}}\mathrm{H_2} \quad (5)$$

The rate of decomposition of Schradan is controlled by reaction (1). The only other slow reaction is the fourth, but this is much slower than the first, so that $(\mathrm{Me_2N})_2\mathrm{PO}\cdot\mathrm{OH}$ can be found in quantity after most of the Schradan has been decomposed.

*Kinetics of Hydrolysis*

Almost invariably the reaction between a phosphorus compound and water, acid or base is found to obey second order kinetics, i.e. the rate is proportional both to the concentration of the phosphorus compound and to the concentration of the catalyst:

$$d[P]/dt = -k_C[P][C].$$

(The constants in this book are all given in the units: litre.mole$^{-1}$.min$^{-1}$ except when the catalyst is water itself. The concentration of water in dilute aqueous solution is always 55·5 moles/litre, so the solvolytic constant $k_W$ is used, where $k_W = k_{H_2O}[H_2O] = 55{\cdot}5\, k_{H_2O}$. The units are taken as understood, and are not repeated.) For the hydrolysis of the same compound $k_C$ measures the activity of the catalyst; and for the hydrolysis of different compounds by the same catalyst the different values of $k_C$ measure the susceptibilities of the compounds to hydrolysis.

In the most general case, a compound may be hydrolysed simultaneously by hydrogen ions, hydroxyl ions, general acids, general bases and water. The experimental constant, $k$, is then the sum of the individual constants:

$$k = k_H[\overset{\oplus}{H}] + k_{OH}[OH^{\ominus}] + k_A[A] + k_B[B] + k_W.$$

The hydrolysis of the P—Cl bond in $(Me_2N)_2PO{\cdot}Cl$ is believed to be exceptional in taking place by a uni-molecular reaction.[6] The rate is controlled by the ionization of the P—Cl bond, and the products then react with water:

$$(Me_2N)_2PO{\cdot}Cl \rightarrow (Me_2N)_2PO^{\oplus} + Cl^{\ominus} \xrightarrow{+H_2O} (Me_2N)_2PO{\cdot}OH + \overset{\oplus}{H} + Cl^{\ominus}.$$

This type of reaction may take place with other amidic chlorides.

*General Acid-Base Catalysis*

The hydrolysis of DFP is catalysed by many bases, i.e. the reaction is subject to general base catalysis. It may be subject to general acid catalysis also.[2] The basic rate constants, $k_B$'s, increase with the strengths of the bases, as measured by their dissociation constants, $K_B$'s, according to the Bronsted relation (Appendix):

$$\log (k_B/q) = \text{const.} + \beta \log (pK_B/q),$$

as shown in Fig. 4.1. Similarly Mipafox is hydrolysed by buffer constituents[7] as shown in Table 4.1. It is not known whether the acid or basic constituents of the buffer were the catalysts, so that $k_A$ and $k_B$ could not be separated. There is much scattered evidence that the hydrolysis of other phosphorus compounds also is catalysed by buffers, and this may be a general phenomenon.

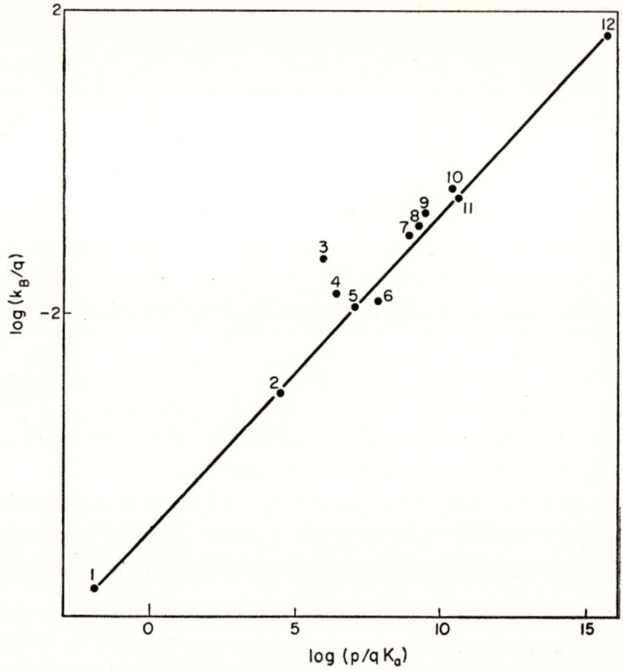


FIG. 4.1. The general base catalysis of the hydrolysis of DFP.[2]

A 'Bronsted plot' is shown. The points represent the following bases: 1, $H_2O$; 2, $MeCO_2^{\ominus}$; 3, $Me_2As \cdot O^{\ominus}$; 4, $HCO_3^{\ominus}$; 5, $HPO_4^{\ominus}$; 6, $(HO \cdot CH_2 \cdot CH_2)_3N$; 7, $(HO \cdot CH_2 \cdot CH_2)_2NH$; 8, $NH_3$; 9, $HO \cdot CH_2 \cdot CH_2 \cdot NH_2$; 10, $EtNH_2$; 11, $MeNH_2$; 12, $OH^{\ominus}$.

After: M. KILPATRICK and M. L. KILPATRICK, *J. Phys. Chem.* **53**, 1371 (1949) (with acknowledgements).

TABLE 4.1. HYDROLYSIS OF MIPAFOX AT pH 5·70 AND 25°[7]

The contributions made to the total rate by the various species in 0·01274 M citrate–phosphate buffer (McIlvaine's) are given, along with the bimolecular rate constants.

| | Water | $\overset{\oplus}{H}$ | $OH^{\ominus}$ | Buffer ions |
|---|---|---|---|---|
| Per cent of total rate | 19 | 23 | 2 | 56 |
| Rate constants | $2·3 \times 10^{-6}$ | 0·66 | 49 | $6·0 \times 10^{-4}$ |

*Effects of Substituents on the Rate of Hydrolysis*

The effects of substituents should be compared in terms of energies and entropies of activation (see Appendix), but these are rarely known accurately enough, so one has to compare rate constants determined at the same temperature. In practice, such comparisons are satisfactory.

*I. Acid rates.* The effects of substituents on rates of acid hydrolysis have not been studied in much detail. From the simplest mechanism (p. 84) the rate should depend upon the ease of protonation of the group removed, being faster the more basic the group. In accordance with this theory, alkoxy groups are only removed slowly—acid catalysis has only been demonstrated in a few instances—whereas amino groups are removed readily, as shown in Table 4.2. It is obvious, however, that the general structure of the

TABLE 4.2. ACID RATE CONSTANTS AT 25° FOR SOME PHOSPHORAMIDIC COMPOUNDS

If there are n amino groups which can be removed from a molecule, then the rate per amino group is $k_H/n$. The basicity of the amino groups is measured by the base constants for the amines released. For dimethylamine and isopropylamine $K_b = 5 \times 10^{-4}$; for diethylamine, $1{\cdot}3 \times 10^{-3}$; and for morpholine, $5 \times 10^{-6}$.

| Compound | $k_H/n$ | Compound | $k_H/n$ |
|---|---|---|---|
| $(Me_2N)_3PO$[3] | $7{\cdot}4 \times 10^{-5}$ | $[(Me_2N)(EtO)PO]_2O$[4] | $2 \times 10^{-3}$ |
| $(Me_2N)_2PO{\cdot}OH$[3] | $1{\cdot}6 \times 10^{-4}$ | $(Pr^iNH)_2PO{\cdot}F$[7] | $2{\cdot}1 \times 10^{-1}$* |
| $(Me_2N)_5P_3O_5$[8] | $7 \times 10^{-4}$ | $(Me_2N)_2PO{\cdot}F$[3] | $2{\cdot}8 \times 10^{-1}$ |
| $(Me_2N)_4P_2O_3$[3] | $9{\cdot}0 \times 10^{-4}$ | $(Me_2N)(morph)PO{\cdot}F$[4] | $7 \times 10^{-1}$ |

* The experimental rate constant is the sum of the constants for P—N and P—F fission. Only the contribution made by P—N fission is given here.

compound greatly influences its rate of hydrolysis. In the compounds in Table 4.2 all the amino groups are of similar basicity, but the rate constants differ by factors of up to 10,000. An explanation is proposed on p. 83.

*II. Basic rates.* Hydroxyl ion catalysis has been studied in more detail, and the effects of many substituents are known. Most obey the electronic theory summarized in Chapter I: the greater the negative charge conferred on the phosphorus atom by the substituents the slower the rate; and the greater the tendency of the most acidic group to separate as an anion the faster the rate.

The effects of all but the acidic group are readily summarized. Thionates hydrolyse slower than their phosphate analogues, because sulphur is less electrophilic than oxygen (Table 4.3A). Electromeric donation of electrons decreases in the order $NR_2 > OR > Ph > R$, so rates increase in the order: $P{-}NR_2 < P{-}OR < P{-}OPh < P{-}R$ (Table 4.3B, C and D). Sulphur atoms are readily polarized during the reaction (a mesomeric effect), so thiolates hydrolyse faster than phosphates (Table 4.3E). Alkyl

TABLE 4.3. THE EFFECTS OF SOME SUBSTITUENTS ON THE RATE OF ALKALINE HYDROLYSIS

Suppose two phosphorus compounds differ only in that one, P—G, contains the group G attached to the phosphorus, while the other, P—G′, contains the group G′. The table contains the change of structure involved in going from one compound to the other, the formula of P—G, and the bimolecular rate constants, $k_G$ and $k_{G'}$, for the alkaline hydrolysis of both compounds. The effects of the change of structure on the rate are given by $k_G/k_{G'}$. Sometimes more than one equivalent substitution has been made, e.g. $(Me_2N)_2PO{\cdot}F \rightarrow (Et_2N)_2PO{\cdot}F$ involves four substitutions of ethyl for methyl groups. As each substitution should have the same effect, the effect of one is given by $\sqrt[n]{k_G/k_{G'}}$, where n is the number of substitutions made. Some rates were determined in buffers, and may have been affected by general base catalysis.

| Code | Change of structure | Compound | $k_G$ | $k_{G'}$ | $t$°C | $\sqrt[n]{k_G/k_{G'}}$ | Refs. |
|---|---|---|---|---|---|---|---|
| A | P=O → P=S | $(EtO)_2PO{\cdot}O{\cdot}C_6H_4$-4-$NO_2$ | 0·52 | $5{\cdot}7 \times 10^{-2}$ | 25 | 9·1 | (9) |
| | | $(MeO)_2PO{\cdot}O{\cdot}C_6H_4$-4-$NO_2$ | 7·7 | 0·69 | 37 | 11·2 | (10) |
| | | $(EtO)_2PO{\cdot}S{\cdot}CH_2{\cdot}SO{\cdot}Et$ | 70 | 3·3 | 28 | 21 | (11) |
| B | RO → $R_2N$ | (MeO)(Me)PO·F | $6{\cdot}4 \times 10^3$ | 62 | 25 | 100 | (12, 13) |
| | | $(EtO)_2PO{\cdot}F$ | 110 | $2{\cdot}5 \times 10^{-5}$ | 25 | 2100 | (14, 15) |
| C | Ph → EtO | $(EtO)(Ph)PS{\cdot}O{\cdot}C_6H_4$-4-$NO_2$ | 1·2 | 0·23 | 37 | 5·2 | (10) |
| D | R → RO | $Et_2PO{\cdot}F$ (one Et only) | $5 \times 10^4$ | $1{\cdot}2 \times 10^3$ | 25 | 42 | (14) |
| | | (EtO)(Et)PO·F | $1{\cdot}2 \times 10^3$ | 110 | 25 | 11 | (14) |
| | | $(Pr^iO)(Pr^i)PO{\cdot}F$ | 126 | 50 | 25 | 2·5 | (12, 14) |
| E | EtS → EtO | (EtS)(Me)PO·F | $1{\cdot}2 \times 10^6$ | $3{\cdot}7 \times 10^3$ | 25 | 320 | (13) |
| | | $(EtS)(EtO)PO{\cdot}O{\cdot}C_6H_4$-4-$NO_2$ | 800 | 1·7 | 37 | 470 | (10, 16) |
| F | Me → Et | (MeO)(Me)PO·F (first Me only) | $6{\cdot}4 \times 10^3$ | $3{\cdot}7 \times 10^3$ | 25 | 1·7 | (12) |
| | | $(MeO)_2PS{\cdot}O{\cdot}C_6H_4$-4-$NO_2$ | 0·23 | $5{\cdot}6 \times 10^{-2}$ | 25 | 2·0 | (9, 17) |
| | | $(Me_2N)_2PO{\cdot}F$ | $3{\cdot}4 \times 10^{-3}$ | $2{\cdot}5 \times 10^{-5}$ | 25 | 3·4 | (3, 15) |
| G | Et → $Pr^i$ | (EtO)(Et)PO·F (first Et only) | $1{\cdot}2 \times 10^3$ | 560 | 25 | 2·1 | (12, 14) |
| | | $(EtO)_2PS{\cdot}O{\cdot}C_6H_4$-4-$NO_2$ | $5{\cdot}7 \times 10^{-2}$ | $3{\cdot}3 \times 10^{-3}$ | 25 | 4·1 | (17, 18) |
| | | $(EtNH)(Me_2N)PO{\cdot}F$ | 12·3 | 8·4 | 25 | 1·5 | (15) |
| H | $X{\cdot}CH_2{\cdot}CH_2{\cdot}O$ → EtO | $(Br{\cdot}CH_2{\cdot}CH_2{\cdot}O)(Me)PO{\cdot}F$ | $9{\cdot}7 \times 10^3$ | $3{\cdot}7 \times 10^3$ | 25 | 2·6 | (12, 14) |
| | | $(Me_3\overset{\oplus}{N}{\cdot}CH_2{\cdot}CH_2{\cdot}O)(Me)PO{\cdot}F$ | $5{\cdot}6 \times 10^4$ | $3{\cdot}7 \times 10^3$ | 25 | 15 | (12, 14) |

groups exert inductive effects, which are super-imposed on the electromeric effects, so that rates increase in the order: $Pr^i < Et < Me$ (Table 4.3F and G); and negative substituents in alkyl groups increase the rates still

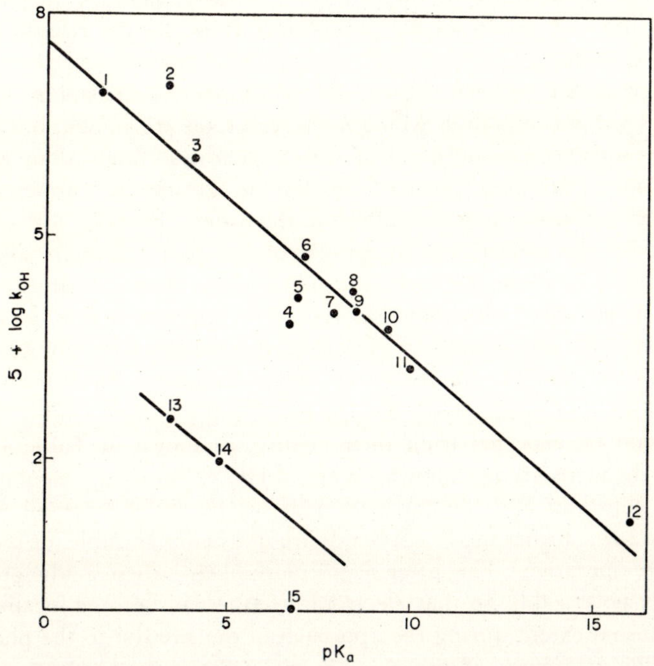


FIG. 4.2. Correlation between the rates of alkaline hydrolysis of some compounds at 25° and the acid strengths of the acids, HX, displaced.

Numerals refer to the compounds as follows: No. 1–12, $(EtO)_2PO{\cdot}X$, where X is: 1, $O{\cdot}PO(OEt)_2$[21]; 2, F[14]; 3, $O{\cdot}C_6H_3$-3:4-$(NO_2)_2$[19]; 4, $O{\cdot}PO(NMe_2)_2$[20]; 5, $O{\cdot}C_6H_2$-2:4:5-$Cl_3$[19]; 6, $O{\cdot}C_6H_4$-4-$NO_2$[9, 19]; 7, $O{\cdot}C_6H_3$-2:4-$Cl_2$[19]; 8, $O{\cdot}C_6H_4$-3-$NO_2$[19]; 9, $O{\cdot}C_6H_4$-2-Cl[19]; 10, $O{\cdot}C_6H_4$-4-Cl[19]; 11, $O{\cdot}Ph$[19]; 12, OEt[22]. No. 13–15, $(Me_2N)_2PO{\cdot}X$ where X is: 13, F[3]; 14, $N_3$[15]; 15, $O{\cdot}PO(NMe_2)_2$[3]. The constants for the aryl phosphates were determined in buffers[19] at 37° but $k_{OH}$ at 25° is known for No. 6, so all the constants found in buffers are corrected by multiplying them by $k_{25}/k_{37}$, where the $k$s are the constants for No. 6 at 25° in hydroxyl ions and 37° in buffer. The constant for No. 15 may be in error by a factor of about 5. The constants for No. 1 and 15 have been halved, as the molecules are symmetrical, so either end may be attacked, and that for No. 12 has been divided by three, as it has three identical groups. The p$K_a$ of ethanol used, 16·0, is a recent value.[23]

more (Table 4.3H). For most acidic groups the rule is obeyed that the rate increases with increasing acidity, as shown in Fig. 4.2, where the rate constants for 12 diethyl phosphate esters and 3 tetramethylphosphorodiamidic compounds are plotted against the dissociation constants of the acids released.* As diethyl 4-nitrophenyl phosphate (No. 6, Fig. 4.2) is hydrolysed faster than its 3-nitrophenyl isomer (No. 8), it is probable that the rates are more affected by the stability of the ion, $X^\ominus$, released than by the electronic effects of the —X group on the phosphorus atom, for the reasons given in Chapter I, p. 26.

The agreement between theory and experiment is satisfactory for about four-fifths of the compounds whose hydrolysis rates are known, but it breaks down for some compounds, all of which hydrolyse faster than expected. Compounds containing primary alkylamino groups are amongst these. According to theory, rates should fall in the order: P—OR $\gg$ P—NHR $>$ P—$NR_2$, i.e. the substitution by an alkoxy group of a primary alkylamino group should increase the rate very much more than substitution by the latter of a secondary alkylamino group. The opposite is true, as shown in Table 4.4. DFP and Mipafox are hydrolysed at similar rates, but substitution of $Me_2N$— by MeNH— increases the rate enormously. Possible reasons are discussed later. Certain acidic groups are also much more readily displaced than would be expected from their acidity, as shown in Table 4.5. The $pK_a$'s of the acids are not known except of HCN, but those calculated from the rate constants and the correlation shown in Fig. 4.2 are all obviously much too low. Either the P—X bonds are unusually unstable, or mesomeric effects are operative. The latter seems more likely—these acidic groups are all highly polarizable, so that electronic charge can become localized in X to an unusual extent during the approach of the catalyst to the phosphorus atom. The mesomeric effects of —SR groups have already been discussed, and should be the same whether the sulphur group is the one displaced or not. In the case of the vinyl compounds the alkanoate released goes mainly to an aldehyde (keto–enol tautomerism):[66]

$$(MeO)_2PO\cdot O\cdot CH{:}CCl_2 \xrightarrow{OH^\ominus} (MeO)_2PO_2^\ominus + OCH\cdot CHCl_2;$$

* If the best straight line is drawn through points 2, 6, 8, 10 and 11, all the deviations from it are in accordance with electronic theory, except that of No. 12. Numbers 3, 5, 7 and 9 represent ortho-compounds, so that the $K_A$'s of the acids released are increased by ortho-effects. These are absent from the phosphorus compounds, so that the rates should be a little lower than the $K_A$'s indicate, as found. Numbers 1, 4 and 5 represent compounds containing pyrophosphate bonding. This is strong in inorganic phosphates, and its strength is usually attributed to an electronic structure approximating to: $\overset{\delta\ominus}{P}\cdots\overset{\oplus}{O}\cdots\overset{\delta\ominus}{P}$.[25] The dipole moment of TEPP suggests that the P·O·P system has a similar structure in organic pyrophosphates also.[26] The rates may therefore be lower than expected because the bonds to be broken are unusually strong. Numbers 13 and 14 should fall on a line parallel to that for the diethyl phosphates, as they do. Why triethyl phosphate, No. 12, should hydrolyse more rapidly than expected cannot be explained.

TABLE 4.4. THE EFFECT OF PRIMARY ALKYLAMINO GROUPS ON RATES OF ALKALINE HYDROLYSIS

| Compound | $k_{OH}$ (25°) |
|---|---|
| 1. $(Pr^iO)_2PO \cdot F$(2) | 50 |
| 2. $(Pr^iNH)_2PO \cdot F$(15) | 49 |
| 3. $(MeNH)(Me_2N)PO \cdot F$(15) | 17·6 |
| 4. $(Me_2N)_2PO \cdot F$(3) | $3{\cdot}4 \times 10^{-3}$ |
| 5. $(MeNH)(Me_2N)PO \cdot O \cdot PO(NMe_2)_2$(24) | $3{\cdot}1 \times 10^{-2}$ |
| 6. $(Me_2N)_2PO \cdot O \cdot PO(NMe_2)_2$(15) | $\sim 1 \times 10^{-5}$ |

TABLE 4.5. SOME COMPOUNDS HYDROLYSING ABNORMALLY RAPIDLY IN ALKALINE SOLUTION

The $pK_a$ for HX was calculated from the rate constant using Fig. 4.2, after making corrections from Table 4.3 for the effects on the rates of substituents other than EtO. Strictly, only rates at 25° should have been taken, but the corrections are relatively trivial.

| Compounds | $k_{OH}$ | t°C | $pK_a$ | | Refs. |
|---|---|---|---|---|---|
| | | | calc. | obs. | |
| $(Me_2N)(EtO)PO \cdot CN$ | 450 | 25 | 5 | 9·1 | (13) |
| $(EtO)_2PO \cdot S \cdot C_6H_4\text{-}4\text{-}NO_2$ | 150 | 37 | 1 | — | (16) |
| $(EtO)_2PO \cdot S \cdot CH_2 \cdot CH_2 \cdot SEt$ | 0·81 | 25 | 7 | — | (27) |
| $(MeO)_2PO \cdot O \cdot CH{:}CCl_2$ | 60 | 30 | 5 | — | (28) |
| $(EtO)_2PO \cdot O \cdot CMe{:}CH \cdot CO \cdot OEt$ | 2 | 28 | 6 | — | (28) |

and the transfer of negative charge is likely to be facilitated by the vinyloxy system, in which the charge can be distributed between three atoms:

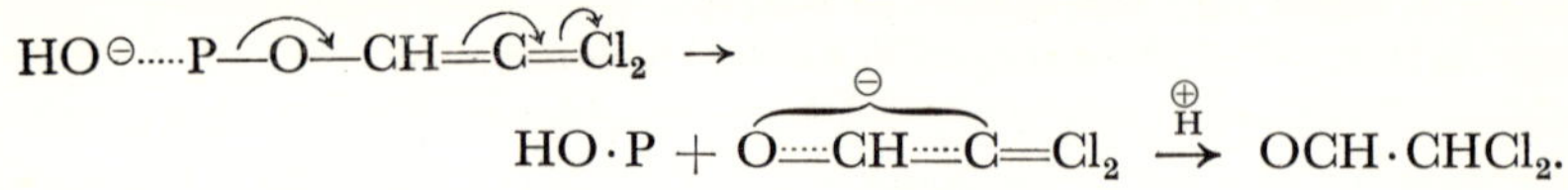

*III. Solvolytic rates.* In neutral solution hydrolysis by water molecules, i.e. solvolysis, may be accompanied by hydrolysis catalysed by acids and bases also present, either as buffer constituents or as products. Most results published do not distinguish between solvolysis and catalysed hydrolysis, and cannot, therefore, be used in the present discussion.

Little is known about the solvolysis of phosphoramidic compounds. Such evidence as there is suggests that water acts mainly as an acid, displacing the alkylamino groups.(7)

Tetra-alkyl pyrophosphates hydrolyse only in the sense:

$$(RO)_4P_2O_3 + H_2O \rightarrow 2(RO)_2PO{\cdot}OH,$$

i.e. only the acid anhydride bonds are broken.(29) The rate constants for this reaction are given in Table 4.6. They can be explained in two ways.

TABLE 4.6. THE RATE CONSTANTS FOR THE HYDROLYSIS OF SOME PYROPHOSPHATES, $(RO)_2PO{\cdot}O{\cdot}PO(R'O)_2$ BY WATER AT 25°(7)

| Compound R | Compound R′ | $10^3k$ | Compound R | Compound R′ | $10^3k$ |
|---|---|---|---|---|---|
| Me | Me | 25 | Et | $Bu^n$ | 0·95 |
| Me | Et | 7·0 | Pr | Pr | 0·65 |
| Me | Pr | 5·6 | Et | $Pr^i$ | 0·28 |
| Et | Et | 1·6 | Pr | $Pr^i$ | 0·20 |
| Me | $Pr^i$ | 1·1 | $Bu^n$ | $Pr^i$ | 0·20 |
| Et | Pr | 1·0 | $Pr^i$ | $Pr^i$ | 0·09 |

Reproduced from *J. chem. Soc.*, 3804 (1956), with permission.

It can be assumed that all the alkyl groups affect the rate constant equally, and to the extent shown in Table 4.3. Then, for example, as the substitution of a methyl group for an ethyl group increases the rate about twofold, tetramethyl pyrophosphate should hydrolyse about $2^4$, i.e. 16 times faster than tetraethyl pyrophosphate, as it does. All the rates have been explained on these lines.(30) A consideration of the mechanism casts some doubt on this theory. If a molecule of water attacks a phosphorus atom, only the alkyl substituents on the atom attacked should affect the rate, the effects of the other two being attenuated by transmission through the P—O—P bonding. If the pyrophosphate oxygen is attacked, then all four groups should affect the rate, but to a lesser extent, as the effects have to be transmitted further. The alternative is to assume that a phosphorus atom is attacked, that only the groups on this phosphorus affect the rate, and to explain the differences between theory and observation in terms of steric hindrance.[7] This, however, requires that attack on a diethyl phosphate group be more hindered sterically than the attack on a dimethyl phosphate group, for which there is no evidence from the relative rates of hydrolysis of other diethyl and dimethyl phosphate esters. Neither theory is, therefore, entirely satisfactory.

*Mechanisms of Hydrolysis*

While it is gratifying that the effects of substituents on rates of hydrolysis agree with the electronic theory of reaction rates, the effects themselves are matters of fact, whereas mechanisms of hydrolysis are still much more matters for speculation. It is obvious why. During hydrolysis any of the three atoms in the ester bonding, P—O—C, may be attacked by acids, bases, or water, and each of these nine basic modes can be formulated in more than one way. Three modes are intrinsically unlikely, and are usually ignored as possibilities. The relative distribution of charge is: $\overset{\delta\oplus}{\mathrm{P}}—\overset{\delta\ominus}{\mathrm{O}}—\overset{\delta\oplus}{\mathrm{C}}$, and the atoms are more likely to be attacked by groups of unlike charge. It is therefore assumed that bases do not attack the oxygen atom, or acids either the phosphorus or carbon atoms.

Acid hydrolysis of phosphorodiamidic compounds yields amines, and of phosphate esters, alkanols. One stage in the reaction is therefore likely to be protonation of the nitrogen or oxygen respectively. For phosphoramidic compounds the following reaction scheme is proposed:

$$\overset{\oplus}{\mathrm{H}} + \mathrm{P—N}< \rightleftharpoons \mathrm{P—}\overset{\oplus}{\mathrm{N}}\mathrm{H}<,$$

$$\mathrm{H_2O} + \mathrm{P—}\overset{\oplus}{\mathrm{N}}\mathrm{H}< \rightarrow \mathrm{H_2O}\cdots\mathrm{P}\cdots\overset{\oplus}{\mathrm{N}}\mathrm{H}< \rightarrow \mathrm{HO—P} + >\overset{\oplus}{\mathrm{N}}\mathrm{H_2}.$$

The first stage is reversible protonation. The resulting cation is then attacked by water in a nucleophilic displacement reaction at the phosphorus atom. This may explain two facts which other theories do not.

(1) With some of the more stable phosphoramidic compounds $k_{\mathrm{H}}$ falls as $[\overset{\oplus}{\mathrm{H}}]$ is raised above 1N. The proportion of each compound protonated at lower hydrogen ion concentrations is obviously small, as the compounds can be extracted from aqueous acids by chloroform with the same efficiency as from neutral solution. Under these conditions the proportion protonated will be proportional to the hydrogen ion concentration, and so will the rate be. At higher hydrogen ion concentrations, however, the proportion protonated may be an appreciable fraction of the whole (this has not been

checked), and then it will increase less rapidly than the hydrogen ion concentration, as observed.

(2) The rate depends on the structure of the molecule as a whole, and not mainly on the basicity of the amine substituents, i.e. protonation is not rate-controlling, as it is on the simplest theory. Suppose two compounds are equally protonated, but one contains more electrophilic substituents than the other. As already shown the one with the more electrophilic substituents will be more susceptible to nucleophilic displacement reactions, and will therefore hydrolyse faster. The results in Table 4.2 are on the whole consistent with this view. A complete theory is hard to develop, as the ease of protonation of one substituent must depend on the other substituents to some extent, but, as interactions between substituents must be transmitted through intervening atoms, this effect may be small.

The rate of removal of alkoxy groups in acid solution may well depend solely on the rate of protonation of the oxygen atoms. This is consistent with the results obtained in neutral solution described below.

Hydrolysis by water or bases could involve attack on either the carbon or phosphorus atoms. In $H_2{}^{18}O$ trimethyl phosphate gives unlabelled methanol on alkaline hydrolysis, but methanol as heavily labelled as the water is obtained on hydrolysis in neutral solution.[31–33] The following equations show that the phosphorus must be attacked during alkaline hydrolysis (either by water, or by the equally labelled hydroxyl ions), and that the carbon must be attacked in neutral solution.

$$\mathrm{H_2{}^{18}O + P{-}O{-}C \rightarrow H_2{}^{18}O \cdots P \cdots O{-}C \rightarrow}$$

$$\mathrm{H^{18}O{-}P + {}^{\ominus}O{-}C \xrightarrow{\overset{\oplus}{H}} HO{-}C;}$$

$$\mathrm{P{-}O{-}C + H_2{}^{18}O \rightarrow P{-}O \cdots C \cdots {}^{18}OH_2 \rightarrow P{-}OH + C{-}{}^{18}OH.}$$

Similarly it is found that tri-2-octyl phosphate prepared from optically active 2-octanol gives back optically active 2-octanol on alkaline hydrolysis, but almost inactive octanol in neutral solution[34]—attack on the phosphorus does not break the C—O bond, so the optical activity is retained, but attack on the carbon does break the C—O bond, and the alcohol racemizes.

Phosphorohalidates are attacked in the same way by both water and hydroxyl ions. The phosphorus is attacked in both, as shown by the labelling of the hydrogen phosphates produced.[35] Acid catalysis of the hydrolysis of DFP can then be explained in terms of a push-pull mechanism:[36]

$$\mathrm{H_2O + P{-}F + \overset{\oplus}{H} \rightarrow H_2O \cdots P \cdots F \cdots \overset{\oplus}{H} \rightarrow HO{\cdot}P + HF + \overset{\oplus}{H}.}$$

In agreement with this, the reaction is slightly slower in $D_2O$ than in $H_2O$.

Attack by bases on the phosphorus atom can be formulated in many ways. Several require a phosphorus atom to have five groups attached to it for much of the time, so that, for example, DFP in water exists as

$$(Pr^iO)_2PO(OH)(F) \quad \text{or} \quad (Pr^iO)_2P(OH)_2(F).$$

These formulations can be rejected on the grounds that the effects of substituents on the rates of hydrolysis of chloroformates and phosphorochloridates are remarkably similar:[37]

| Change of structure | | $10^3 k_G$ | $\sqrt[n]{(k_G/k_{G'})}$ |
|---|---|---|---|
| $Pr^i \rightarrow Et$ | $Pr^iO \cdot CO \cdot Cl$ | 2·5 | 2·2 |
| | $(Pr^iO)_2PO \cdot Cl$ | 3·7 | 1·8 |
| $Et \rightarrow Me$ | $EtO \cdot CO \cdot Cl$ | 5·4 | 1·7 |
| | $(EtO)_2PO \cdot Cl$ | 10·2 | 1·7 |

(the conventions are as in Table 4.3). Carbon cannot be bonded to five atoms. There is other indirect evidence against such formulations.[12] There are four others which are not excluded on these grounds:

$$1.\ HO^{\ominus} + P{-}O{-}C \rightarrow \overset{\delta\ominus}{HO}\cdots\overset{\delta\oplus}{P}\cdots\overset{\delta\ominus}{O}{-}C \rightarrow HO \cdot P + {}^{\ominus}O{-}C.$$

$$2.\ \begin{matrix}H\\H\end{matrix}\!\!>\!O + P{-}O{-}C \rightleftharpoons \begin{matrix}H\\H\end{matrix}\!\!>\!O\cdots P\cdots O{-}C \xrightarrow{+B}$$

(B=base)

$$\begin{matrix}H\\B\cdots H\end{matrix}\!\!>\!O\cdots P\cdots O{-}C \rightarrow HO \cdot P + {}^{\ominus}O \cdot C + {}^{\oplus}BH.$$

$$3.\ H_2O + O{:}P{-}O{-}C \rightleftharpoons (HO)_2P{-}O{-}C \xrightarrow{+B}$$
$$B\cdots H\cdots O\cdots P(OH)(O \cdot C) \rightarrow O{:}P \cdot OH + {}^{\ominus}O \cdot C + {}^{\oplus}BH.$$

$$4.\ H_2O + O{:}P{-}O{-}C \rightleftharpoons [HO \cdot P \cdot O \cdot C]^{\oplus}OH^{\ominus} \rightarrow$$
$$HO\cdots P(OH)\cdots O{-}C \rightarrow (HO)_2P^{\oplus} + {}^{\ominus}O{-}C \rightarrow$$
$$O{:}P{-}OH + HO \cdot C.$$ [38]

The third and fourth mechanisms are unsatisfactory. They require the rate-controlling step to be the combination of the doubly bound atom with a hydrogen atom. Sulphur does not form bonds with hydrogen readily, so P:S compounds should be hydrolysed very much more slowly than P:O compounds. In fact thionates are only hydrolysed a little more slowly than phosphates, less than could be readily accounted for by the lower electronegativity of sulphur. These mechanisms therefore seem less plausible than the first two. The second mechanism cannot be correct in the form put down. The acid formed by hydrolysis in $H_2{}^{18}O$ is only singly labelled, so the first step cannot be reversible.(36) It could be right if the first reaction is not reversible:

2a. 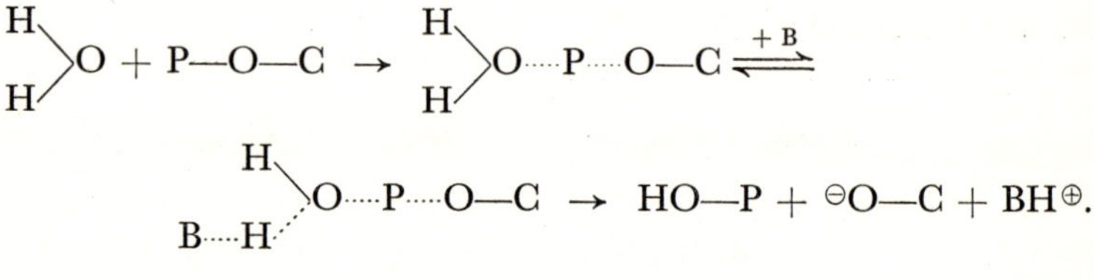

No firm decision can be reached between this mechanism and the first. For hydrolysis by hydroxyl ions the first is favoured by most authors for the following reasons:

(1) The A1 type of reaction: $Y^{\ominus} + (RO)_2PO \cdot X \rightarrow X^{\ominus} + (RO)_2PO \cdot Y$, takes place readily with many ions in non-aqueous solution (Chapter II and the end of this Chapter).

(2) Similar reactions, which must go by this mechanism, take place between many very active anions and phosphorus compounds in water, as shown later. These anions react as rapidly as hydroxyl ions, and it is unlikely that they can react by this mechanism and that the much less bulky hydroxyl ion cannot.

(3) Pyridine, although a weak base, is an effective catalyst for the solvolysis of tetrabenzyl pyrophosphate in isopropanol. Collidine and lutidine although they are stronger bases are much less effective, and their catalytic constants are only one-third as great in $Pr^iOD$. The removal of a deuterium atom is only rate-controlling on the second mechanism, in which the intermediate is:

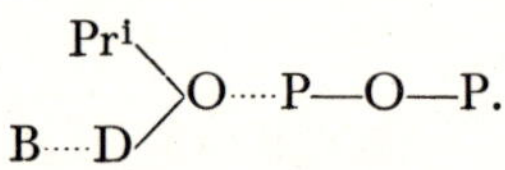

Thus pyridine probably catalyses by the first mechanism, and collidine and lutidine, which are more bulky, by the second. By implication the first mechanism is the normal one.(39)

The main argument against the first mechanism is that the Bronsted relation holds for the hydrolysis of DFP by many bases, including hydroxyl ions, implying that all these bases act by the same mechanism. On the first mechanism hydroxyl ion is unique, in that it is the only base which gives the final product in one step:

$$OH^{\ominus} + (Pr^iO)_2PO \cdot F \rightarrow F^{\ominus} + (Pr^iO)_2PO \cdot OH.$$

Hydrolysis by other bases must take place in two steps:

$$B^{\ominus} + (Pr^iO)_2PO{\cdot}F \rightarrow F^{\ominus} + (Pr^iO)_2PO{\cdot}B \xrightarrow{+H_2O} BH + (Pr^iO)_2PO{\cdot}OH.$$

This two-step mechanism for general base catalysis would not, however, be inconsistent with observation if the intermediate, $(Pr^iO)_2PO{\cdot}B$, is very unstable, so that the rate of reaction is controlled by its rate of formation. This has recently been shown to be so when TEPP is hydrolysed by monohydrogen phosphate ions. The intermediate $(EtO)_2PO{\cdot}O{\cdot}PO_3^{\ominus}$ is formed and breaks down rapidly.[71] The first mechanism is thus much the most plausible on the present evidence, and will be assumed to be correct.

The rapid hydrolysis by water and bases of phosphoramidic compounds containing at least one primary amino group can be explained in two ways. It could be an effect of hydrogen bonding between the attacking group and the hydrogen on the primary amino nitrogen, which constrains the reagents into suitable relative positions for reaction to take place:

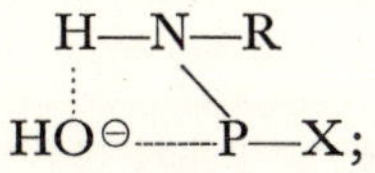

or hydrolysis could take place by a different mechanism:[40]

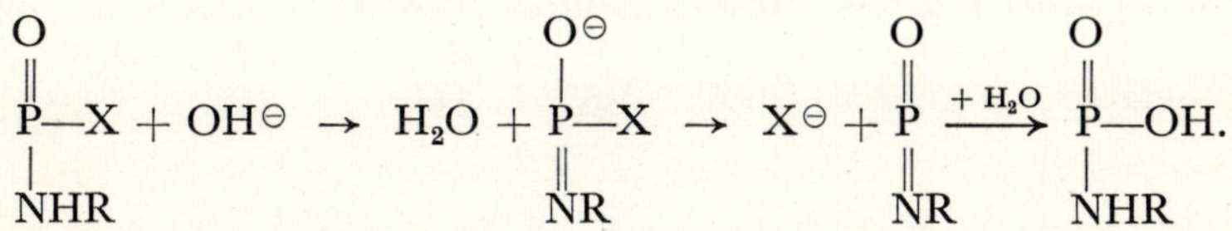

There is no experimental proof of either theory. It has also been suggested that tetra-alkylphosphorodiamidic compounds hydrolyse abnormally slowly because of steric hindrance.[15] The abnormality, however, is in the rates of the trialkyl compounds, which hydrolyse as rapidly as the corresponding dialkyl phosphates contrary to electronic theory, not in the rates of the tetra-alkyl compounds, so this hypothesis is probably incorrect.

The effects of substituents on hydrolysis by hydroxyl ions are explained by electronic theory on the assumption that the charge on the phosphorus atom varies with the substituents, so that the attraction between the phosphorus atom and the hydroxyl ion varies. This is too simple a picture. It implies that the differences in rates can be attributed solely to differences in the energy required from a collision for reaction to take place, i.e. that variations in rate can be entirely accounted for by differences in energies of activation. Table 4.7 shows that differences in entropies of activation are equally important. Thus increasing the negative charge on the phosphorus atom decreases the probability of a collision of the requisite energy breaking the P—X bond, either because the energy is dissipated in other bonds, or because the chances of a collision coinciding with a favourable configuration of the molecule is less, or both. No one has discussed these matters.

TABLE 4.7. RATE CONSTANTS AND ENERGIES AND ENTROPIES OF ACTIVATION FOR THE ALKALINE HYDROLYSIS OF SOME PHOSPHORUS COMPOUNDS

$k_{OH}$, $\Delta S$ and $\Delta E$ are taken or calculated from results in the literature. Values of $\Delta S$ are accurate to better than $\pm$ 2 E.U. and of $\Delta E$ to better than $\pm$ 1·0 kcal/mole. In the offset columns the experimental differences in activation energy are given ($\delta\Delta E$), and the differences calculated on the assumption that the whole change in rate should be accounted for by a difference in activation energy in each case. The discrepancies reveal the importance of changes in $\Delta S$.

| Compound | $k_{OH}$ (25°) | $\Delta S$ | $\Delta E$ | $\delta\Delta E$ Obs. | $\delta\Delta E$ Calc. | Refs. |
|---|---|---|---|---|---|---|
| (MeO)(Me)PO·F | 6360 | −16 | 10·5 | | | (12) |
| | | | | −1·4 | +0·84 | |
| ($Pr^iO$)(Me)PO·F | 1550 | −24 | 9·1 | | | (12) |
| | | | | +2·1 | +2·05 | |
| $(Pr^iNH)_2PO\cdot F$ | 50 | −24 | 11·2 | | | (15) |
| | | | | +3·5 | +5·72 | |
| $(Me_2N)_2PO\cdot F$ | $3{\cdot}4 \times 10^{-3}$ | −31 | 14·7 | | | (3) |
| | | | | +1·9 | +1·47 | |
| $(Me_2N)(Et_2N)PO\cdot F$ | $2{\cdot}9 \times 10^{-4}$ | −30 | 16·6 | | | (15) |
| | | | | +0·4 | +1·46 | |
| $(Et_2N)_2PO\cdot F$ | $2{\cdot}5 \times 10^{-5}$ | −33 | 17·0 | | | (15) |
| | | | Total | 6·5 | 11·5 | |

## HYDROXYLAMINE DERIVATIVES AND HYDROLYSIS

The hydrolysis of DFP, Sarin(41) and TEPP(14) in water is accelerated by hydroxylamine. The dependence of the rate on the pH of the solution shows that the active compound is hydroxylamine itself, not the hydroxylammonium ion, and the reaction kinetics show that the rate-controlling step is probably:

$$(Pr^iO)(Me)PO\cdot F + NH_2\cdot OH \rightarrow (Pr^iO)(Me)PO\cdot O\cdot NH_2 + HF.$$

Fast reactions then give the hydrogen phosphonate, and the overall reaction:

$$(Pr^iO)(Me)PO\cdot F + 3NH_2\cdot OH \rightarrow (Pr^iO)(Me)PO\cdot OH + HF + N_2 + NH_3 + 2H_2O.$$

The hydrolysis of Sarin and DFP is not much faster than is to be expected from the basicity of hydroxylamine, but that of TEPP is (see Table 4.9).

Rates of hydrolysis are more catalysed by many derivatives of hydroxylamine than by hydroxylamine itself. Only a few derivatives are taken as examples here. All are compounds in which the hydroxylamine nitrogen is attached to a group with powerful electron-attracting properties. They are of three types: hydroxamic acids, (—CO·NH·OH), aldoximes, (—CH:N·OH), and ketoximes, (—CR:N·OH). The carbonyl group in the hydroxamic acids is itself electrophilic; in the aldoximes and ketoximes the electrophilic groups are either heterocyclic rings with electrophilic substituents, as in pyridine-2-aldoxime methiodide, $[(C_5H_4\overset{\oplus}{N}Me\text{-}2\text{-}CH{:}N{\cdot}OH)I^{\ominus}$, 2-hydroxyiminomethyl-methylpyridinium iodide, P2-AM], or ketonic groups, as in hydroxyimino-acetone (MeCO·CH:N·OH), and diacetyl monoxime (MeCO·CMe:N·OH, DAM).

The active species is the ion in each case. This is illustrated in Table 4.8, which shows that the rate of reaction is proportional to the concentration of the anion and not to that of the unionized compound.

TABLE 4.8. CONSTANTS FOR THE REACTION OF HYDROXYIMINO-ACETONE WITH SARIN[42] AT 25°C

The bimolecular rate constants are calculated from the apparent rate constants, and the percentages of the compound present as anion calculated from the experimental pH and the $pK_a$ of the compound (8·30).

| pH | 7·4 | 7·6 | 7·8 |
|---|---|---|---|
| Experimental constant | 27·8 | 45 | 57·9 |
| Bimolecular constant | 248 | 271 | 242 |

Reproduced from L. A. Green and B. Saville, *J. chem. Soc.* 3887 (1956), with permission.

The sequence of reactions with hydroxamic acids has been studied in detail. For the reaction between benzohydroxamic acid and Sarin[43, 44] the rate-controlling step is:

$$\text{Ph}\cdot\text{CO}\cdot\text{NH}\cdot\text{O}^{\ominus} + (\text{Pr}^i\text{O})(\text{Me})\text{PO}\cdot\text{F} \rightarrow \underset{\text{I}}{(\text{Pr}^i\text{O})(\text{Me})\text{PO}\cdot\text{O}\cdot\text{NH}\cdot\text{CO}\cdot\text{Ph}} + \text{F}^{\ominus}.$$

This is followed by the rapid reactions:

$$\text{I} + \text{OH}^{\ominus} \rightarrow \underset{\text{II}}{(\text{Pr}^i\text{O})(\text{Me})\text{PO}\cdot\text{O}\cdot\overset{\ominus}{\text{N}}\cdot\text{CO}\cdot\text{Ph}} + \text{H}_2\text{O},$$

$$\text{II} \rightarrow \text{Ph}\cdot\text{NCO} + (\text{Pr}^i\text{O})(\text{Me})\text{PO}_2^{\ominus},$$

$$\text{Ph}\cdot\text{NCO} \xrightarrow{\text{H}_2\text{O}} \text{Ph}\cdot\text{NH}_2 + \text{CO}_2\ (25\%),$$

$$\text{Ph}\cdot\text{NCO} \xrightarrow{\text{Ph}\cdot\text{CO}\cdot\text{NH}\cdot\text{O}^{\ominus} + \overset{\oplus}{\text{H}}} \text{Ph}\cdot\text{NH}\cdot\text{CO}\cdot\text{O}\cdot\text{NH}\cdot\text{CO}\cdot\text{Ph}.$$

The rates of formation and the nature of the products fix the structure of the first intermediate,[44] which is too unstable to study directly. Stable compounds which resemble the intermediate are obtained from some analogues of benzohydroxamic acid: phenyl N-hydroxycarbamate[45] (Ph·O·CO·NH·OH), and benzamidoxime[46], Ph·C($NH_2$):N·OH, and the structures of these products have been proved. The rate-controlling step in the action of the hydroxyimino compounds is also the attack of the anion on the phosphorus atom, which is followed by the rapid hydrolysis of the product:[42]

$$HO^{\ominus} + R\cdot CO{-}CR'{=}N{-}O{-}P \rightarrow R\cdot CO\cdot OH + R'CN + {}^{\ominus}O{-}P.$$

The secondary reactions are more complex with more complex compounds, such as diacetyl monoxime. As the hydroxylamine derivatives are decomposed in these reactions they cannot be described strictly as catalysts.

The intermediates are phosphoric–oxyiminic anhydrides. This indicates that the rate-controlling step is the simple displacement reaction:

$$=NO^{\ominus} + P{-}X \rightarrow =N{-}\overset{\delta\ominus}{O}\cdots P\cdots\overset{\delta\ominus}{X} \rightarrow =N{-}O{-}P + X^{\ominus}.$$

Some anions react as rapidly as hydroxyl ions with phosphorus compounds (Table 4.9). The basic mechanism is likely to be the same, as pointed out

TABLE 4.9. RATE CONSTANTS FOR THE HYDROLYSIS OF FOUR PHOSPHORUS COMPOUNDS BY SEVERAL BASES AT 25°C

| Base | p$K_a$ | Bimolecular rate constants | | | |
|---|---|---|---|---|---|
| | | TEPP | DFP | Sarin | Paraoxon |
| $H_2O$ | −1·7 | $1{\cdot}6 \times 10^{-3}$ | $1{\cdot}2 \times 10^{-4}$ | $10^{-4}$ | $10^{-6}$ |
| $NH_2\cdot OH$ | 6 | 26 | 1·3 | 2·6 | — |
| $Ac_2C$:N·O⊖* | 7·4 | 16 | — | 380 | 0·30 |
| P2-AM⊖† | 7·8 | — | — | 120 | — |
| Ph·CO·NH·O⊖ | 8·8 | 160 | 36 | 1020 | 1·00 |
| OH⊖ | 15·7 | 21‡ | 50 | 2000 | 0·52 |

* Ac = MeCO.

† Pyridine-2-aldoxime methiodide.

‡ A value of 158 has been given for this constant by other workers. The results are as given by Green *et al.*[14] except for P2-AM.[47]

in the previous section. As the oxyiminic anions are much weaker bases than hydroxyl ions the high activity of the hydroxylamine derivatives needs to be explained. Two tentative hypotheses can be proposed. The electrophilic groups attached to the oxyimino groups may confer a partial positive charge on the nitrogen atoms, which can then be attracted to the phosphoryl oxygen atoms of the compounds attacked. Alternatively the attraction may be by way of hydrogen bonding through a water molecule. The two configurations are as follows:

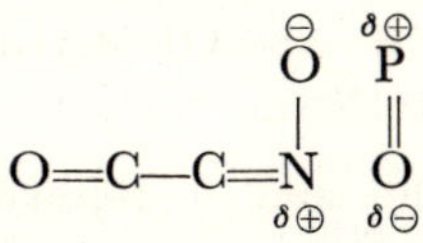

or

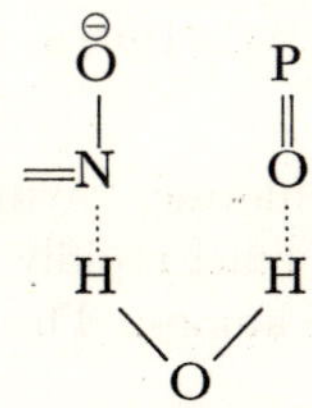

The suggestion that the compounds react in the tautomeric hydroxyimidic form:

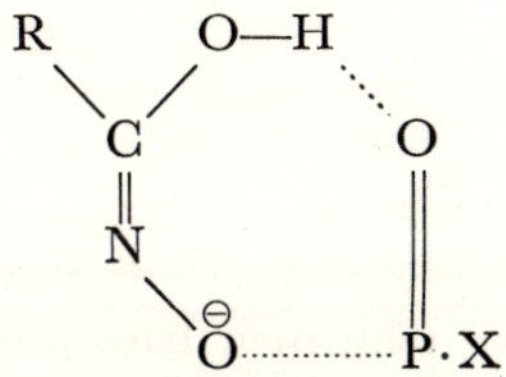

can be ruled out, as N-hydroxyphthalimide ion,

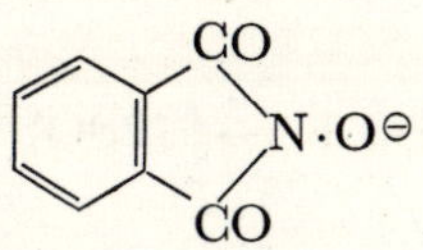

which cannot exist in this form, is active.[14]*

The anions are never much more reactive than hydroxyl ions, so that compounds stable to the latter are stable to the former also. The rate constants increase with the base strengths of the anions according to the Bronsted relation,[14, 48, 49] the slopes varying somewhat with the phosphorus compound.† In consequence, only a few compounds are strong catalysts at physiological pH's, i.e. about pH 7·4, as the anions of strong acids are unreactive, and weak acids are too little ionized. Thus $\beta$ in the Bronsted relation for the reaction of Sarin with benzohydroxamic acids is 0·80, whence the optimum $pK_a$ is 0·6 greater than the pH of the solution.[48] In this case $\beta$ is large, so that acids with $pK_a$'s as high as 10 are still effective at pH 7·4. For Tabun, TEPP and DFP $\beta$ is smaller,[14] so that acids with $pK_a$'s above

* *Added in proof*. G. Aksnes (*Acta chem. Scand.* **14,** 1515 (1960)) has adduced evidence that oxyimino anions and some other exceptionally active nucleophils may be so active because they are readily polarizeable when reacting with phosphate esters —cf. p. 80.

† Individual hydroxyimino compounds deviate a little from the Bronsted relation, probably because of steric effects arising from the attachment of large groups close to the $=NO^{\ominus}$ group.[14, 49]

9 have markedly less than optimum activity. This is relevant to the main use of these compounds, which is the re-activation of phosphorylated enzymes, both *in vivo* and *in vitro*.

## REACTIONS WITH OTHER ABNORMALLY ACTIVE ANIONS

Aqueous chlorine, hydrogen peroxide and 1:2-dihydroxybenzenes (catechols) all react rapidly with many phosphorus compounds. The active species are the anions. The rate constants for reaction with Sarin are given below:

| | $ClO^{\ominus}$ (50) | $HO_2^{\ominus}$ (51) | $Cat^{\ominus}$ (52) | $CrO_4^{2\ominus}$ (53) | $MoO_4^{2\ominus}$ (53) | $WO_4^{2\ominus}$ (53) | $OH^{\ominus}$ |
|---|---|---|---|---|---|---|---|
| $k_O$ | 600 | 80 000 | 589 | 0·1 | 0·14 | 0·15 | 1550 |
| $pK_a$ | 7·4 | 11·7 | 9·5 | 3·5 | 3·9 | 4·1 | 15·7 |

All the ions are more active than their base strengths suggest, unless they react differently in some respects from hydroxyl ions.

Chlorine acts as the hypochlorite ion, and the first step is followed by rapid hydrolysis:

$$ClO^{\ominus} + (Pr^iO)(Me)PO\cdot F \rightarrow \underset{\text{I}}{(Pr^iO)(Me)PO\cdot O\cdot Cl} + F^{\ominus},$$

$$\text{I} + H_2O \rightarrow HClO + (Pr^iO)(Me)PO\cdot OH.$$

It has been suggested that the reaction is rapid because Cl—O is polarized in the sense: $\overset{\delta\oplus}{Cl}—\overset{\delta\ominus}{O}$, and is therefore attracted to the $\overset{\delta\oplus}{P}=\overset{\delta\ominus}{O}$ group.(50) Hypochlorite ion is a catalyst in the strict sense, as it is not destroyed in the reaction. Hypochlorite ions have been used to remove contamination with some toxic phosphorus compounds. As bleaching powder also attacks phosphoramidic compounds (by a different mechanism), there are probably few toxic compounds which it does not eventually destroy, though thionates pass through the more toxic phosphates first.

The first stage of the reaction with peroxidate ion gives a peroxy-acid of phosphorus.(51) The high activity of the ion is attributed to hydrogen bonding:(54)

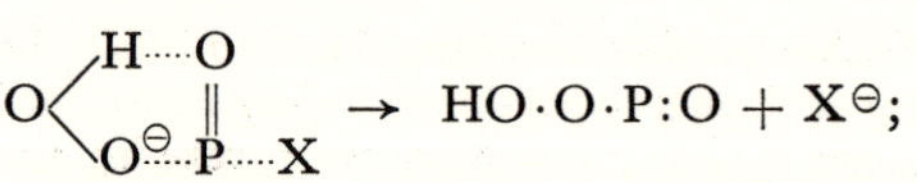


or:

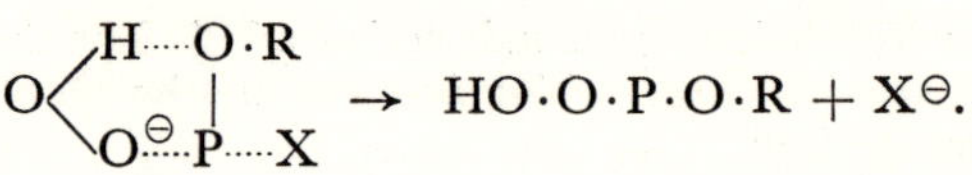

The peroxy-acid is decomposed fairly rapidly by water, hydrogen peroxide and hydroxyl ions, but it can undergo various other reactions. Very sensitive tests for toxic compounds depend on this. For instance, in the Schoenemann reaction[55, 56] the per-acid is made with perborate in the presence of indole, which is rapidly oxidized to indigo via indoxyl and di-indoxyl:

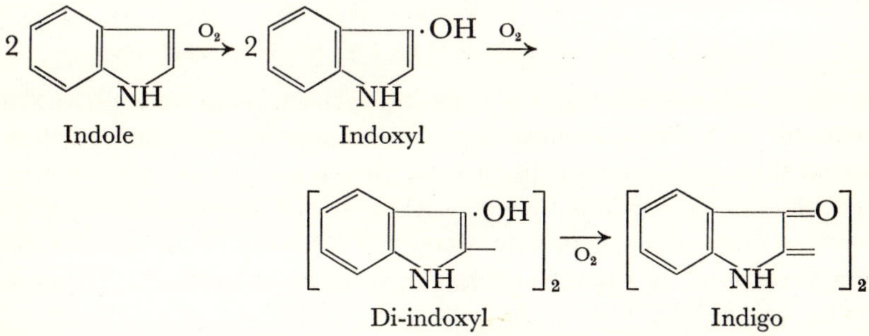


Indoxyl and di-indoxyl are fluorescent, and their formation provides a test sensitive to 1–10 μg of phosphorus compound. The oxidation of benzidine is also used in some variations of the test.[57] As the phosphorus per-acids are about equally unstable whatever the parent compounds from which they are formed, and the fluorescent compounds are unstable, the test is only applicable to compounds which react rapidly with peroxidate ion. This reacts about 100 times more rapidly than an equal concentration of hydroxyl ions, so the test is useful only for compounds which are unstable to alkalis. The limit appears to be reached when $k_{OH}$ is less than 0·1. Compounds hydrolysing more slowly than this are not powerful inhibitors of cholinesterase, so the test is applicable to most poisons which act directly, but not to those which have to be metabolized first.

Catechols give stable hydroxyphenyl phosphate esters.[52, 58, 59] Only the singly ionized catechols react, and hydrogen bonding is postulated to explain the high rates:[52]

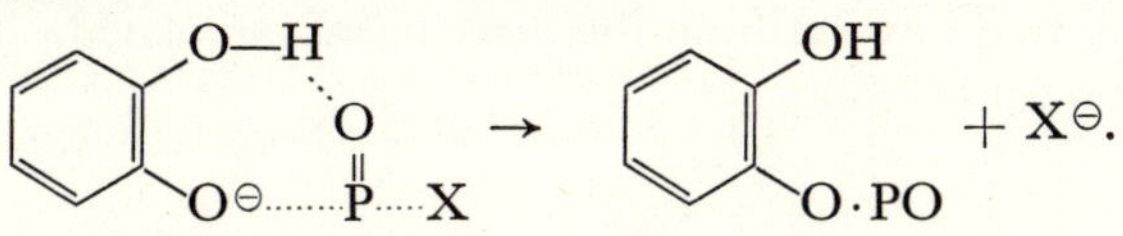


This is consistent with much evidence. Only the 1:2-dihydroxy compounds react rapidly, mono-hydric phenols containing other ortho substituents and 1:3 and 1:4-dihydroxybenzenes reacting at the slow rate of phenol itself. Only one phosphorus group can be introduced. 1:2:3-trihydroxybenzene (pyrogallol) reacts five times faster than catechol with one molecule of phosphorus compound, more slowly with a second, and not at all with a third. Substituents which decrease the basicity of catechols decrease the rate according to the Bronsted relation: $\log k_C = \text{const.} + 0{\cdot}89 \log K_b$.

Chromates, molybdates and tungstates catalyse the hydrolysis of Sarin[53] and these and metavanadates the hydrolysis of Tabun[60] more than other ions of similar basicity. The catalytic effects are attributed to complex formation, but no suggestions have been made as to the structure of the complexes.

## REACTIONS WITH WEAKLY BASIC ANIONS

It can be assumed that an ion cannot displace a more basic group from phosphorus in a solvolytic medium. Accordingly fluoride ion, which is a powerful nucleophil, cannot displace an alkoxy group from phosphorus in water. Even chloride ions, however, catalyse the liberation of hydrogen phosphates from simple phosphate esters,[38, 61] as shown by the rate constants for the breakdown of trimethyl phosphate at 80° given below.[61] (For water $k_N = k_{H_2O} = k_W/55{\cdot}5$.)

| Nucleophil: | $H_2O$ | $Cl^\ominus$ | $Br^\ominus$ | $CNS^\ominus$ | $I^\ominus$ | $S_2O_3^{2\ominus}$ | $OH^\ominus$ |
|---|---|---|---|---|---|---|---|
| $k_N$ | $3{\cdot}6 \times 10^{-4}$ | $2{\cdot}4 \times 10^{-2}$ | 0·12 | 0·44 | 0·23 | 23 | 8·3 |

The attack is apparently at the carbon atom, with the formation of products according to the equation:

$$\text{P—O—C} + X^\ominus \rightarrow \text{P—O}\cdots\text{C}\cdots X^\ominus \rightarrow \text{P—O}^\ominus + \text{C—X}.$$

This reaction is not hydrolytic, as the carbon product is not an alcohol. Except for the hydroxyl ion, the rate increases as the redox potential of the attacking ion increases. Thus the rate-controlling factor is the ease with which the ion can lose an electron, as in oxidation: $X^\ominus \leftrightarrows \frac{1}{2}X_2 + e$. The hydroxyl ion, which attacks the phosphorus atom, reacts about 1000 times faster than expected on this correlation. From the very small number of phosphorus compounds which have been studied it appears that the catalysed rates increase with the activity of the phosphorus compounds as alkylating agents, as one would expect from the mechanism postulated. The reaction is likely to be a general one.

## HYDROLYSIS BY SOME METALLIC COMPOUNDS

The hydrolysis of Sarin is catalysed by cupric, ceric and manganous salts at slightly acid pHs.[62] The dependence of rate on pH shows that the active species in each case is the mono-hydroxo ion: $Cu(OH)(H_2O)_3^\oplus$, $Ce(OH)(H_2O)_5^{2\oplus}$ and $Mn(OH)(H_2O)_5^\oplus$. These ions are even more active than the hydroxyl ion, but the greatest experimental rate constant found was only 0·11 min$^{-1}$ (with copper at pH 6·0), as only low concentrations of the mono-hydroxo ions can be obtained. In acids they are protonated, and in alkaline solution the metal hydroxides precipitate. The mechanism is supposed to involve either bi-polar attack, e.g.

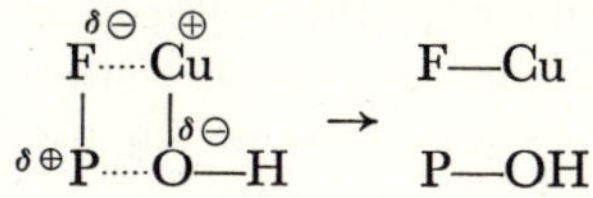


or a push-pull intermediate, such as:

$$\overset{\oplus}{Cu}\cdots\overset{\delta\ominus}{O}H\cdots\overset{\delta\oplus}{P}\cdots\overset{\delta\ominus}{F}\cdots\overset{\oplus}{Cu}\cdot OH, \quad \text{or} \quad HO^{\ominus}\cdots\overset{\delta\oplus}{P}\cdots\overset{\delta\ominus}{F}\cdots\overset{\oplus}{Cu}\cdot OH.$$

Uranyl salts also catalyse Sarin hydrolysis to some extent in neutral solution, and calcium and magnesium salts in alkaline solution, in which, however, catalysis by hydroxyl ion obscures the catalytic effects of the mono-hydroxo ions.

The catalytic effects of cupric salts can be increased, and extended to higher pHs, by the addition of chelating agents. The chelates of other metals are only poor catalysts. $2{\cdot}3 \times 10^{-2}$ M 2:2′-dipyridyl chelate of copper increases the rate of hydrolysis of DFP about 600-fold in bicarbonate buffer at pH 7·6 and 38°. β-Alanine, glycine, aspartic acid, glutamic acid, arginine, lysine, threonine ethylenediamine, imidazole, 2-phenanthroline, 4:4′-dimethyl-2:2′-dipyridyl and L-histidine are nearly as effective.[63] The catalytic effects are destroyed if more than two groups are co-ordinated to the copper. Thus monoethylenediamine chelate is effective:

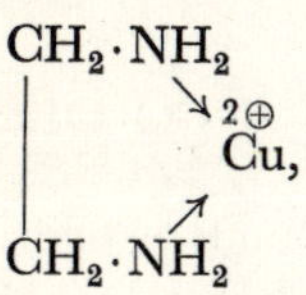


but the dichelate is not:

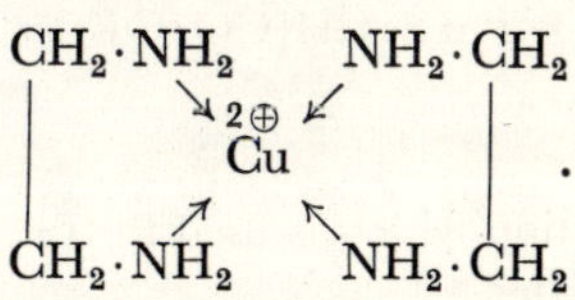


For the same reason the chelate with ethylenediaminetetra-acetic acid is not a catalyst. The kinetics of the reactions are consistent, in the case of the dipyridyl chelate, with the catalyst being any of the species:

$$\text{dipyr.Cu}(H_2O)_2^{2\oplus}, \ \text{dipyr.Cu}(OH)(H_2O)^{\oplus} \ \text{or dipyr.Cu}(OH)_2.$$

Owing to a dimerization reaction the concentration of the second is very small, so that the rate constant calculated on the assumption that it is the catalyst is unreasonably great. It is therefore best to regard the first as the only catalyst in the pH range 6–8, and the last as the catalyst in alkaline solution.[64] The mechanism probably involves bipolar attack, which increases the charge on the phosphorus atom, and lays it open to attack by hydroxyl ions.

Copper salts greatly increase the rate of hydrolysis of the phosphorothionates Parathion and EPN at pH 7·6–8·6.[65] They do not catalyse the hydrolysis of Paraoxon, the phosphate analogue of Parathion. The thionates are not hydrolysed by copper when it is complexed with ethylenediaminetetra-acetic acid. The catalytic activity is therefore attributed to P═S→Cu complex formation, which increases the charge on the phosphorus atom. In the Paraoxon experiments the catalyst concentration, $3{\cdot}2 \times 10^{-6}$ M, was too low to produce a significant increase in rate from the catalytic effects of cupric salts on phosphates already described.

## MISCELLANEOUS REACTIONS IN WATER

When a group attached to a phosphorus atom consists of a chain: $-\mathrm{S}\cdot\overset{|}{\mathrm{C}}\mathrm{H}\cdot\overset{|}{\mathrm{C}}\mathrm{H}-$, with a polar substituent containing a doubly bound atom on one of the carbons, the product of treatment with alkalis may not be a thiol but a vinyl compound.[66] Thus the sulphone of Demeton-S gives ethyl vinyl sulphone:

$$(\mathrm{EtO})_2\mathrm{PO}\cdot\mathrm{S}\cdot\mathrm{CH_2}\cdot\mathrm{CH_2}\cdot\mathrm{SO_2}\cdot\mathrm{Et} + \mathrm{OH}^{\ominus} \rightarrow$$
$$(\mathrm{EtO})_2\mathrm{POS}^{\ominus} + \mathrm{H_2O} + \mathrm{CH_2{:}CH}\cdot\mathrm{SO_2}\cdot\mathrm{Et}.$$

The sulphide (Demeton-S) and the sulphoxide do not react in this way, but are hydrolysed normally. Malathion and its analogues give dialkyl hydrogen dithioates and fumaric ester:

$$(\mathrm{RO})_2\mathrm{PS}\cdot\mathrm{S}\cdot\mathrm{CH(CO}\cdot\mathrm{OEt)}\cdot\mathrm{CH_2}\cdot\mathrm{CO}\cdot\mathrm{OEt} + \mathrm{OH}^{\ominus} \rightarrow$$
$$(\mathrm{RO})_2\mathrm{PS_2}^{\ominus} + \mathrm{H_2O} + (\mathrm{:CH}\cdot\mathrm{CO}\cdot\mathrm{OEt})_2,$$

a reaction which is used for the estimation of these compounds as hydrogen dithioates.[67]

## SOLVENT EFFECTS

Mixtures of water and dioxane or ethanol have often been used for experiments on hydrolysis because of the low solubilities of many of the compounds in water alone. Usually the rates are similar to those in water, and the products are the same. Some compounds, however, give different products. Thus $(\mathrm{Pr^iO})_2\mathrm{PO}\cdot\mathrm{Cl}$ in 80% aqueous ethanol gives both the ethyl ester, $(\mathrm{Pr^iO})_2\mathrm{PO}\cdot\mathrm{OEt}$, and the hydrogen phosphate. The products are in the ratio: 1·22:1, i.e. the ethanol reacts somewhat more rapidly than the water.[37] Unexpected products are also obtained from the hydrolysis of some dialkyl aryl phosphates and thionates in aqueous ethanol. Although the aryl groups are the most acidic, the alkyl groups are in some cases removed preferentially,[68] as shown in Table 4.10. Increasing the hydroxide concentration, or decreasing the ethanol concentration, reduces the extent of the abnormal reaction.[68, 69] The reaction is more rapid with dimethyl than with diethyl compounds, which suggests that the reaction is not a hydrolysis but a de-alkylation:

$$(RO)_2PO\cdot X + EtOH \rightarrow (RO)(HO)PO\cdot X + Et\cdot O\cdot R,$$

and that one product is an ethyl ether. Although no attempt has been made to identify the compounds in which the alkyl groups appear, there is some evidence that such reactions take place. Trimethyl phosphate yields some dimethyl ether on hydrolysis in both acid and alkaline solution.

TABLE 4.10. ABNORMAL PRODUCTS IN 95% AQUEOUS ETHANOL[68]

The percentage of the products from the alkaline hydrolysis of some dialkyl aryl esters appearing as alkyl aryl hydrogen esters was determined chromatographically. Hydrolysis was carried out with equimolar KOH (0·025 M) for 20 hr at room temperature. The total percentage of each parent compound decomposed is given in brackets.

| Common name | Formula | Abnormal products (%) | |
|---|---|---|---|
| Trolene | $(MeO)_2PS\cdot O\cdot C_6H_2$-2:4:5-$Cl_3$ | 64 | (46) |
| — | $(MeO)_2PO\cdot O\cdot C_6H_2$-2:4:5-$Cl_3$ | 49 | (69) |
| Dicapthon | $(MeO)_2PS\cdot O\cdot C_6H_3$-2-Cl-4-$NO_2$ | 46 | (35) |
| Chlorthion | $(MeO)_2PS\cdot O\cdot C_6H_3$-3-Cl-4-$NO_2$ | 35 | (35) |
| Parathion-Me | $(MeO)_2PS\cdot O\cdot C_6H_4$-4-$NO_2$ | 46 | (33) |
| Trolene-Et | $(EtO)_2PO\cdot O\cdot C_6H_2$-2:4:5-$Cl_3$ | 29 | (38) |
| Parathion | $(EtO)_2PS\cdot O\cdot C_6H_4$-4-$NO_2$ | 11 | (9) |
| Paraoxon | $(EtO)_2PO\cdot O\cdot C_6H_4$-4-$NO_2$ | 12 | (41) |
| Diazinon | 4-$(EtO)_2PS\cdot O\cdot$pyrimidinyl-6-Me-2-$Pr^i$ | 9 | (13) |

Reactions in non-protolytic solvents are mainly relevant to methods of preparation of toxic compounds. A systematic study carried out on the reactions of some chloridates and fluoridates with ethanoate, fluoride and phenate ions and some uncharged bases is, however, relevant here.[35, 37] The rates are strictly second order in all cases. The reactions with anions are not affected by the addition of salts to the reaction mixtures, but the reactions with amines are greatly accelerated by the addition of salts, and go faster the more polar the solvent. The only reasonable conclusion is that both types of reaction are nucleophilic displacements:

$$\mathrm{B}^{\ominus} + \mathrm{P{-}X} \rightarrow \overset{\delta\ominus}{\mathrm{B}}\cdots\overset{\delta\oplus}{\mathrm{P}}\cdots\overset{\delta\ominus}{\mathrm{X}} \rightarrow \mathrm{B{-}P} + \mathrm{X}^{\ominus},$$

$$\mathrm{HN} + \mathrm{P{-}X} \rightarrow \overset{\delta\oplus}{\mathrm{H}}\cdots\overset{\delta\ominus}{\mathrm{N}}\cdots\overset{\delta\oplus}{\mathrm{P}}\cdots\overset{\delta\ominus}{\mathrm{X}} \rightarrow \mathrm{N{-}P} + \mathrm{HX}.$$

In the reaction with anions the transition complex is less polar than the reactants, so the rate is not increased by adding salts. In the reaction with uncharged compounds, the transition state is more polar than the reactants, so that the rate of reaction is highly dependent on the dielectric constant of the medium. These mechanisms are analogous to the first of those postulated to explain alkaline hydrolysis, and the results here show that this mechanism cannot be opposed on steric grounds. The reactions are all slower than hydrolysis—ethanoate ion reacts in anhydrous alcohol at about one-hundredth the rate at which water reacts in 80% aqueous ethanol—but this may be a solvent effect, as the activation energies are similar.

Very little has been published on a type of reaction which is of great practical importance—that of an otherwise pure compound with traces of water. The compound $(Me_2N)(Pr^iNH)PO{\cdot}F$ reacts quantitatively with water in a few days at 76° according to the equation:[4]

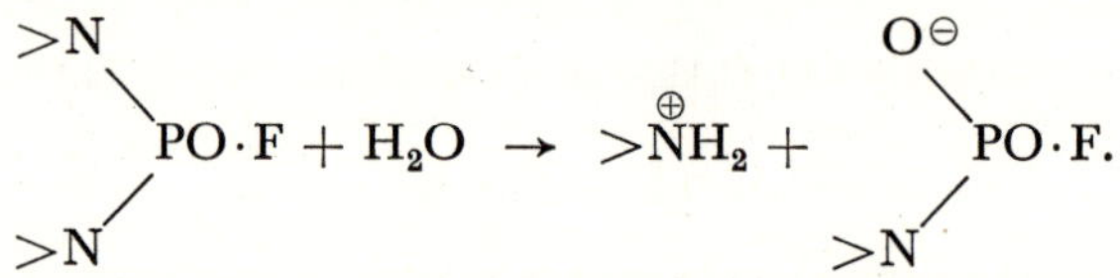

Probably some of each type of amino group is removed. Mipafox undergoes several reactions, including one which is probably identical with the above, and condensation reactions in which P—N—P bonding is produced. Mipafox is a crystalline compound, but it is hygroscopic, and small samples are rapidly destroyed by these reactions when they are exposed to a humid atmosphere.[70] DFP and TEPP are rapidly hydrolysed to the hydrogen phosphates, as described in the last Chapter.

## REFERENCES


1. W. A. Waters and C. G. M. de Worms. *J. chem. Soc.* 926 (1949)
2. M. Kilpatrick and M. L. Kilpatrick. *J. phys. Chem.* **53,** 1371 (1949)
3. D. F. Heath and P. Casapieri. *Trans. Faraday Soc.* **47,** 1093 (1951)
4. D. F. Heath, unpublished
5. L. Larsson. *Acta chem. scand.* **7,** 306 (1953)
6. H. K. Hall, Jr. *J. org. Chem.* **21,** 248 (1956)
7. D. F. Heath. *J. chem. Soc.* 3804 (1956)
8. G. S. Hartley, D. F. Heath, J. M. Hulme, D. W. Pound and M. Whittaker. *J. Sci. Food and Agric.* **2,** 303 (1951)
9. E. F. Williams. *Ind. eng. Chem.* **43,** 950 (1951)

10. R. L. Metcalf and R. B. March. *Science* **117,** 527 (1953)
11. J. S. Bowman and J. E. Casida. *J. econ. Ent.* **51,** 838 (1958)
12. L. Larsson. *Acta chem. scand.* **11,** 1131 (1957)
13. L. Larsson. *Acta chem. scand.* **12,** 783 (1958)
14. A. L. Green, G. L. Sainsbury, B. Saville and M. Stansfield. *J. chem. Soc.* 1583 (1958)
15. D. F. Heath. *J. chem. Soc.* 3796 (1956)
16. W. N. Aldridge and A. N. Davison. *Biochem. J.* **52,** 663 (1952)
17. J. A. A. Ketelaar. *Rec. Trav. Chim.* **69,** 649 (1950)
18. B. Topley. *Chem. and Ind.* 859 (1950)
19. T. R. Fukuto and R. L. Metcalf. *J. Agric. Food Chem.* **4,** 930 (1956)
20. J. E. Casida, T. C. Allen and M. A. Stahmann. *Nature, Lond.* **172,** 243 (1953)
21. J. A. A. Ketelaar and A. H. Bloksma. *Rec. Trav. Chim.* **67,** 665 (1948)
22. P. Cavalier. *Ann. Chim. Phys.* **18,** 449 (1899)
23. E. E. Caldin and G. Long. *J. chem. Soc.* 3737 (1954)
24. D. F. Heath, D. W. J. Lane and P. O. Park. *Phil. Trans.* **239B,** 191 (1955)
25. L. Pauling. *The Nature of the Chemical Bond*, Cornell University Press (1940)
26. G. Speroni. *Chim. e Industr.* **33,** 543 (1951)
27. K. Gardner and D. F. Heath. *Analyt. Chem.* **25,** 1849 (1953)
28. P. E. Gatterdam, J. E. Casida and D. W. Stoutamire. *J. econ. Ent.* **52,** 270 (1959)
29. A. D. F. Toy. *J. Amer. chem. Soc.* **70,** 3882 (1948); **72,** 2065 (1950)
30. F. H. Brock. *J. org. Chem.* **22,** 114 (1957)
31. J. B. M. Herbert and E. Blumenthal. *Nature, Lond.* **144,** 248 (1939)
32. E. Blumenthal and J. B. M. Herbert. *Trans. Faraday Soc.* **41,** 611 (1945)
33. P. W. C. Bernard, C. A. Bunton, D. R. Llewellyn, K. G. Oldham, B. L. Silver and C. A. Vernon. *Chem. and Ind.* 760 (1955)
34. W. Gerrard, W. J. Green and R. A. Nutkins. *J. chem. Soc.* 4076 (1952)
35. I. Dostrovsky and M. Halmann. *J. chem. Soc.* 1004 (1956)
36. M. Halmann. *J. chem. Soc.* 305 (1959)
37. I. Dostrovsky and M. Halmann. *J. chem. Soc.* 502, 508, 511, 516 (1953)
38. E. Cherbuliez and J-P. Leber. *Helv. Chim. Acta* **35,** 2589 (1952)
39. G. O. Dudek and F. H. Westheimer. *J. Amer. chem. Soc.* **81,** 2641 (1959)
40. F. H. Westheimer. *Chem. Soc.* special publ. **8,** 181 (1957)
41. B. J. Jandorf. *J. Amer. chem. Soc.* **78,** 3686 (1956)
42. A. L. Green and B. Saville. *J. chem. Soc.* 3887 (1956)
43. B. E. Hackley, Jr., R. Plapinger, M. Stolberg and T. Wagner-Jauregg. *J. Amer. chem. Soc.* **77,** 3651 (1955)
44. R. Swidler and G. M. Steinberg. *J. Amer. chem. Soc.* **78,** 3594 (1956)
45. G. M. Steinberg and J. Bolger. *J. org. Chem.* **21,** 660 (1956)
46. R. E. Plapinger and O. O. Owens. *J. org. Chem.* **21,** 1186 (1956)
47. A. L. Green and H. J. Smith. *Biochem. J.* **68,** 32 (1958)
48. R. Swidler, R. E. Plapinger and G. M. Steinberg. *J. Amer. chem. Soc.* **81,** 3271 (1959)
49. B. E. Hackley, Jr. *Dissertation Abstr.* **18,** 1255 (1958)
50. J. Epstein, V. E. Bauer, M. Saxe and M. M. Demek. *J. Amer. chem. Soc.* **78,** 4068 (1956)
51. L. Larsson. *Acta chem. scand.* **12,** 723 (1958)
52. J. Epstein, D. H. Rosenblatt and M. M. Demek. *J. Amer. chem. Soc.* **78,** 341 (1956)
53. L. Larsson. *Acta chem. scand.* **12,** 1226 (1958)

54. J. Epstein, M. M. Demek and D. H. Rosenblatt. *J. org. Chem.* **21,** 796 (1956)
55. B. Gehauf and J. Goldenson. *Analyt. Chem.* **29,** 276 (1957)
56. R. H. Cherry, G. M. Foley, C. O. Badgett, R. D. Eanes and H. R. Smith. *Analyt. Chem.* **30,** 1239 (1958)
57. B. Gehauf, J. Epstein, G. B. Wilson, B. Witten, S. Sass, V. E. Bauer and W. H. C. Rueggeberg. *Analyt. Chem.* **29,** 278 (1957)
58. K-B. Augustinsson. *Acta chem. scand.* **6,** 959 (1952)
59. B. J. Jandorf, T. Wagner-Jauregg, J. J. O'Neill and M. A. Stolberg. *J. Amer. chem. Soc.* **74,** 1521 (1952)
60. K-B. Augustinsson. *Acta chem. scand.* **12,** 1286 (1958)
61. R. F. Hudson and D. C. Harper. *J. chem. Soc.* 1356 (1958)
62. J. Epstein and D. H. Rosenblatt. *J. Amer. chem. Soc.* **80,** 3596 (1958)
63. T. Wagner-Jauregg, B. E. Hackley, Jr., T. A. Lies, O. O. Owens and R. Roper. *J. Amer. chem. Soc.* **77,** 922 (1955)
64. F. W. Fowkes, G. S. Ronay and L. B. Ryland. *J. phys. Chem.* **62,** 867 (1958)
65. J. A. A. Ketelaar, H. R. Gersmann and M. M. Beck. *Nature, Lond.* **177,** 392 (1956)
66. R. Mühlmann and G. Schrader. *Z. Naturf.* **12b,** 196 (1957)
67. M. V. Norris, W. A. Vail and P. R. Averall. *J. Agric. Food Chem.* **2,** 570 (1954)
68. F. W. Plapp and J. E. Casida. *J. econ. Ent.* **51,** 800 (1958)
69. H. R. Kreuger, J. E. Casida and R. P. Niedermeier. *J. Agric. Food Chem.* **7,** 182 (1959)
70. I. K. H. Otter, unpublished.
71. D. M. Brown and N. K. Palmer. *J. chem. Soc.* 1155 (1960)

# Part II

# BIOCHEMISTRY

## CHAPTER V

# Some Enzymes

THE SUCCEEDING Chapters are much concerned with cholinesterases and A-esterases (phosphorylphosphatases), especially the former. Cholinesterases, unlike phosphorylphosphatases, can be described without reference to phosphorus compounds, and, as some knowledge of them has to be assumed in later Chapters, their characteristics are described here. The naming of phosphorylphosphatases and some other esteratic enzymes is also dealt with, as this has caused confusion in the past.

### CHOLINESTERASES

*Function and kinetics.* Each mammalian species contains two cholinesterases (ChE's), i.e. enzymes which hydrolyse rapidly choline esters of fatty acids:

$$RCO\cdot O\cdot CH_2\cdot CH_2\cdot \overset{\oplus}{N}Me_3 + H_2O \rightarrow RCO\cdot OH + HO\cdot CH_2\cdot CH_2\cdot \overset{\oplus}{N}Me_3.$$

These enzymes differ in function, distribution and substrate specificity.

One has the specific function of hydrolysing acetylcholine (ACh) at myoneural junctions and some synapses* and will be called acetylcholinesterase (AChE). In the literature it is often called specific ChE, but names based upon presumed substrate specificities are unsatisfactory for reasons given later. The other ChE fulfils no known function, but often masquerades as AChE in experiments, and so will be called pseudo-cholinesterase (ΨChE).

In mammals AChE's are most concentrated at myoneural junctions and the synapses where they are effective, but high concentrations are also found in nerve cells, and, in most mammals, in erythrocytes. ΨChE's are more widely and evenly distributed throughout the body, but erythrocytes usually contain little or none, and serum a high concentration, especially in the horse and dog.[1]

The catalysis of ACh hydrolysis by AChE is inhibited by quite low concentrations of ACh,[2] so that at about $2 \times 10^{-3}$ M there is a concentration

* A brief description of the way the nervous system functions in mammals is given in Chapter XI, and includes the evidence for this function of AChE.

at which the enzyme activity is a maximum (Fig. 5.1). The activity–substrate concentration curve can be accounted for quantitatively by the reaction scheme:

$$\begin{array}{c} E + S \rightleftharpoons ES \rightleftharpoons ES_2 \\ \downarrow \\ E + \text{products.} \end{array}$$

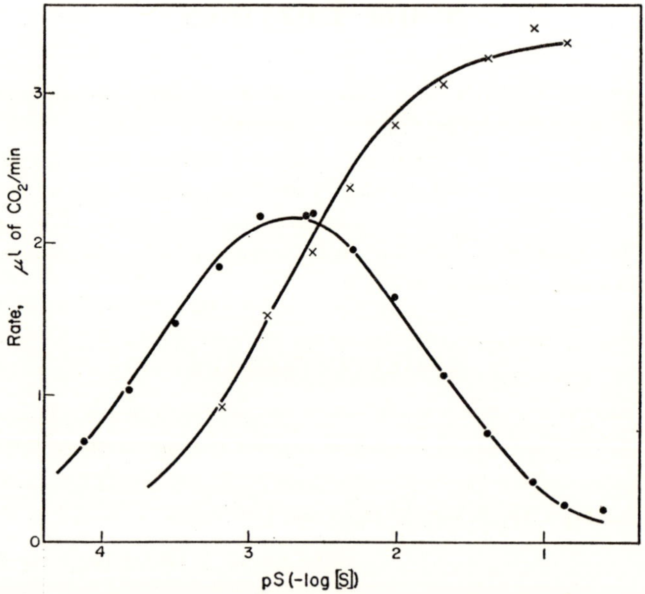


FIG. 5.1. The dependence on concentration of the hydrolysis rate of ACh by human AChE and ΨChE.[3, 4]

Runs were carried out in bicarbonate buffer, and the rates are expressed in $\mu$l of $CO_2$ displaced by the acetic acid released. The rates are plotted against pS, i.e. $-\log$ [ACh]. The curves are theoretical. For ΨChE, $v = V[S]/([S] + K_m)$, (the Michaelis–Menton equation), where

$$V = 3{\cdot}73\ \mu\text{l/min},\ K_m = 1{\cdot}64 \times 10^{-3}\ \text{M};$$

for AChE, $v = V[S]/([S] + K_m + [S]^2/K')$, which is the equation for the reaction scheme given in the text, with the values: $V = 2{\cdot}73\ \mu$l/min, $K_m = 2{\cdot}31 \times 10^{-4}$ M, and $K' = 1{\cdot}22 \times 10^{-2}$ M. $K'$ is the dissociation constant of $ES_2$. ΨChE, ×; AChE ●.

in which the $ES_2$ complex cannot decompose to products, but the ES complex can. Hydrolysis by ΨChE is not inhibited by ACh. The reaction is apparently:

$$E + S \rightleftharpoons ES \rightarrow E + \text{products.}$$

and obeys the Michaelis–Menton equation (Fig. 5.1). The activity of AChE varies with pH differently from the activity of ΨChE. AChE activity is a

TABLE 5.1. RATES OF HYDROLYSIS OF VARIOUS SUBSTRATES BY AChE AND ΨChE*

The letters in brackets refer to sources of the ChE's as follows (a) ox, (b) dog, (c) horse, (d) human.

| Name | Substrate | | Enzyme | | Ref. |
|---|---|---|---|---|---|
| | Formula | Abbreviation | AChE | ΨChE | |
| Acetylcholine | $MeCO\cdot O\cdot CH_2\cdot CH_2\cdot \overset{\oplus}{N}Me_3$ | ACh | 100 (a) | 100 (c) | |
| Acetylthiocholine | $MeCO\cdot S\cdot CH_2\cdot CH_2\cdot \overset{\oplus}{N}Me_3$ | AThCh | 149 (a) | 430 (c) | (7, 8) |
| Acetyl-2-methylcholine | $MeCO\cdot O\cdot CH(Me)\cdot CH_2\cdot \overset{\oplus}{N}Me_3$ | AMeCh | 18 (a) | 0 (c) | (8) |
| Benzoylcholine | $PhCO\cdot O\cdot CH_2\cdot CH_2\cdot \overset{\oplus}{N}Me_3$ | BzCh | 0 (a) | 67 (c) | (8) |
| 2-acetoxy-tetraethylammonium ion | $MeCO\cdot O\cdot CH_2\cdot CH_2\cdot \overset{\oplus}{N}Et_3$ | | 94 (b) | 26 (c) | (9) |
| Ethyl acetate | $MeCO\cdot OEt$ | | 2 (d) | 1 (d) | (10, 11) |
| Triacetin | $(Glyceryl)(MeCO)_3$ | TA | 42 (d) | 14 (d) | (10, 11) |
| 3:3-dimethylbutyl acetate | $MeCO\cdot O\cdot CH_2\cdot CH_2\cdot CMe_3$ | Carbachol | 60 (d) | 35 (d) | (10, 11) |
| 2-chloroethyl acetate | $MeCO\cdot O\cdot CH_2\cdot CH_2\cdot Cl$ | | 37 (d) | 10 (d) | (12) |

* Relative rates, such as those shown in this and the following 3 tables, depend, of course, on the concentrations of substrates. Ideally comparisons should be made in terms of $K_m$ and maximum rates, but this is often not possible. In practice, most comparisons were made at concentrations giving nearly the maximum rates.

maximum at pH 8–8·5, and falls off rapidly as the pH is raised or lowered. This dependence on pH can be explained by assuming that the enzyme can exist in either of three forms, un-, mono- or di-protonated, all in equilibrium;

$$E^{\ominus} + H^{\oplus} \underset{}{\overset{K_1}{\rightleftharpoons}} \boxed{EH + H^{\oplus} \overset{K_2}{\rightleftharpoons} EH_2^{\oplus}}$$

and that only the EH form can react with substrate.[5] $K_1$ and $K_2$ are the acid dissociation constants of EH and $EH_2^{\oplus}$ respectively, with $pK_1 = 9{\cdot}3$ and $pK_2 = 7{\cdot}2$. The optimum substrate concentration varies a little with pH. The activity of ΨChE rises rapidly from pH 6–8, when it reaches a maximum, and does not fall again on increasing the pH further. For this enzyme: $HE \rightleftharpoons H^{\oplus} + E^{\ominus}$, and $E^{\ominus}$ reacts. In practice a fall above pH 8 is often observed because the enzyme is de-natured, but this can be avoided with proper techniques.[6]

Both AChE and ΨChE hydrolyse many esters, including some which are not esters of choline. A few examples are given in Table 5.1, in which the abbreviations should be noted. The choline esters inhibit their own hydrolysis by AChE,[13] though the optimum concentrations of some, especially AMeCh, are quite high—about 0·05 M. Neutral esters do not inhibit their own hydrolysis.

The effects on the rates of hydrolysis of varying the acidic moieties in the esters are illustrated in Table 5.2, and of varying the alcoholic moieties in Table 5.3.

On AChE rates of hydrolysis fall in the order: Acetyl > propionyl > butyryl. This order is reversed on ΨChE, which hydrolyses even iso-amyl hexanoate rapidly. On both enzymes the rates increase as the structure of the alcohol approaches that of choline. The ester of 3:3-dimethylbutanol, which is iso-electronic with choline, is hydrolysed the fastest.

Substrate specificities vary with the species from which the enzymes are taken (Table 5.4), but not with the part of the animal from which they are taken (Table 5.5), nor do they differ from animal to animal in the same species. Some small differences between ChE's from different parts of the same animal have been reported, but probably only represent the effects of impurities on activity. Thus the species must be specified but the source within the species need not be. The differences from species to species are very marked, especially amongst insects. Most insect AChE's differ from mammalian AChE's in hydrolysing nitrophenyl acetate more rapidly than ACh.[19] They also show greater species differentiation amongst themselves.

Both AChE's and ΨChE's are to some extent stereo-specific.[21, 22] They hydrolyse L-AMeCh about 20 times slower than D-AMeCh (ΨChE's hydrolyse even the D-isomer very slowly) but each hydrolyses the corresponding L- and D-acetylthiomethylcholines at the same rate.[22]

TABLE 5.2. THE DEPENDENCE ON THE ACYL GROUP OF THE RATE OF HYDROLYSIS OF SUBSTRATES BY ERYTHROCYTE ACHE AND PLASMA ΨCHE

Rates are given relative to ACh (100). All the results for ΨChE were obtained on horse ΨChE. The sources of AChE are given in brackets.

| Compound | Source | | Refs. |
|---|---|---|---|
| | AChE | ΨChE | |
| Acetylcholine, ACh | 100 | 100 | |
| Propionylcholine, PrCh | 80 (man) | 170 | (13, 14) |
| Butyrylcholine, BuCh | 2·5 (man) | 250 | (10, 14) |
| Acetylthiocholine, AThCh | 149 (ox) | 407 | (7, 8) |
| Butyrylthiocholine, BuThCh | 0 (ox) | 590 | (8) |
| iso-Amyl acetate | 24 (man) | 7 | (10, 15) |
| iso-Amyl propionate | 10 (man) | 13 | (10, 15) |
| iso-Amyl butyrate | 1 (man) | 14 | (10, 15) |
| iso-Amyl hexanoate | — | 5 | (15) |

TABLE 5.3. EFFECTS OF VARYING THE STRUCTURE OF THE ALKYL GROUPS IN ALKYL ACETATES AND BUTYRATES ON THE RATES OF HYDROLYSIS BY PURIFIED HUMAN ACHE AND PURIFIED HUMAN ΨCHE

Acetates were used for AChE and butyrates for ΨChE.

| Structure of alkyl group | Rates | |
|---|---|---|
| | AChE[10] | ΨChE[11] |
| $—CH_2 \cdot CH_2 \cdot CH_3$ | 9 | 3 |
| $—CH_2 \cdot CH_2 \cdot CH_2 \cdot CH_3$ | 16 | 10 |
| $—CH_2 \cdot CH_2 \cdot CH_2 \cdot CH_2 \cdot CH_2 \cdot CH_3$ | 7 | 2 |
| $—CH_2 \cdot CH_2 \cdot CH(CH_3)_2$ | 24 | 27 |
| $—CH_2 \cdot CH_2 \cdot C(CH_3)_3$, 3:3-dimethylbutyl | 60 | 35 |
| $—CH_2 \cdot CH_2 \cdot \overset{\oplus}{N}(CH_3)_3$, cholinyl | 100 | 100 |

TABLE 5.4. SUBSTRATE SPECIFICITIES OF CHE'S FROM DIFFERENT SPECIES

The rates are given relative to the rate of hydrolysis of ACh (100). The following abbreviations are used: PrCh = propionylcholine; BuCh = butyrylcholine; AMeCh = acetyl-2-methylcholine; BzCh = benzoylcholine; TA = triacetin.

| Source | PrCh | BuCh | AMeCh | TA |
|---|---|---|---|---|
| AChE's | | | | |
| Horse[16] | 87 | 2 | 33 | 73 |
| Cobra venom[17] | 48 | 1 | 33 | 22 |
| *Torpedo*[13] | 32 | 0 | 27 | — |
| *Lebistes*[13] | 75 | 12 | — | 10 |
| Housefly[18] | — | 52 | 29 | 60[19] |
| Leafhopper[19] | — | — | 121 | 63 |
| ΨChE's[14] | PrCh | BuCh | BzCh | |
| Dog | 160 | 270 | 60 | |
| Horse | 170 | 250 | 30 | |
| Man | 160 | 230 | 30 | |
| Pigeon | 120 | 125 | 15 | |
| Rat | 220 | 130 | 30 | |
| Chicken | 150 | 90 | 10 | |
| Mouse | 140 | 80 | 15 | |
| Pig | 90 | 120 | 15 | |

TABLE 5.5. THE IDENTITY OF ACHE'S FROM HUMAN RED BLOOD CELLS, BRAIN AND MUSCLE[20]

The identity is illustrated by each responding in the same way to four inhibitors.

| Compound | $I_{50}$(M) r.b.c. | Brain | Muscle |
|---|---|---|---|
| Sarin | $3 \times 10^{-9}$ | $3{\cdot}3 \times 10^{-9}$ | $3{\cdot}6 \times 10^{-9}$ |
| Tabun | $1{\cdot}5 \times 10^{-8}$ | $1{\cdot}5 \times 10^{-8}$ | $2 \times 10^{-8}$ |
| TEPP | $3{\cdot}5 \times 10^{-8}$ | $3{\cdot}2 \times 10^{-8}$ | $3{\cdot}5 \times 10^{-8}$ |
| DFP | $4 \times 10^{-7}$ | $3 \times 10^{-7}$ | $2{\cdot}5 \times 10^{-7}$ |

Quoted from D. Grob and J. C. Harvey. *J. clin. Invest.* **37,** 350 (1958) (with acknowledgements).

To summarize, AChE's and ΨChE's can be distinguished by kinetic experiments in the following ways:

(1) Hydrolysis of ACh by AChE is inhibited by excess substrate, by ΨChE it is not.

(2) AChE's hydrolyse ACh faster than BuCh; ΨChE's hydrolyse either BuCh or PrCh faster than ACh.

(3) Most AChE's hydrolyse AMeCh rapidly, and most ΨChE's hydrolyse it very slowly.

(4) Most AChE's hydrolyse BzCh slowly, and most ΨChE's hydrolyse it rapidly.

These distinctions are maintained in spite of the gross impurity of almost all ChE preparations. The most highly active specimen of AChE ever obtained was from electric eels.[(23)] It hydrolysed 1·25 g ACh/min/mg protein, and sedimented homogeneously in an ultra-centrifuge. Assuming that this was pure, then a highly purified specimen from bovine erythrocytes contained about 1% AChE[(24)] (the units of activity are accidentally multiplied by 1000 in the original paper). The purest ΨChE's have only contained about 11% (from horse serum[(25)]) and 30% (from human serum[(26)]) of the enzyme. Most experiments have been carried out on preparations such as red cells, brain homogenates and serum, in which about 0·001% of the protein was ChE, and the ChE concentrations were about $2 \times 10^{-10}$ M. Even these low concentrations are similar to the effective concentrations of some inhibitors, which may also react with other proteins, so that results obtained with very active inhibitors on most ChE preparations must be treated with caution.

In some preparations it is not certain that only one enzyme able to hydrolyse the substrate is present. I have tried to indicate this possibility wherever necessary, but more work will doubtless reveal some errors.*

*The active centres*. ChE's are inhibited by many simple molecules and ions and the way the degree of inhibition varies with the structure of the inhibitor leads to a more specific picture of the active centres in these enzymes.

* The test most commonly used until recently to show that only one enzyme was present was the summation experiment. The rates of hydrolysis of two substrates are determined, and then the rate of hydrolysis of a mixture of the two at the same final concentrations. If two enzymes are present, each hydrolysing one substrate, the total rate of hydrolysis of the mixture should be equal to the sum of the rates of hydrolysis of the individual substrates. If one enzyme only is present, then, as two substrates are competing for it, the rate of hydrolysis of the mixture should be between the individual rates. When the mixed rate is the sum of the individual rates, the presence of two enzymes is nearly certain, but results are rarely so clear cut, and can be misleading. For example, if a mixture of AChE and ΨChE were to be tested with ACh and BuCh, it is probable that the evidence would indicate the presence of one enzyme only, partly because ΨChE hydrolyses both substrates, but mainly because BuCh inhibits AChE.

AChE is inhibited by simple tetra-alkylammonium ions and by choline esters. It is not inhibited by substrates without positively charged side-chains. From this it seems that the active part of AChE consists of two sites,[2] an anionic site, which attracts positively charged ions, and an esteratic site, at which esters are hydrolysed:

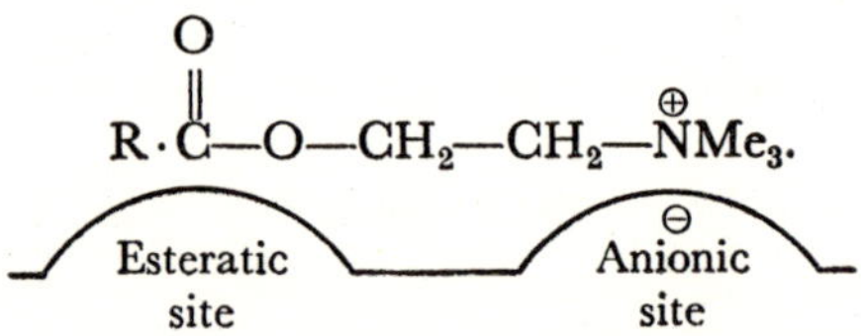


A choline ester cannot be hydrolysed if the anionic site is occupied by another molecule, as it cannot take up the right position. If the molecule attracted to the anionic site is fairly large, it will also hinder the hydrolysis of uncharged esters by hindering their approach to the esteratic site.

The charge on the anionic site of AChE is probably one electron. The argument in principle is that ΨChE is uncharged, and has unit charge less than AChE. The proof is complex, because Van der Waals' dispersion forces attract molecules towards both AChE and ΨChE, and the neglect of these forces leads to a different answer. Two types of observation show that ΨChE probably has no negative charge. Firstly, choline esters do not inhibit their own hydrolysis by ΨChE. Secondly, $K_I$ for the competitive inhibition of ΨChE by 3:3-dimethylbutyl acetate is the same as $K_m$ for the hydrolysis of ACh by ΨChE.[27] The inhibitor and substrate are sterically alike, and $K_I$ and $K_m$ both measure the attraction between enzyme and compound. The positive charge on ACh does not, therefore, increase the attraction. The difference in charge between AChE and ΨChE was shown as follows. Equal concentrations of choline inhibited equally the hydrolysis of ACh by AChE and ΨChE. The $K_m$'s for ACh on AChE and ΨChE respectively were, under the conditions of those experiments, $4{\cdot}4 \times 10^{-5}$ M and $1{\cdot}47 \times 10^{-3}$ M: i.e. the ACh–AChE complex was 30 times as stable as the ACh-ΨChE complex. Hence choline must have been 30 times more attracted to AChE than to ΨChE to produce equal inhibition.[27] Simple electrostatic theory showed that this was the factor expected if AChE has a charge of one electron greater than ΨChE. This has been confirmed by experiments in which AChE[28] and ΨChE[29] were inhibited by methylammonium ions. The $K_I$'s decreased in the order: $Me\overset{\oplus}{N}H_3 > Me_2\overset{\oplus}{N}H_2 > Me_3\overset{\oplus}{N}H = Me_4\overset{\oplus}{N}$, i.e., the energies of attraction increased in this order. The free energy of attraction is given by the equation: $\Delta F = -RT \ln K_I$, and increases by about 1·2 kcal per methyl group up to 3. Only 3 methyl groups can make contact with the enzyme surface unless the enzyme is wrapped round the inhibitor, and this is improbable because the much larger ion,

$Bu_4\overset{\oplus}{N}$, was a more powerful inhibitor than $Me_4\overset{\oplus}{N}$.[30] When ACh is adsorbed on to AChE, 2 methyl groups should make contact and contribute a binding energy of 2·4 kcal/mole. The total binding energy is about 4·5 kcal/mole, leaving 2·1 kcal/mole to be explained by electrostatic attraction, which is consistent with the hypothesis of unit charge.*

The theory leads to the conclusion that both AChE and ΨChE have sites able to attract molecules and ions by dispersion forces, and that these are equally attracting and are positioned at about the same distances from the esteratic sites in both enzymes. In AChE this site has in addition a single negative charge which about doubles its attraction for ACh.

A similar type of argument shows that the energy of attraction at the esteratic site is 2·7 kcal/mole, which might well be due to hydrogen bonding.[29]

When AChE is fully complexed by substrate the rate is a maximum. The ratios of the maximum rates for ACh and ethyl acetate agree precisely with the ratios of their rate constants for hydrolysis by hydroxyl ions,[29] which suggests that reaction at the esteratic site is closely analogous to alkaline hydrolysis, and that ChE's can be regarded as nucleophilic reagents.

Ions of the type $Me_3\overset{\oplus}{N}\cdot(CH_2)_n\cdot\overset{\oplus}{N}Me_3$ inhibit both AChE and ΨChE. The $K_I$'s for inhibition of ΨChE decrease geometrically with n, i.e. the energy of attraction increases by the same amount for each $—CH_2—$ group, presumably due to the operation of dispersion forces. The energy of attraction to AChE, calculated in the same way, increases at the same rate until $n = 6$, but from $n = 6$ to $n = 10$ it increases much more rapidly with chain length.[31] This can be held to show that the anionic site of AChE has two, closely spaced, charges, to which both positive ions can only be attracted when the intervening chain is long enough to allow them to lie close together. The argument appears to be unsound, as the total energy of attraction of two (mutually repelling) cations to two anions close together is very little more than the energy of one cation to two anions. An alternative explanation is that there is a second negatively charged site some way from the anionic site, and that the two can be bridged only when n is 6 or more.[29]

The distance between the esteratic and anionic sites can be calculated in four ways. If ACh, the favoured substrate, is stretched between the two sites, then the distance is that between the ester oxygen and quaternary nitrogen, i.e. about 4·4 Å. In eserine, which is very strongly adsorbed, ($K_I = 6 \times 10^{-8}$ M), the corresponding distance is 5·5 Å.[29, 32] If, however, ACh takes up its most probable position, the distance is only 2·3 Å, a little shorter than

* An alternative theory[31] ignores the attraction between the $\overset{\oplus}{N}Me_3$ group in ACh and ΨChE arising from dispersion forces, and so assigns unit charge to ΨChE. Although some evidence for this interpretation is given it is less likely to be true on the above showing, and some experiments are wholly inconsistent with it—see footnote on p. 138.

the corresponding distance in a group of powerful inhibitors in which $\overset{\oplus}{N}Me_3$ and O·CO·R groups are attached to a cyclopentane ring at rigidly fixed distances from each other.(33) The longer distances are more consistent with other results. Ammonium salts, for example, reduce inhibition by eserine by competing for the anionic site, but they do not affect inhibition by DFP. The N·····F distance between adsorbed ammonium ions and reacting DFP cannot be under about 3·5 Å, the sum of the non-bonded radii. Larger ions than ammonium do slow inhibition by DFP,(34) as expected even if the distance is about 5 Å.

Positive ions of all sizes inhibit the hydrolysis of choline esters by AChE, but only ions larger than $NH_4^{\oplus}$ inhibit the hydrolysis of neutral esters. The effects of positive ions on Ψ'ChE are, however, more complex, some catalysing the hydrolysis.(35)

From the above discussion the active centres are concluded to have the following properties.

(1) Both AChE and Ψ'ChE possess esteratic sites, at which substrates are hydrolysed. The sites react like nucleophilic reagents.

(2) Both active centres also have stereospecific sites, probably about 5 Å away from the esteratic sites, which attract particularly branched chains such as $—CMe_3$ and $—\overset{\oplus}{N}Me_3$. Affinity increases up to at least $—\overset{\oplus}{N}Bu_3$.(30)

(3) In AChE the stereospecific site carries unit negative charge, and is called the anionic site. Ψ'ChE carries no charge.

(4) AChE possesses another negatively charged group displaced some way, probably at least 10 Å, from the anionic site.

This picture of the two enzymes, although a theoretical construct, will be assumed to be basically correct. It is wholly consistent with the reactions of the compounds with phosphorus inhibitors.

*Determination of ChE activity.* All methods depend on determinations of the rate of formation of products or the rate of disappearance of substrate at its optimum concentration. For AChE, the preferred substrates are usually ACh and AMeCh, and for Ψ'ChE, BuCh or BzCh. These are specific for AChE or Ψ'ChE from some sources, but there are some cases where they are not, and where, for example, AMeCh cannot be used to estimate AChE in the presence of Ψ'ChE.

The precision of each method summarized is given in brackets, and is the precision on that ChE level which can be determined most accurately. At one-tenth of this optimum level the precision is usually only about one-tenth as great. Thus suppose in an experiment two ChE samples are compared, one a control at the optimum concentration, and the other initially at this concentration but 90% inhibited before determination. Then with a precision of 2% the second sample would be found to be about $10 \pm 2\%$ of the first.

The most common methods depend on estimating the acid released by the hydrolysis:

$$R\cdot CO\cdot O\cdot CH_2\cdot CH_2\cdot \overset{\oplus}{N}Me_3 + H_2O \rightarrow HO\cdot CH_2\cdot CH_2\cdot \overset{\oplus}{N}Me_3 + RCO\cdot OH,$$

allowance being made for spontaneous hydrolysis of the substrate. When determining AChE with ACh the optimum substrate concentration is only about $3 \times 10^{-3}$ M, so there is some risk of using up much of the substrate in the course of the experiment, and hence of underestimating the enzyme concentration. 0·035 M $Mg^{2\oplus}$ is sometimes included in the medium for this reason,[35] as, by competing for the anionic site, it increases the optimum ACh concentration to about $1{\cdot}5 \times 10^{-2}$ M. Most buffer systems, however, precipitate $Mg^{2\oplus}$. The acid liberated can be estimated in any of the ways listed below.

(1) In a Warburg apparatus[36] by the volume of $CO_2$ released from a bicarbonate buffer, pH 7·7 (2%). $Mg^{2\oplus}$ can be included in this buffer.

(2) By the change in pH in standard time of a phosphate-barbiturate or similar buffer, initially at pH 8·0. For rough tests (20%) the pH change can be followed with an indicator visually,[37] or, more accurately, spectrophotometrically[38, 39] (3–5%); or pH can be determined with a glass electrode[40, 41] (3–5%).

(3) The rate of consumption of alkali required to maintain the pH constant can be determined, preferably with a glass electrode[42, 43] (1–3%). In one method with an automatic titrator no buffer is required.[6]

(4) The $CO_2$ evolved from a bicarbonate buffer can bc cstimated in a Cartesian diver.[44, 45, 46] This method is sensitive enough to determine the AChE in a single nerve cell, but great skill is required to achieve high precision (5%).

The rate of disappearance of substrate may be estimated. ACh is estimated as its ferric hydroxamate[47, 48] (5–10%), or acetyl- or butyryl-thiocholine are estimated spectrophotometrically at 250 m$\mu$[49] (3–5%).

Finally, the alcohol liberated can be estimated if it is a coloured compound, such as indophenol[50] (3–5%), or a compound readily converted to one.[51] The substrates for these methods are, however, structurally very different from choline esters, and the chances of other enzymes interfering are high.

Histochemical methods depend very much on the use of selective phosphorus inhibitors, so their description is relegated to Chapter VIII.

## A-ESTERASES AND OTHERS

Enzymes are usually named by chemical function and natural substrate, as 'acetylcholinesterase' for the enzyme hydrolysing acetylcholine. There is, however, a group of enzymes without a known function which hydrolyses phosphorus inhibitors rapidly. Their natural substrates are unknown, so they must be named in terms of their reactions with phosphorus compounds. They hydrolyse many compounds, so that specific names, such as Sarinase and DFP-ase, are misleading. Two methods of naming have been suggested. The most rational is based upon a three-fold division of esterases[52, 53] into those hydrolysing phosphorus compounds but not inhibited by them (A-esterases), those not hydrolysing phosphorus compounds but inhibited by them (B-esterases), and those not reacting with phosphorus compounds in any way (C-esterases). This nomenclature has the advantage that if natural functions and substrates are discovered naming will present no difficulties. Enzymes hydrolysing phosphorus compounds will, therefore, be called A-esterases throughout. This name has only the trivial disadvantage that it is possible on cursory reading to confuse A-esterase with Ali-esterase. The latter is the name given to an enzyme when it hydrolyses simple aliphatic esters, such as triacetin or phenyl acetate, and has no known natural function. It may be an A-, B- or C-esterase. The alternative name given to an enzyme hydrolysing phosphorus compounds is phosphorylphosphatase and is used in American Chemical Abstracts and many journals. Although a very general name, it still appears to imply that nerve gases are natural substrates.

## REFERENCES


1. E. Kolb. *Z. Veterinarmed.* **4,** 967 (1957)
2. A. Zeller and A. Bisegger. *Helv. chim. Acta.* **26,** 1619 (1943)
3. C. I. Wright and J. C. Sabine. *J. Pharmacol.* **93,** 230 (1948)
4. J. C. Sabine. Atomic Energy Comm. Techn. Inform. Service, AECU-1346 (1952)
5. I. B. Wilson and F. Bergmann. *J. biol. chem.* **186,** 683 (1950)
6. E. Heilbronn. *Acta chem. scand.* **13,** 1255 (1959)
7. G. B. Koelle and J. S. Friedenwald. *Proc. Soc. exp. Biol., N.Y.* **70,** 617 (1949)
8. G. B. Koelle. *J. Pharmacol.* **100,** 158 (1950)
9. P. Holton and H. R. Ing. *Brit. J. Pharmacol.* **4,** 190 (1949)
10. D. H. Adams. *Biochim. biophys. Acta.* **3,** 1 (1949)
11. D. H. Adams and V. P. Whittaker. *Biochim. biophys. Acta.* **3,** 358 (1949)
12. R. A. McNaughton and E. A. Zeller. *Proc. Soc. exp. Biol., N.Y.* **70,** 165 (1949)
13. K-B. Augustinsson. *Arch. Biochem.* **23,** 111 (1949)
14. D. K. Myers. *Biochem. J.* **55,** 67 (1953)
15. L. M. Sturge and V. P. Whittaker. *Biochem. J.* **47,** 518 (1950)
16. L. A. Mounter and V. P. Whittaker. *Biochem. J.* **47,** 525 (1950)
17. L. A. Mounter. *Biochem. J.* **50,** 122 (1951)
18. B. N. Smallmann and L. S. Wolfe. *J. cell. comp. Physiol.* **48,** 197 (1956)
19. J. E. Casida. *Biochem. J.* **60,** 487 (1955)
20. D. Grob and J. C. Harvey. *J. clin. Invest.* **37,** 350 (1958)
21. D. Glick. *J. biol. Chem.* **125,** 729 (1938)
22. K-B. Augustinsson and T. Isacshen. *Acta. chem. scand.* **11,** 750 (1957)
23. M. A. Rothenberg and D. Nachmansohn. *J. biol. Chem.* **168,** 223 (1947)
24. M. G. P. J. Warringa and J. A. Cohen. *Biochim. biophys. Acta.* **16,** 300 (1955)
25. F. Strelitz. *Biochem. J.* **38,** 86 (1944)

26. D. M. Sturgenor and D. Ellis. *J. Amer. chem. Soc.* **76,** 6049 (1954)
27. D. H. Adams and V. P. Whittaker. *Biochim. biophys. Acta.* **4,** 543 (1950)
28. I. B. Wilson. *J. biol. Chem.* **197,** 215 (1952)
29. I. B. Wilson. *J. biol. Chem.* **208,** 123 (1954)
30. F. Bergmann and M. Wurzel. *Biochim. biophys. Acta.* **11,** 440 (1953)
31. F. Bergmann and R. Segal. *Biochem. J.* **58,** 692 (1954)
32. S. L. Friess and W. J. McCarville. *J. Amer. chem. Soc.* **76,** 1363 (1954)
33. S. L. Friess and H. D. Baldridge. *J. Amer. chem. Soc.* **78,** 2482 (1956)
34. D. K. Myers. *Arch. Biochem.* **31,** 29 (1951)
35. D. Nachmansohn and M. A. Rothenberg. *J. biol. Chem.* **158,** 653 (1945)
36. K-B. Augustinsson. *Methods of Biochemical Analysis* **5,** 1 (1957) Intersci. N.Y.
37. A. Meyer and W. Wilbrandt. *Helv. physiol. Acta.* **12,** 206 (1954)
38. J. Gregoire, J. Gregoire and N. Limozin. *Bull. Soc. Chim. Biol.* **37,** 81 (1955)
39. W. T. Caraway. *Amer. J. clin. Path.* **26,** 945 (1956)
40. H. O. Michel. *J. Lab. clin. Med.* **34,** 1564 (1949)
41. L-E. Tammelin. *Scand. J. clin. Lab. Invest.* **5,** 267 (1953)
42. M. Sanz. *Helv. physiol. pharm. Acta.* **2,** C29, (1944)
43. J. C. Sabine. Atomic Energy Comm. Techn. Rept., AECU-2575 (1953)
44. K. Linderstrom-Lang. *Nature, Lond.* **140,** 108 (1937)
45. E. Giacobini and J. Zajicek. *Nature, Lond.* **177,** 185 (1956)
46. J. Zajicek and E. Zeuthen. *Exp. Cell Res.* **11,** 568 (1956)
47. S. Hestrin. *J. biol. Chem.* **180,** 249 (1949)
48. R. L. Metcalf. *J. econ. Ent.* **44,** 883 (1951)
49. I. I. A. Tabachnick, J. Mershon, M. E. Grelis and A. A. Rubin. *Arch. int. Pharmacodyn.* **114,** 351 (1958)
50. D. N. Kramer and R. M. Gamson. *Analyt. Chem.* **30,** 251 (1958)
51. D. E. McOsker and L. J. Daniel. *Arch. Biochem.* **79,** 1 (1959)
52. F. Bergmann, R. Segal and S. Rimon. *Biochem. J.* **67,** 481 (1957)
53. W. N. Aldridge. *Ann. Rep. chem. Soc.* **53,** 294 (1956); *Biochem. J.* **53,** 111 (1953)

CHAPTER VI

# The Inhibition of Enzymes by Phosphorus Compounds *in vitro*

TABLE 6.1 demonstrates the wide variety of enzymes inhibited by organophosphorus compounds. Most of these enzymes are esterases or proteases. A few, of low susceptibility, are oxidases or dehydrogenases. These will not be considered further as little work has been done on them.

## THE NATURE OF THE INHIBITORY ACTION

Proteases and esterases are inhibited by the phosphorylation of their esteratic sites:

$$(RO)_2PO\cdot X + HE \rightarrow (RO)_2PO\cdot E + HX.^*$$

The evidence for this is of two kinds—stoicheiometric and kinetic.

*Stoicheiometry.* When chymotrypsin is inhibited by a dialkyl phosphate ester, one molecule of acid, HX, is liberated per molecule of chymotrypsin inhibited, and each molecule of inhibited enzyme contains two alkoxy groups and one phosphorus atom, as shown in Table 6.2. The protein is not destroyed by inhibition—the product from chymotrypsin and DFP is a crystalline compound of the same molecular weight as chymotrypsin itself, within experimental error.[20, 24] Similar, though less complete studies have been carried out on all the enzymes asterisked in Table 6.1.

Direct stoicheiometric evidence cannot be obtained for the reaction with ChE's, as the molecular weights of the enzymes are unknown. It can, however, be shown in the following way that the active sites of AChE's are probably phosphorylated. A very active specimen of electric eel AChE[27] was inhibited with $DF^{32}P$, and the $^{32}P$ content of the inhibited enzyme determined.[28] On the assumption that one atom of phosphorus was bound to each inhibited site, the number of sites was calculated. From the rate of hydrolysis of ACh by the specimen before inhibition the number of molecules

* The equation:

$$(RO)_2PO\cdot X + E^{\ominus} \rightarrow (RO)_2PO\cdot E + X^{\ominus},$$

may, of course, be equally valid. The same conclusions are drawn from both equations, so which is used is a matter of choice.

of ACh hydrolysed per site per minute (the turnover number) was calculated to be 420 000. A similar turnover number was found for bovine erythrocyte AChE by the following method,[10, 29] less direct than for eel AChE because the bovine AChE was less pure. The AChE was treated with butyrylcholine (BuCh), and then with unlabelled DFP. Both reagents were then washed

TABLE 6.1. ENZYMES INHIBITED BY PHOSPHORUS COMPOUNDS

The name of the enzyme and the $pI_{50}$ of the most active inhibitor tried are given. ($pI_{50} = -\log I_{50}$, where $I_{50}$ is the molar concentration inhibiting 50% of the enzyme.) For ChE's the results obtained by different authors differ too much for the most active inhibitors to be chosen. The source of the enzyme is also given where necessary.

| Enzyme and source | $pI_{50}$ of the most active inhibitor | Refs. |
|---|---|---|
| AChE (various)* | several at 10–11 | |
| ΨChE (various)* | several at about 9 | |
| Chymotrypsin (α)* | $(Bu^nO)_2PO \cdot F$, 5·8 | (1) |
| (β* and γ*) | DFP, ~5 | (2) |
| Trypsin* | $(Bu^nO)_2PO \cdot F$, 6·1 | (1, 26) |
| Plasmin | $(Bu^nO)_2PO \cdot F$, 4·4 | (3) |
| Thrombin | $(Bu^nO)_2PO \cdot F$, 4·0 | (1) |
| Proteinase* (bacterial) | DFP, 4·5 | (4) |
| Haemolytic activity of complement (guinea-pig) | DFP, ~4 | (5) |
| Lipases (wheat germ) | DDVP, 7·1 | (1) |
| (milk) | DFP, 5·2 | (6) |
| (pancreas) | Paraoxon, 5·2 | (7) |
| Acid phosphatase (rabbit) | DFP, ~3 | (6) |
| Pyrophosphatase (yeast) | DFP, ~3 | (6) |
| Ali-esterases | | |
| (livers of various species)* | DFP, ~6·5 | (6, 8, 9) |
| (bovine red cell stroma)* | DFP, ~6 | (10) |
| Esterases | | |
| (sera of various species) | Paraoxon, ~8 | (11) |
| (insects, several species) | $(EtO)(EtS)PO \cdot O \cdot C_6H_4\text{-}4\text{-}NO_2$, ~9 | (12, 13) |
| (insect eggs) | TEPP, ~6 | (14) |
| (saprophytic bacteria) | Paraoxon, ~6 | (15) |
| (citrus fruit) | TEPP, ~6 | (16) |
| Succinoxidase (housefly, mouse liver) | $(MeO)(MeS)PO \cdot S \cdot CH(CH_2 \cdot CO \cdot OEt) \cdot CO \cdot OEt$, ~3·5 | (17) |
| Ascorbic acid oxidase | Demeton-S, ~3 | (18) |
| Dehydrogenase (brain) | DFP, ~4 | (19) |
| Sulphydryl enzymes (bromelain, papain and ficin) | DFP, 3·4 | (115) |

Several of the results given above are unconfirmed.

TABLE 6.2. THE STOICHEIOMETRY OF THE REACTION BETWEEN α-CHYMOTRYPSIN AND PHOSPHORUS INHIBITORS

The results have been recalculated on the assumption that the molecular weight of chymotrypsin is 24 000, which is about the average of various values given in the references below and other literature.[20, 21]

| Inhibitor | Per cent inhibited | Product estimated | Mole product per mole inhibited | Refs. |
|---|---|---|---|---|
| Paraoxon | 50 | 4-nitrophenol released | 0·97 | (22) |
| | 100 | 4-nitrophenol released | 1·10 | (22) |
| DFP | > 95 | Acid released | 1·02 | (23) |
| DFP | > 95 | P in enzyme | 0·96–1·07 | (24) |
| DFP | > 95 | $Pr^iO$ groups in enzyme | 1·95 | (23) |
| $(PhO)_2PO \cdot Cl$ | > 99 | P in enzyme | 0·98 | (25) |
| $(EtO)_2PS \cdot Cl$ | > 99 | P in enzyme | 0·97 | (25) |
| $(Pr^iO)_4P_2O_3$ | > 99 | P in enzyme | 1·02 | (25) |
| Paraoxon | > 99 | P in enzyme | 0·98 | (25) |

off, and the enzyme activity shown to be the same as before treatment. Thus the BuCh had blocked the active sites, so that they were not inhibited by DFP; and any other groups which it was possible to phosphorylate had been phosphorylated. The enzyme was then inhibited with $DF^{32}P$, and the turnover number calculated as before. The only assumption made in this procedure is that BuCh does not protect parts of the enzyme other than the active sites from unlabelled DFP. If it does, then these take up $^{32}P$ in the second treatment, so that the $^{32}P$ content of the inhibited enzyme is too high, and the turnover number calculated is too low. The turnover number found by this method was 278 000, and by a modification of it 372 000, both of which are in reasonable agreement with the turnover number of eel AChE. Unless the inhibition of both AChE's involves a stoicheiometric reaction of the enzymes with the inhibitor in which the active sites are phosphorylated, the chances of obtaining any agreement at all would be remote.

The experiments already described are equally consistent with phosphorylation of the esteratic site or with phosphorylation elsewhere, with the consequent destruction of the enzyme. Three lines of evidence indicate that the point of attack is the active site:

(1) Normal substrates stop or slow inhibition.

(2) Many inhibited enzymes can be re-activated by water or anions which catalyse the hydrolysis of phosphate esters.

(3) The rates of inhibition by phosphate esters and the rates of hydrolysis of substrates depend in the same way on pH.

The evidence for these statements is given later.

*The kinetics of the reaction.* The kinetics of the reaction confirm the conclusions reached by studies of the stoicheiometry, extend them to cover ChE's, and give more insight into the nature of the reaction. Much preliminary work was carried out on enzymes other than ChE's, and may be found in the papers already referred to. The mass of quantitative data on the inhibition of ChE's is, however, more convenient to use for a systematic treatment.

The reaction between an inhibitor and a ChE obeys first order kinetics (Fig. 6.1), and the rate is proportional to the concentration of inhibitor

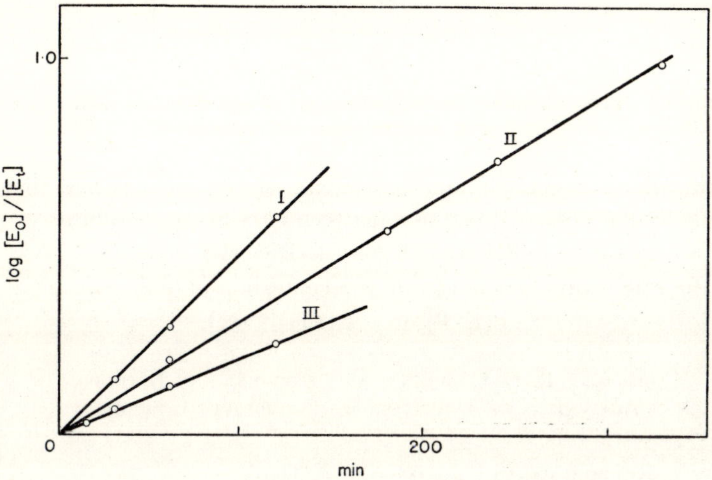


FIG. 6.1. The rates of reaction between phosphorus inhibitors and AChE's at 37°C.

I. Paraoxon ($1{\cdot}04 \times 10^{-8}$ M) with goat red cell AChE.[30]
II. Diethyl 8-quinolyl phosphorothionate* ($1{\cdot}53 \times 10^{-4}$ M) with sheep red cell AChE.[31]
III. Diethyl 4-chlorophenyl phosphate ($3{\cdot}97 \times 10^{-4}$ M) with sheep red cell AChE.[32]

* The active compound may have been the corresponding phosphate present as an impurity.

(Fig. 6.2). The reactions increase in rate rapidly with temperature, corresponding to energies of activation similar to those of alkaline hydrolysis (Table 6.3). This shows that a definite chemical reaction takes place. The inhibited enzymes are not re-activated by the addition of substrate. Inhibition is thus progressive, chemical, and irreversible, unlike inhibition by reversible inhibitors, which is not progressive and only involves physical adsorption on the active centre. Although there are other explanations—and

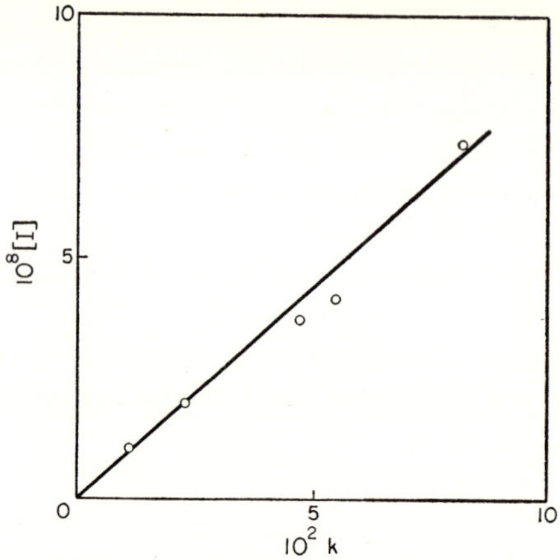


FIG. 6.2. Proportionality between the rate of inhibition of goat red cell AChE and concentration of Paraoxon.[30]

a modification is suggested later—by far the simplest is to assume that there is a bimolecular chemical reaction between enzyme and inhibitor:

$$EH + P\cdot X \rightarrow E\cdot P + HX.$$

Then the rate is proportional to the concentration of both reactants:

$$d[E\cdot P]/dt = k_I[EH][P\cdot X].$$

TABLE 6.3. RATE CONSTANTS AND ENERGIES AND ENTROPIES OF ACTIVATION OF INHIBITION BY PHOSPHORUS COMPOUNDS

| Reacting system | $k$ (37°) | $E$ kcal/mole | $\Delta S$ (E.U.) |
|---|---|---|---|
| Paraoxon + rabbit AChE[33] | $2\cdot6 \times 10^6$ | 10·6 | − 3·6 |
| L-(EtO)(Et)PO·S·$CH_2$·$CH_2$·SEt + electric eel AChE[34] | $3\cdot0 \times 10^3$ | 11·5 | − 7·2 |
| TIPA* + rat ΨChE[35] | $5\cdot1 \times 10^4$ | 14·5 | + 1·0 |
| Paraoxon + trypsin[26] | 3·1 | 15·5 | − 15 |
| Paraoxon + chymotrypsin[26] | 21 | 21·6 | + 9·6 |
| Paraoxon + $OH^{\ominus}$[36, 37] | 1·14 | 12·4 | − 27 |

* $(Pr^iNH)_4P_2O_3$.

The inhibitor is in vast excess in most experiments, so that at any given inhibitor concentration,

$$d[E\cdot P]/dt = k[EH],$$

where $k = k_I[P\cdot X]$, whence

$$\log_e [EH]_0/[EH]_t = kt.$$

Thus inhibition obeys first order kinetics, with $k$ proportional to $[P\cdot X]$.

The inhibited enzyme is a phosphate ester, which has to be hydrolysed if the enzyme is to be re-activated. Hydrolysis is slow, so inhibition is substantially irreversible.

Both substrates[38] and reversible inhibitors[39] of AChE and ΨChE greatly reduce the rate of inhibition by phosphorus compounds, indicating that the active centres are attacked. Thus at concentrations of DFP and Paraoxon which cause less than 50% inhibition of goat AChE in 20 min the addition of ACh to a final concentration of 0·14 M reduces the rate of inhibition to less than 2% of its original rate.[30] Some results illustrating the protective action of the reversible inhibitor eserine* are given in Table 6.4.

TABLE 6.4. PROTECTION OF HUMAN RED CELL AChE FROM TEPP BY ESERINE[39] AT 37°C

Concentrations were as follows: substrate, 0·03 M; TEPP, $4{\cdot}6 \times 10^{-8}$ M; eserine, $3{\cdot}1 \times 10^{-8}$ M. The compounds were added in the order given, and each inhibitor was incubated for 20 min before the addition of the next or the addition of the substrate.

| Control | Percentage inhibition |
|---|---|
| TEPP | 94 |
| Eserine | 54 |
| Eserine + TEPP | 63 |

It may appear anomalous that a concentration of reversible inhibitor which only inhibits 54% of the enzyme should reduce by over 90% the rate of inhibition by TEPP. The explanation reveals the essential difference between competitive, reversible inhibitors, such as eserine, and non-competitive irreversible inhibitors. A reversible inhibitor is in equilibrium with enzyme and enzyme–inhibitor complex:

$$K_I = [E][I]/[EI]$$

Suppose for eserine the constant is about $10^{-9}$ M. Then at the concentration used to obtain the results in Table 6.4, $10^{-9} = 3{\cdot}1 \times 10^{-8}\,[E]/[EI]$, whence only about 3% of the enzyme is un-inhibited. This is the only fraction which can be attacked by a phosphorus inhibitor. A phosphorus compound seems to have no great affinity for the enzyme, and cannot displace the

* Inhibitors such as eserine may react like phosphate esters, at least to some extent, but the inhibited enzyme is very unstable, so that the inhibitors behave kinetically as though they adsorbed on the enzyme surface without reacting with it.[54]

reversible inhibitor. Consequently in the presence of eserine very few active sites are exposed with which it can react. When substrate is added, however, the substrate displaces some of the reversible inhibitor, so that a far higher percentage of the enzyme is available to hydrolyse it. An equilibrium state is reached in about an hour.(39) The proportion of reversible inhibitor displaced by the substrate, and hence the rate of hydrolysis of the substrate, depends upon the substrate concentration.

The mechanism of phosphorylation will now be considered in more detail. Table 6.3 shows that the rate constants for enzyme inhibition are very much higher than those for a hydrolytic reaction with a similar energy of activation, i.e. an energetic collision between an active site and a compound is more likely to lead to reaction than a collision of similar energy between a hydroxyl ion and the compound, even when the energy requirements are equal. This is summarized by the statement that $\Delta S$ is more positive for the reaction with the enzyme. There are two possible explanations. Perhaps as the inhibitor approaches the enzyme it is constrained by hydrogen-bonding and Van der Waals' forces into an exceptionally favourable position for reaction to take place. The reaction is then formally very similar to hydrolysis, and can be written:

$$\text{H—E} + \text{P—X} \rightarrow \text{H}\cdots\text{E}\cdots\text{P}\cdots\text{X} \rightarrow \text{E—P} + \text{H—X}.$$

It can, however, be assumed instead that the first stage is reversible adsorption, and that the resulting complex holds the reactants in position until activation takes place from collisions with solvent molecules. The reaction scheme then becomes:

$$\text{P—X} + \text{E—H} \underset{k_2}{\overset{k_1}{\rightleftharpoons}} [\text{P—X/E—H}] \underset{k_4}{\overset{k_3}{\rightleftharpoons}} \text{H}\cdot\text{E}\cdots\text{P}\cdots\text{X} \xrightarrow{k_4} \text{E}\cdot\text{P} + \text{H}\cdot\text{X},$$

where [P—X/E—H] represents phosphorus inhibitor adsorbed on enzyme. This scheme is only consistent with non-competitive inhibition and with second order kinetics if two conditions are satisfied. The concentration of the adsorption complex must be much smaller than the concentration of enzyme; and $k_2$ must be much greater than $k_3$.*

* This can be shown as follows. As the rates of formation and destruction of the transition complex are very rapid indeed, the reaction scheme reduces to:

$$\text{E} + \text{I} \underset{k_2}{\overset{k_1}{\rightleftharpoons}} \text{I/E} \xrightarrow{k_3} \text{E}_\text{i}.$$

If $k_3 \ll k_2$, [I/E] does not depend on $k_3$, and $[\text{E}][\text{I}]/[\text{I/E}] = k_2/k_1$. Now the rate of inhibition, $\text{d}[\text{E}_\text{i}]/\text{d}t = k_3[\text{I/E}] = \frac{k_1k_3}{k_2}[\text{E}][\text{I}]$, where $[\text{E}_\text{i}]$ is the concentration of the inhibited enzyme. But the original concentration of enzyme, $[\text{E}_\text{o}]$, is given by

$$[\text{E}_\text{o}] = [\text{E}] + [\text{I/E}] + [\text{E}_\text{i}].$$

Only if $[\text{E/I}] \ll [\text{E}]$, does $[\text{E}] = [\text{E}_\text{o}] - [\text{E}_\text{i}]$, $\text{d}[\text{E}]/\text{d}t = -\text{d}[\text{E}_\text{i}]/\text{d}t$, and $\text{d}[\text{E}_\text{i}]/\text{d}t = \text{Const.}\,[\text{E}][\text{I}]$, which is the equation for a bimolecular reaction.

A rough calculation shows that the formation of an adsorption complex could explain most of the differences between the entropies of activation of inhibition and of hydrolysis. Assume that two inhibitors of similar shape and size have a similar tendency to form a complex. The assumption is plausible, as rates of inhibition are closely correlated with rates of hydrolysis, which is not likely to be true if the tendency to form complexes is different. Diethyl phenyl phosphate and Paraoxon inhibit AChE additively(31) (Table 6.5), so

TABLE 6.5. THE ADDITIVITY OF INHIBITION OF SHEEP ACHE BY DIETHYL PHENYL PHOSPHATE AND PARAOXON AT 37°C(31)

Various concentrations of the two inhibitors were incubated separately and together with the enzyme for 30 min, and the activities determined as percentages of the control rates. If the two inhibitors separately, at concentrations I and I′ respectively, reduce the activity to $b$ and $b'$% it can be shown that mixed at the same concentrations the inhibition will be reduced to $b''$, where $\log b'' = \log b + \log b' - 2$. The found and calculated percentage activities are given for the mixture only.

| Concentrations | | % activity remaining | |
|---|---|---|---|
| Paraoxon | Diethyl phenyl phosphate | obs. | calc. |
| $2 \times 10^{-8}$ M | $9{\cdot}8 \times 10^{-4}$ M | 41·7 | 40·1 |
| $2 \times 10^{-8}$ M | $8{\cdot}4 \times 10^{-2}$ M | 12·0 | 10·0 |
| $1{\cdot}25 \times 10^{-8}$ M | $4{\cdot}0 \times 10^{-3}$ M | 31·1 | 31·7 |

Taken from W. N. Aldridge and A. N. Davison. *Biochem. J.* **52** (1952) with acknowledgements.

the less active inhibitor cannot form a complex at its highest concentration with more than a small fraction of the enzyme, say 0·2. The chance that a collision between diethyl phenyl phosphate and enzyme will be energetic enough for reaction to take place is proportional to the product of the concentrations of the reactants: $8{\cdot}4 \times 10^{-2}$[E]. Similarly the chance of a collision between water and complex are: $55 \times 0{\cdot}2$[E] $= 11$[E]. Thus the activation of a complex may be over 100 times more probable than activation by direct collision. Even this small factor is equivalent, however, to a change in $\Delta S$ of $-$ 10 E.U., and would reduce the entropy of activation of inhibition by Paraoxon to $-$ 14 E.U., similar to the entropies of activation of some of the more rapid hydrolyses (Table 4.8).

There is no more direct evidence that a complex is formed, and as the affinity postulated is very low ($K_I = 0{\cdot}34$) it is difficult to devise ways of getting any. The formation of a complex with a $K_I$ in this range is, however, not unlikely. Tributyl phosphate inhibits purified human AChE reversibly[40] with $K_I = 0{\cdot}03$ (calculated from the percentage inhibition).

To summarize—phosphorus inhibitors react chemically with the enzymes they inhibit, phosphorylating them. The reaction is progressive, non-competitive and not readily reversed. Reaction may be preceded by the formation of an adsorption complex, but, if so, the complex is very unstable.

## REVERSAL OF INHIBITION

Phosphorylated enzymes are themselves phosphate esters, and can be hydrolysed by nucleophilic reagents such as water:

$$(RO)_2PO \cdot E + H_2O \rightarrow EH + (RO)_2PO \cdot OH.$$

This type of reaction restores the activity of the enzyme, and is called re-activation.

Re-activation of inhibited ChE's is described first. The reaction obeys first order kinetics[33] (Fig. 6.3 and 6.4), and the rates increase rapidly with temperature, giving energies of activation of about 14 kcal/mole.

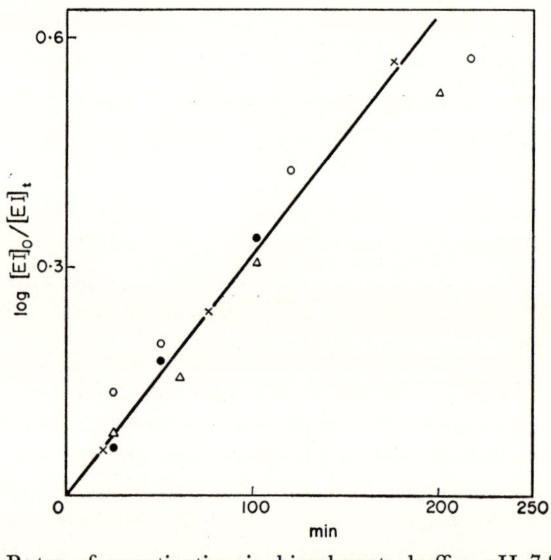


FIG. 6.3. Rates of re-activation in bicarbonate buffer, pH 7·8, at 37°C of rabbit AChE inhibited by various dimethyl phosphate esters.[41]

Let the concentrations of inhibited enzyme be $[EI]_0$ and $[EI]_t$ at $t = 0$ and $t$. Then for a first order reaction, $\log [EI]_0/[EI]_t = kt$. This linear relationship is demonstrated for AChE inhibited by:

$(MeO)_2PO \cdot O \cdot C_6H_4\text{-}4\text{-}NO_2$, ×; $(MeO)_2PO \cdot O \cdot PO(OMe)_2$, ●; $(MeO)_2PO \cdot F$, ○; $(MeO)_2PO \cdot S \cdot C_6H_4\text{-}4\text{-}NO_2$, Δ.

The enzyme was treated with inhibitor at 37°C for 10 min, and the excess inhibitor remaining was then destroyed by an A-esterase.

Rates depend on the species from which the ChE is taken (top of Table 6.6), but probably not upon the part of the animal. Small apparent differences between rates of re-activation of ΨChE's from different parts of the same animal may well be due to impurities in the crude preparations used.

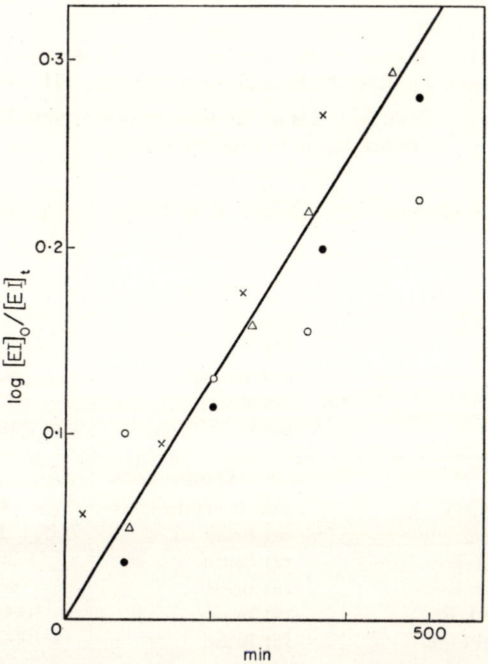


FIG. 6.4. Rates of re-activation in bicarbonate buffer, pH 7·8, at 37°C of rat serum ΨChE inhibited by various diethyl phosphate esters.[41]

The conditions were substantially as in Fig. 6.3, and the types of points are for the corresponding diethyl phosphate esters.

The rate of re-activation does not depend upon the acidic group in the original inhibitor, as shown in Figs. 6.3 and 6.4. This finding confirms the equation for inhibition:

$$(RO)_2PO{\cdot}X + EH \rightarrow (RO)_2PO{\cdot}E + HX,$$

according to which all compounds containing the same 'R' groups give the same inhibited enzyme, whatever X is. All diethyl phosphate esters, for example, give diethylphosphorylated enzymes.

Rates of re-activation vary with the basic groups on the phosphorus. Those of phosphorylated AChE's fall in the order:

$$(MeO)_2PO{\cdot}E \gg (EtO)_2PO{\cdot}E > (PrO)_2PO{\cdot}E \gg (Pr^iO)_2PO{\cdot}E,$$
$$(Pr^iNH)_2PO{\cdot}E,$$

(Table 6.6). The same order is found for phosphorylated ΨChE's, except that $(MeO)_2PO \cdot$ΨChE is unexpectedly stable. This exception may have the special explanation given later. Otherwise the rates fall as expected on electronic theory if re-activation is a nucleophilic displacement reaction in which water displaces the acidic group, E, from the phosphorus.

TABLE 6.6. RATES OF RE-ACTIVATION OF PHOSPHORYLATED ENZYMES AT 37°C IN BICARBONATE BUFFER, pH 7·8[35]

The rates are given in terms of the time in hours taken for 25% re-activation to take place ($t_{0 \cdot 25}$).

| Formula of enzyme | Source of enzyme | $t_{0 \cdot 25}$(hr.) |
|---|---|---|
| | (ΨChE) | |
| $(EtO)_2PO \cdot E$ | hen | 0·9 |
| | rat | 2·1 |
| | guinea pig | 30 |
| | horse | 82 |
| | man | 310 |
| | (AChE) | |
| $(MeO)_2PO \cdot E$ | rabbit erythrocyte | 0·66 |
| | rat brain | 1·3 |
| $(EtO)_2PO \cdot E$ | rat brain | 20 |
| $(PrO)_2PO \cdot E$ | rat brain | 40 |
| $(Pr^iO)_2PO \cdot E$ | rat brain | > 1000 |
| $(Pr^iNH)_2PO \cdot E$ | rat brain | > 1000 |
| | ΨChE (various) | |
| $(MeO)_2PO \cdot E$ | rat | > 80 |
| $(EtO)_2PO \cdot E$ | rat | 2·1 |
| $(PrO)_2PO \cdot E$ | rat | 7·1 |
| $(Pr^iO)_2PO \cdot E$ | rat | > 80 |
| $(Pr^iNH)_2PO \cdot E$ | rat | > 80 |

AChE inhibited by metabolites and oxidation products of tetramethylphosphorodiamidic compounds, such as Schradan and Dimefox, cannot be re-activated by water,[42] whereas horse ΨChE inhibited by these compounds is re-activated[43] with a half-life of 28 hr at 37°C. The nature of the phosphorylated enzyme is not known for certain in these cases, as dimethylamino groups must be partly oxidized before inhibition can take place.

Rates of re-activation depend on pH in the way shown in Fig. 6.5. The re-activation of ΨChE's is acid catalysed, but the rates of re-activation of two AChE's are greatest at about pH 7·8. These results still await an explanation.

Water is a weak nucleophilic reagent, and re-activation is more rapid by those derivatives of hydroxylamine, described in Chapter IV, which are outstanding catalysts of hydrolysis in neutral solution. The other catalysts described there are either less active (tungstate and molybdate) or change the behaviour of enzymes in undesirable ways, e.g. cupric chelates, and will not be considered.

Some hydroxylamine derivatives re-activate certain phosphorylated ChE's in a few minutes at low concentrations. Often these re-activators fail

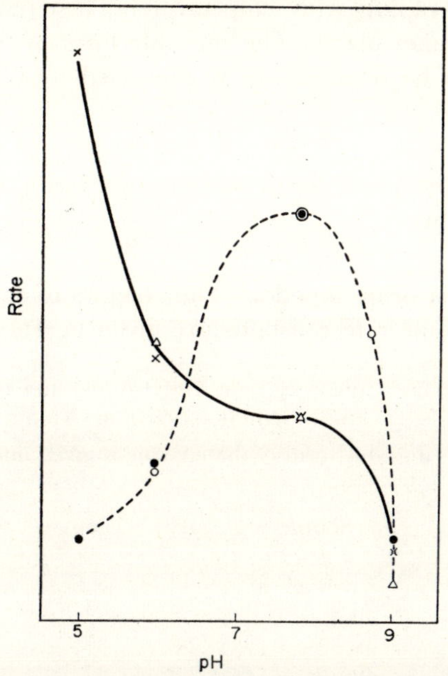


FIG. 6.5. Dependence on pH of the rates of re-activation of inhibited ChE's in water.[35]

The relative rates are given on a linear scale. Diethylphosphorylated ΨChE's: rat, ×; chicken, Δ; dimethylphosphorylated AChE's: rat, ●; rabbit, ○. The curves are drawn as an aid to the eye. There are not enough results to show whether they are correct in detail.

to restore all the enzymic activity. A part of the enzyme is re-activated rapidly, but the rest cannot usually be re-activated by any means. The fraction which can be re-activated is less the longer the phosphorylated enzyme is stored before re-activation is attempted. Allowance has to be made for this in re-activation experiments, and is best considered before continuing the description of re-activation. (In the experiments already described with water as a re-activating agent most of the original enzyme activity was recovered, so that the findings to be described do not alter the conclusions already reached.)

The fraction of phosphorylated enzyme in the stable form is determined as follows. A specimen of enzyme of known activity is stored with enough inhibitor to give nearly complete inhibition rapidly. At set times the inhibitor is removed, in the case of red cell AChE by washing, and in other cases by hydrolysis by an A-esterase or the hydroxylamine derivative. The enzyme is then re-activated as far as possible, and its activity determined.

Such experiments show that the fraction which can be re-activated falls exponentially with the time of storage, i.e. at a first order rate, and that the rate of fall increases rapidly with temperature. These results indicate that a chemical reaction takes place. The unstable form of the phosphorylated enzyme, which can be re-activated, is converted to a stable form which cannot be:

$$E \cdot P_u \rightarrow E \cdot P_s,$$

denoting unstable and stable forms by the subscripts u and s. The rate of conversion is given by:

$$d[E \cdot P_s]/dt = k[E \cdot P_u],$$

as shown by the first order kinetics. The energies of activation are about 15 kcal/mole.[48] Some rate constants are given in Table 6.7. The rates

TABLE 6.7. RATE CONSTANTS FOR THE CONVERSION OF INHIBITED ChE'S TO FORMS WHICH CANNOT BE RE-ACTIVATED

The initial, unstable form of the enzyme is given under $E.P_u$.

| $E.P_u$ | Source | pH | Temp. | $k$ (min$^{-1}$) | Refs. |
|---|---|---|---|---|---|
| $(EtO)_2PO \cdot AChE$ | man | 7·8 | 37° | $3 \times 10^{-4}$ | (45) |
| | ox | 7·8 | 37° | $3 \times 10^{-4}$ | (45) |
| | mouse | 7·8 | 37° | $3{\cdot}7 \times 10^{-4}$ | (46) |
| $(MeO)_2PO \cdot AChE$ | electric eel | 7·0 | 25° | $\sim 6 \times 10^{-4}$ | (47) |
| $(Pr^iO)(Me)PO \cdot AChE$ | man | 8·0 | 37° | $3{\cdot}0 \times 10^{-3}$ | (48) |
| $(Pr^iO)_2PO \cdot AChE$ | man | 7·8 | 37° | $2{\cdot}6 \times 10^{-3}$ | (45) |
| | ox | 7·8 | 37° | $2{\cdot}5 \times 10^{-3}$ | (45) |
| | mouse | 7·8 | 37° | $2{\cdot}4 \times 10^{-3}$ | (46) |
| $(EtO)_2PO \cdot \Psi ChE$ | man | 7·8 | 37° | $7 \times 10^{-3}$ | (44) |
| $(Pr^iO)_2PO \cdot \Psi ChE$ | man | 7·8 | 37° | $2{\cdot}5 \times 10^{-2}$ | (44) |

depend both on the type of phosphorylation, and on the type and source of the enzyme. Inhibited human ΨChE's go to stable forms about ten times faster than inhibited human AChE's. This may be a general difference between AChE's and ΨChE's. If so, it explains the very slow rate of re-activation by water of $(MeO)_2PO \cdot \Psi ChE$ (Table 6.6), which would be converted almost completely to the stable form before re-activation could be started. There are discrepancies in the literature which could be explained if rat $(MeO)_2PO \cdot AChE$ goes to the stable form faster than rabbit $(MeO)_2PO \cdot AChE$, but no rate constants for these conversions have been determined. Rates also depend on pH—see Fig. 6.6.

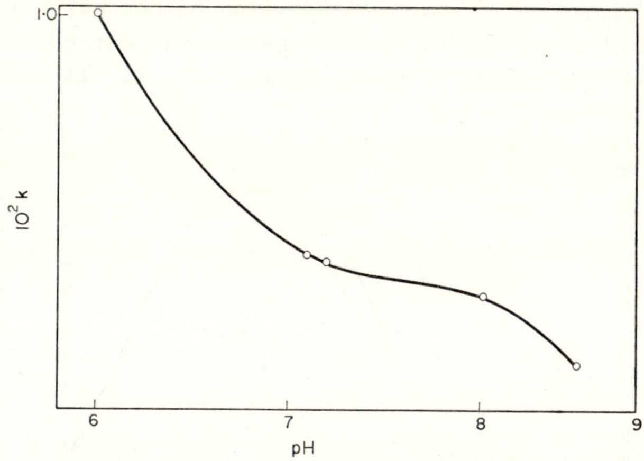


FIG. 6.6. Variation with pH of the rate of conversion of human AChE inhibited by Sarin to the stable form.[48]

The drawn curve is only an aid to the eye.

Several hypotheses on the nature of this reaction have been put forward, but only one is consistent with the present evidence.[117] On this hypothesis conversion to the stable form consists of the removal by hydrolysis of one alkoxy group:

$$(RO)_2PO \cdot E + H_2O \rightarrow (RO)(HO)PO \cdot E + ROH.$$

Horse ΨChE was inhibited by $DF^{32}P$, and the degree of conversion at various times after inhibition found in the usual way. Portions of the inhibited enzyme were also subjected to peptic digestion at the same times, and the phosphorylated peptides obtained analysed. Immediately after inhibition nearly all the phosphorylated peptides were di-isopropylphosphorylated, but mono-isopropylphosphorylated peptides were found as conversion proceeded, and the fraction of the total bound $^{32}P$ in this form agreed well at every stage with the fraction of the enzyme which had been converted to its stable form. The results were confirmed using $DF^{32}P$ in which the isopropyl groups were also labelled with $^{14}C$. As the enzyme underwent conversion $^{14}C$ was lost from it, although its $^{32}P$ content remained unchanged; and $^{14}C$-isopropanol was identified in the supernatant. The hypothesis explains the stability of the stable form. The stable form is a hydrogen phosphate, and therefore resistant to nucleophilic reagents (Chapter IV).

Re-activation is now considered in more detail. It has been studied mainly on inhibited AChE's rather than on inhibited ΨChE's, as the inhibited

AChE's are more tractable, tend to go to stable forms more slowly, and are more relevant to toxic action.

The main features in the action of re-activating agents can be illustrated by results obtained with hydroxyimino-acetone, MeCO·CH:N·OH. This compound hydrolyses inhibited enzymes according to first order kinetics[48] at rates which vary with pH in the way shown in Fig. 6.7. The whole of this

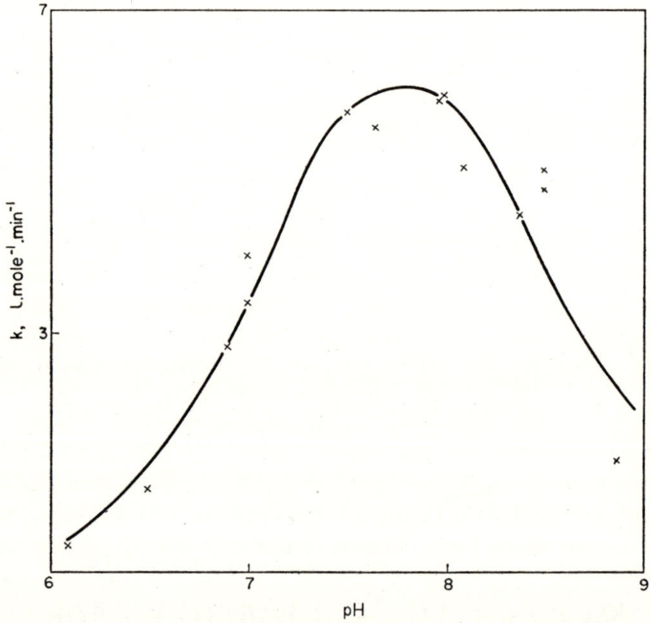


FIG. 6.7. Dependence on pH of the rate of re-activation of $(EtO)_2PO$·AChE (human erythrocytes) by hydroxyimino-acetone at 25°C.[48]

After D. R. Davies and A. L. Green. *Biochem. J.* **63,** 529 (1956), with acknowledgements.

curve can be accounted for quantitatively by assuming that only the oxy-imino anion* can react (as in oxime-catalysed hydrolysis), and that it can react only with a protonated form of the inhibited enzyme, which has a $pK_a$ of 7·3. The primary reactions taking place can probably, therefore, be written:

$$:N\cdot OH \rightleftharpoons :N\cdot O^{\ominus} + \overset{\oplus}{H}.$$

$$E\cdot P^{\ominus} + \overset{\oplus}{H} \rightleftharpoons HE\cdot P,$$

$$HE\cdot P + :N\cdot O^{\ominus} \rightarrow HE^{\ominus} + :N\cdot O\cdot P.\dagger$$

* As the most reactive compounds all contain the group :N·OH, it is convenient to call them hydroxyimines when referring to them as a class, and to call the corresponding anions oxyimino ions, even though hydroxamic acids are included.

† Only the differences in charge on the enzymes are necessarily correct in these equations.

The $pK_a$'s of the phosphorylated enzymes average about 7·6,[48] and this value is taken in calculations made later.

At low concentrations of hydroxyimino-acetone the rate constant, $k$, is proportional to the concentration of re-activator, [A]. At high concentrations it is not—a graph of $1/k$ against $1/[A]$ does not pass through the origin (Fig. 6.8). This suggests that an adsorption complex is formed between the

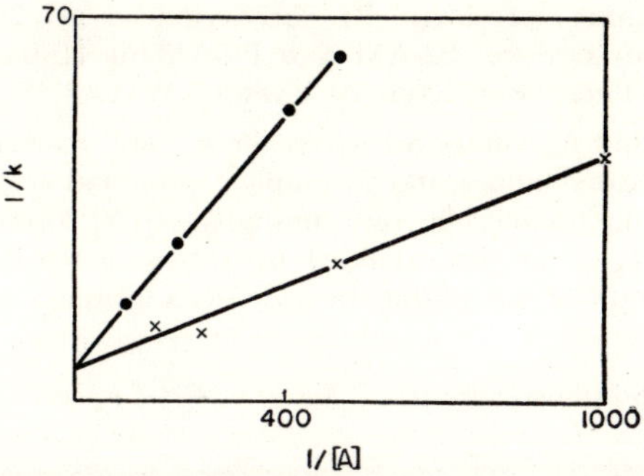


FIG. 6.8. Re-activation by hydroxyimino-acetone of $(EtO)_2PO \cdot AChE$ and $(Pr^iO)(Me)PO \cdot AChE$ (human) at pH 7·4 and 25°C.[51]

The reciprocal of the bimolecular rate constants are plotted against the reciprocal of the concentration of re-activator. $(EtO)_2PO \cdot AChE$, ●; $(Pr^iO)(Me)PO \cdot AChE$, ×.

After A. L. Green and H. J. Smith. *Biochem. J.* **68,** 28 (1958), with acknowledgements.

inhibited enzyme and the re-activator, and that the complex then gives the enzyme:

$$A^{\ominus} + HE \cdot P \underset{k_2}{\overset{k_1}{\rightleftharpoons}} A^{\ominus}/HE \cdot P \xrightarrow{k_3} HE^{\ominus} + A \cdot P.$$

Re-activation is thus characterized by two constants: an equilibrium constant, $K_A$, for the dissociation of the complex, and a rate constant, $k_3$, for its chemical reaction.* Some constants for several re-activations are given in

* The mathematical treatment is as follows.

The only chemical step is the last, so $k_2 \gg k_3$. Then, denoting inhibited enzyme by $E_i$,

$$K_A = [E_i][A]/[A/E_i]; \; K_A[A/E_i] = [E_i][A] \quad (1)$$

The final enzyme concentration is equal to the initial concentration of inhibited enzyme, i.e. $[E_i]_0 = [E]_\infty$. As $[E_i]_t = [E_i]_0 - [E] - [A/E_i]$, from (1) we obtain:

$$[A/E_i] = [A]([E]_\infty - [E])/(K_A + [A]) \quad (2)$$

$$d[E]/dt = k_3[A/E_i] \quad (3)$$

[*Footnote continued on next page*

Table 6.8. The equation of the straight line is: $1/k = K_A/k_3[A] + 1/k_3$. The gradient therefore gives $K_A/k_3$, and the intercept on the $1/k$ axis gives $1/k_3$, so both $K_A$ and $k_3$ can be found. Theoretically, if such a complex is formed the rate should reach a maximum when all the inhibited enzyme is complexed. This cannot be demonstrated with hydroxyimino-acetone, as it can only take place at concentrations at which the rates of re-activation are

TABLE 6.8. $K_A$ AND $k_3$ FOR THE RE-ACTIVATION AT pH 7·4 AND 25°C BY HYDROXYIMINO-ACETONE, P2-AM AND P4-AM OF HUMAN AChE PHOSPHORYLATED IN VARIOUS WAYS[51, 52]

In addition to $K_A$ and $k_3$, values are given for $k_3$ corr. (corrected for incomplete ionization of the hydroxyimines and incomplete protonation of the enzyme), the observed values for the bimolecular rate constants, ($k/[A]$ corrected as before), and the rate constants, $k_{A\ominus}$, for the catalysed hydrolysis of Sarin, $(Pr^iO)(Me)PO·F$. The p$K$ of the inhibited enzyme was assumed to be 7·6.

| Re-activator | Phosphoryl group | $K_A$ | $k_3$ | $k_3$ corr. | $k/[A]$ corr. | $k_{A\ominus}$ |
|---|---|---|---|---|---|---|
| Hydroxyimino-acetone | $(Pr^iO)(Me)PO$— | 0·01 | 0·24 | 3·5 | 350 | 250 |
| | $(EtO)_2PO$— | 0·02 | 0·16 | 2·3 | 110 | |
| P2-AM | $(EtO)_2PO$— | 0·00014 | 0·081 | 0·65 | 4700 | 120 |
| | $(Pr^iO)_2PO$— | 0·0008 | 0·015 | 0·12 | 150 | |
| P4-AM | $(Pr^iO)(Me)PO$— | 0·0008 | 0·031 | 0·69 | 880 | 380 |
| | $(EtO)_2PO$— | 0·0031 | 0·034 | 0·76 | 240 | |

too fast to follow. Pyridine-2-aldoxime methiodide (P2-AM), however, forms a more stable complex, as shown by the smaller value of $K_A$ in Table 6.8, and the maximum rate is readily achieved.

Complex formation is confirmed by the results of experiments in which hydroxyimino-acetone and P2-AM were allowed to re-activate an inhibited enzyme simultaneously. The rates varied with the concentrations of the two compounds in the way calculated on the assumption that both compete for the same sites.[52] Competition between nicotinohydroxamic acid and its methiodide has also been demonstrated.[53]

The constant, $k_3$, varies with pH, as the fractions of hydroxyimine ionized

---

so, for any constant value of [A], from (2),

$$d[E]/dt = k([E_0] - [E]) \quad (4)$$

which is the equation for a first order reaction.

The variation of $k$ with [A] is given by:

$$k = k_3[A]/(K_A + [A]) \quad (5)$$

or:

$$1/k = K_A/k_3[A] + 1/k_3 \quad (6)$$

and of inhibited enzyme protonated vary. A rate constant for the reaction between the active species, $:NO^{\ominus}$ and $HE \cdot P$, can be derived from $k_3$ by correcting for incomplete dissociation of the hydroxyimine and incomplete protonation of the inhibited enzyme. The new constant, $k_3$ corr., is:

$$k_3([AH] + [A^{\ominus}])([EI] + [EI \cdot H])/[A^{\ominus}][EI \cdot H],$$

where [AH], $[A^{\ominus}]$, [EI] and $[EI \cdot H]$ are the concentrations of undissociated hydroxyimine, oxyimino ion, unprotonated inhibited enzyme and protonated inhibited enzyme respectively. The concentrations can be calculated from the known total concentrations of hydroxyimine and enzyme and their $pK_a$'s. The new constant is a direct measure of the rate of reaction between the active species, and is therefore more fundamental than $k_3$. Values are given in Table 6.8 under '$k_3$ corr.' They are constants for the catalysed hydrolysis of phosphate esters, and should therefore vary with structure in the same way as the rate constants for the hydrolysis of phosphate esters catalysed by the same ions. The catalytic constants for the hydrolysis of Sarin are given under '$k_{A^{\ominus}}$' in Table 6.8 All three ions are about equally active as re-activating agents and as catalysts. Unfortunately it is difficult to determine $k_3$ accurately, and the activity of a re-activator is usually expressed in a different way, which obscures any structure-activity relationship there may be. At very low concentrations the observed rate constant, $k$, which is given by: $k_3[A]/(K_A + [A])$, reduces to $k_3[A]/K_A$, as $[A] \ll K_A$. Hence the activity can be expressed in terms of the pseudo-bimolecular constant, $k/[A]$, which is equal to $k_3/K_A$. This constant can, of course, be corrected for pH as before. It includes, however, both $k_3$ and $K_A$, whereas true constants for the catalysed hydrolysis of phosphorus compounds are only comparable with $k_3$. The lack of correlation between $k/[A]$ and $k_{A^{\ominus}}$ can be seen by comparing the last two columns of Table 6.8.

Table 6.9 lists a few of the many compounds tested as re-activators. It includes most of those which have any therapeutic value, and others which are of theoretical interest. Pseudo-bimolecular rate constants are given for the re-activation of $(EtO)_2PO \cdot AChE$ at pH 7·4, and also the constants for the reaction between the anions and the protonated enzyme. The former constants characterize the rates under normal conditions of use, but the latter must be used for discussing the effects of structure.

The most powerful re-activators of AChE are compounds containing a quaternary nitrogen, especially when it is part of a pyridine ring. The high activity is due to the formation of stable complexes, as already shown (Table 6.8), and not to any unusually high intrinsic activity of the anions of such compounds, as measured by $k_{A^{\ominus}}$ or $k_3$. It is found that any factor which makes complex formation difficult greatly reduces activity. The action of quaternary compounds therefore, differs from that of the rest in the ways listed below. Comparisons are in terms of the pseudo-bimolecular constants

TABLE 6.9. RATE CONSTANTS FOR THE RE-ACTIVATION OF DIETHYLPHOSPHORYLATED AChE BY SEVERAL COMPOUNDS

For each compound both $k/[A]$ is given and $k/[A]$ corr., the latter corrected for incomplete dissociation of the re-activator and incomplete protonation of the enzyme. Most p$K_a$'s were taken from Green and Saville.[54]

| Compound | Formula | p$K_a$ | $k/[A]$ | $k/[A]$ corr. | Refs. |
|---|---|---|---|---|---|
| Hydroxyimino-acetone | MeCO·CH:N·OH | 8·3 | 7·0 | 110 | (48, 51) |
| Dihydroxyimino-acetone | $CO(CH:N·OH)_2$ | 7·5? | 8·4 | 30 | (48) |
| Diacetyl monoxime | MeCO·CMe:N·OH | 9·3 | 0·1 | 13 | (55) |
| Hydroxyimino-acetylacetone | $(MeCO)_2C$:N·OH | 7·4 | 0·7 | 2·2 | (48) |
| Hydroxyimino-acetophenone | PhCO·CH:N·OH | 8·25 | 10·7 | 140 | (48) |
| Picolinohydroxamic acid |  ·CO·NH·OH | 8·5 | 2·9 | 100 | (48) |
| Nicotinohydroxamic acid |  ·CO·NH·OH | 8·4 | 0·01 | 0·2 | (53, 56) |
| Nicotinohydroxamic acid methiodide | 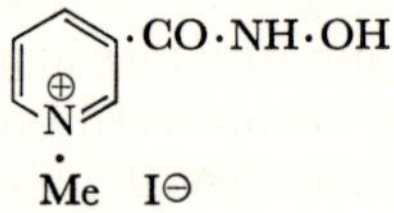 ·CO·NH·OH, Me I⊖ | 6·5 | 0·3 | 0·6 | (48) |
| Nicotinoyl formaldoxime | 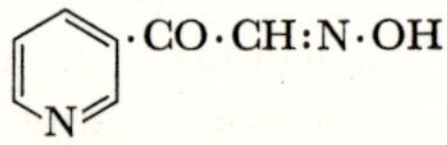 ·CO·CH:N·OH | 7·8 | 80 | 450 | (50) |
| Nicotinoyl formaldoxime methiodide | 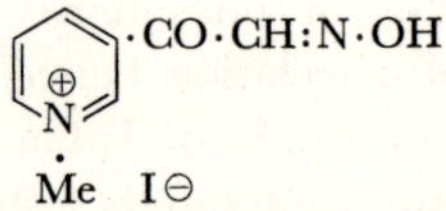 ·CO·CH:N·OH, Me I⊖ | 7·2 | 2800 | 7500 | (50) |
| isoNicotinoyl formaldoxime | 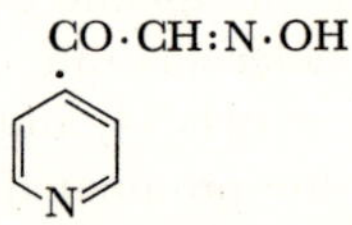 CO·CH:N·OH | 7·8 | 21 | 120 | (50) |
| isoNicotinoyl formaldoxime methiodide | 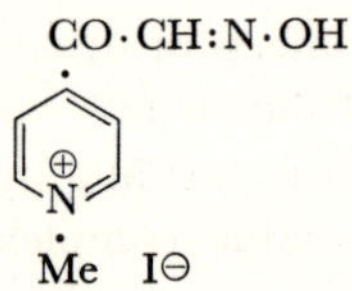 CO·CH:N·OH, Me I⊖ | 7·1 | 2500 | 6000 | (50) |
| Pyridine-2-aldoxime (2-hydroxyimino-methylpyridine) | 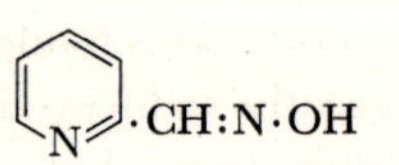 ·CH:N·OH | 10·1 | 0·34 | 270 | (50) |

Table 6.9. contd.

| Compound | Formula | p$K_a$ | $k$/[A] | $k$/[A] corr. | Refs. |
|---|---|---|---|---|---|
| Pyridine-2-aldoxime methiodide (2-hydroxyimino-methyl methylpyridinium iodide) P2-AM | 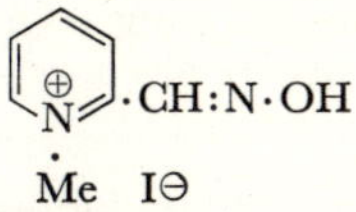 | 8·0 | 20 000 | 160 000* | (50) |
| Pyridine-4-aldoxime | 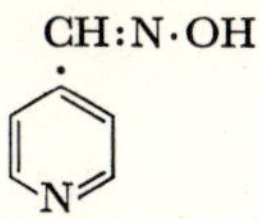 | 10·2 | 0·80 | 800 | (50) |
| Pyridine-4-aldoxime methiodide | 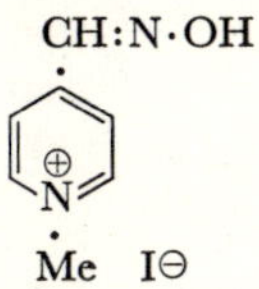 | 8·3 | 1400 | 20 000* | (52) |
| Choline | $Me_3\overset{\oplus}{N}\cdot CH_2\cdot CH_2\cdot OH$ | 12 | 0·1 | 0·28 | (57) |
| Dimethylaminoethanol | $Me_2N\cdot CH_2\cdot CH_2\cdot OH$ | 12 | < 0·001 | < 0·001 | (57) |

* These values, obtained on electric eel AChE, are very much bigger than those given in Table 6.8, which were determined on a different AChE.

corrected for pH, so that the differences are often less than would be observed in practice, as comparisons between different compounds are usually carried out at the same pH.

(1) With very few exceptions quaternary compounds are much more active than their ternary analogues (Table 6.9), although the anions of the ternary analogues are more basic, and hence more reactive (Chapter IV, p. 91).

(2) Compounds without a quaternary group hydrolyse $(EtO)_2PO\cdot AChE$ about five times more rapidly than $(Pr^iO)_2PO\cdot AChE$. This is in accordance with the inductive effects of alkoxy groups on the affinity of the phosphorus atom for other nucleophilic reagents. Quaternary compounds, however, hydrolyse $(EtO)_2PO\cdot AChE$ about fifty times more rapidly (Table 6.10), probably because the isopropyl groups in $(Pr^iO)_2PO\cdot AChE$ interfere with the approach of the quaternary nitrogen atom to the anionic site, and hence with the formation of the adsorption complex.

(3) Simple cations ($K^{\oplus}$, $NH_4^{\oplus}$,) reduce the activities of quaternary compounds by competing for the anionic site,* but they tend, for some unknown reason, to increase the activities of uncharged compounds.[58]

* This is the usual explanation given and may well be basically correct. It is, however, difficult to see why e.g. P2-AM should compete with hydroxyimino-acetone, while a concentration of cation capable of inhibiting the action of the former should catalyse the action of the latter.

TABLE 6.10. RELATIVE RATES OF RE-ACTIVATION OF DIETHYLPHOSPHORYLATED- AND DI-ISOPROPYLPHOSPHORYLATED-ACHE'S

The ratios of the rate constants for the re-activation of $(EtO)_2PO \cdot AChE$ and $(Pr^iO)_2PO \cdot AChE$ are given for some compounds containing quaternary nitrogen atoms and for others without this group. Only a selection of the available results are given.

| | Ratios of constants | Refs. |
|---|---|---|
| Quaternary compounds | | |
| Nicotinohydroxamic acid methiodide | 30 | (47) |
| isoNicotinohydroxamic acid methiodide | 90 | (47) |
| Picolinohydroxamic acid methochloride | >120 | (47) |
| Betainehydroxamic acid | 60 | (47) |
| Choline | >100 | (47) |
| Other compounds | | |
| Hydroxyimo-acetone | 9 | (48) |
| Dihydroxyimino-acetone | 10·5 | (48) |
| Hydroxyimino-acetophenone | 2·0 | (48) |
| Nicotinohydroxamic acid | 3·5 | (47) |
| isoNicotinohydroxamic acid | 2 | (47) |
| Picolinohydroxamic acid | 6 | (47) |

(4) The activities of pyridine-aldoxime methiodides depend very much on their steric configurations.[50] Thus P2-AM and P4-AM are both very active, while P3-AM is almost without activity (Fig. 6.9). (The explanation given here is a much modified version of the one put forward by the original authors.) The double bond in the hydroxyimino group conjugates with the aromatic ring. This tends to hold the molecules planar, in the case of P2-AM in the two configurations shown in Fig. 6.9 as syn-P2-AM and anti-P2-AM.* Compounds of this type are therefore most likely to react when planar. Assuming that the quaternary nitrogen atom is adsorbed firmly on to the anionic site, one can calculate the position of the phosphorus atom which most

* These stereo-isomers were originally believed to be separable, and a compound described as syn-P2-AM was prepared.[50] J. Poziomek, D. N. Kramer, B. W. Fromm and W. A. Mosher (*J. org. Chem.* **26,** 423 (1961)) have shown recently that the product obtained from the attempted preparation was not syn-P2-AM, but was a carbinolamine, i.e. it contained a —CH(OH)·NH·OH group instead of a —CH=N·OH group. This of course explains the very low rates of re-activation got with the so-called 'syn-P2-AM'. The present evidence suggests that the syn and anti forms do not exist separately as stable compounds. This removes the need for the hypothesis that the effects of conjugation are unusually great in these molecules—i.e. that conjugation alone could so restrict rotation about the C-ring bond that the two isomers could be separated.

favours reaction. During the reaction the O—P bond is expected to be formed in that plan perpendicular to the molecule which contains the N, O and P atoms, with the N—O—P angle and the O · · · P distance roughly

FIG. 6.9. Rates of re-activation of inhibited enzymes by isomers of PAM.[50]

| | Name | AChe | ψChE | Chymotrypsin |
|---|---|---|---|---|
| 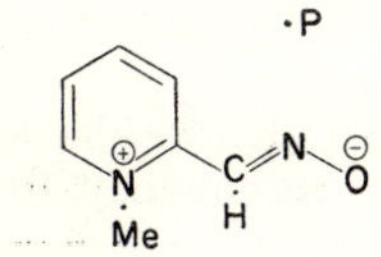 | syn-P2-AM | | | |
| 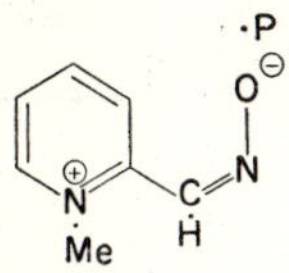 | anti-P2-AM | 160 000 | 12 | 3 |
| 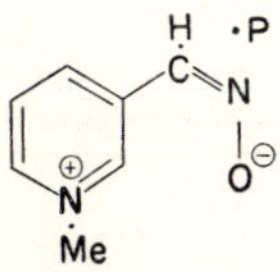 | P3-AM (one form only) | <1 | 1 | 3 |
| 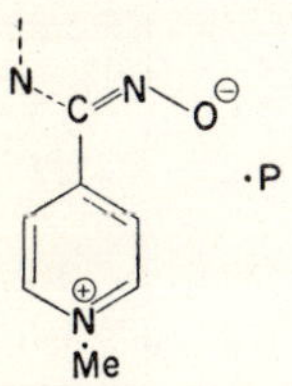 | P4-AM (one form only) | 20 000 | 3 | 3 |

as in the final product, i.e. about 110° and 3 Å. If P2-AM reacts in the 'syn' form, then neither P3-AM nor P4-AM can find suitable positions to react. For reaction with anti-P2-AM the best position of the phosphorus atom is about 2·5 Å out of the plane of the molecule from the position marked 'P' in Fig. 6.9, in which bonds and inter-bond angles are drawn to scale. From the figure it is obvious that there is no configuration of P3-AM which can react, even though the molecule can rotate in its plane about its quaternary nitrogen atom. P4-AM, however, has nearly as favourable a configuration as P2-AM.

Two observations confirm this interpretation. As P2-AM reacts in the 'anti' form, any substituent in the 3-position on the ring should lower the activity, as steric hindrance will force the molecule into a non-planar form. Correspondingly it is found that 3-methyl-P2-AM is only one-sixteenth as active as P2-AM itself. Secondly, when the enzyme has no anionic site all the compounds are about equally active, as demonstrated by the rates of re-activation of $(EtO)_2PO$·chymotrypsin and ·ΨChE (Fig. 6.9).*

The most powerful re-activators yet discovered are derivatives of P4-AM, not of P2-AM.[59, 60, 61] A few are shown in Table 6.11. A large group is

TABLE 6.11. DERIVATIVES OF PYRIDINE-ALDOXIME WITH TWO QUATERNARY GROUPS

The rate of re-activation of $(EtO)_2PO$·AChE at pH 7·4 and 37°C is given as a multiple of the rate of re-activation by P2-AM under the same conditions.[60, 61]

| Compound | p$K_a$ | Rate |
|---|---|---|
| 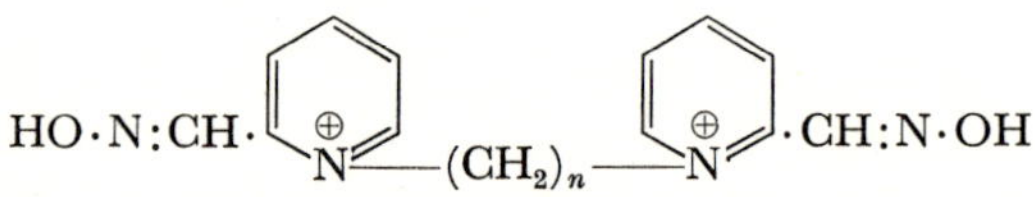 | | |
| $n$ = 5 (best) | 7·8 | 2 |
| 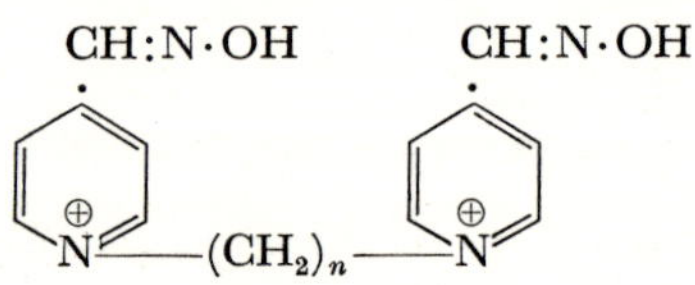 | | |
| $n$ = 1 | 7·6 | 3 |
| $n$ = 3 (Compound 16) | 8·2 | 22 |
| $n$ = 5 | 8·3 | 16 |
| $n$ = 10 | 8·4 | 2 |
| 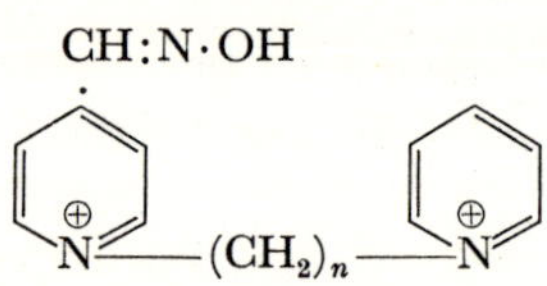 | | |
| $n$ = 3 | — | 8 |
| $n$ = 5 | — | 7 |

* These results, and those given in Table 6.10, are very difficult to understand unless AChE has a negatively charged site and ΨChE has none (p. 110). The 10 000-fold difference between the rates of re-activation of $(EtO)_2PO$·AChE and $(EtO)_2PO$·ΨChE by P2-AM is, however, too great to explain in this way alone. The energy of electrostatic attraction is about 2·4 kcal/mole (p. 110f), only enough to increase the attraction, and hence the rate, by a factor of about 70. Another anomaly is that choline re-activates $(EtO)_2PO$·AChE 500 times more rapidly at pH 7·4 than does dimethylaminoethanol (bottom of Table 6.9), although at this pH both compounds must be equally cationic. No explanation can be offered as yet.

attached to the pyridine nitrogen, and this group itself contains a quaternary nitrogen. In the most active molecules the quaternary nitrogens are separated by aliphatic chains of 3–6 carbon atoms. The most active is $(CH_2)_3(C_5H_4\overset{\oplus}{N}\text{-4-}CH{:}N\cdot OH)_2$, $2Br^{\ominus}$, NN′-trimethylene-bis(4-hydroxy-iminomethyl-pyridinium) dibromide. This is taken as an example of the group in later sections and is referred to as compound 16. The kinetics of re-activation by these compounds is still obscure, so the rates at which they re-activate $(EtO)_2PO\cdot AChE$ are only given in Table 6.11 as multiples of the rate at which the same molar concentration of P2-AM re-activates the same enzyme. Possibly these compounds are more active than P2-AM because the extra cationic groups are attracted to the second negatively-charged site in AChE (p. 111). The increase in activity with short chain lengths does not, however, seem fully consistent with this idea, as the two negatively-charged sites in AChE can apparently only be bridged by a chain of more than 6 carbon atoms.

Two re-activators, choline and hydroxylamine, act in unusual ways. Hydroxylamine is of low activity(62)—most experiments have been carried out with concentrations between 0·1 and 2·0 M. The dependence of the rate on pH shows that the active form is hydroxylamine, not its cation,(47, 63, 64) but, as trimethylamine oxide, $Me_3\overset{\oplus}{N}\text{—}\overset{\ominus}{O}$, is also a re-activator, the free base may act in its zwitterion form, $H_3\overset{\oplus}{N}\text{—}\overset{\ominus}{O}$.(47) The kinetics are abnormal—$k/[A]$ increases with increasing concentration(64) both for chymotrypsin and AChE, instead of decreasing as it should if a complex is formed, or remaining constant if it is not. The effect may be due to the high concentrations used. Choline behaves normally up to pH 8, forming a complex(57) for which $K_A = 0{\cdot}03$. The rate does not, however, drop as the pH is further increased, and from the whole rate–pH curve it is concluded that the reactive species are the unprotonated form of the enzyme and choline itself,(53) i.e. the opposite to usual. In Table 6.9 $k/[A]$ corr. has a different meaning, therefore, for choline and dimethylaminoethanol, with the result that the values shown are very similar to those of $k/[A]$. It is as though for re-activation to take place either the re-activator or the enzyme must be protonated, but not both: $EI\cdot H + [A^{\ominus}]$ or $EI + HA$, but not $EI\cdot H + HA$. Probably the reaction requires the formation of a hydrogen bond.

Only certain inhibited AChE's have been fully investigated—those containing $(EtO)_2PO$—, $(Pr^iO)(Me)PO$—, or $(Pr^iO)_2PO$— groups—and there is little information on the possibility of re-activating others. Apparently some AChE's inhibited with Tabun(65) or Mipafox cannot be re-activated,*

* Electric eel AChE inhibited by Tabun is re-activated fairly fast *in vitro* by compound 16, but only very slowly by P2-AM.(119) This is consistent with earlier statements, as compound 16 is unusually active.

in agreement with electronic theory, according to which

$$(Me_2N)(EtO)PO\cdot AChE \quad \text{and} \quad (Pr^iNH)_2PO\cdot AChE$$

should be more stable than those mentioned previously. AChE inhibited by the metabolite(s) of the tetramethylphosphorodiamidic compound, Schradan, cannot be re-activated,[42] but inhibited horse ΨChE can be.[43]

The catalytic effects of oxyimino anions on the hydrolysis of phosphorus compounds show that rates of re-activation should fall in the order:

$$(Pr^iO)(Me)PO\cdot AChE > (EtO)_2PO\cdot AChE \gg (Pr^iO)_2PO\cdot AChE$$

were no complexes formed. The last two are always in this order (see Table 6.10), but only some re-activators hydrolyse $(Pr^iO)(Me)PO\cdot AChE$ faster than $(EtO)_2PO\cdot AChE$ (e.g. hydroxyimino-acetone, dihydroxyimino-acetone, hydroxyimino-acetylacetone and nicotinohydroxamic acid methiodide), while others reverse the order (e.g. hydroxyimino-acetophenone and picolinohydroxamic acid). As many factors may influence the formation of complexes, no explanation of such differences in terms of structure can be expected yet.

Re-activation of AChE inhibited by methylfluorophosphinylcholine, $(Me)(F)PO\cdot O\cdot CH_2\cdot CH_2\cdot \overset{\oplus}{N}Me_3$, is interesting. The inhibited enzyme presumably has the structure: $(Me_3\overset{\oplus}{N}\cdot CH_2\cdot CH_2\cdot O)(Me)PO\cdot AChE$. It is re-activated fairly rapidly by water, as expected from the very electrophilic nature of the phosphorus atom; but it is not re-activated by P2-AM,[66] presumably because the anionic site is already occupied by the $-\overset{\oplus}{N}Me_3$ group, so that P2-AM cannot adsorb on it.

Re-activation of some other inhibited enzymes, e.g. of inhibited chymotrypsin, has already been referred to. Re-activation by water is very slow indeed,[16] and many enzymes have not been fully re-activated even by the most powerful re-activators. There is, however, no evidence that these inhibited enzymes go to stable forms. $(EtO)_2PO$- and $(Pr^iO)(Me)PO$-chymotrypsin have been completely re-activated;[49, 50] and $(Pr^iO)_2PO$-(horse liver ali-esterase) is re-activated at the same slow rate by hydroxyimino-acetone whether re-activation is attempted immediately, or after two days' storage.[9] In the last case one molecule of $(Pr^iO)_2PO\cdot OH$ has been recovered for each molecule of enzyme re-activated—good evidence that re-activation is by hydrolysis, and that conversion to a stable form does not take place.

Re-activators can have secondary effects, which are of little importance in *in vitro* experiments, but which may be important in some *in vivo* experiments when high doses are given. Firstly, quaternary re-activators are themselves weak, reversible inhibitors of AChE, as expected from their capacity to form complexes.[67] P2-AM concentrations above $5 \times 10^{-4}$ M should protect AChE against inhibition by phosphorus compounds to a significant extent.

Secondly, a re-activator may form a powerful anticholinesterase from the original inhibitor. Thus P4-AM reacts with Sarin as follows:

$$C_5H_4\overset{\oplus}{N}Me\text{-}4\text{-}CH{:}NO^{\ominus} + (Pr^iO)(Me)PO\cdot F \rightarrow$$
$$F^{\ominus} + C_5H_4\overset{\oplus}{N}Me\text{-}4\text{-}CH{:}N\cdot O\cdot PO(Me)(Pr^iO).$$

The product has a half life in water of about 700 min at pH 7·6, and is a more powerful inhibitor than Sarin itself.[68] At the concentrations used in most experiments the effect should be small, but it may be more important in other cases.

## MEASURES OF ANTI-ENZYME ACTIVITY

The obvious measure of the activity of an inhibitor towards an enzyme is the bimolecular rate constant for the reaction between the two. Unfortunately, rate constants take too long to determine when many compounds are to be screened, and there are in some instances practical and theoretical obstacles in the way of their determination which cannot easily be overcome. In this section these obstacles are discussed, and the significance of the alternative measure, the $pI_{50}$, is assessed.

In practice the use of grossly impure enzyme preparations is unavoidable. These may either adsorb or decompose inhibitors, with the result that the inhibitor concentration is not constant during the reaction, and that the effective rate constant, $k_I[I]$, varies as inhibition proceeds. Some inhibitors are completely decomposed within a few minutes of their introduction into the enzyme system. The same problem arises when the inhibitor is unstable in the medium.

Some inhibited enzymes, notably some $(MeO)_2PO\cdot$AChE's and $(EtO)_2PO\cdot\Psi$ChE's, are re-activated by the medium, so that inhibition obeys the kinetics of two successive reactions:

$$EH + I \xrightarrow{k_I} EI \xrightarrow{k_R} EH + \text{products},$$

and not first order kinetics (Fig. 6.10). The mathematical analysis of this system is quite simple, but even slight interference from adsorption or decomposition of inhibitor increases its complexity enormously, as exemplified by the kinetics of inhibition of AChE by $(MeO)_2PO\cdot O\cdot CH{:}CCl_2$, DDVP.[69]

Most authors have, therefore, not determined bimolecular rate constants, but have measured activity by the molar concentration required to inhibit half the enzyme, the $I_{50}$. For convenience this is best expressed as its logarithm with the sign changed, the $pI_{50}$. The concept was taken from the study of reversible inhibitors, where it has a specific meaning, as there is equilibrium between enzyme, substrate and inhibitor, which can be characterized by an equilibrium constant. The value of the constant depends only on temperature and substrate concentration, so that, if these are held constant, the activities of different compounds can be compared. Used uncritically for non-competitive, irreversible inhibitors, the concept can lead to absurdity. Thus, if any two inhibitors are perfectly stable in the enzyme preparation, they will continue to react with the enzyme until either the enzyme or the inhibitors have totally reacted. Under these conditions the $I_{50}$'s of both inhibitors are the same—half the enzyme concentration—even though one may react much faster than the other. Such obvious absurdities are avoided

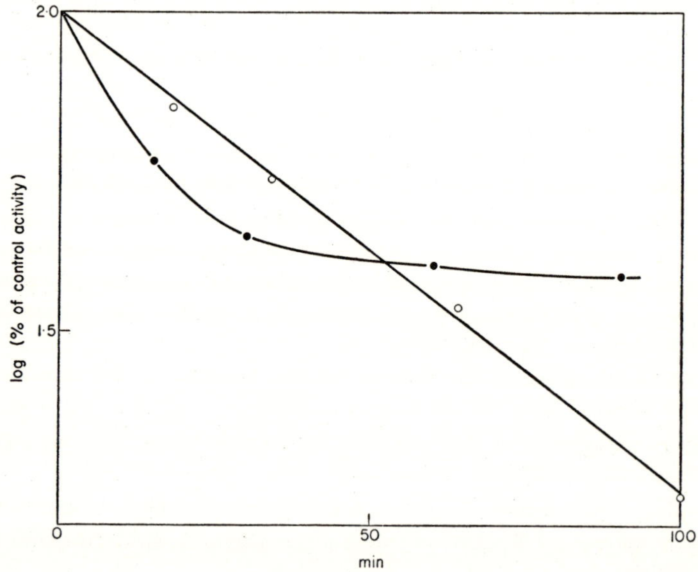


FIG. 6.10. Rates of inhibition of AChE by a dimethyl phosphate and a diethyl phosphate inhibitor.[33]

Rabbit AChE. Paraoxon-Me, ●; Paraoxon, ○. Temperature, 37°C.

After W. N. Aldridge. *Biochem. J.* **54,** 442 (1953), with acknowledgements.

by considering only the inhibition taking place in a set interval of time, but $I_{50}$'s then mean different things in different systems. Three cases can be distinguished:

(1) When the inhibition obeys first order kinetics, i.e. when a stable inhibitor reacts with an enzyme to form a stable product, $I_{50} = \ln_e 2/k_I t_{0\cdot5}$, where $t_{0\cdot5}$ is the time taken for half inhibition. Then, if the time is fixed, the $I_{50}$ is inversely proportional to $k_I$, so that $I_{50}$ and rate constant are equally valid as measures of activity. As $k_I$ depends upon temperature, and $I_{50}$ also upon the time the inhibitor and enzyme are incubated together, both time and temperature must be specified.

(2) If the inhibited enzyme is re-activated, a similar relationship between $I_{50}$ and $k_I$ holds, except that a different constant from $\ln_e 2$ is required. Slow

decomposition of the inhibitor has the same effect on the relationship, though the constants are different in each case. When it can be shown that inhibition increases for longer than the time chosen for the determination of $I_{50}$, it is unlikely that a value of $k_I$ calculated from the $I_{50}$ using the constant, $\ln_e 2$, will differ from the right value by a factor of more than 2. This precision is ample for most purposes.

(3) When the reaction is complete within the time interval chosen, because the inhibitor is rapidly decomposed or adsorbed, the $I_{50}$ does not depend on the incubation time. As, also, medium and enzyme are competing for inhibitor, and both reactions may depend similarly on temperature, the $I_{50}$ may be almost independent of temperature, or vary with it in a complex way. What is measured by the $I_{50}$ is the success of the enzyme at competing with the medium for the inhibitor, not the rate of reaction between enzyme and inhibitor. Consequently, $I_{50}$'s of this type cannot be compared with those of the other two types, and are often almost meaningless.

In this book an attempt has been made only to compare $I_{50}$'s, when either inhibition has been shown to be progressive, or, by analogy, it seems likely it would be. Exceptions have been made only when the effects are so striking that, even were the inhibitor to be rapidly decomposed, the results would be significant. Thus if an inhibitor is found to be exceptionally active and it is rapidly decomposed, its $I_{50}$ only underestimates its activity. So much work, however, has been done on systems whose kinetics can only be guessed, that some comparisons made may well be invalid.

## EFFECTS OF STRUCTURE ON ANTI-ENZYME ACTIVITY

The effects of the structures of inhibitors on the rates at which they inhibit enzyme have only been investigated in any detail on ChE's. Inhibition of ChE's is therefore considered first. Few of the published results can be relied upon, because there is no evidence that more active impurities were eliminated.

A phosphorus inhibitor contains three types of group:

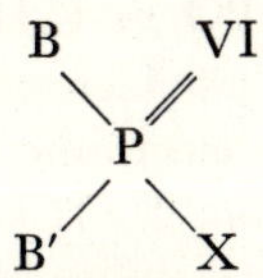


the basic or neutral 'B' groups, the acidic group, X, and the atom from the sixth period of the periodic table, usually oxygen or sulphur, doubly bonded to the phosphorus. Each type of group has a typical effect on activity. In addition there are certain steric and charge effects.

*Basic Groups.* Variation of the basic groups in phosphorus compounds, the structures of which are otherwise the same, produces a complex variation in rates of inhibition, and this variation is different according to whether the enzyme inhibited is an AChE or a ΨChE. Electronic theory, which explains differences in hydrolysis rates, is barely applicable to these effects, although the reactions are analogous. Thus according to electronic theory rates of inhibition should fall in the order:

$$(EtO)(Me)PO{\cdot}X > (MeO)_2PO{\cdot}X > (EtO)_2PO{\cdot}X > (Pr^iO)_2PO{\cdot}X > (Pr^iNH)_2PO{\cdot}X \gg (NMe_2)_2PO{\cdot}X.$$

For AChE's the order may differ considerably from this. For fly brain AChE it is:

$$(EtO)_2PO{\cdot}X > (EtO)(Me)PO{\cdot}X > [(MeO)_2PO{\cdot}X, (Pr^iO)_2PO{\cdot}X] > (Pr^iNH)_2PO{\cdot}X \gg (NMe_2)_2PO{\cdot}X$$

and for horse ΨChE it is:(71)

$$(Pr^iO)_2PO{\cdot}X \approx (EtO)_2PO{\cdot}X > (MeO)_2PO{\cdot}X \approx (Pr^iNH)_2PO{\cdot}X \gg (Me_2N)_2PO{\cdot}X.$$

The actual orders found vary with the enzyme source, but can be summarized as follows:

For AChE's,(70, 71)

$$[(EtO)(Me)PO{\cdot}X \gtrless (EtO)_2PO{\cdot}X] > [(MeO)_2PO{\cdot}X \gtrless (Pr^iO)_2PO{\cdot}X] > (Pr^iNH)_2PO{\cdot}X;$$

and for ΨChE's,

$$[(EtO)(Me)PO{\cdot}X \gtrless (Pr^iO)_2PO{\cdot}X \gtrless (EtO)_2PO{\cdot}X] > [(MeO)_2PO{\cdot}X \gtrless (Pr^iNH)_2PO{\cdot}X] \gg (Me_2N)_2PO{\cdot}X.$$

The failure of electronic theory to account for differences in rates of inhibition has been well demonstrated by finding the rates of inhibition of fly brain AChE by analogues of Paraoxon of the general structure,

$$(EtO)(R)PO{\cdot}O{\cdot}C_6H_4\text{-}4\text{-}NO_2,$$

where R represents a different alkyl group in each compound. Rates of hydrolysis varied according to electronic theory, but were very poorly correlated with rates of inhibition.(70) It is obvious that the effects of the basic groups on the electrophilic activity of the phosphorus atom are outweighed in all but extreme cases by some other effects which they exert, and which are specific to reactions on the surfaces of the enzymes.

It appears that the best way to regard the effects of basic groups is in terms of 'fit'. This concept is simple enough qualitatively, but does not lend itself to quantitative treatment. Any group is attracted by any other group close to it by dispersion forces, but if the array of groups which forms the surface of an enzyme does not conform fairly closely with the shape of the substrate or inhibitor, then only some groups can get close enough to attract, and those further apart will be repelled, so the fit is poor. It follows that the substitution of a hydrogen atom by a methyl group in a group which fits well may destroy the fit completely, as it may make it impossible for most of the atoms to lie close to the enzyme surface.

Evidence that fit is important has been obtained. Suppose that the rate of inhibition of AChE by a given inhibitor is represented by the rate constant $k_A$, and of ΨChE by the same inhibitor by $k_P$. The relative rates are then given by $k_A/k_P$. A similar ratio can be obtained for the rates of hydrolysis of a substrate by the two enzymes. Now there is a close resemblance in shape between the alkyl groups in the acid radical in a choline ester and the alkoxy groups in a phosphate ester. A methoxy group in a dimethyl phosphate ester, for example, is like the ethyl group in propionylcholine:

$$CH_3{-}O{-}P{=}O \quad \text{and} \quad CH_3{-}CH_2{-}C{=}O.$$

If fit is important the ratios of the rates of inhibition by phosphorus compounds should, therefore, be correlated with the ratios of the rates of hydrolysis of substrates. A general correlation is found, as shown in Table 6.12. The change from a dimethyl to a diethyl phosphate ester increases the values of $k_P/k_A$, and so on down the table. A particularly interesting case is Sarin, $(Pr^iO)(Me)PO{\cdot}F$, which inhibits AChE and ΨChE at similar rates.[73, 75] A tentative explanation is as follows. Either the MeP:O or the $Pr^iO{\cdot}P{:}O$ group may be adsorbed on the enzyme surface. The MeP:O group resembles the acetyl group in ACh, and the $Pr^iO{\cdot}P{:}O$ group the isovaleryl group in isovalerylcholine. ACh is a favoured substrate for AChE and isovalerylcholine for ΨChE. If then the MeP:O group is adsorbed on AChE and the $Pr^iO{\cdot}P{:}O$ group on ΨChE the high rates of inhibition of both enzymes are understandable.

The analogy with substrate specificity is not close. In two compounds: $(MeO)_2PO{\cdot}X$ and $(EtO)_2PO{\cdot}X$, where X is the same acidic group, $k_P/k_A$ for the latter is always greater than $k_P/k_A$ for the former. The absolute values of $k_P/k_A$, however, may vary with X, and certainly vary with sources of the enzymes, even though the ratios of the constants for the hydrolysis of substrates do not vary. Table 6.13 shows that two di-isopropylphosphorodiamidic compounds always inhibit ΨChE preferentially, but do so to very different degrees according to the species the enzyme is taken from. The difference between horse and man is particularly marked, although the ΨChE's of these two species have almost identical substrate specificities.*

* Much use has been made of specific inhibitors of ΨChE's in histochemistry, where AChE and ΨChE cannot be separated. One practical conclusion from Table 6.13 is that TIPA at about $3 \times 10^{-5}$ M can be used as a specific inhibitor on tissues from the horse, dog, guinea pig and monkey, but would be barely specific enough for rat tissues and almost useless on human tissues.

TABLE 6.12. RELATIVE RATES OF INHIBITION OF ACHE AND ΨCHE BY PHOSPHORUS INHIBITORS CONTAINING DIFFERENT GROUPS[71, 72]

| Inhibitor | Formula | $pI_{50}$'s horse[71] AChE | $pI_{50}$'s horse[71] ΨChE | $pI_{50}$'s rat[35] AChE | $pI_{50}$'s rat[35] ΨChE | Inhibition ($k_P/k_A$) horse | Inhibition ($k_P/k_A$) rat | Hydrolysis Substrate | Hydrolysis Rate ratios AChE/ΨChE horse |
|---|---|---|---|---|---|---|---|---|---|
| Paraoxon-Me | $(MeO)_2PO \cdot O \cdot C_6H_4$-4-$NO_2$ | 6·27 | 6·1 | 7·40 | 5·19 | 0·67 | 0·006 | $MeCH_2 \cdot CO \cdot Ch$ (propionylcholine) | 1·9 |
| Paraoxon | $(EtO)_2PO \cdot O \cdot C_6H_4$-4-$NO_2$ | 6·38 | 6·85 | 7·80 | 7·58 | 2·9 | 0·6 } | $EtCH_2 \cdot CO \cdot Ch$ (butyrylcholine) | 94 |
| TEPP | $(EtO)_2PO \cdot O \cdot PO(OEt)_2$ | 6·52 | 8·39 | 7·85 | 7·92 | 74 | 1·2 } | | |
| | $(Pr^nO)_2PO \cdot F$ | — | — | 7·16 | 8·00 | — | 5·5 | $Pr^nCH_2 \cdot CO \cdot Ch$ | 44 |
| DFP | $(Me_2CH \cdot O)_2PO \cdot F$ | — | — | 6·14 | 7·80 | — | 45 } | | |
| Paraoxon-Pr[i] | $(Me_2CH \cdot O)_2PO \cdot O \cdot C_6H_4$-4-$NO_2$ | 5·50 | 6·50 | 6·49 | 7·15 | 10·0 | 4·5 } | $Me_2CH \cdot CH_2 \cdot CO \cdot Ch$ (isovalerylcholine) | 263 |
| Mipafox | $(Me_2CH \cdot NH)_2PO \cdot F$ | 3·82 | 7·42 | 4·35 | 6·77 | 3950 | 254 } | | |
| TIPA | $(Me_2CH \cdot NH)_4P_2O_3$ | 2·51 | 6·48 | 3·57 | 6·34 | 9300 | 590 } | | |
| Sarin* | $(Pr^iO)(Me)PO \cdot F$[73, 75] | 8·6 | 8·4 | — | — | 0·6 | | | |

* Human ChE's.

The dependence of anti-esterase activity on fit could be explained in at least two ways. The most plausible explanation is that the compounds form adsorption complexes with the enzyme, in agreement with the conclusion reached earlier in this Chapter. Their different affinities for the enzyme then correspond to different values of the Michaelis constant $K_m$. The differences are never so great that very low values of $K_m$ have to be postulated, so that there is no inconsistency between this hypothesis and the observation that direct inhibitors do not inhibit reversibly to a measurable extent.

TABLE 6.13. RELATIVE RATES OF REACTION OF AChE AND ΨChE FROM DIFFERENT SPECIES WITH TWO INHIBITORS

Rates are given as $pI_{50}$'s, determined at 37°C after 30 min incubation. The ratios of the rates are given as $k_P/k_A$.

| | Mipafox | | | TIPA | | | |
|---|---|---|---|---|---|---|---|
| | ΨChE | AChE | $k_P/k_A$[71] | ΨChE | AChE | $k_P/k_A$[71] | $k_P/k_A$[74] |
| Man | 6·41 | 4·66 | 56 | 5·27 | 3·52 | 56 | 50 |
| Horse | 7·42 | 3·80 | 4200 | 6·52 | 2·47 | 11 300 | 6500 |
| Dog | 7·29 | 4·06 | 1700 | 6·54 | 2·42 | 13 200 | — |
| Rat | 6·75 | 4·22 | 340 | 6·32 | 3·60 | 530 | — |
| Guinea pig | 6·51 | 4·21 | 200 | 5·89 | 2·34 | 3600 | 1200 |
| Monkey | — | — | — | — | — | — | 5000 |

The examplcs so far given do not show the extent of the differences between the rates of reaction of a given inhibitor with two enzymes of the same type. DDVP, for example, inhibits fly AChE about 50 times more rapidly than mouse AChE.[69]

A special case of good fit is provided by those compounds which contain choline, or choline-like basic groups. These are considered later.

*Acidic groups.* When changes in structures are confined to the acidic group, X, there is generally close correlation between rates of hydrolysis and rates of inhibition. An example of the correlation between inhibition of AChE and hydrolysis in phosphate buffer is shown in Fig. 6.11. More extensive, though less direct, correlations may be found in the literature.[79] Rates of hydrolysis depend primarily on the strength of the acid, HX. Exceptions are cyanidates, thiolates, and phosphate esters containing vicinyl double bonds (P·O·C:C), which hydrolyse abnormally rapidly (Chapter IV). Some of these compounds are shown in Fig. 6.12, which is similar to Fig. 6.11, except that the hydrolysis constants used are those for hydroxyl ion catalysis. The line in Fig. 6.12 was drawn from the results on all those compounds which correlated well in Fig. 6.11, excepting only those for which $k_{OH}$ is unknown.

The two thiolates $(EtO)_2PO \cdot S \cdot C_6H_4$-4-$NO_2$ (No. 7) and Demeton-S (A) fall on the line, i.e. the high rate of hydrolysis is reflected in high anti-ChE activity. Of the doubly-bonded compounds (B—E) one is as active, one is more active, and two are less active than expected. This lack of correlation was further shown on the 23 compounds described in the paper from which

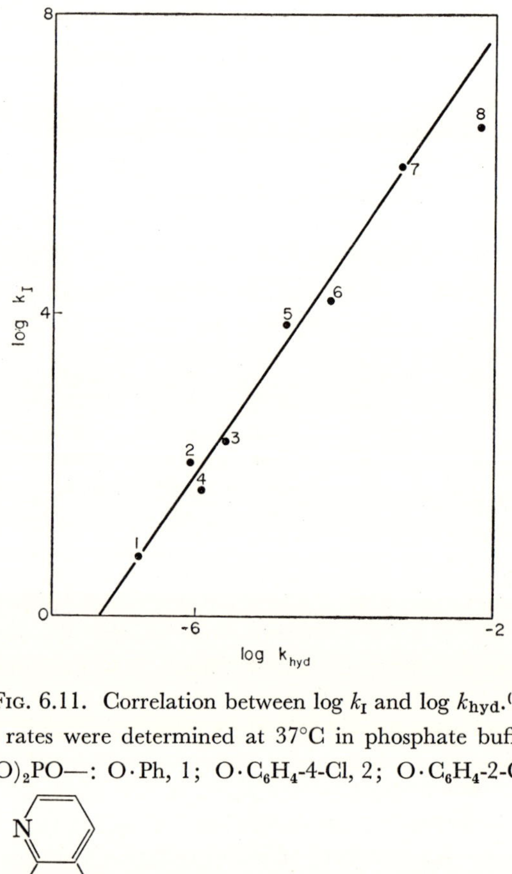


FIG. 6.11. Correlation between log $k_I$ and log $k_{hyd}$.[32]
Hydrolysis rates were determined at 37°C in phosphate buffer, pH 7·6: $(EtO)_2PO$—: O·Ph, 1; O·$C_6H_4$-4-Cl, 2; O·$C_6H_4$-2-Cl, 3;

O·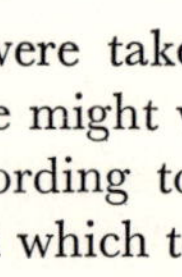, 4; O·$C_6H_4$-3-$NO_2$, 5; O·$C_6H_4$-2-$NO_2$, 6;
S·$C_6H_4$-4-$NO_2$, 7; O·$PO(OEt)_2$, 8.

these results were taken. Hydrogen bonding of such compounds at the enzyme surface might well take place, and would then either aid or hinder inhibition according to whether it led to good or bad 'fit'. There is no information on which to decide whether the correlation holds for cyanidates or not. They are active inhibitors [the $pI_{50}$ for $(EtO)_2PO \cdot CN$ is 7·1 with human AChE],[80] and readily hydrolysed, so probably it holds fairly well.

$(EtO)_2PO \cdot F$ is a surprisingly weak inhibitor. This appears to be a characteristic of fluoridates as inhibitors of AChE's. They inhibit Ψ'ChE's as fast as expected. The abnormally high activity of Paraoxon (G) has not been explained. It does not inhibit fly AChE faster than other compounds with similar hydrolysis rates.[79]

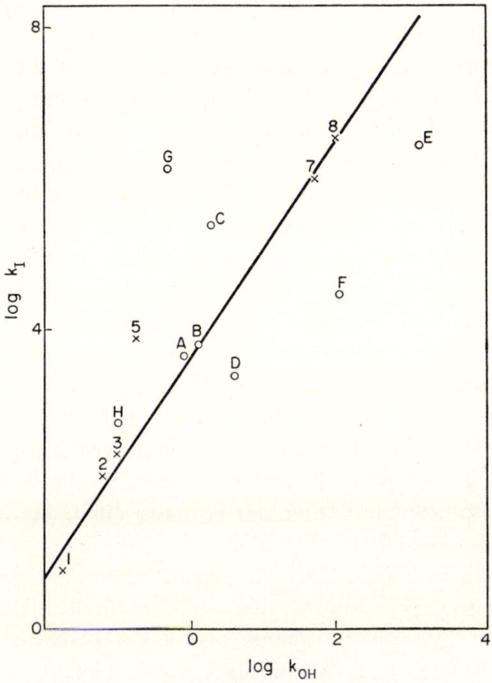


FIG. 6.12. Correlation between $k_I$ and $k_{OH}$.

Numbers refer to the compounds already shown in Fig. 6.11. Letters denote compounds as follows: $(EtO)_2PO$— : $S \cdot CH_2 \cdot CH_2 \cdot SEt$, A;[77] $O \cdot CH{:}CHCl$, B; $O \cdot CMe{:}CH \cdot CO \cdot OEt$, C; $O \cdot C{=}C \cdot CO \cdot OEt$ (ring: $CH_2$ $CH_2$ $CH_2$), D; $O \cdot C(CO \cdot OEt){:}CH \cdot CO \cdot OEt$, E;[78] F, F;[71] $O \cdot C_6H_4$-4-$NO_2$, G;[32] and $(EtO)_2PS \cdot O \cdot C_6H_4$-4-$NO_2$, H;[31]

Phosphorochloridates inhibit ChE's more slowly than expected from their hydrolysis rates. The discrepancy is, however, less than at first appears. The $I_{50}$ as usually determined of $(Pr^iO)_2PO \cdot Cl$ is greater than that of $(Pr^iO)_2PO \cdot F$, i.e. it appears a weaker inhibitor. In fact the chloridate reacts 30 times faster with bovine AChE than the fluoridate. The $I_{50}$ is greater because the chloridate is completely hydrolysed by the medium in three minutes.[81] This is a clear example of the errors which can arise when the $I_{50}$'s of stable and unstable inhibitors are compared.*

P:O *and* P:S *groups.* Thionates isomerize or are oxidized readily to more active compounds. Consequently there is only one trustworthy value for the rate of inhibition of any ChE by any thionate, that of sheep AChE by Parathion. A carefully purified specimen was hydrolysed, and the rate of fall in Parathion concentration determined both chemically and by anti-AChE activity.[31] The results were in good agreement, and the $pI_{50}$ found was 3·72. The $pI_{50}$ of Parathion is thus about 4 less than the $pI_{50}$ of its oxygen analogue, Paraoxon, i.e. Paraoxon reacts about 10 000 times faster. Several thionates certainly have $pI_{50}$'s at least 3 less than their oxygen analogues, and probably the difference is rarely less. As phosphates only hydrolyse about 10 times faster than the corresponding thionates, most thionates are less active than expected. Perhaps the P═O oxygen in phosphates is hydrogen-bonded to the enzyme surface, and this facilitates the reaction. Sulphur has much less ability to form hydrogen bonds, so the thionates react more slowly. It is interesting in this respect that the thiono-analogues of ACh and related esters are not hydrolysed by AChE and other esterases.[82]

The rate of inhibition of AChE by Parathion is that expected from its hydrolysis constant (point H, Fig. 6.12). The agreement is probably fortuitous—Paraoxon is unexpectedly active, so that, as the difference between the $pI_{50}$'s of phosphates and thionates is maintained, Parathion is an unusually active thionate.

*The effects of structural isomerism.* As already described, both AChE and ΨChE hydrolyse one optical isomer of acetyl-2-methylcholine faster than the other. Inhibition by optical isomers also takes place at different rates. L-(EtO)(Et)PO·S·$CH_2$·$CH_2$·SEt inhibits AChE (electric eel, horse and ox) and ΨChE (horse) about twenty times more rapidly than the D-isomer does.[34] The rate of inhibition of one AChE by the L-isomer is given in Table 6.3.

* It has been argued that if rates of inhibition and of alkaline hydrolysis are directly related, then a phosphorochloridate should always be a much stronger inhibitor than the corresponding fluoridate, however the $I_{50}$'s are determined. This contention is wrong. The correlation between $k_{hyd}$ and $k_I$ must break down when $k_{hyd}$ is very great. Thus, for hydrolysis:

$$k_{hyd} = K.e^{\Delta S_h/R}.e^{-\Delta E_h/RT}, \text{ and for inhibition: } k_I = K.e^{\Delta S_I/R}.e^{-\Delta E_I/RT}.$$

$\Delta S_I$ is presumably nearly independent of $k_I$, and is about zero (Table 6.3). Thus, in order to hold the ratio $k_I/k_{hyd}$ constant, i.e. to maintain the correlation, one may have to assume that $\Delta E_I$ is negative, which is absurd. This is equivalent to the statement that there is a top limit to the possible value of $k_I$, as some energy must always be needed to break the P—X bond. A calculation on these lines shows that $\Delta E_I$ for $(Pr^iO)_2PO$·Cl would have to be under 4 kcal/mole for the correlation to hold.

Amongst phosphate esters with vicinal double bonds in the acidic group the α-isomers (probably the *cis*-isomers) react about 50 times faster with AChE than the β-isomers.[78] The α-isomers are only a little less stable in alkaline solution than the β-isomers, so they are probably more active because they fit the enzyme surface better. The rates of inhibition and of hydrolysis of some α-isomers are included in Fig. 6.12. Compounds such as DDVP, $(MeO)_2PO \cdot O \cdot CH:CCl_2$, exist, of course, in only one form, as the $:CCl_2$ group is symmetrical.

*Compounds containing choline-like groups*. The most active anti-AChE'S known contain groups with a structural resemblance to choline. In Table 6.14 some of them are compared to TEPP and Sarin. Sarin is probably the most active compound known without a choline-like group. The compounds containing choline-like groups all resemble each other in several ways, but it is convenient to sub-divide them into three groups for discussion.

(1) *Dialkoxyphosphinylthiocholines and sulphur analogues*. These, along with some uncharged analogues, are numbers 1–10 in Table 6.14. The cationic head is separated from the phosphorus atom by the system: $—S \cdot (CH_2)_n—$, where n is 2 or 3. In the nitrogen series the ternary compounds such as No. 3 are mainly cationic at pH 7·4 and are not much weaker inhibitors of ChE's than their quaternary analogues. They are somewhat weaker. This is reminiscent of the finding that N-dimethylaminoethyl acetate is hydrolysed slower than ACh. The effects of introducing a positive charge are clearly demonstrated by the sulphur compounds, amongst which the sulphonium compounds are about 1000 times more active than their uncharged analogues. The attachment of a large aliphatic group to the ternary sulphur also appears to increase the activity, as the two most active compounds contain an $\overset{\oplus}{S}(Et) \cdot CH_2 \cdot CH_2 \cdot SEt$ group (only one, No. 8, is shown in Table 6.14).

The ionic compounds are more stable than TEPP, but are more active. During hydrolysis the P—S bond is broken. The same bond is likely to be broken during inhibition of AChE, so that inhibition by these compounds closely resembles the first stage in the hydrolysis of acetylthiocholine. The high activity of the compounds can therefore be attributed to their excellent fit on the enzyme surface. Acetylhomocholine,

$$MeCO \cdot O \cdot CH_2 \cdot CH_2 \cdot CH_2 \cdot \overset{\oplus}{N}Me_3,$$

is hydrolysed by ChE's much more slowly than ACh,

$$MeCO \cdot O \cdot CH_2 \cdot CH_2 \cdot \overset{\oplus}{N}Me_3,$$

because the introduction of the third $CH_2$ group into the chain spoils the fit. This may explain why the phosphorus compound with a 3-carbon chain, No. 7, is 30 times less active than its homologue with a 2-carbon chain, No. 6. The arguments on this point are complex, as may be seen by comparing references (76) and (84).

TABLE 6.14. THE ANTI-CHE ACTIVITY OF COMPOUNDS CONTAINING CHOLINE-LIKE GROUPS
The $pI_{50}$'s found with human ACHE and human plasma ΨChE are quoted, along with either $k_{OH}$, or the half life in water relative to TEPP.

| Compound | $pI_{50}$'s AChE | $pI_{50}$'s ΨChE | $k_{OH}$ | $\frac{t_{0.5}\text{ compound}}{t_{0.5}\text{ TEPP}}$ | Ref. |
|---|---|---|---|---|---|
| Phosphinylthiocholines and analogues | | | | | |
| 1. $(EtO)_2PO\cdot S\cdot CH_2\cdot CH_2\cdot \overset{\oplus}{N}Me_3$ | 8·4 | 8·9 | 40 | — | (83) |
| 2. $(EtO)_2PO\cdot S\cdot CH_2\cdot CH_2\cdot \overset{\oplus}{N}Et_3$ | 8·3 | — | — | — | (84) |
| 3. $(EtO)_2PO\cdot S\cdot CH_2\cdot CH_2\cdot NMe_2$ | 7·9 | 8·0 | — | — | (83) |
| 4. $(Pr^iO)(Me)PO\cdot S\cdot CH_2\cdot CH_2\cdot \overset{\oplus}{N}Me_3$ | 8·4 | 7·1 | — | — | (83) |
| Sulphur analogues | | | | | |
| 5. $(MeO)_2PO\cdot S\cdot CH_2\cdot CH_2\cdot \overset{\oplus}{S}EtMe$ | 7·2* | — | — | 100 | (77) |
| 6. $(EtO)_2PO\cdot S\cdot CH_2\cdot CH_2\cdot \overset{\oplus}{S}EtMe$ | 8·3* | — | — | > 100 | (77) |
| 7. $(EtO)_2PO\cdot S\cdot (CH_2)_3\cdot \overset{\oplus}{S}EtMe$ | 6·8* | — | — | — | (84) |
| 8. $(EtO)_2PO\cdot S\cdot CH_2\cdot CH_2\cdot \overset{\oplus}{S}Et\cdot CH_2\cdot CH_2\cdot SEt$ | ~ 9·1* | — | — | 0·4 | (84) |
| 9. $(EtO)_2PO\cdot S\cdot CH_2\cdot CH_2\cdot SEt$ | 5·4* | — | — | — | (84) |
| 10. $(EtO)_2PO\cdot S\cdot (CH_2)_3\cdot SEt$ | 4·7* | — | — | — | (84) |
| Phosphostigmines and analogues | | | | | |
| 11. 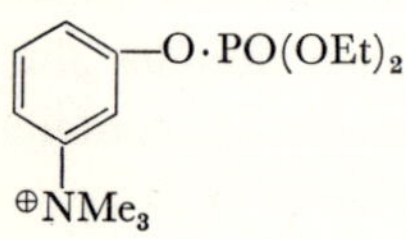 | 7·1 | 8·9 | — | 200 | (85) |
| 12. 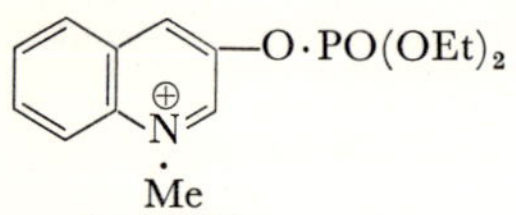 Ro3-0422 | 9·5 | — | — | 5 | (86) |
| 13. Ternary analogue of 12 | 7·1 | 8·4 | — | — | (86) |
| Fluoromethylphosphinylcholines | | | | | |
| 14. $MePO(F)—O\cdot CH_2\cdot CH_2\cdot \overset{\oplus}{N}Me_3$ | 10·0 | 8·4 | $5{\cdot}6 \times 10^4$ | — | (73) |
| 15. $—O\cdot CHMe\cdot CH_2\cdot \overset{\oplus}{N}Me_3$ | 8·4 | 8·4 | $2{\cdot}3 \times 10^4$ | — | (73) |
| 16. $—O\cdot (CH_2)_3\cdot \overset{\oplus}{N}Me_3$ | 11·0 | 8·4 | $1{\cdot}8 \times 10^4$ | — | (73) |
| 17. TEPP | 7·9 | 9·1 | 160 | 1 | (85) |
| 18. Sarin | 8·6 | 8·4 | 1560 | — | (75) |

* Sheep red cell.

The phosphates corresponding to the thiolates are much less active (only those with nitrogen in the side-chain have been tested).[83] They are also much more difficult to hydrolyse.

(2) *Phosphostigmines and analogues.* These compounds differ from the preceding group in that the relative positions of the cationic groups and phosphorus atoms in them are nearly fixed by aromatic rings. The compounds were designed to resemble neostigmine:

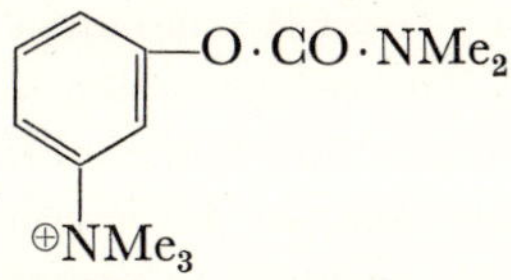

The quinolinium compounds such as No. 12 are more active than the trimethylanilinium compounds, perhaps because the $\overset{\oplus}{N}$—P distance in them makes a better fit possible. It is certainly closer to the $\overset{\oplus}{N}$—P distance in the phosphinyl-thiocholines. Compounds with a ternary nitrogen atom instead of a quaternary one are markedly less active towards AChE. They are not appreciably charged at the pH of inhibition.

It is interesting that the quinolinium compounds obviously fit well on to AChE, but that the quinolinium analogue of P2-AM is a poor re-activator of inhibited AChE's.

(3) *Fluorophosphinylcholines.* These are No. 14–16 in the table. They differ from the rest in that the choline-like residue is not the group displaced during hydrolysis or inhibition; the fluorine atom is displaced instead. It therefore seems unlikely that their high activity should be explained by good fit on the anionic site, as this appears to put the P—F bond in an unfavourable position for reaction to take place. Two sorts of evidence confirm this view. Firstly, the homocholine analogue is the most active compound, although presumably it does not fit so well as the choline analogue. Secondly the compounds are not much more active than expected from their very fast hydrolysis rates.[84]

Other compounds which are more active than the $k_{OH}$–$k_I$ correlation indicates contain groups with some resemblance to choline. Examples are: $(EtO)_2PO \cdot O \cdot C_6H_4$-3-$NMe_2$ and $(EtO)_2PO \cdot O \cdot C_6H_4$-3-$CMe_3$, both of which are about 1000 times more active against fly AChE than other compounds with similar hydrolysis rates.[79]

*Summary and extension.* There is no general relationship between the rates at which compounds are hydrolysed and the rates at which they inhibit ChE's, because the effects of the basic groups on hydrolysis rates are mainly electronic, and on rates of inhibition mainly steric. When only the acidic groups are varied, the two rates are correlated, and deviations from this correlation can be attributed to special factors. Thionates inhibit both AChE and ΨChE more slowly than expected from their hydrolysis rates.

One broad generalization can be made. Compounds which are difficult to hydrolyse are weak inhibitors. There is no compound known with a $pI_{50}$ (AChE) greater than 4 for which $k_{OH}$ is less than $0{\cdot}1[OH^{\ominus}]$/min at 25°C. Thus all trialkyl phosphates and tetramethylphosphorodiamidic compounds are very weak inhibitors. Some early reports of active trialkyl phosphates certainly arose from the use of compounds containing small proportions of tetra-alkyl pyrophosphates.*

*Enzymes other than ChE's.* Structure-activity relationships cannot be developed for lack of data. The results shown in Table 6.15 demonstrate that the effects of varying the structure of the inhibitor on rates of inhibition are not the same for these enzymes. As 'fit' is so important, and its requirements are likely to vary from enzyme to enzyme, this is not surprising. Cationic inhibitors have only been tested against a few enzymes. They are weak inhibitors, in agreement with the theory that only ChE's have anionic sites or stereospecific sites in the same position.

## THE NATURE OF THE INHIBITED SITE

Our understanding of the nature of the esteratic sites in enzymes has been greatly increased by experiments with phosphorus inhibitors. Two main lines have been followed. In the first the dependence of rates of inhibition on pH has been studied; and in the second the nature of the phosphorylated peptides obtained on degrading inhibited enzymes. Different immediate conclusions are reached, which modern theories have partly reconciled.

*Dependence on pH.* Both rates of hydrolysis of substrates and rates of inhibition vary in the same way with pH. For AChE's and some other enzymes the rates are a maximum at about pH 8, and are less at both higher and lower pH's. The plot of rate against pH is then a typical bell-shaped curve like that shown in Fig. 6.13. The accepted interpretation is that the enzyme exists in three forms, E, EH and $EH_2$, of which only EH can react. The fraction of enzyme present as EH is then given by:

$$F = 1/(1 + K_1/[\overset{\oplus}{H}] + [\overset{\oplus}{H}]/K_2),$$

where $K_1$ and $K_2$ are the dissociation constants of EH and $EH_2$. The rates are proportional to $F$. The curve shown in Fig. 6.13 is a theoretical one

* Several phosphonates without acid anhydride bonds have been reported as inhibitors. Mostly, either impurities have not been excluded, or, as in the case of Dipterex, conversion to a phosphate inhibitor takes place rapidly in the medium (p. 66). One compound, $(MeO)_2PO{\cdot}CH(O{\cdot}CO{\cdot}Pr){\cdot}CCl_3$, has however been obtained pure, and its bovine $pI_{50}$ is 5·0.[116] It is not known whether this indicates that a phosphonate can be an inhibitor if the P—C bond is sufficiently weakened by electrophilic substituents on the carbon atom, or that the AChE first hydrolyses the aliphatic ester to give Dipterex, which then re-arranges to DDVP. The kinetics of inhibition should be very different in the two cases.

TABLE 6.15. THE INHIBITION OF SEVERAL ENZYMES

The $pI_{50}$'s are given.

| Compounds | | ChE's | | Lipases | | Esterases | | | | | | | Chymo-trypsin | Tryp-sin | Thrombin |
|---|---|---|---|---|---|---|---|---|---|---|---|---|---|---|---|
| Type | Specific | AChE | ΨChE | A | B | A | B | C | D | E | F | G | | | |
| Diethoxy compounds | Paraoxon | 7·8 | 6·9 | 5·2 | 4·9 | 7·6 | 7·7 | 7·5 | 6·6 | 8·6 | 8·6 | 7·4 | < 2·5 | < 3 | < 3 |
| | TEPP | 7·9 | 7·9 | — | 3·5 | — | 7·3 | — | 6·7 | 7·8 | 8·4 | 6·1 | 4·7 | < 3 | < 3 |
| | $(EtO)_2PO \cdot S \cdot C_6H_4\text{-}4\text{-}NO_2$ | 7·7 | — | — | — | — | — | 7·3 | 5·9 | 7·7 | 8·6 | 7·6 | — | — | — |
| Dipropoxy compounds | $(PrO)_2PO \cdot O \cdot CH{:}CCl_2$ | 7·8 | 9·3 | — | — | — | 7·3 | — | — | — | — | — | 5·6 | — | — |
| | $(PrO)_4P_2O_3$ | 6·0 | 7·7 | — | < 2 | — | — | — | — | — | — | — | 4·8 | 2 | 3 |
| Di-isopropoxy compounds | DFP | 6·1 | 7·8 | < 3 | 3·9 | 6·7 | — | — | — | — | — | — | 5·9 | 4·2 | 4 |
| | Paraoxon-$Pr^i$ | 6·5 | 7·2 | 2·6 | — | 8·4 | — | — | — | — | — | — | — | — | — |
| $(Pr^iO)(Me)$ compound | Sarin | 8·6 | 8·4 | — | — | — | — | — | — | — | — | — | 3·3 | — | — |
| Di-isopropyl-amino compound | Mipafox | 4·4 | 6·8 | < 3 | — | 5·5 | — | — | — | — | — | — | — | — | — |
| Dimethoxy compound | DDVP | 6·3 | 7·8 | — | 7·1 | — | — | — | — | — | — | — | 3·3 | < 2 | < 2 |
| Cationic compound | Ro3-0422 | 9·5 | > 9 | — | 3·5 | — | — | — | — | — | — | — | < 2·5 | < 2·5 | < 3 |
| References | | (1, 31, 35, 75, 86, 87) | | (7) | (1) | (7) | (11) | (13) | (12) | | | | (1, 25, 87) | (1) | (1) |

Ro3-0422 is compound 12 in Table 6.14. Lipases: A, rat pancreas; B, wheat germ. Esterases: A, rat pancreas; B, rat serum, C, locust; D, *Tribolium castaneum*; E, *Tenebrio molitor*; F, *Blatella germanica*; G, *Dysdercus fasciatus*.

which obeys this equation. It does not fit the experimental results precisely, probably because the tendency of EH to form a complex with substrate, as measured by $K_m$, varies with pH.[89] Rough values of $K_1$ and $K_2$ can, however, be obtained from the experimental curves. The rates for other enzymes, e.g. trypsin and chymotrypsin,[88, 89] increase up to pH 8, and then remain constant, i.e. the enzymes exist in only two forms, of which the

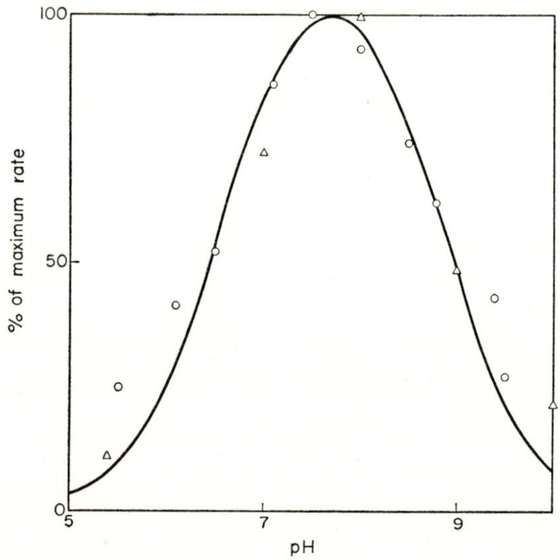


FIG. 6.13. The dependence on pH of the rate of hydrolysis of ethyl bromoacetate by AChE[90] and the rate of inhibition of AChE by TEPP.[107] Purified AChE from *Electrophorus electricus* was used. The published results for inhibition by TEPP[107] have been converted to rates of inhibition. The curve shown was calculated from the theoretical equation, with $K_1 = 1{\cdot}4 \times 10^{-9}$ and $K_2 = 3 \times 10^{-7}$.

Hydrolysis rates, ○; inhibition rates, Δ.

unprotonated reacts. Some values of $pK_1$ and $pK_2$ are given in Table 6.16. They are very similar whether determined from rates of hydrolysis or from rates of inhibition, in agreement with the theory that the controlling steps in hydrolysis and inhibition are similar.

The function of the group with a p$K$ of about 9·3 is indicated by the fact that the hydrolysis of acetylthiocholine by AChE does not slow above pH 8, whereas that of ACh does.[82] If the group forms a hydrogen bond with the ester oxygen, then the rate of hydrolysis of a thio-ester should be unaffected. The only group likely to be present in an enzyme to which this p$K$ can be assigned is the phenolic group of tyrosine,

$$HO \cdot C_6H_4\text{-}4\text{-}CH_2 \cdot CH(NH_2) \cdot CO \cdot OH.^{(90)}$$

Presumably it is absent from enzymes such as trypsin.

TABLE 6.16. DISSOCIATION CONSTANTS OF ENZYMES

The p*K*'s have been calculated from the published pH–rate curves. When the rates do not fall above pH 8, or it is not known whether they do or not (asterisked), there is no p*K* corresponding to p$K_1$.

| Enzyme | Dissociation constants | | | | | Refs. |
|---|---|---|---|---|---|---|
| | From hydrolysis of substrates | | Inhibitor | From inhibition | | |
| | p$K_1$ | p$K_2$ | | p$K_1$ | p$K_2$ | |
| AChE (ox) | 9·1 | 6·7 | DFP | 9·3 | 6·7 | (106) |
| AChE (electric eel) | 8·9 | 6·5 | TEPP | 8·9 | 6·6 | (107) |
| ΨChE (human) | 9·3† | 6·7 | DFP | 9·3† | 6·7 | (106) |
| α-Chymotrypsin | — | 6·6 | DFP | — | 6·6 | (106) |
| Trypsin | — | 6·7 | DFP | — | 6·7 | (106) |
| Thrombin* | — | 7·1 | DFP | — | 6·9 | (106) |
| Lipase* (wheat germ) | — | 6·7 | $(EtO)_2PO{\cdot}F$ | — | 6·7 | (106) |
| A-esterase* (hog kidney) | — | 6·9 | — | — | — | (106) |
| A-esterase* (rabbit serum) | — | 6·8 | — | — | — | (106) |

† This finding is probably incorrect, as ΨChE de-natures at high pH's (see p. 106).

Two groups may have p*K*'s of about 6·7: the imidazole nitrogen of histidine[91] or the nitrogen of $\Delta^2$-oxazoline:[92]

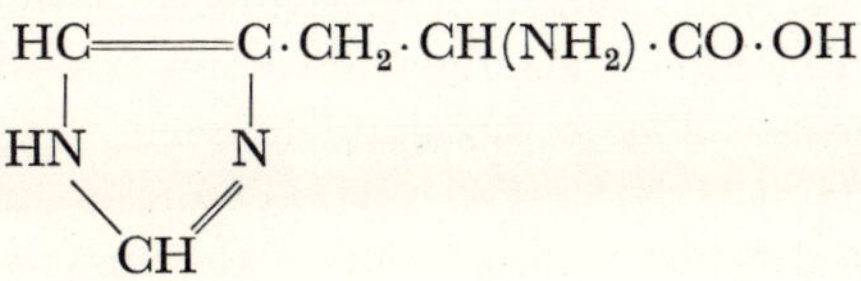

histidine

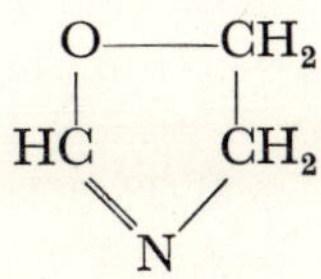

$\Delta^2$-oxazoline

The p*K*'s of imidazole derivatives (7–7·5) are more consistent with the values of p$K_2$ than those of the few $\Delta^2$-oxazoline derivatives studied (4–5·5), but the structure cannot be ruled out.

The oxazoline theory has only been put forward recently, and it is not known to what extent the following evidence, usually held to support the imidazole theory, is consistent with it.

(1) The only amino-acids both found in proteins and attacked by DFP are histidine and tyrosine.[93, 94] These are also the most rapidly attacked by other phosphorus compounds.[95]

(2) Imidazoles catalyse the hydrolysis of DFP and several aliphatic esters.[82] The rate increases with pH, showing that the unprotonated form

is the catalyst. The catalysed hydrolysis of DFP involves transient phosphorylation,[96] and of aliphatic esters transient acylation.[97]

(3) Cautious oxidation of chymotrypsin by light and methylene blue destroys one of the two histidine residues in the enzyme and, simultaneously, the capacity of the enzyme to react with DFP.[98]

(4) If trypsin is either acetylated or treated with formaldehyde its activity is unchanged, showing that the active sites do not contain $—NH_2$ groups.[99]

(5) Imidazole compounds have been prepared which have both the same rate constants for hydrolysis and the same dependence of rate on pH as the hydrolysis of substrates by chymotrypsin.[100, 101] An example is:

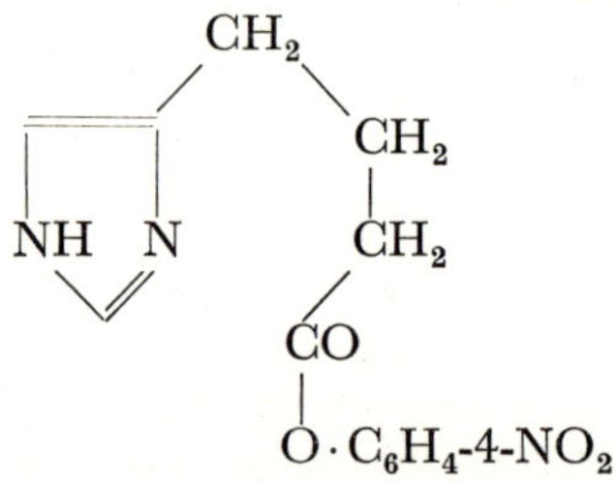


In this compound the group to be hydrolysed is held in the neighbourhood of the catalyst. The compound therefore resembles a reacting enzyme, in which the formation of the adsorption complex fulfils the same function.

To conclude, the active sites of esterases contain a molecule of histidine (or, perhaps, of $\Delta^2$-oxazoline), and some also contain tyrosine.

*Degradation of inhibited enzymes.* When many enzymes (α-chymotrypsin, electric eel AChE, trypsin, horse ΨChE, liver ali-esterase and bacterial proteinase) inhibited with DFP are refluxed with hydrochloric acid for several hours to degrade them to amino-acids, the only amino-acid to which phosphorus is attached is serine. The compound obtained is O-serine phosphoric acid, $H_2PO_3 \cdot O \cdot CH_2 \cdot CH(NH_2) \cdot CO \cdot OH$. This does not prove that the phosphorus moiety is attached to this group in the enzyme, as in acids N-(di-isopropoxyphosphinyl)-serine methyl ester is converted rapidly to serine phosphoric acid:[102]

$$(Pr^iO)_2PO \cdot NH \cdot CH(CH_2 \cdot OH) \cdot CO \cdot OMe \xrightarrow[100^\circ]{HCl}$$

$$H_2PO_3 \cdot O \cdot CH_2 \cdot CH(NH_2) \cdot CO \cdot OH + 2Pr^iOH + MeOH.$$

Similar $N \rightarrow O$ migration might take place during the degradation of an enzyme, even if the phosphorus moiety were to be attached to the nitrogen atom of a different amino-acid. The phosphorus is still found attached to serine, however, after degradation under the very mild conditions of tryptic digestion for a few minutes at 37°C. This has been demonstrated on electric eel AChE,[103, 104] chymotrypsin, trypsin, human ΨChE and liver ali-esterase.[104] The serine is di-isopropylphosphorylated, as shown by the recovery of di-isopropyl hydrogen phosphate on mild alkaline hydrolysis of the peptides.[9, 104] The phosphorus is apparently attached to the oxygen atom of serine, as reduction, followed by hydrolysis, yields one molecule of alanine per atom of phosphorus bound.[105]

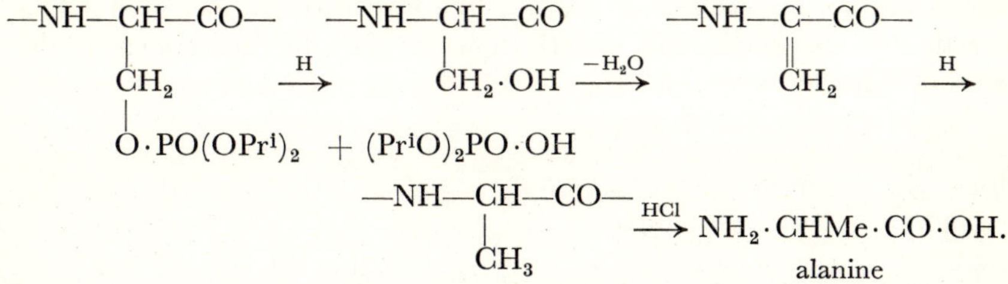

It is thus very probable that the phosphorus moiety is attached to the serine oxygen atom very shortly after inhibition takes place, which is in apparent contradiction to the evidence that the primary reaction is with histidine.

The contradiction can be resolved if imidazole catalyses the phosphorylation of serine. This is the present view. As shown below, this theory does not necessarily imply that the imidazole group is phosphorylated. Detailed analysis of the peptides obtained on degrading inhibited enzymes shows that the serine to which the phosphorus is attached is not near a histidine residue (Table 6.17). They may be neighbours in the whole enzyme because of the tertiary coiling of the protein molecule. There is independent evidence that coiling of the protein is essential for enzymic activity. Uncoiling the trypsin

TABLE 6.17. AMINO-ACID SEQUENCES ABOUT THE PHOSPHORYLATED SERINE IN SEVERAL ENZYMES

The di-isopropylphosphoryl group is abbreviated to *P*.

| Enzyme | Sequence |
|---|---|
| Trypsin[108] | Asp–Ser–Cys–Glu–Gly–Gly–Asp–Ser–Gly–Pro–Val–Cys–Ser–Gly–Lys<br>\|<br>*P* (on the second Ser) |
| α-Chymotrypsin[109] | Gly–Asp–Ser–Gly–Gly–Pro–Leu<br>\|<br>*P* (on Ser) |
| ΨChE (horse)[118] | Phe–Gly–Glu–Ser–Ala–Gly–($Ala_2$, Ser)<br>\|<br>*P* (on Ser) |
| Ali-esterase[9] (horse liver) | Gly–Glu–Ser–Ala–Gly–Gly–(Glu, Ser)<br>\|<br>*P* (on Ser) |
| Thrombin[110] | Gly–Asp–Ser–Gly<br>\|<br>*P* (on Ser) |

molecule with 8 M urea removes its enzymic activity, but it behaves normally again when the urea is removed.[111]

The kinetics of hydrolysis of acetyl-L-tyrosine ethyl ester by chymotrypsin suggest that an imidazole ring catalyses the hydrolysis without itself acting directly.[112] Inhibition and re-activation can then be represented by the series of reactions:

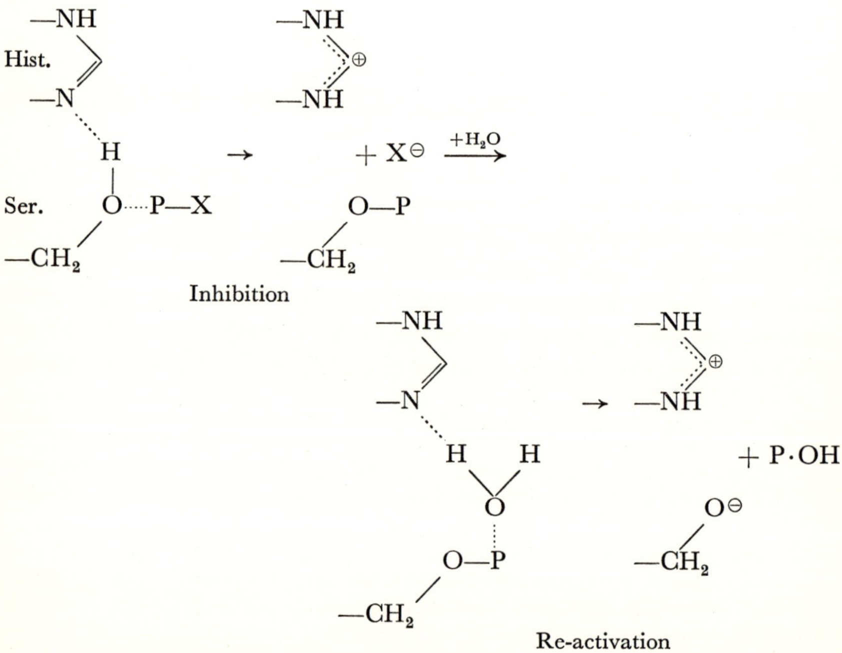


An alternative approach assumes that an oxazoline ring is formed from serine and the amino-acid next to it, and that this then reacts as below:[93, 113]

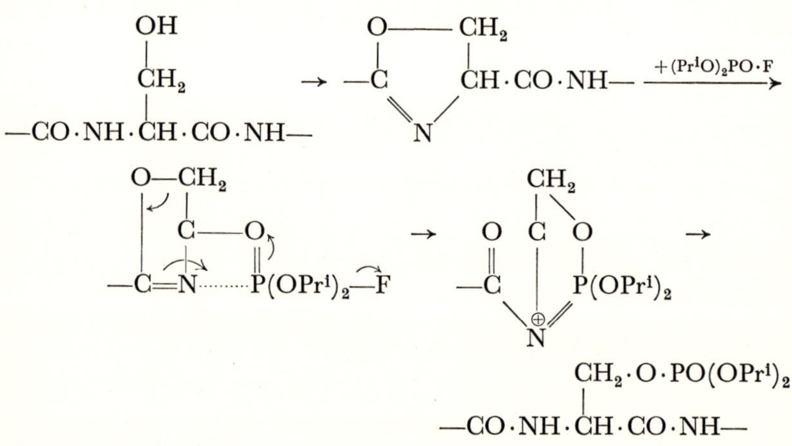

This scheme still requires the presence of an imidazole ring or some other group of similar p*K* to explain the kinetics of re-activation, and of de-acylation during hydrolysis of substrate.[112]

Neither theory accounts for many details of the behaviour of esterases,[114] but both account for the major features. Probably this is as much as can be expected of any theory on the action of these very complex systems at present.

## REFERENCES


1. L. A. Mounter, K. D. Tuck, H. L. Alexander, 3rd and L. T. H. Dien. *J. biol. Chem.* **226,** 873 (1957)
2. E. F. Jansen and A. K. Balls. *J. biol. Chem.* **194,** 721 (1952)
3. L. A. Mounter and B. A. Shipley. *J. biol. Chem.* **231,** 855 (1958)
4. H. Matsubara and S. Nishimura. *J. Biochem., Tokyo* **45,** 503 (1958)
5. E. L. Becker. *J. Immunol.* **77,** 462 (1956)
6. E. C. Webb. *Biochem. J.* **42,** 96 (1948)
7. D. K. Myers, A. Schotte, H. Boer and H. Borsji-Bakker. *Biochem. J.* **61,** 521 (1955)
8. D. K. Myers and B. Mendel. *Biochem. J.* **53,** 16 (1953)
9. H. S. Jansz, C. H. Posthumus and J. A. Cohen. *Biochim. biophys. Acta* **33,** 387, 396 (1959)
10. J. A. Cohen, R. A. Oosterbaan and M. G. P. J. Warringa. *Biochim. biophys. Acta* **18,** 228 (1955)
11. W. N. Aldridge. *Biochem. J.* **53,** 110 (1953)
12. K. A. Lord and C. Potter. *Chem. and Ind.* 1214 (1954)
13. A. Stringer and A. H. Fielding. *Ann. appl. Biol.* **44,** 626 (1956)
14. K. A. Lord and C. Potter. *Ann. appl. Biol.* **38,** 495 (1951)
15. D. K. Myers, J. W. Tol and M. H. T. de Jonge. *Biochem. J.* **65,** 223 (1957)
16. E. F. Jansen, M. D. F. Nutting and A. K. Balls. *J. biol. Chem.* **175,** 975 (1948)
17. R. D. O'Brien. *J. econ. Ent.* **49,** 484 (1956)
18. L. P. Bocharova and N. S. Ukrainets. *Biokhimiya* **23,** 388 (1958); (*Chem. Abstr.* **52,** 17340e, 1958)
19. M. Michaelis, N. I. Arango and R. W. Gerard. *Amer. J. Physiol.* **157,** 463 (1949)
20. E. I. Smith and D. M. Brown. *J. biol. Chem.* **195,** 525 (1952)
21. P. E. Wilcox, J. Kraut, R. D. Wade and H. Neurath. *Biochim. biophys. Acta* **24,** 72 (1957)
22. B. S. Hartley and B. A. Kilby. *Nature, Lond.* **166,** 784 (1950)
23. E. F. Jansen, M. D. F. Nutting, R. Jang and A. K. Balls. *J. biol. Chem.* **185,** 209 (1950)
24. E. F. Jansen, M. D. F. Nutting and A. K. Balls. *J. biol. Chem.* **179,** 201 (1949)
25. E. F. Jansen, A. L. Curl and A. K. Balls. *J. biol. Chem.* **190,** 557 (1951)
26. B. A. Kilby and G. Youatt. *Biochem. J.* **57,** 303 (1954)
27. M. A. Rothenberg and D. Nachmansohn. *J. biol. Chem.* **168,** 223 (1947)
28. H. O. Michel and S. Krop. *J. biol. Chem.* **190,** 119 (1951)
29. J. A. Cohen and M. G. P. J. Warringa. *Biochim. biophys. Acta* **11,** 52 (1953)
30. W. N. Aldridge. *Biochem. J.* **46,** 451 (1950)
31. W. N. Aldridge and A. N. Davison. *Biochem. J.* **52,** 663 (1952)
32. W. N. Aldridge and A. N. Davison. *Biochem. J.* **51,** 62 (1952)
33. W. N. Aldridge. *Biochem. J.* **54,** 442 (1953)

34. H. S. Aaron, H. O. Michel, B. Witten and J. I. Miller. *J. Amer. chem. Soc.* **80,** 456 (1958)
35. A. N. Davison. *Biochem. J.* **60,** 339 (1955)
36. J. A. A. Ketelaar. *Rec. Trav. Chim.* **69,** 649 (1950)
37. E. F. Williams. *Ind. eng. Chem.* **43,** 950 (1951)
38. K-B. Augustinsson and D. Nachmansohn. *J. biol. Chem.* **179,** 543 (1949)
39. A. S. V. Burgen. *Brit. J. Pharm.* **4,** 219 (1949)
40. J. C. Sabine and F. N. Hayes. *Arch. indust. Hyg.* **6,** 174 (1952)
41. W. N. Aldridge and A. N. Davison. *Biochem. J.* **55,** 763 (1953)
42. H. Kewitz. *Arch. Biochem.* **66,** 263 (1957)
43. A. N. Davison. *Biochem. J.* **61,** 203 (1955)
44. F. Hobbiger. *Brit. J. Pharmacol.* **10,** 356 (1955)
45. F. Hobbiger. *Brit. J. Pharmacol.* **11,** 295 (1956)
46. F. Hobbiger. *Brit. J. Pharmacol.* **12,** 438 (1957)
47. I. B. Wilson. *Disc. Faraday Soc.* **20,** 119 (1955)
48. D. R. Davies and A. L. Green. *Biochem. J.* **63,** 529 (1956)
49. A. L. Green and J. D. Nichols. *Biochem. Proceedings,* 11.12. (1958)
50. I. B. Wilson, S. Ginsburg and C. Quan. *Arch. Biochem.* **77,** 286 (1958)
51. A. L. Green and H. J. Smith. *Biochem. J.* **68,** 28 (1958)
52. A. L. Green and H. J. Smith. *Biochem. J.* **68,** 32 (1958)
53. I. B. Wilson, S. Ginsburg and E. K. Meislich. *J. Amer. chem. Soc.* **77,** 4286 (1955)
54. I. B. Wilson, M. A. Hatch and S. Ginsburg. *J. biol. Chem.* **235,** 2312 (1960)
55. B. M. Askew. *Brit. J. Pharmacol.* **11,** 417 (1956)
56. I. B. Wilson. *J. Amer. chem. Soc.* **77,** 2383 (1955)
57. I. B. Wilson. *J. biol. Chem.* **199,** 113 (1952)
58. A. L. Green and H. J. Smith. *Biochim. biophys. Acta* **27,** 212 (1958)
59. I. B. Wilson and S. Ginsburg. *Biochem. Pharm.* **1,** 200 (1958)
60. W. K. Berry, D. R. Davies and A. L. Green. *Brit. J. Pharmacol.* **14,** 186 (1959)
61. F. Hobbiger and P. W. Sadler. *Brit. J. Pharmacol.* **14,** 192 (1959)
62. I. B. Wilson. *J. biol. Chem.* **190,** 111 (1951)
63. L. W. Cunningham, Jr. and H. Neurath. *Biochim. biophys. Acta* **11,** 310 (1953)
64. L. W. Cunningham, Jr. *J. biol. Chem.* **207,** 443 (1954)
65. I. B. Wilson and F. Sondheimer. *Arch. Biochem. Biophys.* **69,** 468 (1957)
66. I. Enander. *Acta chem. scand.* **12,** 780 (1958)
67. R. Holmes and E. L. Robins. *Brit. J. Pharm.* **10,** 490 (1955)
68. B. E. Hackley, Jr., G. M. Steinberg and J. C. Lamb. *Arch. Biochem.* **80,** 210 (1959)
69. K. van Asperen and H. M. Dekhuijzen. *Biochim. biophys. Acta* **28,** 603 (1958)
70. T. R. Fukuto and R. L. Metcalf. *J. Amer. chem. Soc.* **81,** 372 (1959)
71. W. N. Aldridge. *Biochem. J.* **53,** 62 (1953)
72. W. N. Aldridge. *Chem. and Ind.* 473 (1954)
73. L-E. Tammelin. *Arkiv. Kemi.* **12,** 287 (1958)
74. L. Austin and W. K. Berry. *Biochem. J.* **54,** 695 (1953)
75. D. Grob and J. C. Harvey. *J. clin. Invest.* **37,** 350 (1958)
76. N. K. Fruentov. *Bull. exp. Biol. Med.* **48,** 1492 (1959)
77. D. F. Heath and M. Vandekar. *Biochem. J.* **67,** 187 (1957)
78. P. E. Gatterdam, J. E. Casida and D. W. Stoutamire. *J. econ. Ent.* **52,** 270 (1959)
79. T. R. Fukuto and R. L. Metcalf. *J. Agric. Food Chem.* **4,** 930 (1956)
80. B. Holmstedt. *Acta physiol. scand.* **25,** Suppl. 90 (1951)

81. J. R. Bocquet. *Ann. Roy. Soc. Med. et Nat., Bruxelles* **9,** 161 (1956)
82. F. Bergmann, S. Rimon and R. Segal. *Biochem. J.* **68,** 493 (1958)
83. L-E. Tammelin. *Acta chem. scand.* **11,** 1340 (1957)
84. D. F. Heath. *Biochem. Pharmacol.*, in press
85. A. S. V. Burgen and F. Hobbiger. *Brit. J. Pharmacol.* **6,** 593 (1951)
86. F. Hobbiger. *Brit. J. Pharmacol.* **9,** 159 (1954)
87. B. J. Jandorf. *J. Agric. Food Chem.* **4,** 853 (1956)
88. H. Gutfreund. *Trans. Faraday Soc.* **51,** 441 (1955)
89. B. R. Hammond and H. Gutfreund. *Biochem. J.* **61,** 187 (1955)
90. F. Bergmann, R. Segal, A. Shimoni and M. Wurzel. *Biochem. J.* **63,** 684 (1956)
91. I. B. Wilson and F. Bergmann. *J. biol. Chem.* **186,** 683 (1950)
92. G. R. Porter, H. N. Rydon and J. A. Schofield. *Nature, Lond.* **182,** 927 (1958)
93. R. F. Ashbolt and H. N. Rydon. *Biochem. J.* **66,** 237 (1957)
94. R. F. Ashbolt and H. N. Rydon. *J. Amer. chem. Soc.* **74,** 1865 (1952)
95. T. Wagner-Jauregg and B. E. Hackley, Jr. *J. Amer. chem. Soc.* **75,** 2125 (1953)
96. D. M. Brouwer, M. J. Vlugt and E. Havinga. *Koninkle Ned. Proc. ser. B,* **60,** 275 (1957)
97. H. Gutfreund and J. M. Sturtevant. *Biochem. J.* **63,** 656 (1956)
98. L. Weil, S. James and A. R. Buchert. *Arch. Biochem.* **46,** 266 (1953)
99. H. Gutfreund. *Disc. Faraday Soc.* **20,** 270 (1955)
100. T. C. Bruice and J. M. Sturtevant. *Biochim. biophys. Acta* **30,** 208 (1958)
101. T. C. Bruice and J. M. Sturtevant. *J. Amer. chem. Soc.* **81,** 2860 (1959)
102. R. E. Plapinger and T. Wagner-Jauregg. *J. Amer. chem. Soc.* **75,** 5757 (1953)
103. N. K. Schaffer, S. C. May, Jr. and W. H. Summerson. *J. biol. Chem.* **206,** 201 (1954)
104. J. A. Cohen, R. A. Oosterbaan, M. G. P. J. Warringa and H. S. Jansz. *Disc. Faraday Soc.* **20,** 114 (1955)
105. R. A. Oosterbaan, P. Kunst, J. van Rotterdam and J. A. Cohen. *Biochim. biophys Acta* **27,** 549, 556 (1958)
106. L. A. Mounter, H. C. Alexander, 3rd, K. D. Tuck and L. T. H. Dien. *J. biol. Chem.* **226,** 867 (1957)
107. I. B. Wilson and F. Bergmann. *J. biol. Chem.* **185,** 479 (1950)
108. G. H. Dixon, D. L. Kauffman and H. Neurath. *J. biol. Chem.* **233,** 1373 (1958)
109. R. A. Oosterbaan, P. Kunst and J. A. Cohen. *Biochim. biophys. Acta* **16,** 299 (1955)
110. J. A. Gladner and K. Laki. *J. Amer. chem. Soc.* **80,** 1263 (1958)
111. G. H. Dixon and H. Neurath. *Biochim. biophys. Acta* **20,** 572 (1956)
112. T. Spencer and J. M. Sturtevant. *J. Amer. chem. Soc.* **81,** 1874 (1959)
113. H. N. Rydon. *Nature, Lond.* **182,** 928 (1958)
114. M. A. Marini and G. P. Hess. *Nature, Lond.* **184,** 113 (1959)
115. R. M. Heinicka and R. Mori. *Science* **129,** 1678 (1959)
116. B. W. Arthur and J. E. Casida. *J. Agric. Food Chem.* **6,** 360 (1958)
117. F. Berends, C. H. Posthumus, I. v.d. Sluys and F. A. Deierkauf. *Biochim. biophys. Acta* **34,** 576 (1959)
118. H. S. Jansz, D. Brons and M. G. P. J. Warringa. *Biochim. biophys. Acta,* **34,** 573 (1959)
119. J. H. Fleisher, H. O. Michel, L. Yates and C. S. Harrison. *J. Pharmacol.* **129,** 31 (1960)

CHAPTER VII

# Metabolism of Phosphorus Compounds *in vitro*

MOST toxic phosphorus compounds are attacked by at least one enzyme *in vitro*. According to their structure, they may be either hydrolysed, oxidized or reduced. Some aspects of enzyme-catalysed hydrolysis have been investigated fully, quite independently of any effects the reactions taking place may have on the toxic action of the compounds hydrolysed, and these are described in detail in this Chapter. Other hydrolytic reactions, and oxidation and reduction reactions have been studied mainly in order to explain toxic effects. To avoid repetition many of these are described in Chapter IX, and only their main features are illustrated here.

## HYDROLYTIC REACTIONS

The hydrolysis of a phosphorus compound usually gives an acid of phosphorus as one of the products, but some compounds have side-chains containing carboxylic ester or amide groups which can also be attacked. Enzymes acting in the first way are A-esterases (phosphorylphosphatases), those in the second ali-esterases or amidases.

*1. A-esterases.* The wide distribution of A-esterases capable of hydrolysing DFP, Paraoxon and Tabun is illustrated in Table 7.1. Only a few have been studied in detail, all of which appear to hydrolyse the acid anhydride bond:[1, 4, 7, 14]

$$(Pr^iO)_2PO \cdot F + H_2O \rightarrow (Pr^iO)_2PO \cdot OH + HF,$$
$$(EtO)_2PO \cdot O \cdot C_6H_4\text{-}4\text{-}NO_2 + H_2O \rightarrow (EtO)_2PO \cdot OH + HO \cdot C_6H_4\text{-}4\text{-}NO_2,$$
$$(Me_2N)(EtO)PO \cdot CN + H_2O \rightarrow (Me_2N)(EtO)PO \cdot OH + HCN.$$

Some idea of the range of compounds hydrolysed by these enzymes can be got from Table 7.2. There are undoubted differences in specificity. Rabbit serum, for example, hydrolyses Paraoxon, whereas hog kidney does not. Enzymes from different species also differ in substrate specificity: human, but not rabbit, serum, for example, hydrolyses 3-diethoxyphosphinyloxy-methylquinolinium iodide;[23] and Sarin and Tabun are hydrolysed by the same enzyme but at different relative rates according to the species and the organ from which the enzyme is prepared.[12]

Rates of hydrolysis do not depend in any simple way on structure. The only generalizations which can be made are that compounds fairly stable to alkalis are not hydrolysed by A-esterases. The dividing line corresponds closely to that separating powerful and weak inhibitors of AChE. The converse is not true. The very unstable fluorophosphinylthiocholines, $MePO(F) \cdot S \cdot CH_2 \cdot CH_2 \cdot \overset{\oplus}{N}Me_3$ and its homologues, are not hydrolysed by human serum,[24] and Paraoxon is hydrolysed by some enzymes and not by others. There is not as yet enough information for further discussion.

TABLE 7.1. SOURCES OF ENZYMES HYDROLYSING DFP, PARAOXON AND TABUN

| Compound | Source |
|---|---|
| DFP | Rabbit (8 organs),[4] hog (liver and kidney),[2, 3] man (8 organs,[2, 4] rat (12 organs),[2] cat (9 organs), guinea pig (9 organs), pigeon (7 organs), turtle (5 organs),[2] 14 micro-organisms (5 others contained little or none),[2] sera of man,[2, 4] rabbit,[5] and guinea pig,[6] and red cells of man.[4] |
| Paraoxon | Rat (5 organs), rabbit (12 organs),[7] guinea pig liver,[8] bee (head and abdomen), cockroach nerve cord,[9] sera of rat,[10] guinea pig, rabbit, ferret, sheep, goat, man, horse and mouse,[7] and rumen fluid of cow.[11] |
| Tabun | Rabbit (12 organs), cow, pig and rat (2 organs each), and livers of cat and monkey,[12, 13] and sera of rabbit, man, dog, horse, rat, cat, cow, and guinea pig.[13] |

How many enzymes are present in any one animal species is still uncertain. The present situation is shown in Table 7.3, which is based on three types of evidence:

(1) Summation experiments. For a criticism of these see the footnote on p. 109.

(2) Inhibition and activation experiments. If, for example, $Mn^{2\oplus}$ accelerates the hydrolysis of DFP and inhibits the hydrolysis of Tabun, it can be argued that two enzymes must be present. Experiments described later show that such results can be misleading.

(3) The enzyme can be purified, and shown to have the same relative activities towards two or more substrates at different stages of purification. A positive result is convincing, but a negative one is not, as impurities may affect the hydrolysis of one substrate more than the other.

The activities of the A-esterases from hog kidney[26] and rabbit serum[7] depend on pH in much the same way as that of AChE, passing through a maximum at pH 8. This suggests that the enzyme exists in three forms,

TABLE 7.2. COMPOUNDS HYDROLYSED BY A-ESTERASES FROM DIFFERENT SOURCES

The rates of hydrolysis depend upon the purity of the enzymes, and upon whether activators are present or not. The rates found without added activators are given relative to DFP. Compounds hydrolysing 4–10 times faster are marked (++), 2–4 times faster (+), 0·5–2 times faster (O), 0·2–0·5 times as fast (−), and 0·05–0·2 times as fast (=). Those hydrolysed still more slowly are entered as 'not hydrolysed'.

| Enzyme | Hydrolysed | Not hydrolysed |
|---|---|---|
| Rabbit serum | DFP, $(MeO)_2PO{\cdot}F$ (++), $(EtO)(MeO)PO{\cdot}F$ (O), $(EtO)_2PO{\cdot}F$ (++)[4] $(BuO)_2PO{\cdot}F$ (+), DDVP (O), $(EtO)_2PO{\cdot}O{\cdot}CH{:}CCl_2$ (+), $(Cl{\cdot}CH_2{\cdot}CH_2{\cdot}O)_2PO{\cdot}O{\cdot}CH{:}CCl_2$ (O),[3] $(MeO)_2PO{\cdot}O{\cdot}C(OMe){:}CCl_2$ (O?),[17] $(Me_2N)(MeO)PO{\cdot}F$ (−), Tabun (O), $(Me_2N)(MeO)PO{\cdot}CN$ (+), $(Me_2N)(Pr^iO)PO{\cdot}CN$ (−), Ro3-0422† (=),[16] TEPP (O), $(EtO)_2PO{\cdot}NH{\cdot}CH_2{\cdot}CO{\cdot}OEt$ (=), $(PrO)_4P_2O_3$ (=),[3] Paraoxon, Paraoxon-Me (O),[5] $(MeO)_2PO{\cdot}S{\cdot}C_6H_4\text{-}4\text{-}NO_2$ (O?), $(EtO)_2PO{\cdot}S{\cdot}C_6H_4\text{-}4\text{-}NO_2$ (O?).[18] | Paraoxon-Pr$^i$,[5] Parathion,[19] Chlorthion, Malathion, Diazinon, Demeton-O, Demeton-S,[16] Schradan,[19] $(C_8H_{17}O)_2PO{\cdot}O{\cdot}CH{:}CCl_2$, $(EtO)_2PO{\cdot}O{\cdot}C_6H_4\text{-}3\text{-}\overset{\oplus}{N}Me_3$ and 3 analogues,* $(PhO)_2PO{\cdot}NH{\cdot}CH_2{\cdot}CO{\cdot}OEt$, $(EtO)_2PO{\cdot}NH{\cdot}CO{\cdot}O{\cdot}C_6H_4\text{-}4\text{-}NO_2$ and an analogue,[3] $(EtO)_3PO$, $(EtO)_2PO_2^{\ominus}$ and similar compounds.[19] |
| Hog kidney | DFP, $(EtO)_2PO{\cdot}F$ (+), $(BuO)_2PO{\cdot}F$ (O),[3] $(EtO)(Me)PO{\cdot}F$ (O), $(PrO)(Me)PO{\cdot}F$ (+), $(Pr^iO)(Me)PO{\cdot}F$ (O), $(Me_3C{\cdot}CH_2{\cdot}O)(Me)PO{\cdot}F$ (+) $(Me_3C{\cdot}CHMe{\cdot}O)(Me)PO{\cdot}F$ (++), $(C_6H_{11}O)(Me)PO{\cdot}F$ (++), $(Pr^iO)(Et)PO{\cdot}F$ $(Pr^iO)(Pr^i)PO{\cdot}F$ (O), $(Pr^iO)(C_6H_{11})PO{\cdot}F$ (+),[15] $(Bu^iO)_2PO{\cdot}F$ (O) $(AmO)_2PO{\cdot}F$ (−), $(EtO)_2PO{\cdot}CN$ (=),[21] DDVP (O), $(EtO)_2PO{\cdot}O{\cdot}CH{:}CCl_2$ (=) TEPP (O), $(PrO)_4P_2O_3$ (−), $(EtO)_2PO{\cdot}NH{\cdot}CH_2{\cdot}CO{\cdot}OEt$ (=).[3] | $(Am^iO)_2PO{\cdot}F$, $(OctO)_2PO{\cdot}F$,[21] $(Cl{\cdot}CH_2{\cdot}CH_2{\cdot}O)_2PO{\cdot}O{\cdot}CH{:}CCl_2$ and 2 analogues, $(EtO)_2PO{\cdot}O{\cdot}C_6H_4\text{-}3\text{-}\overset{\oplus}{N}Me_3$ and 3 analogues, $(PhO)_2PO{\cdot}NH{\cdot}CH_2{\cdot}CO{\cdot}OEt$ $(EtO)_2PO{\cdot}NH{\cdot}CO{\cdot}O{\cdot}C_6H_4\text{-}4\text{-}NO_2$ and 1 analogue,[3] $(Pr^iO)_4P_2O_3$, Schradan,[21] Paraoxon,[3] Parathion, Dimefox, hexose diphosphate, and several trialkyl phosphates and phosphonates.[21] |
| Escerii coli[22] | DFP, $(EtO)_2PO{\cdot}F$ (O), $(BuO)_2PO{\cdot}F$ (O), TEPP (−), $(PrO)_4P_2O_3$ (=) DDVP (−), $(EtO)_2PO{\cdot}CN$ (=). | $(EtO)_3PO$, Dimefox, Schradan, Parathion, Chlorthion, Dipterex, Paraoxon. |

* One of these, Ro3-0422, is shown by another worker to be slowly hydrolysed.[16]

† 3-(diethoxyphosphinyloxy)-methylquinolinium methosulphate, Ro3-0422.

TABLE 7.3. A-ESTERASES PRESENT IN SOME SPECIES

| Species | Organ | A-esterases |
|---|---|---|
| Rat[2] | Liver | I. Water-soluble, activated by $Mn^{2\oplus}$ and $Co^{2\oplus}$, hydrolysing DFP and $(BuO)_2PO \cdot F$ at the same rates.<br>II. Water-insoluble, inhibited by $Mn^{2\oplus}$, $Co^{2\oplus}$ and $Mg^{2\oplus}$, activated by $Ca^{2\oplus}$, and hydrolysing $(BuO)_2PO \cdot F$ very much faster than DFP. |
| | Kidney | I. Very similar to rat liver I. |
| Hog | Liver | I. Very like rat liver I.[2]<br>II. Very like rat liver II.[21] |
| | Kidney | I. Like rat kidney I. Extensive purification increased activity towards TEPP and DFP equally.[21]<br>II? Even in the purified enzyme I Tabun seems to be hydrolysed by a second enzyme.[15] |
| Rabbit | Serum | I. Hydrolyses DFP and Tabun, not Paraoxon or TEPP.<br>II. Hydrolyses Paraoxon (? and TEPP). The differentiation depends entirely on summation experiments.[16] |
| Man | Serum | I. Hydrolyses DFP and Paraoxon,[16] inhibited by $Sr^{2\oplus}$.[25]<br>II. Hydrolyses Tabun, catalysed by $Sr^{2\oplus}$.[25]<br>III. Hydrolyses TEPP, and perhaps the others to a lesser extent.[25] |

E, EH and $EH_2$, of which only EH is active. For the rabbit enzyme,[3, 7] $pK_1 = 8{\cdot}5$, and $pK_2 = 6{\cdot}9$, and for the hog kidney enzyme,[3, 26] $pK_1 = 9$ and $pK_2 = 6{\cdot}7$, as compared with 9·3 and 7·2 for AChE. For this reason it is generally held that there is no fundamental difference between hydrolysis of ACh by AChE, inhibition of AChE by a phosphorus inhibitor, and hydrolysis by an A-esterase of a phosphorus inhibitor. The last two reactions can both be written:

$$P{-}X + H{\cdot}E \xrightarrow{k_1} H{\cdot}X + \boxed{P{-}E + H_2O \xrightarrow{k_2} P{\cdot}OH + H{\cdot}E}\,.$$

For inhibition $k_1$ is very much greater than $k_2$, whereas for hydrolysis the rates of the two reactions are comparable. As dimethylphosphorylated AChE is hydrolysed, AChE can be regarded as an A-esterase of low activity.

The A-esterases differ from AChE in forming relatively stable complexes with phosphorus compounds. $K_m$ for Paraoxon and rabbit serum A-esterase-II is $4{\cdot}5 \times 10^{-4}$ M,[7] and for Tabun and rabbit serum A-esterase-I is $2{\cdot}15 \times 10^{-3}$ M.[1] The maximum rate of hydrolysis of most compounds is reached below $5 \times 10^{-2}$ M, implying that the $K_m$'s are similarly low. The low $K_m$'s explain how very high rates of hydrolysis can be combined with high energies of activation—15 kcal/mole for Sarin and Tabun on bovine plasma A-esterase.[12]

Some A-esterases are markedly stereospecific for some substrates. When Tabun is added to hog kidney A-esterase in bicarbonate buffer, carbon dioxide equivalent to half of it is evolved rapidly, and the rest at about 0·05 times the rate. This is repeated on the addition of more Tabun. Apparently half the Tabun is less stable than the other half,[28] suggesting that one optical isomer is hydrolysed more rapidly than the other. This has been confirmed with rat serum A-esterase. The mixture becomes slightly laevo-rotary during the rapid initial reaction, and then slowly optically neutral as the slow stage of hydrolysis is completed, indicating that the D-isomer is the less stable.[29] The stereospecificity depends both on the source of enzyme and the phosphorus compound. For Tabun, hog kidney and rat serum enzymes are highly specific, hog liver and rabbit plasma enzymes moderately so, and human serum enzyme is not specific at all.[28] Rat serum enzyme, which is highly stereospecific for Tabun, is not stereospecific for Sarin.[31]

Metallic ions influence the activity of some A-esterases, as illustrated in Table 7.3. Work on hog kidney A-esterase-I illustrates the complexity of metal-enzyme interaction. $Mn^{2\oplus}$ and $Co^{2\oplus}$ activate the hydrolysis of DFP and Tabun, and the activation is increased by the addition of an equimolar concentration of a bi-dentate complexing agent such as 2:2-dipyridyl. Higher concentrations of bi-dentate agents, reduce the action of the ions,[30] as do quadridentate agents such as ethylenediamine tetra-acetic acid at all concentrations.[21] There is an obvious similarity to the effects of complexing agents on hydrolysis catalysed by $Cu^{2\oplus}$ (p. 95). Less pure A-esterase contains naturally occurring complexing agents, removable by electrophoresis, which can increase the effects of $Mn^{2\oplus}$ and $Co^{2\oplus}$. The activating effects of these ions, with or without complexing agents, *varies with the compound hydrolysed.* The rates of hydrolysis of DFP and Tabun are increased 10–30 times, but the rates of hydrolysis of Sarin and eight analogues are increased very little, and in some cases are actually reduced.[15]

Heavy metals, $Ni^{2\oplus}$, $Cu^{2\oplus}$, $Ag^{\oplus}$, $Zn^{2\oplus}$, $Ca^{2\oplus}$, $Ba^{2\oplus}$, $Hg^{2\oplus}$, $Fe^{3\oplus}$, $Pb^{2\oplus}$ and $Pd^{2\oplus}$, inhibit at fairly low concentrations (the $I_{50}$'s are mostly in the range $10^{-4}$–$10^{-6}$ M),[26] as does p-chloromercuribenzoic acid.[7, 26] The inhibition of the last is reversible with cysteine.[15] Although compounds reacting with sulphydryl groups are generally inhibitors, some are very weak (arsenite, iodo-acetate), and it would be premature to assume that the enzyme site contains a sulphydryl group.[15] Ions interfering with enzymic oxidations (azide, fluoride, cyanide) are not inhibitors.[26]

The enzymes hydrolysing Sarin are not confined to any one part of the cell, but are found in nuclei, mitochondria, microsomes and supernatant, roughly in the proportions 5:10:20:65, with some variations according to the animal species and the organ.[32]

The natural function of the enzymes is unknown. The enzyme hydrolysing Paraoxon in rabbit serum (II) also hydrolyses 4-nitrophenyl acetate, and therefore probably possesses general aliphatic esterase activity.[7] Hog kidney A-esterase-I does not hydrolyse triacetin, creatine phosphate, or choline esters.[15]

*II. Ali-esterases and amidases.* Malathion is hydrolysed by an enzyme in rat liver. The acid released appears from its solubility properties and infra-red spectrum to be the half ester obtained on hydrolysing one carbethoxy group:[34]

$$\begin{array}{l}(MeO)_2PS\cdot S\cdot CH\cdot CO\cdot OEt \\ \qquad\qquad\quad | \\ \qquad\qquad\quad CH_2\cdot CO\cdot OEt\end{array} + H_2O \rightarrow \begin{array}{l}(MeO)_2PS\cdot S\cdot CH\cdot CO\cdot OH \\ \qquad\qquad\quad | \\ \qquad\qquad\quad CH_2\cdot CO\cdot OEt\end{array} + EtOH.$$

It is not known which ethoxy group is removed. The enzyme is widely distributed in mammals. It has been found in cows' liver,[35] and in the livers, kidneys, sera, lungs, ilea and spleens of rats, mice, guinea pigs and dogs. In each species the highest activity was always in the liver.[36] It is, however, absent from the heads and bodies of house-flies,[35] an observation which goes far to explain the selective toxicity of the compound. Malaoxon is also hydrolysed readily,[36] the most active tissues hydrolysing about 1 $\mu$g/mg tissue/min. Phosphorus compounds without carboxylic ester groups are not hydrolysed.[37] Acetone-dried liver powders are more stable than tissue minces, and are as active at pH 7·6 in phosphate buffer.

The enzyme hydrolyses rapidly most organic esters with more than one carbethoxy group. All three groups are hydrolysed in triacetin and tributyrin, but only one is hydrolysed in diethyl succinate, malonate and malate, as in Malathion. Dimethyl oxalate, and esters with only one carbethoxy group, including ACh, are hydrolysed slowly or not at all.[27]

The esterase is inhibited progressively by many of the more powerful inhibitors of ChE's such as Paraoxon and DFP. The $pI_{50}$'s are mainly in the range 6–7 (determined after 25 min incubation at 37°C). The enzyme is thus not an A-esterase. The phosphate analogue of EPN,

$$(EtO)(Ph)PO\cdot O\cdot C_6H_4\text{-}4\text{-}NO_2,$$

is one of the more powerful inhibitors. The effects of treating animals with EPN before giving them Malathion are described in the next Chapter.[27, 37]

Similar reactions have been postulated to explain the rapid decomposition to ionic products of Acethion, $(EtO)_2PS\cdot S\cdot CH_2\cdot CO\cdot OEt$, and its analogues by the livers of rats and mice,[38] and of Phosdrin,

$$(MeO)_2PO\cdot O\cdot CMe{:}CH\cdot CO\cdot OMe,$$

by human serum.[39] These compounds are also hydrolysed simultaneously by A-esterases.

*Other enzymes.* Bovine rumen fluid decomposes Trolene,

$$(MeO)_2PS \cdot O \cdot C_6H_2\text{-}2{:}4{:}5\text{-}Cl_3,$$

rapidly.[40] The main product is: $(MeO)(HO)PS \cdot O \cdot C_6H_2\text{-}2{:}4{:}5\text{-}Cl_3$, the same as that formed by the action of sodium hydroxide in aqueous ethanol (p. 97).[40] It is not known whether either reaction involves hydrolysis or demethylation.

Human plasma attacks Delnav to give diethyl hydrogen dithioate, either by hydrolysis of the sulphur-carbon bond, or by a devinylation reaction:[41]

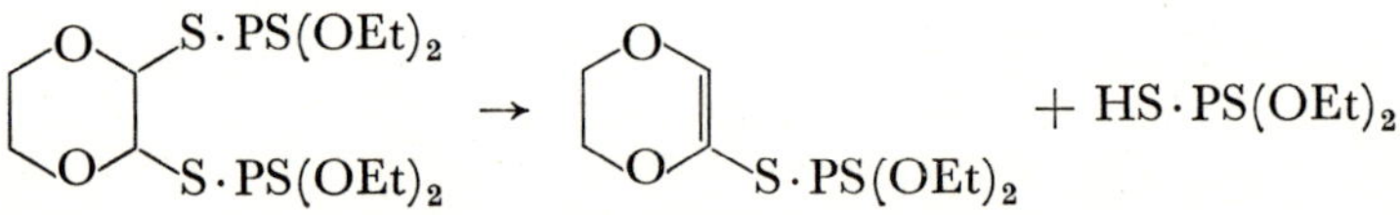


analogous to that which takes place between Malathion and alkalis (p. 96). As *in vivo* some compounds which do not contain $—S \cdot \overset{|}{C}H \cdot \overset{|}{C}H—$ systems, and which cannot, therefore, yield vinyl compounds, also give hydrogen dithioates, hydrolysis certainly takes place. The enzyme involved cannot be classified until more is known of it. It is not an A-esterase, as no bond with the phosphorus atom is broken.

## OXIDATIVE REACTIONS

Four types of compound are oxidized *in vitro*: thionates; compounds containing a sulphur atom in the side-chain (thio-groups); dialkylphosphoramidic compounds; and one thiolate. Many of the compounds studied are listed in Table 7.4.

The identities of the products have only been proved in a few cases, mainly because the yields are so small. More information has been obtained from experiments on whole animals, as described in Chapter IX. If this section is regarded in isolation, the evidence presented will often seem inadequate to support the conclusions drawn.

*Oxidation of thionates.* Liver slices act on several thionates with the formation of much more powerful inhibitors of ChE's. The process is usually called activation. Liver homogenates do not activate thionates, nor do the isolated fractions obtained on centrifugation (supernatant, nuclei, mitochondria and microsomes), or liver slices in an atmosphere of nitrogen.[42] Microsomes plus supernatant are, however, active, and this observation has led to the discovery of microsomal systems which activate thionates reproducibly. The essential additives are a pyridine nucleotide and $Mg^{2\oplus}$; the system must have free access to oxygen or air; and nicotinamide has some catalytic affect. Any of the nucleotides: diphosphopyridine nucleotide

TABLE 7.4. COMPOUNDS ACTIVATED BY SOME BIOCHEMICAL SYSTEMS

| System | Compounds activated | Compounds not activated |
|---|---|---|
| Liver slices in Ringer phosphate | *Thionates:* EPN,[60] Guthion[61] and Parathion[62]<br>*Sulphides:* Demeton-O and -S[53] and Thimet[54]<br>P—$NR_2$ *compounds:* Schradan,[63, 64] Dimefox,[56] $(Me_2N)_3(EtO)P_2O_3$ and $(Me_2N)_2PO \cdot O \cdot C_6H_4$-4-$NO_2$[65] | Malathion[36]<br><br>$(Me_2N)(EtO)_3P_2O_3$, $[(Me_2N)(EtO)PO]_2O$ and $(Me_2N)_2PO \cdot O \cdot PO(OEt)_2$[65] |
| Liver microsomes + DPNH + $Mg^{2\oplus}$ | *Thionates:* Co-ral, Diazinon, Guthion, EPN, Malathion, Parathion, Potasan and Trolene,[43] $(EtO)_2PS \cdot S \cdot CH_2 \cdot CO \cdot NH_2$[38]<br>P—$NR_2$ *compounds:* Schradan,[42] Dimefox,[56] $(Et_2N)_4P_2O_3$ (slight),[43] and $(Me_2N)_3(MeNH)P_2O_3$[66] | Dimethoate,[43] Acethion and several analogues[38] |
| Insect gut | *Thionates:* Malathion,[52] and Parathion[50]<br>*Sulphides:* $(EtO)_2PS \cdot O \cdot C_6H_4$-4-SMe and $(EtO)_2PO \cdot S \cdot C_6H_4$-4-SMe[55]<br>P—$NR_2$ *compounds:* Schradan,[56] Dimefox,[67] and $(Me_2N)_3(MeNH)P_2O_3$[66] | Acethion[38] |

(DPN), triphosphopyridine nucleotide (TPN) or their reduced forms, DPNH or TPNH, may be used, but the reduced forms are more effective.[43, 44] Earlier workers did not obtain satisfactory results with TPN or TPNH.[42] A typical mixture contains the microsomes* from 0·05 g of liver in 2·5 ml of solution containing 0·2 M NaCl, 0·018 M $Mg^{2\oplus}$, 0·01 M nicotinamide, 0·04% nucleotide and $2 \times 10^{-5}$ M thionate. Activation is estimated by the rate at which the product inhibits AChE. The compounds which have been tested with this and similar systems are listed in Table 7.4. Liver homogenates activate thionates when DPNH is added, but they retain their activity for under an hour.[46] They are less stable than microsomes, which are, in turn, less stable than liver slices.

The capacity to activate thionates is not confined to mammalian liver. Small intestines, lungs and kidneys of female rats all activate Parathion, though not so rapidly as the liver.[51] Whether the enzyme acting on other thionates are similarly distributed is not known. Insect fat bodies and gut also activate thionates.[50]

It is generally accepted that the reaction taking place is the oxidation of thionate to phosphate:

$$P{=}S \rightarrow P{=}O.$$

This theory is in accordance with the following observations:

(1) The system requires oxygen.

(2) The phosphate analogues of the thionates inhibit AChE rapidly.

(3) In the few cases where activation is expected but does not take place it has been shown that the enzyme preparations hydrolyse rapidly either the parent compounds or their phosphate analogues.

(4) The enzyme system is inhibited by $10^{-5}$ M 2-diethylaminoethyl diphenylpropylacetate, [$Et_2N \cdot CH_2 \cdot CH_2 \cdot O \cdot CO \cdot C(Ph)_2(Pr)$, SKF 525A] and by cytochrome-*c*, but not by sodium arsenite.[42, 46] It is also more active from rats which have been treated with methylcholanthrene, 3:4-benzpyrene or 1:2:5:6-dibenzanthracene.[47] In all these respects the enzyme system is very similar to those in the microsomal fractions of livers which demethylate or hydroxylate other foreign compounds in mammals.[48, 49] The only major difference is that the enzymes oxidizing thionates can be activated by DPNH or DPN, whereas the others can only be activated by TPNH or TPN.

Certain evidence that phosphate analogues are formed has not been obtained. The active metabolites of Malathion in cockroach guts and liver slices have been separated, and inhibited human ΨChE and fly and mouse AChE's at about the same rates as Malaoxon.[52] The agreement was not, however, good enough to constitute proof of structure. Some quantitative

* 'Microsomes', in the sense used here, are those particles which do not sediment from homogenates at under 8000 *g*, but do sediment in a few hours above 10,000 *g*.

results on the activation of Parathion by liver slices are also difficult to interpret except on the assumption that the active product was Paraoxon. Within an hour the capacity of the products to inhibit AChE reached a maximum, at which it remained constant. Assuming that this represented a state in which Paraoxon was formed by oxidation at the same rate as it was hydrolysed by liver A-esterases, the Paraoxon concentration was calculated from the rate of release of 4-nitrophenol by hydrolysis. The concentration was also calculated from the rate of inhibition of AChE, and agreed within a factor of two.(51)

In spite of the lack of direct evidence, the theory that 'activation' is oxidation of thionate groups is in agreement with so many observations that it is almost certainly correct.

*Oxidation of thio-groups.* Liver slices oxidize the side-chains of Demeton-O and Demeton-S to the corresponding sulphoxides:

$(EtO)_2PS \cdot O \cdot CH_2 \cdot CH_2 \cdot SO \cdot Et$ and $(EtO)_2PO \cdot S \cdot CH_2 \cdot CH_2 \cdot SO \cdot Et$.

Demeton-O also undergoes thionate oxidation. Whether further oxidation to the sulphones takes place cannot be decided on the available data.(53) Similar reactions take place readily in the cockroach gut, but not in mammalian kidney, brain or muscle.(53) Thimet,(54) and $(EtO)_2PS \cdot O \cdot C_6H_4$-4-SMe and its thiolate isomer, $(EtO)_2PO \cdot S \cdot C_6H_4$-4-SMe,(55) are oxidized similarly. The yields of sulphoxide are consistently higher from sulphides than the yields of phosphates from thionates. Sulphides are also oxidized chemically more readily than thionates.

*Oxidation of* P—$NR_2$ *compounds.* The nature of the reaction is considered along with results obtained *in vivo* in Chapter IX.

The enzymes in mammals oxidizing dialkylphosphoramidic compounds are like those oxidizing thionates in most ways.(42, 44, 56) They are in microsomes only, require pyridine nucleotide and oxygen, and are accelerated by nicotinamide and $Mg^{2\oplus}$, and inhibited by SKF 525A, etc. Their activity depends on the species and sex of animal.

Liver activates Dimefox; ileum, colon, kidney and all other tissues tested do not.(56) Parathion is oxidized by tissues other than liver, but the failure to observe Dimefox metabolites may be a consequence of the instability of the Dimefox metabolite. It is completely decomposed in 1 min in the presence of the tissues, and can only be shown to be present by its capacity to inhibit the ChE's endogenous to the tissues. The Parathion metabolite, Paraoxon, is relatively stable.

Many organs from many insect species oxidize Schradan.(57, 58) In only one case have homogenates or cell fractions been investigated. The enzyme system in the gut of the desert locust is in the microsome fraction, but differs in some ways from the mammalian enzymes. It is activated by TPNH, but not by TPN, DPN, DPNH or nicotinamide.(59) The enzyme oxidizing Malathion is activated by DPNH.

The compounds which have been studied are listed in Table 7.4. The only compounds activated contain tetra-alkylphosphorodiamidic groups. The only tetraethyl compound studied was much less readily activated than the rest, which contain tetramethyl groups. The products are all very unstable, with half lives in water of a few hours or less at room temperature, and it is possible that the failure to activate certain compounds is due to the instability in water of their metabolites.

*Oxidation of* $(EtO)_2PO \cdot S \cdot CH_2 \cdot CH_2 \cdot NEt_2$ (*Amiton*). Homogenates and microsomes of the livers from rats, rabbits, mice and guinea pigs, but not those from pigs, dogs, cows, frogs or man, decompose this compound rapidly. The enzyme is in every way similar to that oxidizing thionates, but the products are inert. Probably the first stage of the reaction is oxidation of the thiol sulphur to give $(EtO)_2PO \cdot SO \cdot CH_2 \cdot CH_2 \cdot NEt_2$, which then hydrolyses rapidly.[68]

## REDUCTIVE REACTIONS[11]

Several 4-nitrophenyl phosphate esters (Parathion, Parathion-Me, EPN, Chlorthion, and four others) are reduced to the corresponding 4-aminophenyl compounds by stagnating bovine rumen juices. The reaction is accompanied by enzyme-catalysed hydrolysis. At low concentrations (about 2 p.p.m.) more Parathion is reduced than hydrolysed, but at 300 p.p.m. hydrolysis is faster than reduction. The phosphate analogues give in 3 hr about equal amounts of hydrolysis and reduction products. The toxicological significance of these reactions is discussed in Chapter IX.

## REFERENCES

1. K-B. Augustinsson and G. Heimbürger. *Acta chem. scand.* **8,** 762 (1954)
2. L. A. Mounter and co-workers. *J. biol. Chem.* **215,** 691, 699, 705 (1955)
3. L. A. Mounter, K. D. Tuck, H. L. Alexander, 3rd and L. T. H. Dien. *J. biol. Chem.* **226,** 873 (1957)
4. A. Mazur. *J. biol. Chem.* **164,** 271 (1946)
5. W. N. Aldridge. *Biochem. J.* **54,** 442 (1953)
6. R. Goutier. *Biochim. biophys. Acta* **19,** 524 (1956)
7. W. N. Aldridge. *Biochem. J.* **53,** 117 (1953)
8. F. Heim and V. Neff. *Arzneimittelforsch.* **5,** 587 (1955)
9. R. L. Metcalf, M. Maxon, T. R. Fukuto and R. B. March. *Ann. ent. Soc. Amer.* **49,** 274 (1956)
10. J. Payton. *Nature, Lond.* **171,** 355 (1953)
11. M. K. Ahmed, J. E. Casida and R. E. Nichols. *J. Agric. Food Chem.* **6,** 740 (1958)
12. P. A. Adie. *Canad. J. Biochem. Physiol.* **36,** 15 (1958)
13. K-B. Augustinsson and G. Heimbürger. *Acta chem. scand.* **8,** 753 (1954)
14. L. A. Mounter and L. T. H. Dien. *Proc. Soc. exp. biol., N.Y.* **87,** 40 (1954)
15. J. A. Cohen and M. G. P. J. Warringa. *Biochim. biophys. Acta* **26,** 29 (1957)
16. K-B. Augustinsson and G. Heimbürger. *Acta chem. scand.* **8,** 1533 (1954)
17. B. Herrmann and R. Pulver. *Arch. int. Pharmacodyn.* **117,** 223 (1958)

18. W. N. Aldridge and A. N. Davison. *Biochem. J.* **55,** 763 (1953)
19. L. A. Mounter. *J. biol. Chem.* **209,** 813 (1954)
20. L. A. Mounter and A. Chanutin. *J. biol. Chem.* **204,** 837 (1953)
21. L. A. Mounter and L. T. H. Dien. *J. biol. Chem.* **219,** 685 (1956)
22. L. A. Mounter and K. D. Tuck. *J. biol. Chem.* **221,** 537 (1956)
23. F. Hobbiger. *Brit. J. Pharmacol.* **9,** 159 (1954)
24. L-E. Tammelin. *Arkiv. Kemi.* **12,** 287 (1958)
25. K-B. Augustinsson and G. Heimbürger. *Acta chem. scand.* **9,** 383 (1955)
26. L. A. Mounter, C. S. Floyd and A. Chanutin. *J. biol. Chem.* **204,** 221 (1953)
27. G. Yip and J. W. Cook. *J. Ass. off. agric. Chem., Wash.* **42,** 405 (1959)
28. K-B. Augustinsson. *Acta chem. scand.* **11,** 1371 (1957)
29. F. C. G. Hoskin and G. S. Trick. *Canad. J. Biochem. Physiol.* **33,** 963 (1955)
30. L. A. Mounter and A. Chanutin. *J. biol. Chem.* **210,** 219 (1954)
31. P. A. Adie, F. C. G. Hoskin and G. S. Trick. *Canad. J. Biochem. Physiol.* **34,** 80 (1956)
32. P. A. Adie and J. Tuba. *Canad. J. Biochem. Physiol.* **36,** 21 (1958)
33. W. C. Dauterman, J. E. Casida, J. B. Knaak and T. Kowalczyk. *J. Agric. Food Chem.* **7,** 188 (1959)
34. J. W. Cook and G. Yip. *J. Ass. off. Agric. Chem., Wash.* **41,** 407 (1958)
35. D. E. Weidhaus. *J. Ass. off. Agric. Chem., Wash.* **42,** 445 (1959)
36. S. D. Murphy and K. P. DuBois. *Proc. Soc. exp. Biol., N.Y.* **96,** 813 (1957)
37. J. W. Cook, J. R. Blake, G. Yip and M. Williams. *J. Ass. off. agric. Chem., Wash.* **41,** 399 (1958)
38. R. D. O'Brien, G. D. Thorn and R. W. Fisher. *J. econ. Ent.* **51,** 714 (1958)
39. J. E. Casida, P. E. Gatterdam, L. W. Getzin, Jr. and R. K. Chapman. *J. Agric. Food Chem.* **4,** 236 (1956)
40. F. W. Plapp and J. E. Casida. *J. Agric. Food Chem.* **6,** 662 (1958)
41. B. W. Arthur and J. E. Casida. *J. econ. Ent.* **52,** 20 (1959)
42. A. N. Davison. *Biochem. J.* **61,** 203 (1955)
43. R. D. O'Brien. *Nature, Lond.* **183,** 121 (1959)
44. R. D. O'Brien. *Canad. J. Biochem. Physiol.* **34,** 1131 (1956)
45. R. B. March, T. R. Fukuto, R. L. Metcalf and M. G. Maxon. *J. econ. Ent.* **49,** 185 (1956)
46. S. D. Murphy and K. P. DuBois. *J. Pharmacol.* **119,** 572 (1957)
47. S. D. Murphy and K. P. DuBois. *J. Pharmacol.* **124,** 194 (1958)
48. B. B. Brodie, J. Axelrod, J. R. Cooper, L. Gaudette, B. N. LaDu, C. Mitoma and S. Udenfriend. *Science* **121,** 603 (1955)
49. A. Conney, E. C. Miller and J. A. Miller. *Cancer Res.* **16,** 450 (1956); *J. biol. Chem.* **228,** 753 (1957)
50. G. C. Kok and J. N. Walop. *Biochem. biophys. Acta* **13,** 510 (1954)
51. J. Kubistova. *Arch. int. Pharmacodyn.* **118,** 308 (1959)
52. R. D. O'Brien. *J. econ. Ent.* **50,** 159 (1957)
53. R. B. March, R. L. Metcalf, T. R. Fukuto and M. G. Maxon. *J. econ. Ent.* **48,** 355 (1955)
54. J. S. Bowman and J. E. Casida. *J. econ. Ent.* **51,** 838 (1958)
55. E. Benjamini, R. L. Metcalf and T. R. Fukuto. *J. econ. Ent.* **52,** 94 (1959)
56. M. L. Fenwick, J. R. Barron and W. A. Watson. *Biochem. J.* **65,** 58 (1957)
57. R. D. O'Brien and E. Y. Spencer. *J. Agric. Food Chem.* **3,** 56 (1955)
58. H. Hurst. *Nature, Lond.* **145,** 462 (1940)
59. M. L. Fenwick. *Nature, Lond.* **182,** 607 (1958)
60. S. D. Murphy and K. P. DuBois. *Fed. Proc.* **15,** 462 (1956)

61. K. P. DuBois, D. R. Thursh and S. D. Murphy. *J. Pharmacol.* **119,** 208 (1957)
62. W. M. Diggle and J. C. Gage. *Biochem. J.* **49,** 491 (1951)
63. K. P. DuBois, J. Doull and J. M. Coon. *J. Pharmacol.* **99,** 376 (1950)
64. J. E. Gardiner and B. A. Kilby. *Biochem. J.* **51,** 78 (1952)
65. J. E. Casida, T. C. Allen and M. A. Stahmann. *Nature, Lond.* **172,** 243 (1953)
66. R. D. O'Brien and E. Y. Spencer. *Nature, Lond.* **179,** 52 (1957)
67. M. L. Fenwick. *Biochem. J.* **70,** 373 (1958)
68. J. F. Scaife and D. H. Campbell. *Canad. J. Biochem. Physiol.* **37,** 297 (1959)

CHAPTER VIII

# Inhibition of Enzymes *in vivo*

THIS Chapter is the counterpart of Chapter VI, in which the inhibition of enzymes *in vitro* was described. Far fewer enzymes have been studied *in vivo* because the inhibition of AChE, which is one of the more sensitive ones, kills the animal.

Toxic compounds either inhibit enzymes directly, or their metabolites do. Those which must be metabolized will be called indirect inhibitors. Inhibition by both direct and indirect inhibitors is described.

## INHIBITION OF ENZYMES IN MAMMALS

### CHOLINESTERASES

Inhibition *in vivo* differs from inhibition *in vitro* in one important respect: the concentration of the inhibitor is not constant *in vivo*. This is, of course, also true *in vitro* when particularly unstable compounds are studied, but these cases are unusual.

The natural substrate of AChE is an ion, ACh, so it is safe to assume that the active site is in an aqueous medium. The problem is, therefore, to elucidate those factors which lead to variations in the concentrations of phosphorus compounds in the aqueous part of an animal. Consider first the events following intravenous or arterial injection. Within about a minute the compound is evenly distributed throughout the blood stream. Rather more slowly it then diffuses throughout other aqueous parts of the animal, so that the concentration in the blood falls. During and after this process it is also open to attack by A-esterases, and can phosphorylate B-esterases or be extracted by lipids and fat. The rate of inhibition, therefore, does not depend only on the dose of the compound and the rate at which it reacts with AChE. It also depends on other properties of the compound:

(1) The rate at which it can penetrate membranes, to reach AChE in certain sites.

(2) Its fat–water partition coefficient. An ionic compound cannot be extracted from water by fats and lipids, so that its concentration in the neighbourhood of AChE cannot be reduced in this way.

(3) The rate at which it is hydrolysed by A-esterases.

(4) The rate at which it reacts with other esterases.

After less direct methods of administration the rate of diffusion into the blood-stream may also be important.

The final degree of inhibition depends in addition on the rate of re-activation of the inhibited enzyme *in vivo*, and on the rate of synthesis of new enzyme. It follows that AChE, or any other enzyme, will not be equally inhibited in different parts of the animal, as few of these factors operate uniformly throughout it.

The salient features of inhibition *in vivo* are exemplified by a description of the inhibition of ChE's in blood, brain and some myo-neural junctions. Inhibition in some other tissues is described in Chapters XII–XIV, which are on the detailed pharmacology of the compounds.

*Inhibition of ChE's in Blood*

Blood ChE's are inhibited by direct phosphorus inhibitors in the blood stream.

After intravenous or intra-arterial injection of direct inhibitors maximum inhibition is usually attained in a few minutes (see Fig. 8.5), mainly because the blood concentration falls rapidly.

The rapid fall was demonstrated on rats given three dimethyl phosphate esters intravenously.[1] The rats were killed at different times after receiving 0·7 $LD_{50}$'s of Paraoxon-Me, $(MeO)_2PO\cdot S\cdot CH_2\cdot CH_2\cdot \overset{\oplus}{S}EtMe$ (sulphonium

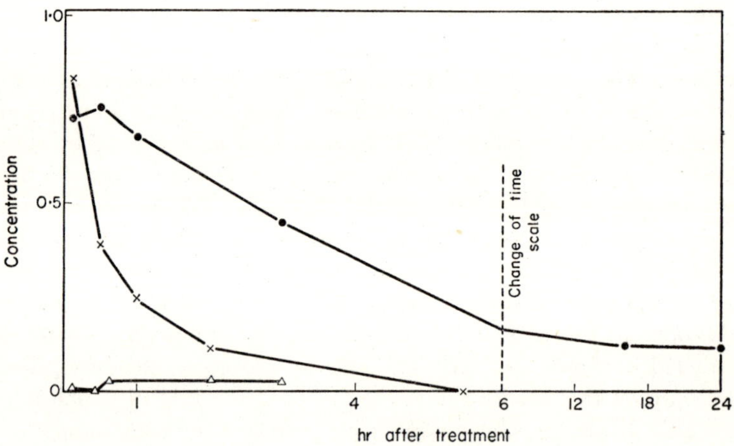


FIG. 8.1. The persistence in rats' blood of three inhibitors *in vivo*.[1]
The concentrations of the inhibitors in rats' serum were determined from the percentage inhibition of rat AChE produced in 1 hr at 37°C. 1·0 on the concentration scale represents the concentration of each compound producing 70% inhibition. This concentration is shown in brackets: Paraoxon-Me, $\Delta$ ($4 \times 10^{-8}$ M); $(MeO)_2PO\cdot S\cdot CH_2\cdot CH_2\cdot \overset{\oplus}{S}EtMe$, $\times$ ($9{\cdot}2 \times 10^{-8}$ M); $(MeO)_2PO\cdot S\cdot CH_2\cdot CH_2\cdot SO_2\cdot Et$, ● ($5{\cdot}4 \times 10^{-4}$ M).

compound), or $(MeO)_2PO \cdot S \cdot CH_2 \cdot CH_2 \cdot SO_2 \cdot Et$ (sulphone), and the concentrations in the serum determined with AChE *in vitro*. When the first samples were taken 5–10 min after injection, the fractions of the initial blood concentrations were already very low: of Paraoxon-Me $< 0 \cdot 0005$; of the sulphonium compound, 0·025; and of the sulphone, 0·04. It is not surprising, therefore, that most of the inhibition took place before this. The fall in Paraoxon-Me concentration should probably be attributed mainly to decomposition by the A-esterase known to be present in rat serum; and the fall in the concentrations of the other two compounds to physical transfer from the blood stream to other parts of the body.

After this initial, rapid fall, the concentrations fell at different rates, as shown in Fig. 8.1. Even 24 hr after injection enough sulphone was present to inhibit 20% of the AChE remaining in 1 hr at 37°C; but inhibition was competing with re-activation of the inhibited enzyme, so that the net effect was some recovery in AChE activity. Probably few water-soluble compounds persist as long *in vivo* as the sulphone.

The concentration in the blood stream is obviously better maintained when the compound is not injected directly into it, but apparently absorption is so rapid after intraperitoneal, intramuscular and subcutaneous injection,

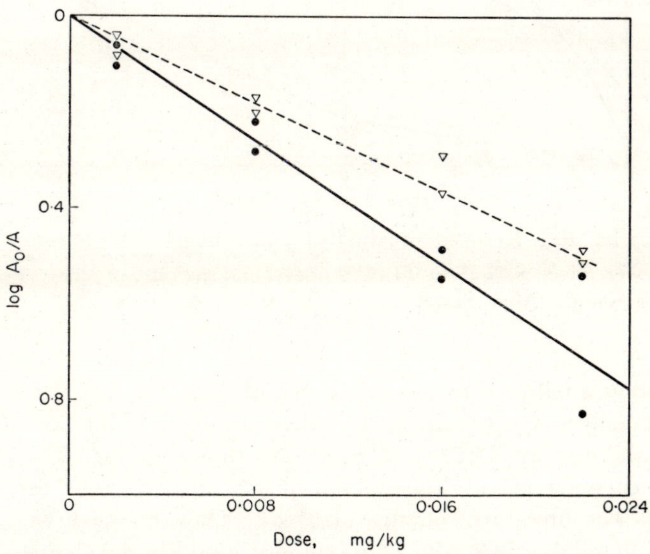


FIG. 8.2. Dependence in man of blood ChE activity on oral dose of Sarin.[4] $Log_{10}(A_0/A)$ is plotted against dose, where $A_0$ is the activity before treatment, and A is the activity 2 hr after treatment. Red cell AChE, ●; serum ΨChE, Δ.

After D. Grob and J. C. Harvey. *J. clin. Invest.* **37,** 350 (1958) (with acknowledgements).

and sometimes after oral dosing, that AChE inhibition normally progresses for under an hour.

Inhibition by indirect inhibitors may progress for a longer time; by Schradan[2] and some thionates[3] for several hours. In these cases inhibitors are being produced continuously, so that the result is not unexpected.

The degree of inhibition increases with the dose. In man the fall in ChE activity (both AChE and ΨChE) is related exponentially to the dose[4] (Fig. 8.2). In small animals there are greater individual variations from animal to animal, so that this precise correlation has not been established.*

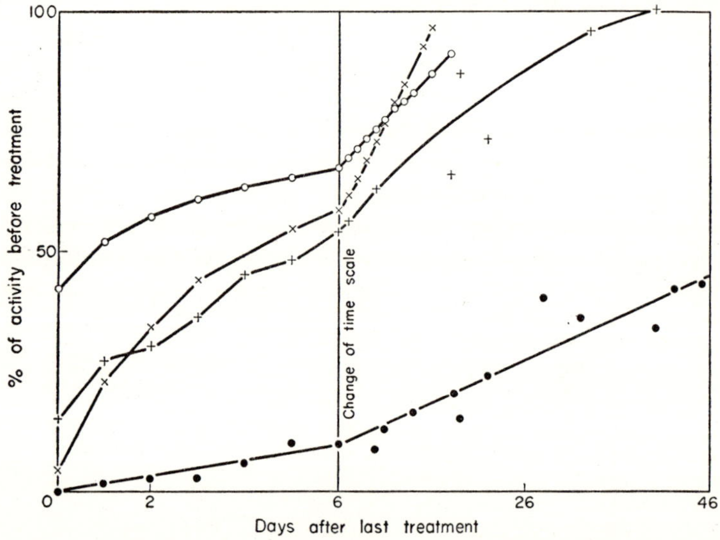


Fig. 8.3. Recovery of inhibited human blood ChE's *in vivo*.

Men were given Sarin[4] or TEPP[5] orally, and the rates of recovery of the ChE's found. After Sarin, ΨChE, +, AChE, ●; after TEPP, ΨChE, ×, AChE, ○.

Inhibition is followed by the return of ChE activity. The results of some very thorough investigations on the rate of recovery of human AChE and ΨChE inhibited by TEPP and Sarin are shown in Fig. 8.3, and some recovery curves, typical of those obtained with small animals, are shown in Fig. 8.4. The literature contains many discrepancies. One worker,[7] for example, found that only 70% of the activity of rat blood AChE inhibited by Paraoxon-Me was recovered in two days, while others[1] found that substantially complete recovery had taken place. Such differences may be due to biological variation, or to differences in technique.

* An exponential fall in activity with dose is the relationship expected if the balance between inhibition and other reactions does not vary with the dose.

Three factors influence the rate of recovery of ChE activity: the rate of synthesis of fresh enzyme, the rate of re-activation of inhibited enzyme, and the rate at which the unstable form of the inhibited enzyme goes to the stable form which cannot be re-activated.

Synthesis of fresh enzyme is most important when the inhibited enzyme is stable. The rate of recovery of activity should then be due entirely to

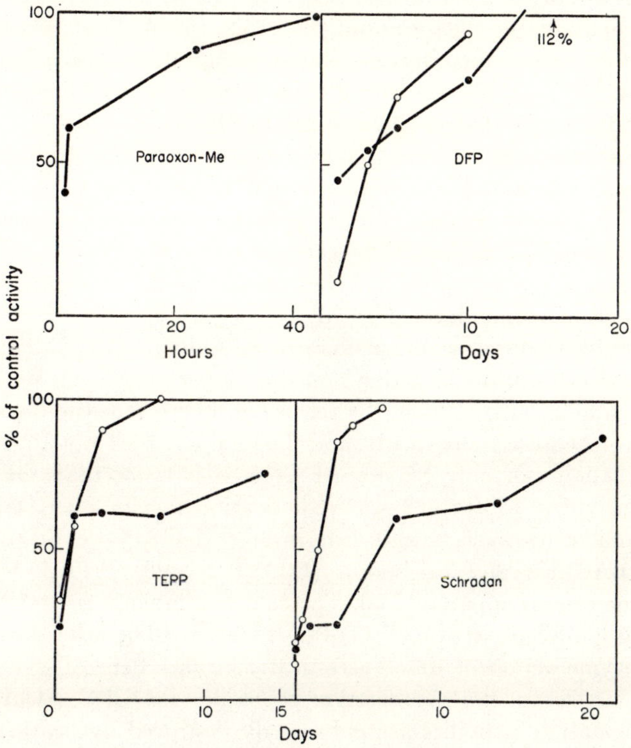


FIG. 8.4. Recovery of blood ChE's in rats treated with four compounds. The ChE's present are expressed as percentages of the original activities. AChE, ●; ΨChE, ○. Refs: Paraoxon-Me,[1] DFP,[6] TEPP,[3] Schradan: AChE,[3] ΨChE.[2]

synthesis of fresh enzyme, and should therefore depend only on the enzyme and species, and not on the inhibitor. This is found to be so. The activity of ΨChE in the rat recovers at the same rate (half life 72 hr) whether it is inhibited with a dimethyl phosphate, a di-isopropyl phosphate or an N,N′-di-isopropylphosphorodiamidate.[7] (These inhibited enzymes all recover much more slowly *in vitro*.[7]) There is also rough agreement between the times taken for 50% recovery of AChE in rat blood inhibited by DFP,[6] Sarin[10] and Schradan metabolite.[3]

The synthesis of fresh enzyme has been shown in several instances to depend on the formation of fresh red cells. The rate of recovery of AChE is the same as the rate of synthesis of red cells in the dog, when AChE is inhibited by Tabun,[12] and the rates of recovery of rat AChE phosphorylated by different compounds all appear fairly consistent with the 59 days' life span of red cells found in other ways.[13, 14] In rabbits treated with $DF^{32}P$ the rate of loss of $^{32}P$ from the red cells runs exactly parallel to the recovery in AChE activity,[15] which can only be explained by the destruction of old cells and their replacement by new, unless, most improbably, only the AChE is labelled. In man the loss of $^{32}P$ from cells labelled in the same way is definitely due to the destruction of old cells.[16]

The appearance of fresh enzyme in the blood may not depend solely on its synthesis. The initial rates of recovery of ΨChE's tend to be faster in the first few days than afterwards; for example, the percentage recovery per day in man of serum ΨChE after DFP treatment is initially 9%/day, falls steadily over 9 days to about 3·5%/day and then remains at this rate until recovery is complete.[5]* The final level of ΨChE activity also tends to become eventually higher in a poisoned animal than in an untreated one.[3, 11] Both the faster initial rates of re-activation and the higher final levels would be explained if inhibition stimulates the release of ΨChE into the blood stream. Recovery of AChE is more steady, and the final levels are not above normal.

Re-activation can only be studied when it is faster than the synthesis of new enzyme. In these cases, re-activation *in vivo* often takes place at the same rates as *in vitro*[7] [rat brain $(MeO)_2PO$-, $(EtO)_2PO$- and $(PrO)_2PO$-AChE's and rat serum $(EtO)_2PO$- and $(PrO)_2PO$-ΨChE's]; but in some cases it apparently takes place rather faster *in vivo*[5, 7, 8] [human serum and guinea pig serum $(EtO)_2PO\cdot$ΨChE's]. The faster rates may be due to enzyme action or differences in the media. Sera of several species markedly accelerate the re-activation of mouse $(EtO)_2PO\cdot$AChE (brain), but their catalytic activities are only partly destroyed by heating them to 100° for 3 min, which suggests that the catalysis is not wholly enzymic.[9]

Conversion of unstable phosphorylated ChE's to their stable forms provides the best explanation of some results already described. Direct evidence of conversion has been obtained. An investigation was carried out on five dimethyl phosphate esters: Demeton-S-Me[$(MeO)_2PO\cdot S\cdot CH_2\cdot CH_2\cdot SEt$], its sulphoxide, sulphone and methylsulphonium derivative, and Paraoxon-Me, and on one diethyl phosphate ester: the methylsulphonium derivatives of Demeton-S [$(EtO)_2PO\cdot S\cdot CH_2\cdot CH_2\cdot \overset{\oplus}{S}EtMe$].[1] By experiments of the

* The initial rapid recovery could not have been due to re-activation of some inhibited enzyme. Human $(Pr^iO)_2PO\cdot$ΨChE goes rapidly to a stable form (Table 6.7), so the rapid rate could not have been maintained for several days: and the DFP was given in several successive doses, so that very little could have been in the unstable form at the beginning of recovery.

type illustrated in Fig. 8.1, the two sulphonium derivatives and Paraoxon-Me were shown to reach low concentrations rapidly in the blood, whereas the other three persisted.* The return of activity *in vivo* was then determined after 0·7 $LD_{50}$'s of each compound had been given intravenously. The results are shown in Fig. 8.5. AChE inhibited by the two unstable phosphate esters, Paraoxon-Me and the sulphonium compound, recovered activity at the same fast rate, i.e. the AChE was re-activated. AChE activity in the animals treated with the persistent dimethyl phosphate esters recovered very

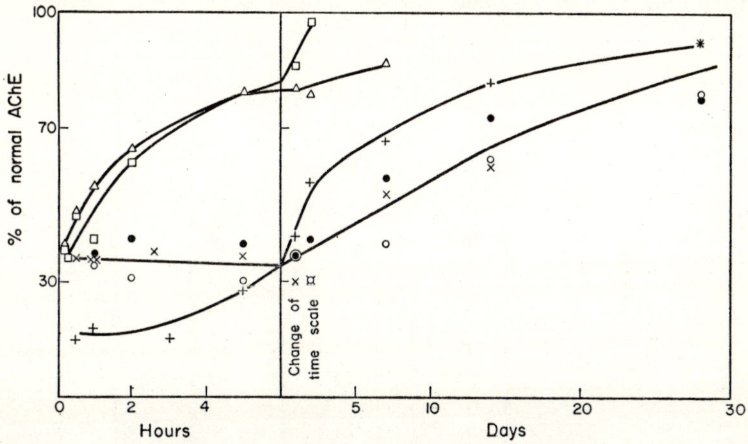


FIG. 8.5. Re-activation *in vivo* of rat red cell AChE.[1]

Most points represent the average AChE levels in two or three rats. Points refer to the following compounds: Demeton-S-Me, ×, its sulphoxide, ○, its sulphone, ●, and its methylsulphonium derivative, □; Paraoxon-Me, Δ; and the methylsulphonium derivative of Demeton-S, +.

slowly. The rate was consistent with the assumption that recovery represented the production of fresh enzyme, i.e. that the enzyme had been irreversibly inhibited. The diethyl ester was even less stable *in vivo* than the other sulphonium compound. In rats treated with it AChE activity rose at an intermediate rate for the first two days, and afterwards only at the rate of AChE synthesis. These results are consistent with the properties of rat $(EtO)_2PO{\cdot}AChE$ *in vitro*. It is re-activated by water more slowly than $(MeO)_2PO{\cdot}AChE$, and goes to a stable form with a half life of 1·5 days (Table 6.7). The conversion of $(Pr^iO)_2PO{\cdot}AChE$ and $(EtO)_2PO{\cdot}AChE$ to stable forms has also been demonstrated in the rabbit, and evidence adduced that $(EtO)_2PO{\cdot}AChE$ goes to the stable form in man.[17]

* Demeton-S-Me and its sulphoxide are probably indirect inhibitors, being converted to the sulphone. The persistence of inhibitor in the blood, i.e. probably of the sulphone, was the same, and is therefore given by the 'sulphone' curve in Fig. 8.1.

If the interpretation of the difference between the action of transient and persistent inhibitors is correct, then any inhibitor should inhibit irreversibly if it can be made to persist long enough. This was demonstrated with Paraoxon-Me, the most transient of the group studied. The compound was transfused into the vena jugularis of rats at constant slow rates. The rats were killed after various periods of infusion, and their brain AChE was split into two portions. One was estimated immediately, to give the total inhibition. The other was incubated for 4 hr at 37° with rabbit serum A-esterase, and then estimated, to give the proportion of AChE in the stable form. The

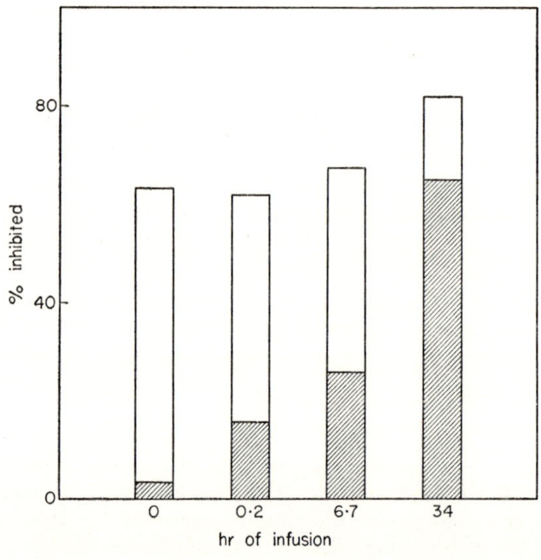


FIG. 8.6. Dependence in rats of the proportion of inhibited AChE in the stable form on the time of contact with Paraoxon-Me.[1]

The total per cent inhibited is shown by open blocks, and the per cent inhibited irreversibly by hatched blocks.

results are shown in Fig. 8.6. The proportion in the stable form increased with the time of perfusion. The prolonged contact required for this effect may also be brought about by oral feeding or dermal application. Thus irreversible inhibition of AChE is produced in monkeys by repeated dermal application of DDVP;[18] and the effects of feeding trials can only be explained in this way.

A steady intake of phosphorus poisons, as in feeding trials, causes a drop in blood ChE's to a level which often remains steady for several weeks[18, 20, 21, 22] (Fig. 8.7A and B). Probably all the enzyme is in the stable form, and the state reached represents equilibrium between inhibition and the synthesis of fresh enzymes.

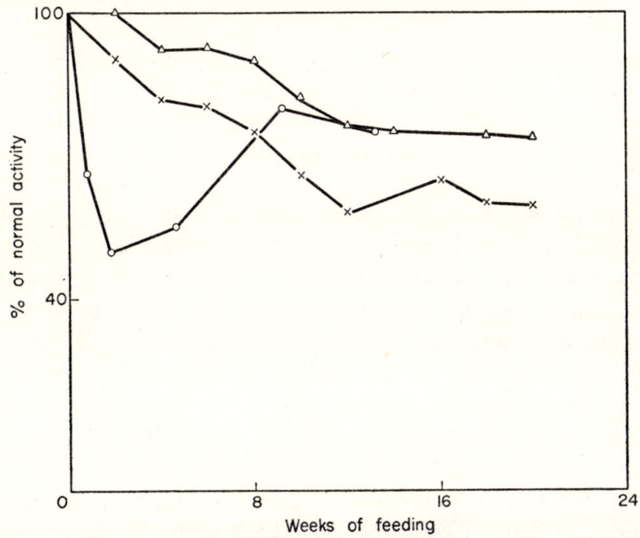


FIG. 8.7A. AChE levels in animals fed various compounds in the diet. 5 p.p.m. Parathion to the dog,[19] ×; 5 p.p.m. Systox to the dog,[19] Δ; 200 p.p.m. DDVP to the rat,[18] ○.

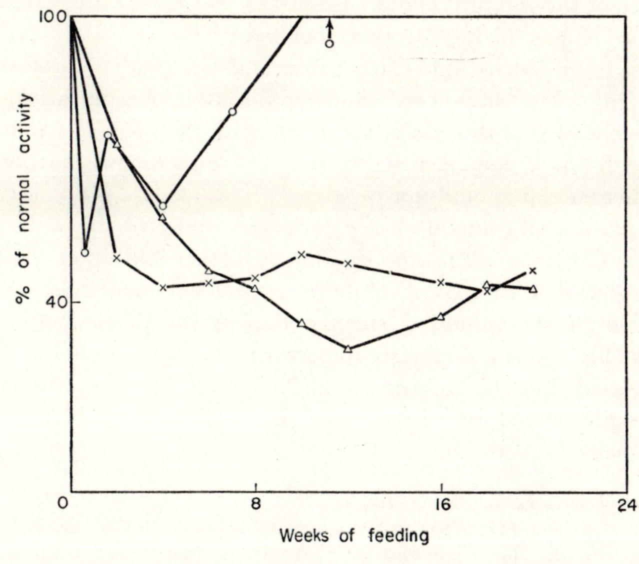


FIG. 8.7B. Ψ'ChE levels in animals fed various compounds in the diet. 2 p.p.m. Parathion to the dog,[19] ×; 5 p.p.m. Systox to the dog,[19] Δ; 200 p.p.m. DDVP to the rat,[18] ○.

*Inhibition of Brain ChE's*

Brain is not inhibited uniformly, though the differences from part to part are not great.[23] In this section only the average inhibition of whole brains is discussed, more detailed studies appearing in Chapter XIII. The effects of inhibitors on the brain is complicated by two factors which do not influence inhibition of blood ChE's. The brain contains a high proportion of lipids, which extract oil-soluble compounds; and it is protected by barriers to diffusion, which limit the penetration of some compounds.

Lipids in the brain extract compounds with high fat–water partition coefficients. In a non-aqueous medium most phosphorus compounds are stable, so they may persist for many hours in the lipids.* This complicates the determination of brain AChE. When the brain is homogenized in the first step in determining the enzyme activity, the oil-soluble inhibitors are released, and inhibit the enzymes *in vitro*.[24] More than 90% of the AChE remaining in the brains of animals treated with Paraoxon or DFP may be inhibited in this way, which leads to spurious results.[25] Four ways of avoiding or correcting for this error have been used.

(1) The brain is homogenized with a solution of ACh. This method is very simple, and in most cases effective.[139]

(2) The brain is halved. The ChE activity of one half is found. The other is homogenized with one half of the brain of an untreated animal, and the activity of the mixture found. From the two values the activity in the brain before homogenizing can be calculated.[24, 26]

(3) * The brain is dried, and extracted several times with chloroform, before homogenizing. The chloroform removes oil-soluble phosphorus compounds quantitatively, so that they do not interfere with the final determination.[27]

(4) The brain is homogenized with an A-esterase preparation, which destroys the remaining inhibitor rapidly.[7] This method is not as general as the others, as not all compounds are destroyed, and probably only a few are destroyed rapidly enough to stop significant further inhibition of ChE's.

When inhibition by excess inhibitor is avoided, one must correct for re-activation of the inhibited enzyme during the determination of the residual AChE. This is obviously important when the enzyme is dimethylphosphorylated, but the correction can be significant when the enzyme is diethylphosphorylated.[139] In most of the work quoted these corrections were not made.†

* *Added in proof.* L. C. Blaber and H. N. Creasey (*Biochem. J.* **77,** 591 (1961)) showed that there is no need to remove excess inhibitor from the brain of sheep one day after treatment. They also showed that considerable re-activation takes place when method (3) is used.

† Such corrections should, of course, be made to all determinations. Unfortunately the papers referred to were published too late for the results in them to be incorporated fully into this Chapter.

Unlike blood ChE's, brain ChE's are very little inhibited by toxic doses of some compounds. The brain is protected by a barrier to diffusion which hinders the passage from the blood-stream to the brain cells of compounds with low fat–water partition coefficients (the blood–brain barrier).[33, 34] Inhibition of brain ChE's therefore depends on the solubility properties of the inhibitors. Ionic compounds cannot be extracted by fats or oils from water, with the result that single doses of compounds containing quaternary ammonium[26, 28, 29] or sulphonium[1] groups never inhibit brain ChE's to a significant extent. The barrier is not complete, and successive doses may inhibit a few per cent. The non-ionic compounds inhibit more the more they partition into oils from water, as shown in Table 8.1. The compounds are in order of increasing oil extractability, and are bracketed where the differences are slight. Compounds with low oil–water partition coefficients, such as the metabolites of Schradan and Dimefox, inhibit more if given in many successive doses, as, for example, with food in feeding trials lasting a few weeks.[3] The ternary analogues of quaternary nitrogen inhibitors have similar partition coefficients to TEPP, and about the same relative effects on blood and brain ChE's. Brain ChE's can be inhibited by ionic compounds by injecting them directly into the brain. The enzymes recover activity at the same rate as enzymes inhibited with their ternary analogues.[23, 26]

TABLE 8.1. THE DEPENDENCE OF BRAIN AChE INHIBITION ON THE CAPACITY OF NON-AQUEOUS SOLVENTS TO EXTRACT THE INHIBITORS FROM WATER

The compounds are in descending order as their oil–water partition coefficients increase. Compounds with similar partition coefficients are bracketed. Indirect inhibitors are placed in the order of their active metabolites, and are asterisked. The percentage inhibition of brain AChE after an $LD_{50}$ dose is compared with the percentage inhibition of blood or skeletal muscle AChE.

| Compound | Species | Route | Percentage Inhibition | | Refs. |
|---|---|---|---|---|---|
| | | | Brain | Elsewhere | |
| Schradan* | rat | i-p. | 5 | 90 | (2) |
| Dimefox* | rat | i-p. | 37 | 90 | (30) |
| $(Me_2N)_2PO \cdot O \cdot PO(OEt)_2$ } | rat | i-p. | 10 | 60 | (31) |
| $[Me_2N)(EtO)PO]_2O$ } | rat | i-p. | 10 | 70 | (31) |
| $(EtO)_2PO \cdot S \cdot CH_2 \cdot CH_2 \cdot NMe_2$ } | rabbit | i-v. | 91 | ? | (26) |
| TEPP } | dog | i-m. | 68 | 95 | (32) |
| | rat | oral | 49 | 71 | (3) |
| Paraoxon } | mouse | s-c. | 65 | 67 | (25, 27) |
| Sarin } | rat | vapour | 97 | 92 | (10) |
| DFP } | mouse | s-c. | 90 | 96 | (25, 27) |

When brain ChE's are inhibited, the reactions take place in substantially the same way as in blood. The fraction inhibited increases with the dose, though some compounds which inhibit brain and blood ChE's equally at toxic doses inhibit brain ChE's less than blood ChE's at low doses. This is true of Sarin in rats.(35) The brain can only receive Sarin via the blood stream, so that low doses may be largely decomposed by the A-esterases present before reaching it. Inhibition usually reaches a maximum within an hour. Enzyme activity then returns, usually at the same rate as in the blood of the same species.(1, 3, 7, 32, 36) There are some minor differences,(8, 10, 37) and one surprising major difference: rat brain $(EtO)_2PO \cdot \Psi ChE$ recovers its activity very much more slowly than $(EtO)_2PO \cdot \Psi ChE$ in other tissues or *in vitro*.(7) Conversion to stable forms of $(MeO)_2PO \cdot AChE$'s takes place in exactly the same way in brain as in blood.(1)

*Inhibition of ChE's at Myoneural Junctions*

Many investigations have shown that the AChE in muscles innervated by the voluntary and parasympathetic systems is concentrated at the myoneural junctions. Junctional AChE can therefore be estimated by carrying out AChE estimations on muscle homogenates. Precautions must be taken against further inhibition *in vitro* by inhibitor stored in the tissues.(27) ΨChE's are more diffusely distributed. Histochemical methods give, however, more information than can be obtained on homogenates, and are generally preferred.

The most common histochemical method(38, 39) and its main variations are as follows. Slices of tissue, 10–20 $\mu$ thick, are frozen, and placed in a maleate buffer, pH 6·0, containing copper glycinate and sodium sulphate. When AChE is to be located, the ΨChE is inhibited by incubation with a specific inhibitor, usually $10^{-7}$ M DFP(39) or $4 \times 10^{-6}$ M Mipafox.(40) Others have been used when these are not specific enough.(41, 42) Acetylthiocholine is then added as substrate. Its hydrolysis leads to the precipitation of a dark brown copper thiocholine salt at, or very near, the enzyme. A darker stain of cupric sulphide can be obtained by treatment with ammonium or hydrogen sulphide, but this is not essential.(40) To find ΨChE's, DFP is omitted, and butyrylthiocholine is used as substrate. Prior treatment with a specific inhibitor of AChE improves the results.(40) Good results are only obtained when the conditions given in the literature are adhered to strictly. Various methods have been suggested for reducing diffusion artifacts,(43) and interference by other esterases.(44) Some other methods have been compared with this, and seem to be somewhat less satisfactory.(45) The thiocholine methods are better for AChE's, which are highly localized, than for ΨChE's. They are, of course, only semi-quantitative. On monochrome photographs of stained tissues, five degrees of staining can usually be distinguished with ease (Plate 8.1), but a skilled microscopist can obtain more precision.(45) The

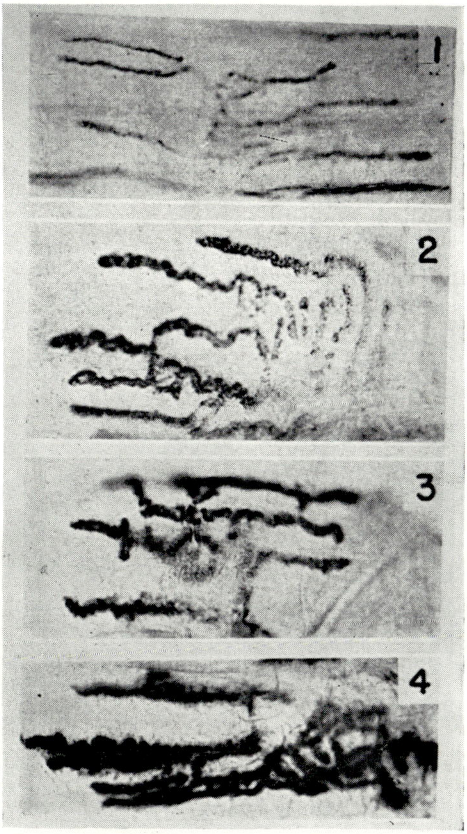


PLATE 8.1. Subneural apparatus in the iliofibularis muscle of the frog stained by the method described in the text. The preparations had been treated with TEPP and P2-AM to give different AChE levels. Four are shown here. The fifth corresponds to complete inhibition.[47]

Reproduced from A. D. Bergner and P. F. Wagley. *Proc. Soc. exp. Biol. N.Y.* **97,** 90 (1958) (with acknowledgements).

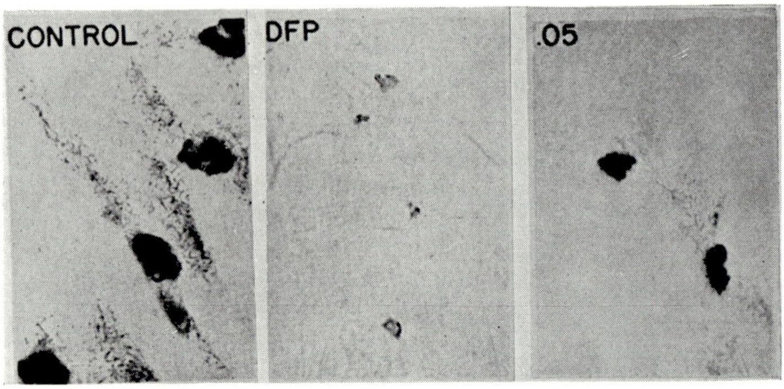


PLATE 8.2. Intercostal muscles in the cat stained as described in the text. End-plates are shown from the muscles of control cats, from cats given 0·02 m-mole/kg of DFP and from cats given 0·02 m-mole/kg of DFP after 0·05 m-mole/kg of WIN-8077, a reversible inhibitor. The AChE level is much higher after WIN-8077 treatment, showing that the AChE is protected against DFP.[50]

Reproduced from G. B. Koelle. *J. Pharmacol.* **120,** 488 (1957) (with acknowledgements).

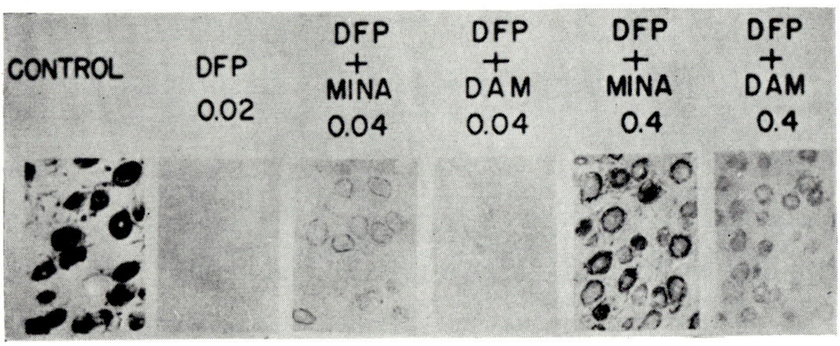


100 1·4 ± 1·9 2·1 ± 2·8 0 5·6 ± 2·7 1·8 ± 0·3

PLATE 8.3. AChE in cat's superior cervical ganglion after various treatments. A control is shown. The remaining cats were given 0·02 m-mole/kg of DFP intravenously, and 20 min later some were given hydroxyimino-acetone (MINA) or diacetyl monoxime (DAM). The doses of the re-activators are shown in m-mole/kg. The percentages of the normal AChE remaining after these treatments were found on homogenates, and are shown, with standard errors, below the histographs.[46]

Reproduced from M. V. Rajapurkar and G. B. Koelle. *J. Pharmacol.* **123,** 247 (1958) (with acknowledgements).

method is, however, particularly valuable in that it is most sensitive to differences when inhibition is nearly complete.[46]

The inhibition of ChE's in muscles has very similar characteristics to inhibition in blood both in dependence on dose and in rate of recovery after inhibition.[36, 44, 48] Recovery of AChE after DFP in rats is, however, somewhat slower than in blood,[6, 49] and after Schradan and some analogues somewhat faster.[2, 31] The degree of inhibition after a given dose varies with the different muscles.

*Protection of ChE's* in vivo

*In vitro* ChE's can be protected from the action of phosphorus compounds either by substrates, or by reversible inhibitors. Protection by substrates cannot be demonstrated *in vivo*, and protection by reversible inhibitors has only rarely been demonstrated. Plate 8.2 shows the effects on the AChE of the intercostal muscles of cats given DFP with and without pre-treatment with the reversible inhibitor WIN-8077.[50] Toxicity studies provide more general evidence.

*Re-activation by Anionic Agents*

Re-activating agents not only re-activate inhibited ChE's, but also catalyse the hydrolysis of phosphate esters, and may protect ChE's from non-competitive inhibitors. Re-activation can only, therefore, be demonstrated convincingly when the compound is given to the animal after the inhibitor. Some results obtained in this way are shown in Table 8.2. On all but inhibited brain AChE they are very similar to those obtained *in vitro* (Chapter VI). $(EtO)_2PO\cdot AChE$ is re-activated at the same rate in the rat whatever diethyl phosphate ester is used to bring about inhibition (No. 5 in Table 8.2); $(EtO)_2PO\cdot AChE$ and $(Pr^iO)(Me)PO\cdot AChE$ are re-activated more rapidly than $(Pr^iO)_2PO\cdot AChE$; and AChE inhibited by Schradan metabolites cannot be re-activated at all. From the original papers it can be concluded that the rates of re-activation depend on the re-activator, falling in the order: compound 16* $>$ P2-AM $>$ hydroxyimino-acetone $>$ diacetyl monoxime. Single doses of P2-AM and compound 16, however, do not re-activate brain AChE,[54, 55] and multiple doses have only a slight effect (No. 7), presumably because these cationic compounds cannot readily penetrate the blood–brain barrier. The other compounds act in the ionized form, but exist partly as uncharged, oil-soluble, free acids at physiological pH's. They can therefore penetrate the blood–brain barrier, and re-activate AChE (No. 11–14).

The ChE activities shown in Table 8.2 were determined on homogenates. Lower degrees of re-activation in tissues can be demonstrated histochemically. The degrees of re-activation in ganglia shown in Plate 8.3 are readily

* See p. 138 for the formula of this ionic re-activator.

TABLE 8.2. RE-ACTIVATION BY HYDROXYIMINO-COMPOUNDS *in vivo*

Species, inhibitors, tissues, re-activators and their doses, and the ChE levels at maximum inhibition as a percentage of the control values are given. The re-activation in the times stated is given as a percentage of the control activity without and with treatment with re-activator. Re-activators were given intraperitoneally, except where stated otherwise, 0·5–1 hr after the inhibitor. When multiple doses were given, they were separated by an hour or more. Asterisked results refer to ΨChE's; and the rest to AChE's.

| Species | Inhibitor | Tissue | Re-activator | Dose (mg/kg) | Min ChE % | % Recovery Without | With | Time | Refs. |
|---|---|---|---|---|---|---|---|---|---|
| 1. Guinea pig | TEPP | blood | Picolinohydroxamic acid | 65 | 60 | ~20 | 40 | 5 days | (51) |
| 2. | | | | 65 | 50* | ~20* | 46* | 5 days* | (51) |
| 3. Man | Parathion | blood | P2-AM | 15 (i-v) | ~5 | 0 | 95 | 20 min | (52) |
| 4. | | | | 15 (i-v) | ~5* | 0* | 40* | 20 min | (52) |
| 5. Mouse | Sev$\cdot(EtO)_2PO\cdot X$'s | blood | P2-AM | 25 | 4 | 6 | 49 | 2 hr | (53) |
| 6. | TEPP | blood | Compound 16 | 4 | 12 | 1 | 40 | 3 hr | (54) |
| 7. | Paraoxon | diaphragm | P2-AM | 25 | 35 | 40 | 62 | 1 day | (27) |
| | | brain | P2-AM | 25 | 35 | 5 | 6 | 1 day | (25) |
| | | | | 5 × 50 | 35 | 5 | 16 | 1 day | (25) |
| 8. | DFP | diaphragm | P2-AM | 25 | 4 | 4 | 24 | 7 hr | (27) |
| | | brain | P2-AM | 2 × 75 + 50 | 6 | 19 | 19 | 2 days | (25) |
| 9. | Schradan | diaphragm | P2-AM | 3 × 50 | 20 | 0 | 0 | 1 day | (27) |
| 10. Rabbit | Parathion | blood | P2-AM | 10 × 15 | 15 | 5 | 65 | 12 hr | (52) |
| 11. Rat | Sarin | blood | Hydroxyimino-acetone | 35 | 6 | 0 | 20 | 1 hr | (55) |
| | | | | 35 | 10* | 0* | 21* | 1 hr* | (55) |
| 12. | | brain | Hydroxyimino-acetone | 35 | 14 | 0 | 25 | 2 hr | (55) |
| 13. | | blood | Diacetyl monoxime | 150 | 15 | 0 | 31 | 2 hr | (55) |
| 14. | | brain | | 150 | 14 | 0 | 9 | 2 hr | (55) |
| 15. | | blood | P2-AM | 100 | 3 | 0 | 32 | 1 hr | (55) |
| 16. | | Skeletal muscle | P2-AM | 100 | 15 | 0 | 51 | 1 hr | (55) |
| 17. | | brain | P2-AM | 100 | 6 | 0 | 3 | 1 hr | (55) |

observed visually, but are too slight to be demonstrated with certainty using homogenates.[46]

The efficacy of a re-activating agent is limited by its toxicity and by its stability *in vivo*. Some $LD_{50}$'s of the more important re-activators are given in Table 8.3. A dose of *x* mg/kg of hydroxyimino-acetone or diacetyl

TABLE 8.3. THE TOXICITIES OF SOME RE-ACTIVATING AGENTS

| Compound | Species | $LD_{50}$(mg/kg) | Refs. |
|---|---|---|---|
| P2-AM | mice | i-p, 260 | (54) |
| | | i-v, 122* | (56) |
| | rats | i-p, 262* | (56) |
| | | i-v, 109* | (56) |
| | | i-p, 305 | (56) |
| | guinea pigs | i-m, 305* | (56) |
| | rabbits | i-v, 147* | (56) |
| | monkeys | i-m, 356* | (56) |
| Compound 16 | mice | i-p, 130 | (54) |
| Hydroxyimino-acetone | mice | i-p, 150 | (57) |
| Dihydroxyimino-acetone | mice | i-p, 20 | (57) |
| Diacetyl monoxime | mice | i-p, 900 | (57) |

* These results were obtained using the methanesulphonate salt instead of the iodide.

monoxime injected intraperitoneally into dogs gives a concentration of about *x* μg/ml in the blood a few minutes later.[58] Safe doses thus give maximum blood concentrations in the range $10^{-3}$–$10^{-4}$ M. The rates of re-activation by hydroxyimino-acetone and diacetyl monoxime *in vivo* referred to in Table 8.2 are about as fast as one would expect from the concentrations present, but P2-AM is considerably less effective than expected from its *in vitro* rate constant[55] (Table 6.9). The blood concentrations fall rapidly. In the dog the concentrations of hydroxyimino-acetone and diacetyl monoxime fall to about half in 20–30 min,[58] and then exponentially with half lives of about 3 hr.[136] P2-AM in rats has a negligible effect 1 hr after injection.[25] Because re-activators persist only a short time, when men poisoned with Parathion are treated with P2-AM the initial, rapid re-activation of both AChE and ΨChE is followed by re-inhibition.[52] This is because Parathion is an indirect inhibitor, owing its toxicity to its metabolic conversion to Paraoxon, which takes place continuously over several hours. To maintain a high level of ChE one must give several successive doses of P2-AM.

When re-activators are given before the inhibitor, i.e. prophylactically, the AChE levels found may be either higher or lower than after therapeutic doses, depending on the stability of the inhibitor and re-activator, and possibly on the ability of the re-activator to decompose the inhibitor. A theoretical analysis of the rates of the various reactions involved leads to the conclusion that only diacetyl monoxime should affect inhibition significantly by hydrolysing the inhibitor, and that even P2-AM should not protect AChE to an appreciable extent.[59] P2-AM does, however, exert a marked protective action on isolated tissues and organs *in vitro* at concentrations only slightly above those tolerated *in vivo*,[47, 60] which casts some doubt on the last conclusion.

### *Inhibition of ChE's and Poisoning*

Poisoning of mammals by phosphorus compounds is generally held to result from AChE inhibition. This is the anti-cholinesterase theory of poisoning. The relationship between inhibition and gross toxic effects goes far to establish this theory, as shown in this section, and is consistent with pharmacological investigations (Part III) which show how phosphorus compounds poison by interfering with the normal functioning of the nervous system.

Early work with specific, reversible inhibitors showed that while inhibition of AChE was correlated with signs of poisoning inhibition of ΨChE was not.[61, 62] This conclusion has been confirmed with phosphorus compounds. ΨChE plays a very minor role in nervous function, and its inhibition is not considered further in this section.

The relationship between inhibition and toxicity is not a simple one, and for this reason the anti-cholinesterase theory has often been contested. Some of the complexities are described first.

Blood AChE is not necessarily inhibited by doses which produce acute signs of poisoning. Myosis follows direct application of most direct inhibitors to the eye with negligible fall in blood AChE[4]; and bronchoconstriction follows the inhalation,[63] and central nervous effects follow intra-cerebral injection[26] of some compounds with only moderate falls in blood AChE. This is not surprising, as these signs are essentially local, and the poison need spend very little or no time in the blood stream before reaching the relevant nerves. Lethal doses, however, always inhibit blood AChE markedly. When an inhibitor can only reach the nervous system via the blood stream, as after intravenous, intraperitoneal, intra-muscular, or subcutaneous injection, dermal application or oral dosing, the first symptoms do not appear until blood AChE is reduced to below 40% of normal levels, and death does not ensue until the levels are still further reduced (Table 8.4). The maximum inhibition after an $LD_{50}$ varies somewhat with species, compound, and route of administration. Successive small doses of DFP, Sarin and TEPP can reduce the AChE in man to under 5% without producing signs, although single doses reducing the level so much are often lethal.[4, 5] This difference in response can be explained by a combination of A-esterase action and adaptation. Probably higher proportions of small doses are destroyed by A-esterases and in other ways during transport in the blood stream to the nervous system; and animals can adapt slowly to low AChE levels. Adaptation is particularly noticeable in feeding trials, or when animals are given small repeated doses in other ways for several weeks. As already described, the AChE's reach a steady low level under these conditions, but the animals only show symptoms for the first few days after the minimum levels are reached. They then apparently recover completely.

TABLE 8.4. BLOOD AChE LEVELS IN POISONED ANIMALS

The levels are expressed as percentages of the levels in control animals, and are those found in animals near death or just showing signs of poisoning.

| Compound | Species | Route | First signs | Near death | Refs. |
|---|---|---|---|---|---|
| DFP | rabbit | i-v | 35 | 3 | (64) |
| | rat | oral | — | 10 | (3) |
| Sarin | man | oral | 22 | — | (4) |
| TEPP | dog | i-v | 38 | 4 | (32) |
| | man | i-m | 19 | — | (5) |
| | rat | oral | — | 29 | (3) |
| EPN* | rat | oral | — | 17 | (3) |
| Parathion* | rat | oral | — | 20 | (3) |
| Schradan* | rat | oral | — | 27 | (3) |

* Indirect inhibitors.

Before and during the presence of signs the animals gain weight more slowly than controls, but as the signs disappear weight gains become normal unless the signs have been very severe.[18, 20, 21, 22] More rapid adaptation also takes place, as shown in Part III.

There is no precise correlation between recovery of blood AChE activity and recovery from toxic doses. This is not unexpected for three reasons. The severity of poisoning increases rapidly with small increases in inhibition (Table 8.4) at AChE levels which are too low to be determined accurately; recovery of blood AChE does not necessarily reflect accurately recovery of AChE in the nervous system; and animals can adapt.

Inhibition and recovery of brain AChE are also not related to toxicity in any simple way. Some compounds kill without inhibiting brain AChE appreciably, while others do not kill until brain AChE is more than 90% inhibited. It has been suggested that only certain sites have to be inhibited in the brain to cause death, and that determinations of AChE inhibition carried out on whole brains may be inadequate to demonstrate the inhibition of these sites if certain compounds attack them selectively. This theory is untenable. Some compounds, Schradan for example, inhibit brain AChE to a negligible extent when given in single lethal doses, but inhibit it considerably when given in several successive non-lethal doses. Pharmacological observations, also, show that the signs of poisoning produced by many compounds can be explained without postulating any direct interference with the central nervous system. Inhibition of brain AChE is only the controlling factor in the action of a few compounds. DFP may be one; and in rats inhibition and recovery of brain AChE are closely correlated with the development of and the recovery from DFP poisoning.[37] In this case there is no correlation with blood AChE, which initially recovers much slower, and eventually somewhat faster than brain AChE.

Inhibition at myoneural junctions is more closely correlated with toxicity. Levels at death are usually below 30% in diaphragms and intercostal muscles, but vary with the muscle.[49] They also vary with the compound in some species (DFP and TEPP in cats;[49] DFP and Paraoxon in mice[25, 27]). Diaphragm AChE levels are, however, closely correlated with death in rats.[48] Thus, when rats are given DFP, TEPP or Paraoxon death takes place when the diaphragm AChE is lowered to 10–15%. By artificial respiration, animals can be kept alive for several hours with still less AChE. When this is done, it is found that rats eventually survive otherwise lethal doses of TEPP and Paraoxon, provided that enough AChE has been re-activated to raise the level to 10–15%. They cannot be made to survive DFP in this way and it is found that the enzyme is not significantly re-activated in a time during which artificial respiration can be maintained.

The failure to obtain close correlations between toxic action and AChE inhibition in the many cases given does not weaken the anti-cholinesterase theory. Pharmacological studies show clearly that different species die in different ways. In some bronchoconstriction is important, in others not, and so on. Many, perhaps most, motor nerves are affected, but to different extents. Which are affected most may well vary from animal to animal, even in the same species, as blood supplies, distribution of fat and other biophysical factors may vary; and will certainly vary with compounds of different oil-solubilities and instabilities in the presence of A-esterases. Death results from the sum of these variable effects. Consequently unless inhibition of AChE in a particular organ is a major factor in bringing about death no close correlation between inhibition in that organ and death can be expected.

Much direct evidence for the theory can be adduced. This will be considered under four heads.

1. *Duration of Signs of Poisoning and Cumulative Action*

The duration of signs of poisoning in survivors of near-lethal doses is precisely as expected when allowance is made for the capacity of animals to adapt slowly. Consider first direct inhibitors which do not persist. Compounds known to be attacked by A-esterases are included in this group. Mammals die quickly, or survive. The duration of symptoms is shorter the faster the rate of re-activation of the inhibited enzyme, i.e. the duration increases in each species in the order:[1, 5, 6, 18, 20, 37, 65, 66]

$(MeO)_2PO \cdot AChE < (EtO)_2PO \cdot AChE <$
$(Pr^iO)(Me)PO \cdot AChE < [(Me_2N)(EtO)PO \cdot AChE, (Pr^iO)_2PO \cdot AChE]$.

The differences in duration can be very great. Thus in rats treated with a dimethyl phosphate ester the duration is under an hour; and in rats treated with a di-isopropyl phosphate ester it is over 8 days. Rats tend to recover more rapidly than cats and dogs. Some dimethyl phosphate esters, as already described, persist in the blood, and inhibit AChE irreversibly. The symptoms are correspondingly prolonged, lasting 5 hr in rats.[1] Indirect inhibitors usually produce signs more slowly than direct inhibitors, and the signs last longer, as inhibitors are produced metabolically over a considerable time. Guthion is an exception. A toxic metabolite is both formed and destroyed rapidly (p. 236), so the symptoms are of short duration.

A compound is said to show cumulative action when a succession of small doses, given at intervals, is as toxic as a single dose equal to the sum of the successive doses. It is obvious from this definition that cumulative action is a matter of degree. Very few, if any, compounds are perfectly cumulative if given in very small doses over a very long period; and every compound is cumulative if successive, fairly large doses are given in rapid succession. Rational indices of cumulative action can be devised, but none which enables the available data to be compared. The following results, however, show that the cumulative action of direct inhibitors increases in the same order as duration of symptoms. Most dimethyl phosphate esters show little cumulative action. Thus rats infused with Paraoxon-Me at less than 1·3 $LD_{50}$'s/hr tolerated about 7 acute $LD_{50}$'s, and rats given 0·5 acute $LD_{50}$'s/day of Phosdrin intraperitoneally for 22 days survived.[20] Diethyl phosphates are more cumulative: few rats survived 0·5 acute $LD_{50}$'s/day of diethyl dichlorovinyl phosphate for more than 4 days;[20] and the $LD_{50}$'s of TEPP and Sarin were less than usual in rabbits whose AChE had been lowered by previous administration of the drugs. Sarin was rather more cumulative than TEPP.[66] In man cumulative effects are measured in terms of the doses producing symptoms of a given severity. A dose of TEPP sensitized men to further doses for 2–3 days, whereas DFP sensitized them to further doses of DFP for 3–4 weeks.[5] Sarin, by the same criterion, was about as cumulative as TEPP.[4]

Dimethyl phosphate esters which persist in the blood should act cumulatively. This has not been checked, but it has been observed that Phosdrin has a cumulative action on rats when given in the food, but none when given intraperitoneally.[20] The compound is likely to be slowly and continuously absorbed when given in the food, and should therefore inhibit AChE irreversibly.

When indirect inhibitors are considered there is, of course, no simple relationship between cumulative action and duration of signs. The duration of signs depends upon how slowly the compound is metabolized, whereas the effects of the second dose depend upon the AChE level when it is given. The AChE level may be much the same a day after equilethal doses of a direct and indirect inhibitor with the same basic groups, unless the two are dimethyl phosphate esters, when the AChE inhibited by the metabolite of the indirect inhibitor is likely to be in the stable form. Thus the cumulative actions of a direct and indirect inhibitor may be the same. This is the case with diethyl dichlorovinyl phosphate and Parathion in rats, when doses are given daily.[20, 67] On the other hand the indirect inhibitor should prove the more cumulative if a second dose is given before metabolism of the first is substantially complete.

Schradan[2] and Dimefox[30] are exceptionally cumulative in rats—0·1 acute $LD_{50}$'s/day for 10 days kills half the rats.* This can be entirely explained by the fact that inhibition of AChE by the metabolites of these compounds is completely irreversible.

The effects of giving two poisons simultaneously at 0·5 $LD_{50}$'s of each are often much as expected. If one acts rapidly, does not persist, and gives a readily re-activated, inhibited enzyme, while the other acts slowly, the combined dose should kill less than half the animals, i.e. the effects should not be additive. This is found to be correct for the pairs: Dipterex-EPN, Dipterex-Systox,† Guthion-Parathion, Guthion-Systox and Malathion-Parathion. The first in each pair is the rapidly-acting compound. When both act slowly, the effects should be additive, as found for Parathion-EPN, Parathion-Systox and Systox-EPN. There are, however, exceptions, Dipterex-Parathion and Guthion-EPN act additively, although one in each pair acts rapidly and the other slowly; and Malathion-Guthion, both of which act rapidly, are less than additive.[68] Some possible explanations are considered later.

* Lower doses of Schradan increase the tolerance of rats to the compound, so that eventually they can withstand indefinitely a daily dose which kills normal rats in 6 days.[69] In the tolerant rats the blood AChE level does not fall continuously as in the normal rats, but reaches a steady level. There are at least three possible causes of this difference in response. Whichever is correct, toxic action and blood AChE levels are closely correlated.

† Systox is a commercial mixture of Demeton-S and Demeton-O.

2. *Protection by Reversible Inhibitors*

*In vitro* eserine and other reversible inhibitors affect the inhibition of AChE by phosphorus compounds. When the AChE is treated with the reversible inhibitor first it is protected from the phosphorus compound; when it is treated with the phosphorus inhibitor first the reversible inhibitor increases the inhibition (Chapter VI). The results obtained *in vivo* are as expected from these observations. Given before, reversible inhibitors have a definite protective effect against otherwise lethal doses of phosphorus compounds; given after, the toxic effects of the two types of compound are roughly additive.[2, 30, 31, 50, 67, 70, 71] The degree of protection conferred varies greatly with the particular reversible inhibitor, phosphorus compound and species. This is not surprising, as protection depends upon both the reversible inhibitor and the phosphorus compound reaching the same sites, which will depend upon the physical properties of the compounds and the biophysical properties of the animal.

3. *Protection and Therapy with Re-activators*

Re-activators are divided into those with cationic groups, e.g. P2-AM and compound 16, which do not penetrate the brain readily, and those without, e.g. hydroxyimino-acetone, which do. The prophylactic and therapeutic effectiveness of some re-activators of the first group are illustrated in Table 8.5, and the therapeutic actions of P2-AM, hydroxyimino-acetone and diacetyl monoxime against Sarin poisoning are compared in Table 8.6.

Therapeutic effects were often determined in animals given atropine before the inhibitor. Atropine is an ACh antagonist, with complex effects described in Chapter XI. It synergizes the action of re-activators, so that results on atropinized animals are much more favourable than on ones given a re-activator only, and the two types of results cannot be compared. If atropine is not given, however, it is often impossible to demonstrate therapeutic action against doses much above the $LD_{50}$, because the animal dies before the re-activator can be given.

Most of the results published can be interpreted in terms of the rates at which the re-activators re-activate the AChE's, and the relative distributions of the inhibitors and re-activators *in vivo*. The action of ionic re-activators (Table 8.5) depend in a complex way upon both factors. These compounds cannot penetrate the blood–brain barrier readily, and must therefore act mainly peripherally. Despite this, they are highly effective therapeutically when given with atropine against TEPP in mice, and the increase in the $LD_{50}$ of TEPP is closely correlated with the rates at which the compounds re-activate $(EtO)_2PO \cdot AChE$ *in vitro*.[72] TEPP does not inhibit brain AChE's much at normal $LD_{50}$'s, and the central effects of the very high doses given in these experiments may have been antagonized by atropine.

TABLE 8.5. THE RELATIVE EFFECTIVENESS OF VARIOUS RE-ACTIVATORS AT COMBATING THE TOXIC ACTION OF SOME PHOSPHORUS INHIBITORS

Both prophylactic (pre-treatment) and therapeutic affects are shown. The index of effectiveness is the factor by which the $LD_{50}$ is raised by the dose of oxime given minus 1. Thus if there is no effect, the $LD_{50}$ is not changed, and the index is zero.

| Re-activator | Dose/kg | Type of effect | Species | Inhibitor: Quat. cpd* | Inhibitor: TEPP | Inhibitor: Sarin | Inhibitor: DFP | Refs |
|---|---|---|---|---|---|---|---|---|
| P2-AM | 9·5 μ-mole | prophylactic | mouse | 0 | 0 | — | 0 | (54) |
| | 95 μ-mole | prophylactic | mouse | — | < 1 | — | 0·3 | (54) |
| 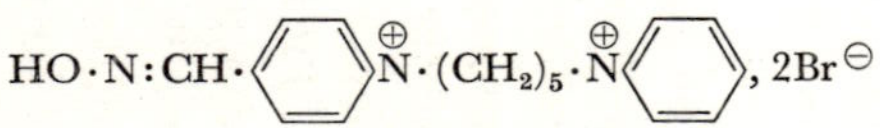 (compound 13) | 9·5 μ-mole | prophylactic | mouse | 15 | 1 | — | — | (54) |
| | 95 μ-mole | prophylactic | mouse | — | 15 | — | 2 | (54) |
|  (compound 16) | 9·5 μ-mole | prophylactic | mouse | 30 | 1 | — | 0·5 | (54) |
| | 95 μ-mole | prophylactic | mouse | — | 20 | — | 5 | (54) |
| P2-AM | 10 mg | therapeutic† | mouse | — | 3 | 1 | — | (72) |
| | 90 mg | therapeutic† | rat | — | 6‡ | — | 2 | (76) |
| Compound 16 | 10 mg | therapeutic† | mouse | — | > 20 | 3 | — | (72) |

* Quat. cpd. = $(EtO)_2PO \cdot O \cdot C_6H_4\text{-}3\text{-}\overset{\oplus}{N}Me_3$, $MeSO_4^{\ominus}$.
† With atropine.
‡ Paraoxon, not TEPP.

TABLE 8.6. THE EFFECTS OF THREE RE-ACTIVATORS ON THE TOXICITY OF SARIN

The re-activators were given intraperitoneally 30 sec after Sarin was given subcutaneously. Therapeutic indices are as in Table 8.5. Doses are in mg/kg[132]

| | | | Species | | |
|---|---|---|---|---|---|
| | | | Rat | mouse | guinea pig |
| $LD_{50}$ of Sarin | | | 0·116 | 0·214 | 0·046 |
| | | Dose | | | |
| Therapeutic indices | P2-AM | 100 | 0 | 0·3 | — |
| | hydroxyimino-acetone | 35 | 2·5 | — | 0·7 |
| | diacetyl monoxime | 150 | 7·3 | — | — |

The protective action of ionic re-activators is, however, very much greater against a quaternary diethyl phosphate inhibitor than against TEPP (Table 8.5), as both the re-activators and the inhibitor are excluded from the brain. They afford much less protection against Sarin and DFP, which have marked central action, though they do afford some.[54, 72, 73, 74, 137] As compound 16 raised the $LD_{50}$ of the reversible inhibitor, neostigmine, a little,[54] they may act partly as competitive AChE inhibitors, and protect the enzyme. The relief of the peripheral effects of Sarin and DFP is, however, probably more important. As shown in Chapter XIV, peripheral effects always contribute appreciably to the toxic action of all compounds, and their relief increases the chances of survival. Other work gives results easier to interpret. P2-AM does not re-activate AChE's inhibited by Schradan metabolite or by Tabun. Correspondingly, it does not protect rats against Schradan,[25] and the slight protection it affords against Tabun[25] can be accounted for by its catalytic effects on Tabun's hydrolysis.[55] Unlike P2-AM, hydroxyimino-acetone and diacetyl monoxime can re-activate brain AChE *in vivo*. Consequently they afford more protection against Sarin (Table 8.6), although they are weaker re-activators *in vitro*. In mice the dodeciodide homologue of P2-AM, $2\text{-HO}\cdot\text{N:CH-C}_5\text{H}_4\overset{\oplus}{\text{N}}\cdot\text{C}_{12}\text{H}_{25}, \text{I}^{\ominus}$, gives more protection against Sarin than P2-AM does, probably because it is more oil-soluble.[75]

The same compound is not equally effective in a given species against different phosphate esters containing the same basic groups.[53] The variations may be due to variations in the persistences of different inhibitors *in vivo*. A persistent compound may continue to inhibit after most of the re-activator has been metabolized. Thus P2-AM reduces the symptoms of Parathion poisoning in man dramatically, but repeated doses are required to maintain the improvement.[52] Similarly a single dose of P2-AM does not protect mice against Parathion, although it protects them against its toxic metabolite, Paraoxon.[73] Sometimes variations could be explained in another way, though, as yet, without experimental evidence. Consider two diethyl phosphate inhibitors. They both give $(EtO)_2PO \cdot AChE$, but to different degrees and in different places, because of differences in stability, oil-solubility, etc. The distribution *in vivo* of a given re-activator is also fixed by its physico-chemical properties, and it may therefore reach some inhibited sites at higher concentrations than others. If this is so, it will not be equally efficacious against both inhibitors. As a result the best re-activator may well be different for each individual inhibitor, even though there is a general fairly close correlation between therapeutic efficiency *in vivo* and re-activation rates *in vitro*.

The evidence for the anti-cholinesterase theory from the action of re-activating agents is particularly strong. The compounds described appear to re-activate inhibited AChE much more readily than other inhibited enzymes (Chapter VI), and it is difficult to account for their success as therapeutic agents except in terms of the theory.

4. *$I_{50}$–$LD_{50}$ Correlations*

Only results for direct inhibitors can be used, for obvious reasons. A precise correlation between $I_{50}$'s and $LD_{50}$'s cannot be expected. Compounds differ in persistence, in the nerve endings they can reach, and in the readiness with which they are absorbed. The last factor can be very important. The sulphonium ion, $(MeO)_2PO \cdot S \cdot CH_2 \cdot CH_2 \cdot \overset{\oplus}{S}EtMe$, is about 500 times more toxic when given intravenously than when given orally.[1] Even among non-ionic compounds, a tenfold variation with route is not uncommon.[77, 78, 79] The sex and age[80, 81] of the animal under test, and even how much exercise it takes,[82] may also influence the toxicity. As, however, both $LD_{50}$'s and $I_{50}$'s of different compounds can vary by factors of over a million, some rough correlation should be found between them on animals all dosed intravenously, so that absorption into the animal is not a controlling factor. This is found to be correct, as shown in Fig. 8.8.* The $I_{50}$'s of No. 2 and 6 are surprisingly low; that of No. 2 is lower in fact than the concentration of AChE usually present in an $I_{50}$ determination, and is therefore suspect. Even including this point, the average deviation of the log $LD_{50}$ from the best

* The figure contains *all* the compounds for which I could find both intravenous $LD_{50}$'s and $I_{50}$'s, though a few of the $I_{50}$'s were determined on AChE's from different species to those on which the toxicity was tested. From various cross-checks it appears that this is not likely to cause any point to fall more than about 0·5 off the best straight line. For three compounds I could not find intravenous $LD_{50}$'s, but it is known that the toxicity of these compounds is nearly independent of the route of administration.

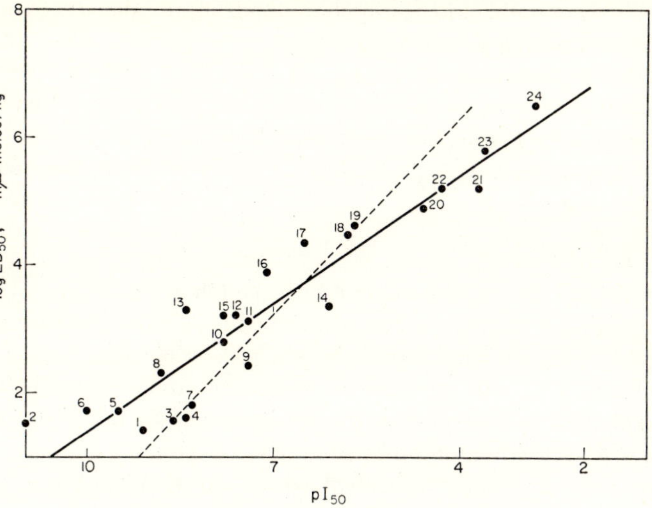


FIG. 8.8. Correlation between $LD_{50}$'s and $I_{50}$'s.

The $LD_{50}$'s are in m$\mu$-moles/kg, and the numerals refer to the following compounds or ions:

1, $(EtO)_2PO \cdot S \cdot CH_2 \cdot CH_2 \cdot \overset{\oplus}{S}Et \cdot CH_2 \cdot CH_2 \cdot SEt$, i-v, rats;(83)
2, $MePO(F) \cdot O \cdot (CH_2)_3 \cdot \overset{\oplus}{N}Me_3$, i-v, rabbits;(84)
3, $(EtO)_2PO \cdot S \cdot CH_2 \cdot CH_2 \cdot \overset{\oplus}{S}Et_2$, i-v, rats;(88)
4, $MePO(F) \cdot O \cdot CHMe \cdot CH_2 \cdot \overset{\oplus}{N}Me_3$, i-v, rabbits;(84)
5, N-methylquinolinium-3-O $\cdot PO(OEt)_2$, i-v, mice;(85)
6, $MePO(F) \cdot O \cdot CH_2 \cdot CH_2 \cdot \overset{\oplus}{N}Me_3$, i-v, rabbits;(84)
7, $(EtO)_2PO \cdot S \cdot CH_2 \cdot CH_2 \cdot \overset{\oplus}{N}Et_3$, i-v, rats;(83)
8, $(Pr^iO)(Me)PO \cdot F$, i-v, rabbits;(84)
9, $(MeO)_2PO \cdot S \cdot CH_2 \cdot CH_2 \cdot \overset{\oplus}{S}EtMe$, i-v, rats;(88)
10, $(EtO_4)P_2O_3$, i-v, mice;(29)
11, $(EtO)_2PO \cdot O \cdot C_6H_4$-4-$NO_2$, i-v, rats;(79)
12, $(EtO)_2PO \cdot S \cdot C_6H_4$-4-$NO_2$, i-v, rats;(79)
13, $(EtO)(Me_2N)PO \cdot CN$, i-v, rabbits;(86) 14, $(Pr^iO)_2PO \cdot F$, i-v, mice;(29)
15, $(MeO)_2PO \cdot O \cdot C_6H_4$-4-$NO_2$, i-v, rats;(88)
16, $(EtO)_2PO \cdot CN$, i-v, mice;(86)
17, $(Pr^iO)_2PO \cdot O \cdot C_6H_4$-4-$NO_2$, i-v, mice;(87)
18, $(EtO)_2PO \cdot O \cdot C_6H_4$-2-$NO_2$, i-v, mice;(79)
19, $(EtO)_2PO \cdot O \cdot C_6H_4$-3-$NO_2$, i-v, mice;(79)
20, $(MeO)_2PO \cdot S \cdot CH_2 \cdot CH_2 \cdot SO_2 \cdot Et$, i-v, rats;(88)
21, $(EtO)_2PO \cdot O \cdot C_6H_4$-4-Cl, i-v, rats;(87) 22, $(Pr^iNH)_2PO \cdot F$, i-p, rats;(87)
23, $(Pr^iNH)_4P_2O_3$, i-p, rats;(87) 24, $(EtO)_2PO \cdot OPh$, s-c, rats.(87)

straight line is only 0·36, i.e. most compounds do not deviate by a factor of more than 2·3, which is surprisingly good. It is obvious, however, that the $LD_{50}$ is not proportional to the $I_{50}$. If the $LD_{50}$'s and $I_{50}$'s were strictly proportional to each other the points should fall about the dotted line. It is clear they do not. There are several possible explanations for this systematic deviation but as there is little evidence for any of them, they will not be discussed.

*General Conclusions*

The anti-ChE theory is consistent with the following facts:

(1) AChE's are the only enzymes involved in nervous function which are always inhibited by toxic doses of phosphorus compounds.

(2) Variations in the duration of symptoms and in cumulative effects can be explained by variations in the persistence of the inhibitors *in vivo* and the rates of re-activation of inhibited AChE's *in vitro*.

(3) Recovery is usually associated with some re-activation of inhibited AChE.

(4) Animals can be protected by reversible inhibitors.

(5) Re-activators of inhibited AChE's are therapeutically effective; and their efficacies depend in the way expected from the re-activation they produce *in vivo*.

(6) There is a rough correlation between the AChE $I_{50}$'s and the $LD_{50}$'s for direct inhibitors.

The theory does not explain why the level of AChE is not reduced to the same extent by lethal doses of different compounds in the same species,* or why an apparently trivial degree of re-activation leads to the disappearance of most signs of poisoning, or why animals can eventually recover in feeding experiments without any apparent rise in AChE levels. It is, perhaps, too much to expect an answer to the first question. A partial one is provided by pharmacological studies, which show that different compounds primarily affect the nerves of different organs, and that these are not equally sensitive to AChE inhibition. Recovery with slight or no rise in AChE levels can only be explained on the theory by the assumption that animals slowly adapt to low AChE levels. Adaptation is not a rare phenomenon and there is pharmacological evidence that it takes place in animals poisoned with phosphorus compounds (Chapters XI–XIV), but very little is known of its mechanism.

Alternative theories must assume either that an, as yet undiscovered, enzyme, essential to the nervous system, is inhibited; or that nervous function is interfered with directly in some biophysical way. In fact the anti-ChE theory proves able to explain most of the pharmacology of phosphorus compounds, and a biophysical theory cannot do this (Part III). The

* Some of the apparent discrepancies are resolved when corrections are made for re-activation during determinations of AChE levels.[139]

undiscovered enzyme theory cannot, of course, be tested (except by finding the enzyme).

## ENZYMES OTHER THAN ChE's

Other than ChE's the only enzyme known to be inhibited by many phosphorus compounds *in vivo* is the B-esterase which hydrolyses carbethoxy groups in Malathion, Malaoxon and other compounds containing two or more carbethoxy groups.*

When liver or serum is taken from rats given EPN 1 hr earlier their capacity to hydrolyse Malaoxon is greatly reduced by an amount roughly

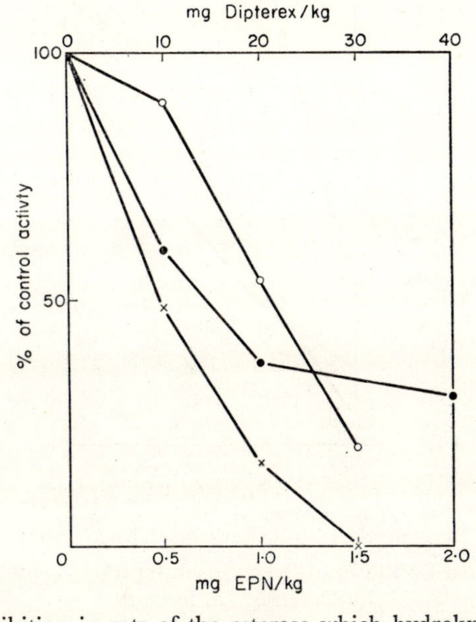


FIG. 8.9. Inhibition in rats of the esterase which hydrolyses Malathion. Inhibition by EPN: in liver, ○; in serum, ×. Inhibition by Dipterex in liver, ●.

proportional to the dose (Fig. 8.9).(89) Inhibition is followed by re-activation (Fig. 8.10). Dipterex also inhibits, but re-activation is faster (Figs. 8.9 and 8.10).(90) *In vitro* the metabolites of Dipterex and EPN (DDVP and the P:O analogue of EPN) inhibit, not the compounds themselves.(91) This is probably true *in vivo* also. Reactions of this enzyme are thus in several ways similar to the reactions of AChE's. The inhibitors are direct inhibitors of AChE's, and inhibition is followed by slow re-activation, which is faster after inhibition by a dimethyl phosphate ester than after inhibition by a

* This enzyme is often called 'Malathionase'. This name is unsatisfactory for the usual reasons.

diethyl phosphate ester. More work is needed to establish that these conclusions are general.

Feeding trials show that EPN almost destroys the enzyme activity at doses which barely affect blood ChE's, and produces 30–40% inhibition at 5 p.p.m. in the food, which is only one-fifth of the lowest concentration producing significant inhibition of blood AChE. Most compounds are not selective in this way. Parathion, Guthion and Dipterex all inhibit blood ChE's at lower dose rates than those at which they inhibit this enzyme.[89]

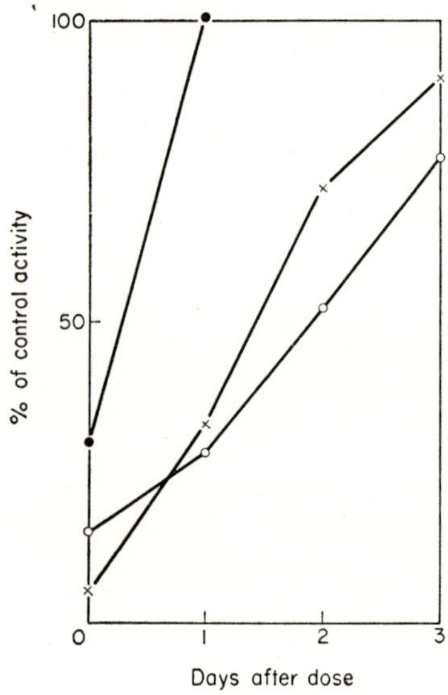


FIG. 8.10. Recovery in rats of the enzyme hydrolysing Malathion. Enzyme inhibited by EPN: in liver, ○; in serum, ×. Enzyme inhibited by Dipterex: in liver, ●.

EPN and Malathion given together are very much more toxic than either alone. The $LD_{50}$ of Malathion to female rats, for example, is 1400 mg/kg, and of EPN is 65 mg/kg; but the $LD_{50}$ of a mixture containing 4% of EPN is 167 mg of Malathion + 6·7 mg of EPN/kg.[92] This effect is called potentiation. The increase in toxicity is paralleled by an increase in the inhibition of AChE. The mechanism of toxic action is, therefore, not changed. EPN potentiates the toxicity of Malathion, not the converse, as shown by experiments on dogs. Mixtures inhibit AChE preferentially to the same extent as Malathion. Thus EPN inhibits the destruction of Malathion, thereby increasing the effect of a given dose. These results can obviously be explained by the inhibition by EPN of the enzyme hydrolysing Malathion. Dipterex, Systox, and Parathion inhibit the esterase at toxic doses. Such inhibition may explain why Dipterex slightly potentiates the action of

Malathion (p. 196). Perhaps some compounds also inhibit other enzymes which decompose phosphorus inhibitors. This could explain the unexpected potentiation of Guthion by Dipterex,[(68)] and other cases mentioned on p. 196.

Certain B-esterases in liver and serum of the rat are inhibited irreversibly by low doses (50–100 mg/kg) of tri-2-tolyl phosphate [tri-ortho-cresyl phosphate, $PO(O \cdot C_6H_4\text{-}2\text{-}Me)_3$, TOCP]. At these doses in this species TOCP has little effect on ChE's and produces no signs of poisoning.[(93)] $DF^{32}P$ also reacts with similar enzymes in the sera of rats,[(93)] guinea pigs, and rabbits,[(94)] and the labelled enzymes have been separated in some instances by electrophoresis.[(94)] In the rat DFP reacts more rapidly with B-esterases than with ΨChE, so that the B-esterases may confer some protection by removing DFP from the system.[(93)]

Proteins in animals poisoned with $^{32}P$ labelled compounds are fairly generally labelled.[(15, 95, 96)] It is likely that this takes place by phosphorylation reactions analogous to those already described.[(95)] It has been shown in the case of DFP that its main metabolite, di-isopropyl hydrogen phosphate, does not react *in vivo* with proteins.[(15)] Whether these reactions lead to toxic effects is not known, but it is very unlikely that they do so.

## INHIBITION OF ENZYMES IN NON-MAMMALIAN SPECIES

### *Insects*

*Inhibition of ChE's*

Most adult insects contain a ChE of the AChE type, which, however, often differs considerably from mammalian AChE's in substrate specificity[(97, 98)] (Chapter V). Some insects also contain ΨChE's,[(99)] and aphids contain only one ChE, which resembles mammalian ΨChE rather than mammalian AChE.[(97)] ChE's are not confined to the nervous system,[(99)] if one can generalize from the few insects in which the distribution of ChE's has been studied.

There are few trustworthy estimates of ChE inhibition in insects. Some workers have either used impure compounds or failed to show conclusively that the enzyme inhibited was ChE. Most workers did not allow for the persistence of the inhibitors in the insects. Persistence is a major problem in insect studies. Paraoxon[(100)] and DDVP[(101)] have been shown to persist in flies (*Musca domestica*), $(EtO)_2PO \cdot S \cdot CH_2 \cdot CH_2 \cdot \overset{\oplus}{N}Et_3\overset{\ominus}{I}$ (Amiton-ethiodide) in locusts[(102)] (*Locusta migratoria migratorioides*), and Schradan metabolite[(104)] and $(EtO)_2PO \cdot S \cdot C_6H_4\text{-}4\text{-}SO \cdot Me$[(103)] in cockroaches (*Periplaneta americana*). The concentration of the sulphinyl compound or its active metabolites in cockroach nerve cord 21 hr after a dose of 3 mg/kg was enough to inhibit the locust nerve cord AChE nearly completely in a few seconds. Inhibition during homogenization is, therefore, likely to take place, and corrections must be made for it,[101, 102, 105] as in the case of mammalian muscle and brain.

When precautions were taken against inhibition during homogenization, inhibition of AChE in the nerve cord of the American cockroach increased with the dose of DFP as expected (Fig. 8.11).[105] Re-activation in several cases has, however, been shown to be much faster *in vivo* in insects than in mammals, or than *in vitro*. Thus in the housefly about 95% inhibition of brain AChE was caused *in vivo* by treatment with nine inhibitors, including members of the $(MeO)_2PO\cdot X$, $(EtO)_2PO\cdot X$ and $(Pr^iO)_2PO\cdot X$ types; but by 22 hr after treatment nearly all the enzyme had been re-activated.

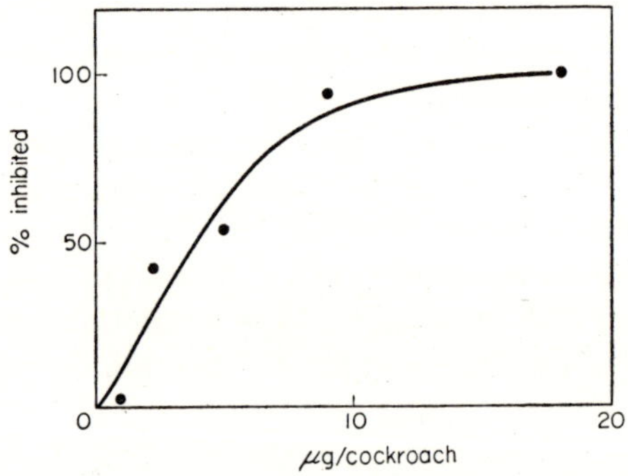


FIG. 8.11. Variation of AChE inhibition in the nerve cord of the American cockroach with dose of DFP.[105]

*In vitro* none of the inhibited enzymes, even $(MeO)_2PO\cdot AChE$, was appreciably re-activated in 24 hr, and the same was found when the inhibited enzymes were taken from poisoned flies. Thus the inhibited insect enzymes were more stable *in vitro* than the corresponding inhibited enzymes from mammals, but were much less stable *in vivo*. The high stability *in vitro* was not due to the conversion of the inhibited enzymes to their stable forms, as they could be re-activated by P2-AM. Results *in vivo* were explained when it was found that a fresh extract of the brains of untreated flies re-activated the inhibited AChE's *in vitro*. Thus in flies natural re-activators are present.[136] This also explains some results obtained on flies with a different group of compounds[106] (when no precautions were taken against inhibition during homogenization) and the rapid re-activation *in vivo* of AChE inhibited by Amiton-ethiodide in the dorsal ganglion of the locust.[102]

Diazinon in the locust (*Schistocerca gregoria Frank*) is cumulative for several days.[108] This may only indicate that Diazinon is unusually persistent, and not that the inhibited enzyme cannot be re-activated.

Toxicity to insects may not be due to inhibition of AChE. Death tends to take place after some recovery of AChE activity, and some flies always die when recovery is nearly complete.[106, 109, 110] Some locusts survive complete inhibition of AChE in their dorsal ganglia by Amiton-ethiodide;[102] but flies die after treatment with DDVP when their brain AChE is only 24% inhibited.[101] The last two observations were made in experiments in which persistence was allowed for.

Other results are in favour of the anti-AChE theory. The $LD_{50}$'s and $I_{50}$'s of the enantiomorphs of $(EtO)(Et)PO \cdot S \cdot CH_2 \cdot CH_2 \cdot SEt$ are closely

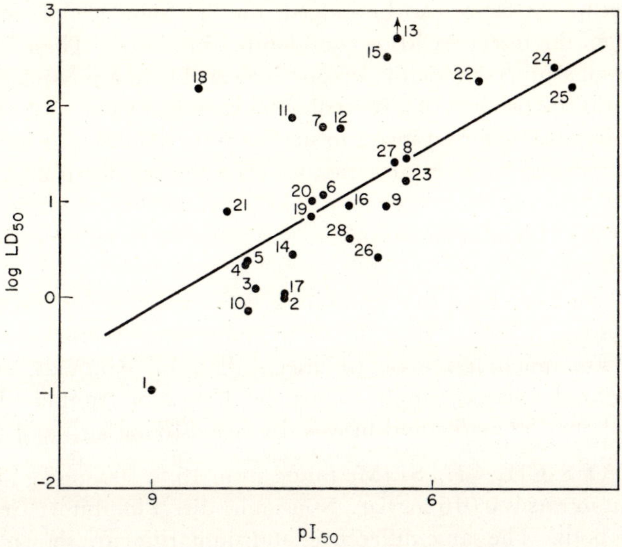


FIG. 8.12. Correlation between $LD_{50}$'s and $I_{50}$'s for houseflies.

$LD_{50}$'s were found after topical application, and are given in $\mu$g/g. $I_{50}$'s were determined on fly brain AChE. Numbers refer to compounds as follows: 1, DDVP;[101] 2–17, $(EtO)(R)PO \cdot O \cdot C_6H_4$-4-$NO_2$[113] where R is: 2, Me, 3, Et; 4, Pr; 5, Bu; 6, Am; 7, Hex; 8, $Pr^i$; 9, $Bu^i$; 10, $Am^i$; 11, $Hex^i$; 12, $Me_3C \cdot CH_2$; 13, cyclohexyl; 14, Ph; 15, $PhCH_2$; 16, $Cl \cdot (CH_2)_3$; 17, EtO; 18–28, $(EtO)_2PO \cdot O \cdot C_6H_n$-$X_{5-n}$,[112] where $X_{5-n}$ is: 18, 2:4-$(NO_2)_2$; 19, 2-$NO_2$; 20, 3-$NO_2$; 21, 2:4:5-$Cl_3$; 22, 2:4:6-$Cl_3$; 23, 2:4-$Cl_2$; 24, 2-Cl; 25, 4-Cl; 26, 4-SO·Me; 27, 3-$NMe_2$; 28, 4-CN.

related when tested on honey bees (*Apis mellifera*), mosquito larvae (*Aedes aegypti*) and houseflies;*[111] a fairly satisfactory $LD_{50}$–$I_{50}$ correlation for flies is found with a considerable range of compounds† (Fig. 8.12);[112, 113]

* The enantiomorphs may be in part indirect inhibitors. This does not destroy the argument. The active metabolites would be formed by oxidation of the ethylthioethyl sulphur atom, which is not optically active, and should react identically in both.

† $LD_{50}$'s of insects vary more with species and route of administration than the $LD_{50}$'s of mammals.[115, 116, 117] The dose–mortality curves also differ from inhibitor to inhibitor,[118] so that a compound with a lower $LD_{50}$ than another may have a higher $LD_{90}$. Effects of temperature[119] and variation with strain may be important also,[120] though these factors are, of course, controlled in laboratory experiments.

the $LD_{50}$'s of Paraoxon to three species of fruit fly are correlated with the $I_{50}$'s of Paraoxon to their AChE's;[98] and compounds which do not inhibit ChE's either directly or indirectly are of very low toxicity.*[114]

There are obviously grounds for dispute. One experimental difficulty does not appear to have been considered in this context. It is often exceedingly difficult to know when an insect dies, or, indeed, to know what constitutes death. As far as can be judged, the first signs of poisoning always appear when the insect AChE is considerably inhibited. These may be the overt signs of a process which is already irreversible in a physiological sense, although it is the product of a reversible biochemical lesion. In other words it may be impossible for an insect to survive if its AChE levels have at some time been reduced by a certain amount, even though death may not ensue until the AChE levels have recovered. The pharmacological studies described in Chapter XVII lend some support to this idea, but there are many remaining difficulties.

Toxicity studies show that inhibitors exert their action on the insect nervous system, whether they do so by inhibiting AChE or not. Ionic inhibitors are much less toxic to insects than to mammals, even when injected into the haemolymph. Thus the $LD_{50}$'s of Amiton-ethiodide in locusts is about 750 mg/kg and in rats (i-v) is 0·018 mg/kg; and the $LD_{50}$'s of $(EtO)_2PO \cdot S \cdot CH_2 \cdot CH_2 \cdot \overset{\oplus}{S}EtMe$ range from 15 to 300 mg/kg† in 5 insect species, but in rats is 0·016 mg/kg. Non-ionic, direct inhibitors are of similar toxicity to both. The same differences and similarities are shown by a wide range of other, non-phosphorus compounds which act on the nervous system; the ionic compounds are much less toxic to insects than to mammals, but the non-ionic compounds are about equitoxic.[121] The results agree with independent evidence that the insect nervous system is protected by a lipoidal membrane which acts as a barrier to the passage of ionic compounds,[122, 123] i.e. it is similar functionally to the blood–brain barrier in mammals. A convincing demonstration of the existence of such a barrier to AChE inhibitors has been carried out on cockroaches treated with Schradan. The metabolite is of low oil-solubility. When it was incubated with an eviscerated cockroach containing an intact cord it caused little inhibition of AChE in the cord, but when the cord was damaged it inhibited much of the AChE.[104] Not surprisingly, the $LD_{50}$ of Schradan to cockroaches is very high.

Evidence has been adduced that the whole of the insect nervous system is surrounded by a lipoidal membrane,[122, 123] but this conclusion has recently

* The few apparent exceptions probably result from the use of compounds containing traces of TEPP or other pyrophosphates.

† In the paper referred to comparisons are made with the i-p. $LD_{50}$ of the compound in rats. Sulphonium compounds are less toxic by this route.

been disproved by the discovery that lipid-insoluble, water-soluble compounds (sugars and ions) can penetrate the nerve sheath and enter the nerve cord of the cockroach.[140] It is, however, only necessary to postulate that the synaptic regions are protected.[141] There is no evidence against this, and it is hard to explain the results quoted in the previous paragraph and in Chapter XVII in any other way.

*Other Enzymes*

In insects, unlike in mammals, inhibition of enzymes besides ChE's has often been demonstrated. The inhibition may be either greater or less, depending on the species, organ and compound. In houseflies, for example, the ali-esterase hydrolysing methyl butyrate was 83% inhibited by lethal doses of DDVP, whereas the AChE was only 27% inhibited.*[101] Precautions were taken to avoid further inhibition during homogenization. Cockroaches also seem to possess both an AChE and an ali-esterase, estimated with nitrophenyl acetate† (NPA). Both were inhibited equally *in vitro* by TEPP.‡[97] Near lethal doses of Paraoxon, DFP, Dipterex or Phosdrin inhibited both esterases in all the organs studied (fat body, gut, muscle, nerve, salivary glands, trachea and blood). Some compounds in some organs inhibited the ali-esterase preferentially, but in the nerve cords all compounds inhibited AChE more than ali-esterase. (No precautions were taken to avoid inhibition during homogenization.) *In vitro* the ali-esterases in *Tribolium castaneum Hbst*, *Tenebrio molitor L*, *Blatella germanica L*, and *Dysdercus fasciatus L* are all about as sensitive as the AChE's in the same species to several phosphorus inhibitors.[126] The $I_{50}$'s of neither type of esterase are closely correlated with $LD_{50}$'s, but such correlation as is obtained for both AChE and ali-esterase would not look out of place in Fig. 8.12.

Dehydrogenases and oxidases may also be inhibited by lethal doses of some compounds. Thus in the rice stem borer (*Chilo simplex Butler*) citric and fumaric dehydrogenase activity is progressively reduced by Parathion and Systox to about 10% at death;[127] and in flies treated with Malathion there is some drop in succinoxidase activity.[128] These losses in activity do

* *Added in proof.* F. W. Plapp and W. S. Bigley (*J. econ. Ent.* **54**, 103 (1961)) found that Parathion and Malathion also inhibited ali-esterase preferentially in houseflies, but that this inhibition was not related to toxic effects, whereas that of AChE was. See also: D. Stegwee (*Can. J. Biochem.* **38**, 1417 (1960)), who confirms the general conclusion.

† Inspection of the results suggests that the AChE has some capacity to hydrolyse NPA, and possibly that the ali-esterase can also hydrolyse ACh, though the effects which indicate this may be due to the crude enzyme preparations used. Crude preparations from some insects contain naturally occurring AChE inhibitors.[124] In many insects the same enzyme hydrolyses both NPA and ACh.[97]

‡ Mammalian ali-esterases are mostly very much less sensitive than AChE's to phosphorus poisons and eserine .[125]

not necessarily imply inhibition of the enzymes by phosphorus compounds; they may be due to the dysfunction of the insect attendant on the effects of the poisons elsewhere.

The eggs of insects at an early stage of embryonic development contain no AChE, but those of several species are killed by TEPP[129, 130] or Parathion.[131] It is obvious in several cases that inhibition of AChE cannot be the cause of death. The eggs contain susceptible ali-esterases, and it has been suggested that the inhibition of these or similar enzymes may be responsible.

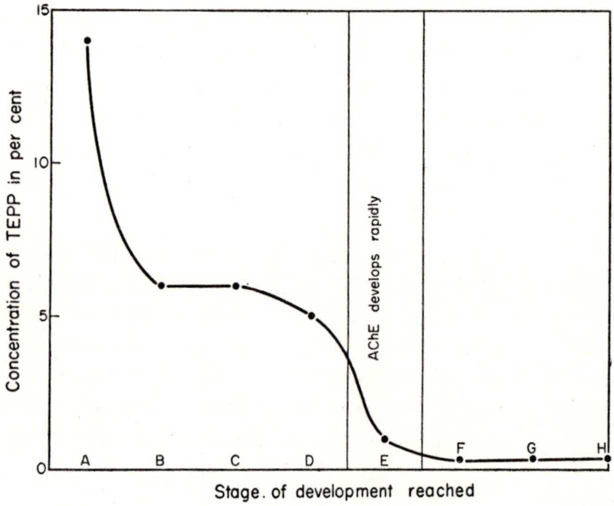


FIG. 8.13. The stage of development reached by *Pieris brassicae* eggs after TEPP treatment at laying.[130]

A detailed investigation of the survival of embryos in *Pieris brassicae* eggs treated with TEPP indicates that there are two mechanisms of death, one of which involves the nervous system.[130] Eggs were treated immediately after laying with various concentrations of TEPP, and the number of embryos reaching different stages* of development determined. The $LC_{50}$ (lethal concentration for 50% kill) of each stage is plotted against the stage in Fig. 8.13. Each stage took about 1 day. The nervous system appeared at stage E, and was accompanied by a rapid increase of AChE activity. Only high concentrations of TEPP stopped development to stage D, but very much lower concentrations stopped development to stage F. The early effect of high concentrations could not have been on the nervous system, which was non-existent, but the later effects of low concentrations probably were, and can be interpreted in two ways. Either TEPP inhibited the

* The stages were those which were morphologically distinguishable. It was not claimed that they are observable in eggs of other species.

formation of AChE, or it persisted, and inhibited the AChE as it was formed. The authors favoured the former theory, but the latter cannot be excluded. The method of treating eggs was likely to give at least as high a concentration of TEPP inside the eggs as was applied. If TEPP were decomposed by aqueous hydrolysis in the eggs, the concentration by stage E would be about $5 \times 10^{-6}$ M, high enough to inhibit most insect AChE's completely in a few seconds. The concentration would be higher were TEPP stored in lipids, or very much less if TEPP were hydrolysed by A-esterases. Only a demonstration that TEPP does not persist would rule this theory out. A similar study was carried out on housefly eggs, which hatch in 12 hr, using Parathion as the poison.[131] The eggs apparently developed normally, but failed to hatch, and no development of AChE activity after the fifth hour was observed in treated eggs, although it was in the control eggs. The ACh content of the treated eggs was significantly higher, as expected.

### *Other Living Things*

Toads (*Bufo viridis*) and frogs (*Rana ridibunda*) proved very resistant to many organophosphorus poisons.[138] Thus the $LD_{50}$'s in toads of some compounds given via the dorsal lymphatic sacs were: DFP, 1450 mg/kg; Dimefox, 1410 mg/kg; Paraoxon, 188 mg/kg; and TEPP, 540 mg/kg. Frogs were somewhat less resistant. The compounds were apparently destroyed in the blood.

Fish (minnow, sunfish and goldfish) were killed by 1 part in $10^9$ of Sarin or cyclohexyl methylphosphonofluoridate in their water. At death their brain AChE levels were reduced to 30–60%. In survivors transferred to fresh water the levels increased to normal over about 2 weeks, during which the fish were more susceptible to further doses.[133]

Ali-esterases in saprophytic mycobacteria were inhibited by Paraoxon and Paraoxon-$Pr^i$ at concentrations varying with species from $10^{-9}$ to $10^{-4}$ M. $I_{50}$'s and $LD_{50}$'s were not significantly correlated.[134]

Esterases hydrolysing phenyl acetate in bean plants and mangolds were inhibited in plants sprayed with Parathion, Dimefox, Systox or Demeton-S. Inhibition was by direct inhibitors of ChE formed metabolically, and may be related to the phytotoxic effects of the compounds.[57]

## REFERENCES


1. M. Vandekar and D. F. Heath. *Biochem. J.* **67,** 202 (1957)
2. K. P. DuBois, J. Doull and J. M. Coon. *J. Pharmacol.* **99,** 376 (1950)
3. J. P. Frawley, E. C. Hagan and O. G. Fitzhugh. *J. Pharmacol.* **105,** 156 (1952)
4. D. Grob and J. C. Harvey. *J. clin. Invest.* **37,** 350 (1958)
5. D. Grob and A. M. Harvey. *Bull. Johns Hopkins Hosp.* **84,** 532 (1949)
6. G. B. Koelle and A. Gilman. *J. Pharmacol.* **87,** 421 (1946)

7. A. N. Davison. *Biochem. J.* **60,** 339 (1955)
8. F. Hobbiger. *Brit. J. Pharmacol.* **6,** 21 (1951)
9. D. Neubert, J. Schaefer and H. Kewitz. *Naturwiss.* **45,** 290 (1958)
10. F. W. Oberst and M. K. Christensen. *J. Pharmacol.* **116,** 216 (1956)
11. A. Locker and H. Siedek. *Experientia* **8,** 146 (1952)
12. F. N. Craig. *J. Pharmacol.* **126,** 174 (1959)
13. L. H. Smith, T. T. Odell, Jr. and B. Caldwell. *Proc. Soc. exp. Biol., N.Y.* **100,** 29 (1959)
14. E. H. Belcher and E. B. Harriss. *J. Physiol.* **146,** 217 (1959)
15. B. J. Jandorf and P. D. McNamara. *J. Pharmacol.* **98,** 77 (1950)
16. J. R. Bove and F. G. Ebaugh, Jr. *J. Lab. clin. Med.* **51,** 916 (1958)
17. F. Hobbiger. *Brit. J. Pharmacol.* **11,** 295 (1956)
18. W. F. Durham, T. B. Gaines, R. H. McCauley, V. A. Sedlak, A. M. Mattson and W. J. Hayes, Jr. *Arch. industr. Hlth.* **15,** 340 (1957)
19. J. P. Frawley and H. N. Fuyat. *J. Agric. Food Chem.* **5,** 346 (1957)
20. J. K. Kodama, M. S. Morse, H. H. Anderson, M. K. Dunlap and C. H. Hine. *Arch. industr. Hlth.* **9,** 45 (1954)
21. T. J. Bombinski and K. P. DuBois. *Arch. industr. Hlth.* **17,** 192 (1958)
22. J. M. Barnes and F. A. Denz. *Brit. J. industr. Med.* **11,** 11 (1954)
23. A. S. V. Burgen and L. M. Chipman. *Quart. J. exp. Physiol.* **37,** 61 (1952)
24. D. Nachmansohn and E. A. Feld. *J. biol. Chem.* **171,** 715 (1947)
25. H. Kewitz and D. Nachmansohn. *Arch. Biochem. Biophys.* **66,** 271 (1957)
26. G. B. Koelle and E. C. Steiner. *J. Pharmacol.* **118,** 420 (1956)
27. H. Kewitz. *Arch. Biochem. Biophys.* **66,** 263 (1957)
28. T. Fredriksson. *Acta physiol. scand.* **42,** Suppl. 145, 47 (1957)
29. A. S. V. Burgen and F. Hobbiger. *Brit. J. Pharmacol.* **6,** 593 (1951)
30. A. J. Okinaka, J. Doull, J. M. Coon and K. P. DuBois. *J. Pharmacol.* **112,** 231 (1954)
31. K. P. DuBois, J. Doull, A. J. Okinaka and J. M. Coon. *J. Pharmacol.* **107,** 464 (1953)
32. C. Dayrit, C. H. Manry and M. H. Seevers. *J. Pharmacol.* **92,** 173 (1948)
33. R. B. Aird and L. Strait. *Arch. Neurol. Psychiat.* **51,** 54 (1944)
34. S. E. Mayer and J. A. Bain. *J. Pharmacol.* **118,** 17 (1956)
35. B. S. Cohen, F. W. Oberst, J. W. Crook and C. Harris. *J. Pharmacol.* **116,** 209 (1956)
36. A. N. Davison. *Biochem. J.* **54,** 583 (1953)
37. A. M. Freedman, A. Willis and H. E. Himwich. *Amer. J. Physiol.* **157,** 80 (1949)
38. G. B. Koelle. *J. Pharmacol.* **100,** 158 (1950)
39. G. B. Koelle. *J. Pharmacol.* **103,** 153 (1951)
40. B. Holmstedt. *Acta physiol. scand.* **40,** 322 (1957)
41. W. J. Pepler and A. G. E. Pearse. *J. Neurochem.* **1,** 193 (1957)
42. G. B. Koelle. *J. Pharmacol.* **114,** 167 (1955)
43. K. Hellmann. *J. Cell comp. Physiol.* **40,** 421 (1952)
44. M. A. Gerebtzoff and E. Hoessels. *C.R. Soc. Biol. Paris* **152,** 546 (1958)
45. F. A. Denz. *Brit. J. exp. Path.* **34,** 329 (1953)
46. M. V. Rajapurkar and G. B. Koelle. *J. Pharmacol.* **123,** 247 (1958)
47. A. D. Bergner and P. F. Wagley. *Proc. Soc. exp. Biol. N.Y.* **97,** 90 (1958)
48. F. A. Denz. *Brit. J. exp. Path.* **35,** 459 (1954)

49. B. P. McNamara, E. F. Murtha, A. D. Bergner, E. M. Robinson, C. W. Bender and J. H. Wills. *J. Pharmacol.* **110,** 232 (1954)
50. G. B. Koelle. *J. Pharmacol.* **120,** 488 (1957)
51. G. Paulet and P. André. *C.R. Soc. Biol. Paris* **150,** 1716 (1956)
52. T. Namba and K. Hiraki. *J. Amer. med. Ass.* **166,** 1834 (1958)
53. F. Hobbiger. *Brit. J. Pharmacol.* **12,** 438 (1957)
54. F. Hobbiger and P. W. Sadler. *Brit. J. Pharmacol.* **14,** 192 (1959)
55. J. P. Rutland. *Brit. J. Pharmacol.* **13,** 399 (1958)
56. D. R. Davies and G. L. Willey. *Brit. J. Pharmacol.* **13,** 202 (1958)
57. L. Dultz, M. A. Epstein, G. Freeman, E. H. Gray and W. B. Weil. *J. Pharmacol.* **119,** 522 (1957)
58. B. M. Askew, D. R. Davies, A. L. Green and R. Holmes. *Brit. J. Pharmacol.* **11,** 424 (1956)
59. A. L. Green. *Biochem. Pharmacol.* **1,** 115 (1958)
60. K. Bethe, W. D. Erdmann, L. Lendle and G. Schmidt. *Arch. exp. Path. Pharmak.* **231,** 3 (1957)
61. R. D. Hawkins and J. M. Gunter. *Biochem. J.* **40,** 192 (1946)
62. R. D. Hawkins and B. Mendel. *Biochem. J.* **44,** 260 (1949)
63. W. K. Berry. *Proc. roy. Soc. Med.* **46,** 801 (1953)
64. A. Mazur and O. Bodansky. *J. biol. Chem.* **163,** 261 (1946)
65. S. Krop and A. M. Kunkel. *Proc. Soc. exp. Biol. N.Y.* **86,** 530 (1954)
66. S. Calloway and D. R. Davies. *Brit. J. Pharmacol.* **12,** 382 (1957)
67. K. P. DuBois, J. Doull, P. R. Salerno and J. M. Coon. *J. Pharmacol.* **95,** 79 (1949)
68. K. P. DuBois. *Arch. industr. Hlth.* **18,** 488 (1958)
69. J. A. Rider, L. E. Ellinwood and J. M. Coon. *Proc. Soc. exp. Biol. N.Y.* **81,** 455 (1952)
70. R. Koster. *J. Pharmacol.* **88,** 39 (1946)
71. J. R. Lewis, W. B. McKeon, Jr. and A. M. Lands. *Arch. internat. Pharmacodyn.* **102,** 371 (1955)
72. W. K. Berry, D. R. Davies and A. L. Green. *Brit. J. Pharmacol.* **14,** 186 (1959)
73. T. O. King and E. Poulsen. *Arch. internat. Pharmacodyn.* **114,** 118 (1958)
74. J. H. Wills, A. M. Kunkel, R. V. Brown and G. E. Groblewski. *Science* **125,** 743 (1957)
75. I. B. Wilson. *Biochim. biophys. Acta* **27,** 196 (1958)
76. H. Kewitz, I. B. Wilson and D. Nachmansohn. *Arch. Biochem. Biophys.* **64,** 456 (1956)
77. R. G. Horton, G. B. Koelle, B. P. McNamara and H. Pratt. *J. Pharmacol.* **87,** 414 (1946)
78. L. Rosival and Fr. V. Selecký. *Chem. Abstr.* **50,** 5174 (1956)
79. W. N. Aldridge and J. M. Barnes. *Nature, Lond.* **169,** 345 (1952)
80. A. M. Freedman and H. E. Himwich. *Fed. Proc.* **7,** 36 (1948)
81. A. M. Freedman and H. E. Himwich. *Amer. J. Physiol.* **153,** 121 (1948)
82. C. L. Punte, E. J. Owens, E. H. Krackow and P. L. Cooper. *Arch. industr. Hlth.* **17,** 34 (1958)
83. D. F. Heath, unpublished.
84. T. Fredriksson. *Arch. internat. Pharmacodyn.* **115,** 474 (1958)
85. F. Hobbiger. *Brit. J. Pharmacol.* **9,** 159 (1954)
86. B. Holmstedt. *Acta physiol. scand.* **25,** Suppl. 90 (1951)
87. J. M. Barnes, unpublished.
88. D. F. Heath and M. Vandekar. *Biochem. J.* **67,** 187 (1957)

89. S. D. Murphy and K. P. DuBois. *Proc. Soc. exp. Biol. N.Y.* **96,** 813 (1957)
90. S. D. Murphy and K. P. DuBois. *Fed. Proc.* **17,** 397 (1958)
91. J. W. Cook, J. R. Blake, G. Yip and M. Williams. *J. Ass. off. agric. Chem. Wash.* **41,** 399 (1958)
92. J. P. Frawley, H. N. Fuyat, E. C. Hagan, J. R. Blake and O. G. Fitzhugh. *J. Pharmacol.* **121,** 96 (1957)
93. D. K. Myers. *Science* **115,** 568 (1952)
94. R. Goutier. *Biochim. biophys. Acta* **19,** 524 (1956)
95. J. A. Cohen and M. G. P. J. Warringa. *J. clin. Invest.* **33,** 459 (1954)
96. D. Hofmann-Credner and H. Siedek. *Arch. internat. Pharmacodyn.* **89,** 74 (1952)
97. J. E. Casida. *Biochem. J.* **60,** 487 (1955)
98. C. C. Roan and S. Maeda. *J. econ. Ent.* **47,** 507 (1954)
99. K. Iyatomi and K. Kanehisa. *Chem. Abstr.* **52,** 14013 (1958)
100. K. A. Lord and S. R. B. Solly. *Chem. and Ind.* 1352 (1956)
101. K. van Asperen. *Nature, Lond.* **181,** 355 (1958)
102. H. S. Hopf and R. T. Taylor. *Nature, Lond.* **182,** 1381 (1958)
103. E. Benjamini, R. L. Metcalf and T. R. Fukuto. *J. econ. Ent.* **52,** 94, 99 (1959)
104. R. D. O'Brien and E. Y. Spencer. *Nature, Lond.* **179,** 52 (1957)
105. L. E. Chadwick and D. L. Hill. *J. Neurophysiol.* **10,** 235 (1947)
106. D. C. Mengle and J. E. Casida. *J. econ. Ent.* **51,** 750 (1958)
107. K. van Asperen and H. M. Dekhuijzen. *Biochim. biophys. Acta* **28,** 603 (1958)
108. R. D. MacCuaig. *Ann. appl. Biol.* **45,** 114 (1957)
109. R. D. O'Brien. *J. econ. Ent.* **50,** 159 (1957)
110. F. P. W. Winteringham and A. Harrison. *Nature, Lond.* **178,** 81 (1956)
111. T. R. Fukuto and R. L. Metcalf. *J. econ. Ent.* **52,** 739 (1959)
112. T. R. Fukuto and R. L. Metcalf. *J. Agric. Food Chem.* **4,** 930 (1956)
113. T. R. Fukuto and R. L. Metcalf. *J. Amer. chem. Soc.* **81,** 372 (1959)
114. G. F. Ludvik and G. C. Decker. *J. econ. Ent.* **44,** 405 (1951)
115. R. L. Beard. *J. econ. Ent.* **42,** 292 (1949)
116. H. Mennsan, Jr. *J. econ. Ent.* **41,** 302 (1948)
117. H. R. Kreuger and J. E. Casida. *J. econ. Ent.* **50,** 356 (1957)
118. G. D. Glynne Jones and J. U. Connell. *Ann. appl. Biol.* **41,** 271 (1954)
119. A. H. McIntosh. *Chem. and Ind.* 2 (1957)
120. F. J. Oppenoorth. *Nature, Lond.* **181,** 425 (1958)
121. R. D. O'Brien and R. W. Fisher. *J. econ. Ent.* **51,** 169 (1958)
122. A. G. Richards. *Ann. N.Y. entomol. Soc.* **51,** 55 (1943)
123. G. Hoyle. *Nature, Lond.* **169,** 281 (1952): *J. exp. Biol.* **30,** 121 (1953)
124. K. A. Lord and C. Potter. *Nature, Lond.* **172,** 679 (1953)
125. D. K. Myers and B. Mendel. *Proc. Soc. exp. Biol. N.Y.* **71,** 357 (1949)
126. K. A. Lord and C. Potter. *J. Sci. Fd. Agric.* **5,** 490 (1954)
127. C. Tomizawa and H. Koike. *Bull. nat. Inst. agr. Sci.* **5C,** 17 (1955)
128. R. D. O'Brien. *J. econ. Ent.* **49,** 484 (1956)
129. K. A. Lord and C. Potter. *Nature, Lond.* **166,** 893 (1950)
130. C. Potter, K. A. Lord, J. Kenten, E. H. Salkeld and D. V. Holbrook. *Ann. appl. Biol.* **45,** 361 (1957)
131. K. N. Mehrotra and B. N. Smallman. *Nature, Lond.* **180,** 97 (1957)
132. B. M. Askew. *Brit. J. Pharmacol.* **12,** 340 (1957)
133. C. M. Weiss. *Ecology* **39,** 194 (1958)
134. D. K. Myers, J. W. Tol and M. H. T. de Jonge. *Biochem. J.* **65,** 223 (1957)
135. K. A. Lord. *Ann. appl. Biol.* **43,** 192 (1955)
136. D. C. Mengle and R. D. O'Brien. *Biochem. J.* **75,** 201 (1960)

137. W. D. Erdmann and O. Latki. *Archiv. für Toxikol.* **18,** 157 (1960)
138. H. Edery and G. Schatzberg-Parath. *Arch. int. Pharmacodyn.* **124,** 212 (1960)
139. W. Schaumann. *Arch. exp. Path.* **239,** 81 *et seq.* (1960)
140. J. Treherne. *Symposium on Radioisotopes and Radiation in Entomology*, 1960. To be published under the auspices of the International Atomic Energy Agency, Vienna.
141. D. F. Heath, *loc. cit.*

CHAPTER IX

# Metabolism in Insects and Mammals

In this Chapter it is shown that phosphorus compounds are metabolized *in vivo* by reactions which are very similar to those shown in Chapter VII to take place *in vitro*. The different types of reaction are described in the same order: hydrolytic, oxidative and reductive, and there is an additional section on the balance between them in some important cases. At each stage the effects of the reactions on toxicity are exemplified.

## HYDROLYTIC REACTIONS

The hydrolysis of a phosphorus compound yields two acids (in the most general sense), one of which contains phosphorus. The other only contains phosphorus in special cases, such as when a pyrophosphate is hydrolysed. The phosphorus products often resemble compounds which occur naturally, and from which they are not easily separated. For this reason hydrolysis *in vivo* has usually been studied with phosphorus compounds labelled with $^{32}P$, with the result that the phosphorus products can be estimated without interference, but the products which do not contain phosphorus are not labelled, and cannot be discovered in most cases. With few exceptions the study of hydrolysis *in vivo* has been the study of those products which contain phosphorus.

The demonstration that a particular hydrolysis product is present does not prove that it is produced directly from the compound given to the animal. In some instances the compound found must have been produced indirectly. Hydrolysis of a thionate, for example, must give a hydrogen thioate:

$$>PS \cdot X + H_2O \rightarrow >PSOH + HX.$$

The presence of a phosphate implies that either the parent compound or the hydrogen thioate has been oxidized:

$$>PS \cdot X \rightarrow >PO \cdot X \rightarrow >PO \cdot OH + HX;$$

or

$$>PS \cdot OH \rightarrow >PO \cdot OH.$$

This example demonstrates both that identification of a product may not identify unambiguously the hydrolytic reaction producing it, and that it may prove that other, non-hydrolytic reactions take place.

Most products of hydrolysis are readily separated from most parent compounds and products from reactions of other types. The hydrolysis products are not extractable from neutral or weakly alkaline aqueous solution by chloroform, whereas most phosphorus compounds, except the cationic inhibitors, are. Much early work stopped at this point. More recently the aqueous fractions have been further separated, and individual products identified by ion exchange chromatography. Some of these identifications are, however, still tentative.

*Mammals*. On mammals the most informative results have been obtained on urine, in which hydrolytic products are concentrated, and which is relatively easy to handle. Analyses of faeces, blood and tissues have usually been less complete because of the low concentrations of $^{32}P$ present. They have always confirmed the results obtained on urine.

Mammals excrete in urine hydrolysis products of every compound studied. As shown in Table 9.1, the rate of excretion, expressed as a percentage of

TABLE 9.1. EXCRETION IN URINE OF HYDROLYSIS PRODUCTS

The compound, species, dose (mg/kg), route of administration, and the percentages of $^{32}P$ given which were excreted as hydrolysis products up to the times shown are given. Most results were read from graphs or obtained from tables by interpolation, and may be about 5% in error, but this is probably within biological variation.

| Code | Compound | Species | Dose | Route | Time in days 0·25 | 1·0 | 5·0 | Refs. |
|---|---|---|---|---|---|---|---|---|
| A | $(EtO)_2PS\cdot O\cdot$ 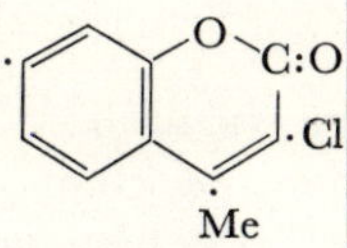 (Co-ral)* | rat | 40 | s-c | 4 | 11 | 31 | (1) |
| | | | 45 | dermal | 1 | 3 | 14 | |
| | | | 50 | oral | 30 | 40 | 50 | |
| | | goat | 30 | dermal | 0·1 | 0·3 | 1·5 | |
| B | $(MeO)_2PS\cdot S\cdot CH_2\cdot CO\cdot NHMe$ (Dimethoate) | rat ♂ | 100 | oral | 53 | 80 | 90 | (2) |
| | | rat ♀ | 100 | oral | 38 | 46 | 55 | |
| C | Parathion | rat | 10 | oral | — | 66 | — | (3) |
| | | cow | 6·7 | oral | — | — | 13 | |
| D | $(MeO)_2PS\cdot O\cdot C_6H_2\text{-}2{:}4{:}5\text{-}Cl_3$ (Trolene) | rat | 100 | oral | 15 | 56 | 65 | (4) |
| | | cow | 100 | oral | 5 | 20 | 50 | |
| E | $(EtO)_2PS\cdot O\cdot$ 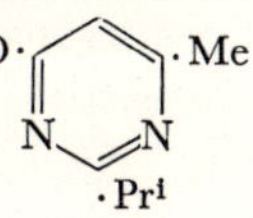 (Diazinon) | rat | 100 | oral | — | 97 | — | (5) |
| | Dimefox | rat | 2·5 | oral | 14 | 23 | 24 | (6) |
| | Schradan | rat | 2·5 | oral | 14 | 23 | 24 | |
| | Dipterex | cow | 25 | oral | 45 | 70 | — | (7) |
| | Phosdrin | cow | 2 | oral | 40 | 51 | 55 | (8) |

* Diethyl 3-chloro-4-methylumbelliferonyl phosphorothionate.

the $^{32}P$ given, depends on the route of administration (A-code in the table), sex (B), species (A, C and D), and the compound given. Dimefox and Schradan are not hydrolysed *in vitro* by any enzyme system studied so far, and there is very strong evidence, given in the section on oxidation reactions, that they are oxidized first.

The variety of products (Table 9.2) reveals the multiplicity of reactions taking place. DFP and Phosdrin are readily hydrolysed *in vitro* to hydrogen phosphates, and these are the only products found in the urine. The thionates give both hydrogen thioates and hydrogen phosphates, the latter showing that hydrolysis is in part either preceded or succeeded by oxidation. The dithioates, Delnav and Thimet, give these products and hydrogen dithioates. Hydrogen dithioates cannot be formed by hydrolytic attack on the phosphorus atom, which gives a hydrogen monothioate:

$$HO^{\ominus} + {>}PS\cdot S\cdot Y \rightarrow {>}PS\cdot OH + SY^{\ominus}.$$

From Delnav the dithioate may be formed by a de-vinylation reaction (p. 96):

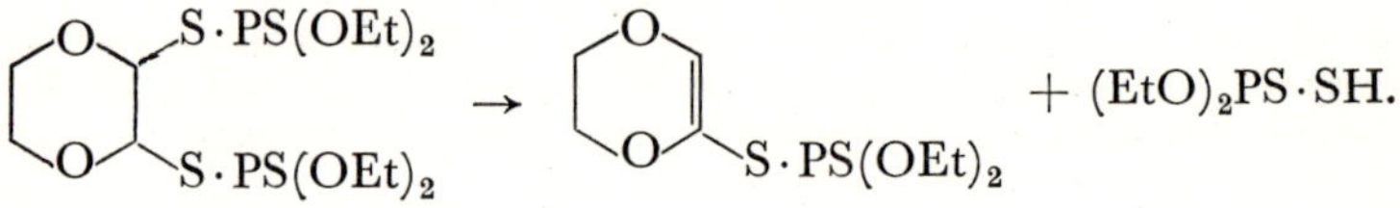


Thimet has the structure: $(EtO)_2PS\cdot S\cdot CH_2\cdot SEt$, and so cannot react in this way. The reaction is probably the enzymic hydrolysis of the S—C bond:

$$(EtO)_2PS\cdot S\cdot CH_2\cdot SEt + H_2O \rightarrow (EtO)_2PS\cdot SH + HO\cdot CH_2\cdot SEt.$$

This reaction has not been brought about chemically. The thionates all give some products formed by de-alkylation, i.e. compounds of the type: (RO)(HO)P(O or S)·X. Dimethyl esters give high yields of these and diethyl esters low ones. The products are the same, and represent about the same fractions of the total hydrolysis products, as those obtained by the action of alkalis in solvents containing a high proportion of ethanol (p. 97). Dipterex is a phosphonate, $(MeO)_2PO\cdot CH(OH)\cdot CCl_3$. In alkaline solution it is readily de-hydrochlorinated, and the product rapidly or simultaneously re-arranges to DDVP (p. 47). This is attacked by A-esterases to give $(MeO)_2PO\cdot OH$ and dichloroacetaldehyde. In cow urine very little $(MeO)_2PO\cdot OH$ is found. The major product is a stronger acid, and may be a compound in which the —$CCl_3$ group is partly hydrolysed and a methyl group is lost.(7) In the dog trichloroethanol glucuronide, equivalent to 66% of the dose given (150 mg/kg orally), is excreted within a day; the same yield is obtained when chloral is given.(11) In both cases the main products are not those obtained by dehydrochlorination followed by hydrolysis. It is not known what the reactions taking place are. They appear to be different in the cow and the dog. As little DDVP is formed in either

TABLE 9.2. HYDROLYSIS PRODUCTS IN URINE

The compound, species and the percentages of various products in urine samples are given, estimated on the total hydrolysis products excreted. The urine analyses were carried out on samples collected in the first 24 hr; later samples usually had a different constitution. X refers to the acidic groups.

| Compound | Species | Products | Refs. |
|---|---|---|---|
| DFP§ | man | $(Pr^iO)_2PO \cdot OH$, 60% | (80) |
| Phosdrin | cow | $(MeO)_2PO \cdot OH$, 100% | (8) |
| Delnav | rat | $(EtO)_2PO \cdot OH$, 11·5%; $(EtO)_2PSOH$, 79%; $(EtO)_2PS \cdot SH$, 9·5% | (9) |
| Thimet* | rat | $(EtO)_2PO \cdot OH$, 17%; $(EtO)_2PSOH$, 80%; $(EtO)_2PS \cdot SH$, 3% | (10) |
| Chlorthion† | rat | $(MeO)_2PO \cdot OH$, 29%; $(MeO)_2PSOH$, 33%; $(MeO)(HO)PO \cdot X$, 13%; $(MeO)(HO)PS \cdot X$, 25% | |
| Dicapthon‡ | rat | $(MeO)_2PO \cdot OH$, 37%; $(MeO)_2PSOH$, 26%; $(MeO)(HO)PO \cdot X$, 10%; $(MeO)(HO)PS \cdot X$, 27% | |
| Trolene | rat | $(MeO)_2PO \cdot OH$, 15%; $(MeO)_2PSOH$, 18%; $(MeO)(HO)PO \cdot X$, 15%; $(MeO)(HO)PS \cdot X$, 42% | (5) |
| Parathion-Me | rat | $(MeO)_2PO \cdot OH$, 60%; $(MeO)_2PSOH$, 27%; $(MeO)(HO)PO \cdot X$, 4%; $(MeO)(HO)PS \cdot X$, 9% | |
| Parathion | rat | $(EtO)_2PO \cdot OH$, 16%; $(EtO)_2PSOH$, 78%; $(EtO)(HO)PO \cdot X$, 2%; $(EtO)(HO)PS \cdot X$, 4% | |
| Diazinon | rat | $(EtO)_2PO \cdot OH$, 24%; $(EtO)_2PSOH$, 71%; $(EtO)(HO)PO \cdot X$, 1%; $(EtO)(HO)PS \cdot X$, 4% | |
| Co-ral§ | rat | $(EtO)_2PO \cdot OH$, 17%; $(EtO)_2PSOH$, 29%; $(EtO)(HO)PO \cdot X$ and $(EtO)(HO)PS \cdot X$, 18%; $H_3PO_4$, 36% | (1) |
| Dipterex | cow | $(MeO)_2PO \cdot OH$ and/or $(MeO)_2PO \cdot H$, 17%; unknown, 77% | (7) |

* $(EtO)_2PS \cdot S \cdot CH_2 \cdot SEt$, 2 day sample.
† $(MeO)_2PS \cdot O \cdot C_6H_3$-3-Cl-4-$NO_2$.
‡ $(MeO)_2PS \cdot O \cdot C_6H_3$-2-Cl-4-$NO_2$.
§ 7 day samples.

species, and Dipterex is not itself a powerful anti-AChE, the low toxicity ($LD_{50} > 200$ mg/kg in several species) is explained.

Surprisingly little work has been done on the hydrolysis products *in vivo* of powerful direct inhibitors such as Sarin and Paraoxon. They are hydrolysed rapidly by A-esterases *in vitro*, and it has been assumed that they are hydrolysed similarly *in vivo*, and that the hydrogen phosphates produced are excreted in the urine. There is much indirect evidence for this theory. Hydrogen phosphates, when given as such, are rapidly excreted in the urine,[12, 13, 14] and certain observations, summarized below, are most easily explained by postulating A-esterase action on toxic phosphates.

(1) The $LD_{50}$ of Sarin given intraperitoneally to rabbits is closely related to the levels in their livers of A-esterases hydrolysing Sarin.[15] The enzyme activity may differ by a factor of four in different rabbits, and the i-p $LD_{50}$ by a factor of about two. The higher the activity the higher the $LD_{50}$. This relationship is not observed when Sarin is given subcutaneously, which is reasonable on physiological grounds. The viscera are rich in A-esterases, so that intraperitoneal injection places the compound in the region where its decomposition is fastest.

(2) DFP has much more effect on the ChE's of an eviscerated rabbit than of a whole one, evisceration removing the organs which are particularly rich in A-esterases.[16]

(3) Rabbit serum A-esterase (purified), when injected into the blood stream of rats, doubles the resistance of the latter to Paraoxon given intravenously.[17]

(4) The effect of liver A-esterases in rats and mice on the toxicity of Paraoxon has been fully investigated.[18] Rats were given a dose via the vena cava which produced 70% inhibition of brain AChE. The same dose given via the portal vein, when the whole dose passed through the liver immediately, caused no inhibition of brain AChE at all. The $LD_{50}$ to mice was also four times greater by intraperitoneal injection than by intramuscular injection. Amiton, $(EtO)_2PO \cdot S \cdot CH_2 \cdot CH_2 \cdot NEt_2$, is only decomposed relatively slowly by the liver, and correspondingly was only a little more toxic to mice by intramuscular injection than by intraperitoneal injection. Amiton was also more toxic than Paraoxon to rats, although it is a much less active anti-ChE *in vitro*.

Not all direct inhibitors are attacked by A-esterases. The fairly close correlation found between the $LD_{50}$'s and $I_{50}$'s of a wide variety of compounds (Fig. 8.8) therefore suggests that the effects of A-esterases on toxicity are rarely very great.

The hydrolytic reactions considered so far all yield acids of phosphorus. Compounds containing esters and amides of carboxylic acids in the side-chain are also hydrolysed by esterases *in vivo*. Dimethoate in rats, for example, gives products which suggest that every type of reaction so far mentioned takes place simultaneously.[2] Urine taken 2 hr after an oral dose of 100 mg/kg contained 'hydrolysis' products in the proportions:

$(MeO)_2PSOH$, 20·5% (P-X hydrolysis);
$(MeO)_2PS\cdot SH$, 23·4% (S-C 'hydrolysis');
$(MeO)(HO)PS\cdot S\cdot CH_2\cdot CO\cdot NHMe$, 12% (de-alkylation);
$(MeO)_2PO\cdot OH$, 6% (hydrolysis and oxidation);
$(MeO)_2PS\cdot S\cdot CH_2\cdot COOH$, 32·4% (amide hydrolysis);

and two unidentified compounds, 5·5%. The urine of steers contained the same products, though in different proportions. Mice and hens treated with Malathion excreted 8 metabolites, all of which were probably acidic. Although these products were not identified, the number is suggestive, as it is only one less than the number produced by the following reactions:[19]

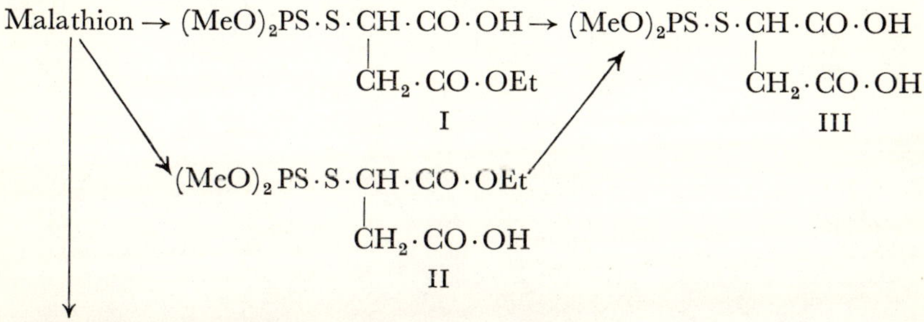

Malaoxon → analogous -oxon products, $PO\cdot S$—, IV, V, VI.
Malathion → $(MeO)_2PS\cdot SH$ and $(MeO)_2PSOH$.
VII VIII

Malaoxon → $(MeO)_2PO\cdot OH$, IX.

It is difficult to devise reactions giving as many products as were found without assuming that the carbethoxy groups were hydrolysed. The number separated may well have been less than the number formed—the chances of two compounds not being distinguished on so crowded a chromatogram are quite high. Butyrylated Dipterex, $(MeO)_2PO\cdot CH(O\cdot CO\cdot Pr)\cdot CCl_3$ is rapidly hydrolysed in rats to Dipterex[14] (analysis of liver). Probably acetylated Dipterex,[11] and Acethion, $(EtO)_2PS\cdot S\cdot CH_2\cdot CO\cdot OEt$, and its analogues[20] are hydrolysed in the same way.

*Insects.* The analysis of excreta is less important in the study of insect metabolism, analyses being carried out on homogenates of whole insects. Direct inhibitors have not been studied much, as, except for a few oil-insoluble compounds, they are about equally toxic to insects and mammals, and it has been assumed that their fate is similar.

Insects often decompose indirect inhibitors more slowly than mammals do, and give fewer products. Thus in cockroaches Trolene,[4] Dicapthon[5] and Chlorthion[5] yield only traces of demethylated products of the (MeO)(HO)PS·X type; and Malathion yields only one ionic metabolite, as against the eight or more produced by mice and hens.[19] All these compounds and Delnav[9] are also attacked much more slowly by insects. Dipterex and butyrylated Dipterex give the same compounds in flies, cockroaches and honeybees as in mammals, but more slowly; and butyrylated Dipterex gives higher concentrations of Dipterex in insects.[14] As the authors point out, these observations go far towards explaining why these compounds are markedly more toxic to insects than to mammals, as illustrated in Table 9.3. The great selectivity of Dimethoate has also been

TABLE 9.3. TOXICITIES OF SOME COMPOUNDS TO MAMMALS AND HOUSEFLIES

All $LD_{50}$'s are in mg/kg.

| Compound | Mammal | | | Fly (topical) | Refs. |
|---|---|---|---|---|---|
| | Species | Route | $LD_{50}$ | | |
| Direct Inhibitors | | | | | |
| Paraoxon | rat | s-c | 0·5 | 0·5 | (21) |
| DDVP | rat | oral | 6 | 0·6 | (11) |
| Indirect Inhibitors | | | | | |
| Acethion | mouse | i-p | 1280 | 9·4 | (20) |
| Chlorthion | rat | oral | 806 | 10·5 | (78, 79) |
| Butonate* | rat | oral | > 3000 | 8 | (14) |
| Dipterex | rat | i-p | 400 | 11 | (11) |
| Dimethoate | rat | oral | 600 | 0·43 | (2) |
| Malathion | rat | s-c | 1000 | 30 | (21) |
| Trolene | cow | oral | 500 | 1·5 | (4, 22) |

* Butyrylated Dipterex.

explained in the same way.[2] In rats' urine a major product is $(MeO)_2PS \cdot S \cdot CH_2 \cdot CO \cdot OH$. By analogy with Malathion it is likely that the hydrolytic reaction yielding this product does not take place in insects, so that Dimethoate persists longer, and more is oxidized to the phosphate analogue, to which the toxic action is attributed.

Dipterex and its acylated derivatives are the only examples up to the present of useful insecticides which do not possess an acid anhydride structure. They appear to have some capacity to inhibit AChE (p. 154), but they are considerably weaker anti-AChE's than any other directly acting insecticides of comparable toxicity. It is therefore reasonable to suppose that they are converted in insects to DDVP by reactions analogous to those which take place in weakly alkaline solution in the absence of enzymes. In accordance with this a compound with partitioning properties and $I_{50}$ similar to those of DDVP has been separated in 5% yield from flies treated with Dipterex.[23] Other workers failed to confirm this finding.[14] It seems, however, intrinsically probable that the theory given for the action of Dipterex is correct. In mammals DDVP has not been found. This is not surprising. DDVP is hydrolysed by A-esterases in mammals, and only a tiny fraction of the Dipterex given need be converted to DDVP to account for the toxic effects. The toxicity of some of the acylated derivatives and homologues to mammals may possibly be accounted for by the direct action of the parent compounds, but not that of Dipterex itself.

## OXIDATION REACTIONS

Many compounds inhibit ChE's much more rapidly *in vivo* than *in vitro*, as illustrated in Table 9.4. The inhibition of blood AChE in rabbits was

TABLE 9.4. ACTIVATION *in vivo* OF SOME PHOSPHORUS COMPOUNDS[24]

The experiments were carried out on rabbits and rabbit AChE. All injections were intravenous.

| Compound | Dose (mg/kg) | Per cent Inhibition | | |
|---|---|---|---|---|
| | | Time (min) | *in vitro* | *in vivo* |
| Parathion | 10 | 10 | 8 | 83 |
| Parathion-Me | 15 | 10 | 6 | 80 |
| Parathion-Pr$^{i}$ | 15 | 10 | 21 | 34 |
| Potasan* | 15 | 20 | 26 | 94 |
| Dimefox | 20 | 7 | 31 | 88 |

* Potasan = diethyl 4-methyl-7-hydroxycoumarinyl phosphorothionate.

determined at set times after dosing with the compounds.[24] The inhibition of rabbit blood AChE was then determined *in vitro* by incubating it for the same time at the same (body) temperature with the same concentration of each compound. The concentrations to use *in vitro* were calculated on the assumption that the whole dose *in vivo* was confined to the blood. This must make the apparent differences less than the real ones, as *in vivo* compounds rapidly leave the blood stream (p. 178f), so the concentrations used *in vitro* were higher than *in vivo*. The apparent differences were still very great. In a somewhat different way, the activation of Schradan in rabbits was also demonstrated.[25] Direct comparisons of *in vivo* and *in vitro* inhibition have not been made with other compounds, but by comparing $I_{50}$'s with doses producing greater than 50% inhibition *in vivo* all the following compound—animal combinations plainly show a similar effect: Delnav (rats);[9] Dimethoate (cows and rats);[2] Disyston (rats);[26] Guthion (rats);[27] Parathion (rats[28] and insects[29]); Parathion-Me (rats);[30] Schradan (several mammals[25] and insects[31]); tri-2-tolyl phosphate (chicken and rabbits)[32] and Trolene (cow and flies);[4] and nearly all thionates are more toxic than their $I_{50}$'s indicate, even allowing for the fairly great deviations from a strict $I_{50}$–$LD_{50}$ relationship which are known to exist.

Most compounds of these types are converted to more powerful anti-ChE's by oxidative enzymes *in vitro* (Chapter VII), so that it is natural to attribute *in vivo* 'activation' to similar enzyme systems. This is plausible on general grounds, as oxidation makes groups more negative, and hence the phosphorus atom more electrophilic and more readily attacked by nucleophilic reagents of all kinds, including ChE's. The evidence for this theory is given in turn for each type of compound attacked.

*Thionates.* The anti-ChE activity of thionates *in vivo* can, theoretically, be attributed either to oxidation to the corresponding phosphates or to isomerization to thiolates. There is no other type of reaction most thionates can be supposed to undergo which could produce the observed effects. That 'activation' *in vivo* involves oxidation and not isomerization has only been demonstrated convincingly on two compounds.[33] Two thionates, their S-alkyl isomers and their phosphate analogues were injected singly into groups of rats, and their effects on AChE and ali-esterase in brains compared. The results are shown in Table 9.5. *In vivo*, as *in vitro*, Paraoxon inhibited AChE more than ali-esterase, while the S-ethyl isomer inhibited both about equally. Parathion acted like Paraoxon, not like its S-ethyl isomer. Methyl

TABLE 9.5. EVIDENCE THAT TWO THIONATES ARE ACTIVATED BY OXIDATION AND NOT BY ISOMERIZATION IN RATS[33]

Times are hours after injection.

| Compound | $pI_{50}$'s | | Per cent Inhibition | | | | | |
|---|---|---|---|---|---|---|---|---|
| | AChE | Ali-E | Time | AChE | Ali-E | Time | AChE | Ali-E |
| Parathion | — | — | 1 | 86 | 36 | | | |
| Paraoxon | 8·0 | 6·7 | 1 | 78 | 26 | | | |
| S-ethyl isomer | 6·4 | 6·4 | 1 | 69 | 62 | | | |
| $MeO \cdot PS(O \cdot C_6H_4\text{-}2\text{-}NO_2)_2$ | — | — | 4 | 72 | 56 | 24 | 92 | 75 |
| $MeO \cdot PO(O \cdot C_6H_4\text{-}2\text{-}NO_2)_2$ | 7·5 | 6·5 | 4 | 80 | 62 | 24 | 93 | 79 |
| $MeS \cdot PO(O \cdot C_6H_4\text{-}2\text{-}NO_2)_2$ | 4·9 | 6·1 | 4 | 40 | 78 | 24 | 16 | 30 |

From: D. K. Myers, H. R. Gersmann and J. A. A. Ketelaar, *Nature, Lond.* **170,** 805 (1952), with acknowledgements.

di-2-nitrophenyl phosphate inhibited AChE more than ali-esterase both *in vivo* and *in vitro*, and *in vivo* the inhibition was greater at 24 hr than at 4 hr after injection. The S-methyl isomer inhibited ali-esterase more than AChE, and the activities of both enzymes were partly restored by 24 hr. The effects of the thionate were identical within experimental error to those of the phosphate, and totally at variance with those of its S-methyl isomer. The difficulty of obtaining pure specimens of S-alkyl isomers has probably dissuaded workers from carrying out analogous experiments with other compounds. It is therefore an assumption that isomerization does not contribute to the toxic effects of any thionates *in vivo*. The assumption is, however, a very plausible one. Firstly, there is no positive evidence that isomerization takes place *in vivo*, or that enzymes can catalyse isomerization *in vitro*; secondly, no enzymes are known which bring about related reactions with any type of substrate; and thirdly, some S-alkyl isomers are not toxic enough to explain the toxic effects of the thionates from which they are derived.[34] The oxidation theory is not open to any of these objections.

Evidence that thionates are converted to phosphates has been obtained with several compounds. The presence of Paraoxon in the livers of rats poisoned with Parathion was shown by extracting them with acetone, and, after various separations, demonstrating that the compounds extracted had the same u-v absorption spectrum, hydrolysis rate and $I_{50}$ to AChE as Paraoxon.[35] There is also chromatographic evidence for the presence of the phosphate analogues in animals poisoned with Parathion (cow);[3] Co-ral (cow and goat);[1] Demeton-O (mouse and cockroach);[9] Thimet sulphoxide and/or sulphone (*Prodenia eridania Cram. larvae*); [10] and, more doubtfully, Diazinon (cow)[36] and Dimethoate (cow).[37]

The toxicities of thionates are to some extent related to the rates at which the thionates are oxidized to phosphates by liver preparations *in vitro* (Chapter VII). Some results are shown in Table 9.6. Both rates of oxidation and toxicity were affected in the same way by pre-treating rats with SKF 525A or methylcholanthrene. SKF 525A inhibited oxidation and reduced toxicity; methylcholanthrene increased both. Again, female rats were more susceptible to Parathion, and their liver preparations oxidized Parathion faster *in vitro*.

Some discrepant results were, however, obtained with EPN and Guthion —both were oxidized faster by liver preparations from male rats although the females were more susceptible. These results may be explained by the action of testosterone in male rats. This inhibited oxidation by preparations of the livers from adult males, but not oxidation by those from females. *In vivo* testosterone may have been present in adult males in sufficient amounts to inhibit oxidation, so that oxidation in males was actually less than in females, in agreement with the toxic effects. The following observations are consistent with this hypothesis:[38]

(1) Weanling rats of both sexes were equally susceptible, and their liver preparations oxidized EPN at the same rate.

(2) In these respects adult females resembled weanlings closely.

(3) When adult males were treated for several weeks with stilboestrol their livers became as inactive *in vitro* as those of normal females; and when adult females were treated for several weeks with testosterone their livers became more active *in vitro*.

TABLE 9.6. INTRAPERITONEAL TOXICITY OF SOME THIONATES TO RATS, AND THEIR ACTIVATION *in vitro*

| Compound | Sex and treatment | $LD_{50}$ mg/kg | Activation* |
|---|---|---|---|
| Guthion[38] | adult ♂ | 11·6 | 10·3 |
| | adult ♀ | 5·7 | 4·2 |
| | adult ♀ pre-treated with Me-cholanthrene† | 2·8 | 10·5 |
| | adult ♀ pre-treated with SKF 525A (20 mg/kg) | > 10 | < 1 |
| EPN[38] | adult ♂ | 26 | 9·6 |
| | adult ♀ | 7·8 | 4·2 |
| | weanling ♂ | 5·9 | 4·7 |
| | weanling ♀ | 5·0 | 4·7 |
| Parathion[39] | adult ♂ | 7 | 7·5 |
| | adult ♀ | 4 | 58 |

* The activation is in arbitrary units, different for each compound, measuring the rate of production of anti-ChE's by liver homogenates or microsomes as described in Chapter VII.

† 25 mg/kg i-p 24 hr before Guthion.

Other observations may have similar explanations. Thus male rats became more susceptible to Parathion when they were treated with stilboestrol;[40] and male and female rats differed in their susceptibilities to Disyston, although the product of oxidation, Demeton-S, was equally toxic to both.[26]

The evidence that oxidation takes place *in vivo* is very strong, and it appears that the effects of sex and of some treatments on toxicity can be explained by differences in the oxidase activities observed in isolated systems. There are still, however, some discrepancies, and more results are needed.

Certain thionates, such as Demeton-O and Demeton-O-Me, isomerize very much more readily than most by a different mechanism (p. 47).

There is some evidence which may indicate that Demeton-O is partly converted to Demeton-S in the mouse.[41] There is some indirect evidence that Demeton-O-Me does not isomerize to any great extent in the rat. As already described (p. 58) isomerization of this compound is accompanied by the formation of very toxic sulphonium compounds. The i-v $LD_{50}$ is, however, very high (216 mg/kg) and death ensues from a narcotic, not from an anti-AChE action (p. 339).[42] There is, in fact, little evidence that the isomerization of any thionate is catalysed *in vivo*.

*Compounds containing sulphide groups*. The three compounds of greatest importance are Demeton-O, Demeton-S and Thimet. In mice and cockroaches both Demeton-S and Demeton-O yielded some products which separated chromatographically as their sulphoxides or sulphones,

$$(EtO)_2PO \cdot S \cdot CH_2 \cdot CH_2 \cdot (SO \text{ or } SO_2) \cdot Et \quad \text{or}$$
$$(EtO)_2PS \cdot O \cdot CH_2 \cdot CH_2 \cdot (SO \text{ or } SO_2) \cdot Et,$$

or as the sulphoxide or sulphone of the phosphate analogue of Demeton-O.[41] The separation was not good enough for all six possible products to be distinguished. The $I_{50}$'s of the chromatographic fractions were consistent with the assumption that the main products were the sulphoxides of the thionate, thiolate and phosphate, and that relatively little if any sulphone was produced. Thimet in cows gave some of the sulphoxide and sulphone, $(EtO)_2PS \cdot S \cdot CH_2 \cdot (SO \text{ and } SO_2) \cdot Et$, mainly in the faeces.[10] Metabolites with the same chromatographic and anti-ChE properties were obtained with mammalian liver slices, and, in the Demeton-O and Demeton-S experiments, with various cockroach tissues, including the nerve cord.

Although the evidence for sulphide oxidation is based mainly on chromatographic evidence, as there was only rough agreement between the $I_{50}$'s to AChE of the metabolites and of the pure sulphoxides and sulphones, there seems little doubt that oxidation does take place *in vivo*. Whether, in fact, one should regard all sulphides of this type as indirect inhibitors *in vivo* is, however, an open question. Demeton-S, for example, is itself a powerful inhibitor, ($I_{50}$ about $10^{-6}$ M), and its sulphoxide and sulphone are not much more powerful or more toxic.[41] It is probable that this and similar compounds act both directly and indirectly.

*Tetramethylphosphorodiamidic compounds*. Most work on compounds of this type has been carried out on Schradan. Schradan approaches the ideal of an indirect inhibitor. It is exceptionally stable, miscible with water, easily purified, and only inhibits AChE at concentrations greater than 0·1 M. It is highly toxic to all mammals on which it has been tested, and produces typical symptoms of anti-AChE poisoning. It is, however, only toxic to some insects (Table 9.7).

*In vitro* Schradan is activated by oxidases in liver microsomes, which are inhibited by SKF 525A (p. 173). Activation *in vivo* is apparently similar. Schradan is antagonized by SKF 525A, Iproniazid and Isoniazid, all three inhibitors of liver microsome oxidases. Pre-treatment of mice with any of these compounds increases the survival times after 50 mg/kg of Schradan by amounts proportional to the protecting dose;(46) and pre-treatment with

TABLE 9.7. $LD_{50}$'S OF SCHRADAN IN DIFFERENT SPECIES
i-h = into haemolymph, t = topical.

| Species | Route | mg/kg | Refs. |
|---|---|---|---|
| Mammals | | | |
| Man | oral | ~ 5 | |
| Rat ♂ | i-p | 8·5 | (43) |
| | | (7·0) | (39) |
| Rat ♀ | i-p | 8·0 | (43) |
| | | (27) | (39) |
| Mice | i-p | 17·0 | (43) |
| Guinea pigs | i-p | 10·0 | (43) |
| Dogs | i-v | 5–10 | (43) |
| Insects | | | |
| American cockroach (*Periplaneta americana*) | i-h | > 100 | (45) |
| German cockroach (*Blatella germanica*) | t | > 1000 | (21) |
| Mealworm larva (*Tenebrio molitor*) | i-h | > 100 | (45) |
| Blowfly (*Phormia regina*) | i-h | 175 | (45) |
| Mourning cloak larva (*Vanessa antiopa*) | i-h | > 400 | (45) |
| Squash bug (*Anasa tristis*) | i-h | 16 | (45) |
| Milkweed bug (*Oncopeltus fasciatus*) | i-h | 30 | (45) |
| Willow aphid (*Lachnus saligenus*) | t | 22 | (31) |
| Pea aphid (*Macrosiphum pisi*) | t | 20 | (21) |

SKF 525A or Iproniazid increases the $LD_{50}$ in mice, SKF 525A by fourfold when given at 20 mg/kg.(47) In confirmation, SKF 525A confers on mice some resistance to $(EtO)(Me_2N)PO \cdot S \cdot CH_2 \cdot CH_2 \cdot NEt_2$, but not to $(EtO)_2PO \cdot S \cdot CH_2 \cdot CH_2 \cdot NEt_2$ (Amiton)—only the phosphoramidic compounds can be activated by oxidation, and hence antagonized by an inhibitor of oxidation. The oxidation of Schradan is nearly confined to the liver, any damage to which greatly reduces the toxicity of Schradan. The liver may be damaged with carbon tetrachloride,[48] but the most convincing results are obtained when the liver is removed.[49] Rats survived total hepatectomy 363 ± 47 min (± standard error). Given 30 mg Schradan/kg intravenously after total hepatectomy they survived almost as long—312 ± 20 min—and showed only slight symptoms. The same dose given to sham-operated rats brought about death in 17·5 ± 0·9 min.

There is still some controversy over the structure of the active metabolite. The facts, and main arguments, are as follows. The characteristics of activation by liver preparations *in vitro* explain the toxic effects of Schradan very well. It is therefore assumed that the same toxic metabolite is formed *in vivo* as *in vitro*.* Liver slices partly convert Schradan to a metabolite, which can be extracted from water by chloroform, and partially separated from Schradan by counter-current extraction[50] (methylene chloride and water as solvents) or by chromatography.[51] The separation is not as good as expected from the differences between the partition coefficients of Schradan and the metabolite, for reasons given later. The metabolite yields formaldehyde on heating with 2·5 N HCl, and the quantity of formaldehyde released is proportional to the rate at which the specimen of metabolite inhibits AChE and chymotrypsin. At the tail of the metabolite peak, where the separation is best, the mole/atom ratio of formaldehyde to phosphorus is 0·96/2. From this it is concluded that the metabolite is a pyrophosphate derivative containing one group which yields formaldehyde quantitatively with acid, and, consequently, that it can be estimated as formaldehyde.[50]

No other peak gives formaldehyde with acid. The compound has the properties listed in Table 9.8. It is a powerful anti-ChE, is unstable in water and more so in acid and alkaline solution, and partitions more into water from organic solvents than Schradan. The yields on Schradan are about 0·3%. Oxidation of Schradan either biochemically with cockroach guts (*Blatella germanica*) or chemically with potassium permanganate or sodium hypochlorite in neutral solution[51] gives better yields (up to 5%) of a compound with the same physico-chemical and biochemical properties (Table 9.8). The identity of the liver metabolite and the permanganate oxidation product has been demonstrated in another way.[51] Both are converted rapidly by acids, and more slowly in water or chloroform to the same product. This has the same properties as one obtained on treating Schradan with peracetic acid (and is usually a by-product of oxidation with permanganate).[58, 59] It is more stable, partitions more into chloroform from water, and is a weaker anti-ChE than the metabolite (Table 9.8). The properties of the metabolite can, therefore, be studied on the product of permanganate oxidation.

* Claims that the active metabolite has been extracted from poisoned insects, separated, and shown to have the same properties as the product from *in vitro* oxidation have not been supported by the publication of figures, so they cannot be assessed.

TABLE 9.8. METABOLITES AND OXIDATION PRODUCTS OF SCHRADAN

| Property | Schradan | Liver metabolite | Cockroach metabolite | Permanganate product | Peracetic acid product | Heptamethylpyro-phosphoramide |
|---|---|---|---|---|---|---|
| Partition coefficients: | | | | | | |
| $CHCl_3/H_2O$ | 7·0 | 1·4–1·6 | 1·72 | 1·2–1·77 | 8–10 | 1·43 |
| $CH_2Cl_2/H_2O$ | 2·0 | 0·65 | — | — | — | — |
| $C_6H_6/H_2O$ | — | — | 0·130 | 0·124 | — | — |
| $pI_{50}$'s: fly AChE (60 min at 37°C) | 0·8 | 6·4 | — | 6·4 | 3·2 | — |
| human ΨChE (30 min at 25°C) | ~1 | — | 6·45 | 6·36 | — | ~4 |
| chymotrypsin (30 min at 25°C) | – 0·2 | 4·6 | — | 4·6 | — | — |
| Decomposition constants in water: | | | | | | |
| $k_{OH}$ at 25°C | $\sim 10^{-5}$ | > 100 | > 100 | $1·65 \times 10^4$ | $\sim 1·5 \times 10^2$ | $3·1 \times 10^{-2}$ |
| $k_H$ at 25°C | $3·4 \times 10^{-3}$ | — | 26 | 24 | — | — |
| $k$ at pH 7·0 | $< 10^{-8}$ | $\sim 5 \times 10^{-4}$ | $1·0 \times 10^{-3}$ | $1·2 \times 10^{-3}$ | $< 10^{-4}$ | — |
| Ratio of residues on degradation: | | | | | | |
| $CH_2O/MeNH_2/Me_2NH/P$ | 0/0/2/1 | 0·48/–/–/1 | — | 0·51/0·60/ 1·48/0·95 | — | 0/1/3/2 |
| Typical lines in infra-red spectrum (mμ) | none at 5·8–6·2 | — | — | 5·93, 6·13 | 5·85, 6·13 | — |

The values are taken from the sources as follows: Schradan,[50, 55, 56, 57] liver metabolite,[25, 50] cockroach metabolite,[54] permanganate oxidation product,[50, 52] peracetic acid oxidation product,[51, 53] and heptamethylpyrophosphoramide.[52, 55]

The product inhibits chymotrypsin irreversibly, and the inhibited enzyme can be purified.[51] On degradation the inhibited enzyme releases formaldehyde, methylamine, dimethylamine and phosphorus in the molar ratios: $CH_2O/MeNH_2/Me_2NH/P$/chymotrypsin of 0·51/0·60/1·48/0·95/1.*

The molar ratios of the metabolite's groups have been found in another way also. The metabolite was separated, and stored in water at pH 8 for 6 hr to hydrolyse it. The aqueous solution was then extracted with chloroform, leaving only the products of hydrolysis in the water layer.[52] On acid treatment these products were found to be in the ratio: $CH_2O/MeNH_2/Me_2NH/P$ of 0·99/1·01/3·10/2. The chloroform layer contained heptamethylpyrophosphoramide, as shown by the ratio: 0/0·82/3·08/2. Heptamethylpyrophosphoramide is not the product of peracetic acid oxidation, which has different physico-chemical properties (Table 9.8).

The above observations are explained if the metabolite has the structure:

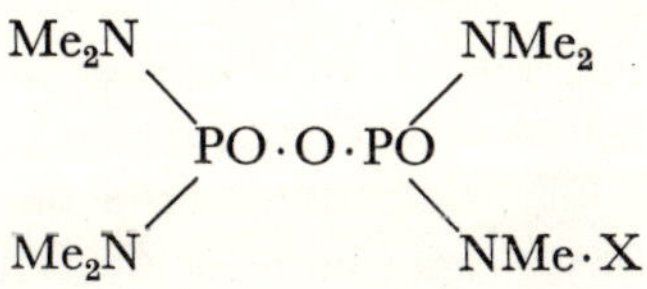


where the group —NMe·X gives formaldehyde with acids.

Two views have been put forward as to the nature of the —NMe·X group. Some think it is an amide oxide group, $—Me_2N\rightarrow 0$. In confirmation, the compound absorbs light at 1690 $cm^{-1}$ like trimethylamine oxide dihydrate (1660 $cm^{-1}$).[50] The spectroscopic evidence can be contested. Thus trimethylamine oxide does not absorb at this frequency when dehydrated, nor does N-dimethyltryptamine N-oxide.[52] The frequency is therefore probably a characteristic of the hydrated group. The metabolite, however, absorbs at this frequency when carefully dehydrated.[52] The N-oxide structure also presents difficulties when attempts are made to assign a structure to the peracetic acid oxidation product (Table 9.8). This absorbs at the same frequency, which suggests that it contains a similar bond. On the N-oxide theory, its only possible structure is: $(Me_2N)_2PO\cdot O\cdot PO(NMe_2)(NMe\cdot OMe)$. This compound has been synthesized, and is much more stable.[52] Others think that the metabolite is $—NMe\cdot CH_2\cdot OH$,[52, 55, 58] the hydroxymethylamide, perhaps formed via the amide oxide. The compound does not absorb at the frequency expected of a compound containing an O—H group.

* These ratios are very surprising, as they imply that both halves of the oxidized Schradan molecule can inhibit the active site of chymotrypsin, although only one half should be greatly affected by the introduction of an electrophilic group. It must be assumed that the electronic effects of the substituent are transmitted with remarkable efficiency through the P—O—P bonding. This is some justification for the similar assumption made in one theory which explains the relative rates of hydrolysis in water of tetra-alkyl pyrophosphates (p. 82).

Otherwise, the structure is more consistent with most observations than the amide oxide theory, as shown below.

(1) Heptamethylpyrophosphoramide is an impurity in most preparations. Its formation by loss of formaldehyde from a hydroxymethylamino compound would be analogous to the known reactions of dialkyl hydroxymethylamines:[60]

$$R_2N \cdot CH_2 \cdot OH \rightarrow R_2NH + CH_2O.$$

(2) The effects on partition coefficients of changes in structure of phosphoramidic compounds are very similar to the effects of the same changes in the structure of amines, as shown in Table 9.9. The difference between the partition coefficients of heptamethylpyrophosphoramide and the metabolite is very like the difference between those of diethylamine and its hydroxymethyl derivative; but the difference between that of Schradan and the metabolite is very unlike the difference between those of an amine and its amine oxide.

(3) The low yields of metabolite are most easily explained if it is not the primary product. The amide oxide can hardly be anything else.

(4) The conversion of the metabolite to the product of peracetic acid oxidation of Schradan (PAO) appears to be faster the higher the concentration. PAO also gives formaldehyde with acids, but a smaller amount. Both observations can be explained if the metabolite has the hydroxymethylamide structure and reacts with heptamethylpyrophosphoramide in the same way as hydroxymethyldialkylamines react with dialkylamines:[60]

$$R_2N \cdot CH_2 \cdot OH + R_2NH \rightarrow CH_2O + R_2N \cdot CH_2 \cdot NR_2 \xrightarrow{\text{acid}} CH_2O + 2R_2NH.$$

The products give formaldehyde with acids. As heptamethylpyrophosphoramide is, on this theory, formed from the hydroxymethylamide, the rate of reaction should increase roughly as the square of the hydroxymethylamide concentration. The proposed sequence of reactions is therefore:

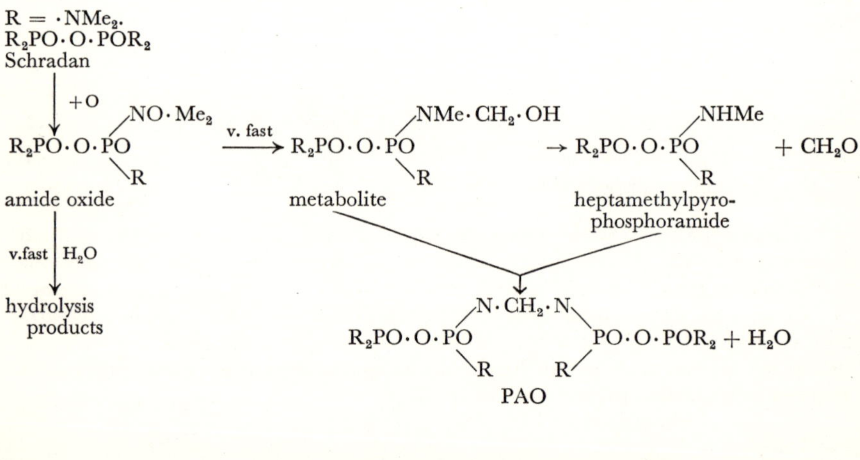

TABLE 9.9. VARIATION OF CHLOROFORM/WATER PARTITION COEFFICIENTS WITH STRUCTURE[55]

The constants for aliphatic amines are corrected for dissociation; $R = \cdot NMe_2$.

| Substitution | Phosphorus compounds | | | Other compounds | | |
|---|---|---|---|---|---|---|
| | Compounds | Coefficients | Ratios | Compounds | Coefficients | Ratios |
| $\cdot Me$ for $\cdot H$ | $R_3P_2O_3\cdot NMe_2/R_3P_2O_3\cdot NHMe$ | 7·1/1·43 | 5·0 | $Me_3N/Me_2NH$ | 3·6/0·80 | 4·5 |
| | $R_2PO\cdot NMe_2/R_2PO\cdot NHMe$ | 6·7/0·46 | 14·6 | $Me_2NH/MeNH_2$ | 0·80/0·098 | 8·2 |
| $\cdot OEt$ for $\cdot NMe_2$ | $R_2PO\cdot OEt/R_2PO\cdot NMe_2$ | 60/6·7 | 9·0 | $RCO\cdot OEt/RCO\cdot NMe_2$ | 98/13·8 | 7·1 |
| $\cdot CH_2\cdot OH$ for $\cdot H$ | metabolite/$R_3P_2O_3\cdot NHMe$ | 1·5/1·43 | 1·05 | $Et_2N\cdot CH_2\cdot OH/Et_2NH$ | 3·7/3·9 | 0·95 |
| $>N$ for $>NO$ | $R_4P_2O_3$/metabolite | 7·1/1·5 | 4·7 | $Et_3N/Et_3NO$ | 29/0·01 | 2900 |

*Quoted from:* D. F. Heath, D. W. Lane and P. O. Park. *Phil. Trans. Roy. Soc.* **239B,** 191 (1955) with acknowledgements.

The rates of formation of active metabolite by various tissues are found by estimating the rates at which the products inhibit AChE. The livers of mammals oxidize at rates varying with the species, but the variation is not great.[61] In insects the organ which converts Schradan most rapidly is the fat body,[45] which fulfils several of the functions in insects performed in mammals by the liver. In at least some insects, however, it is probable that metabolite formed in the fat body does not poison, as the metabolite cannot penetrate the sheath surrounding the nerve cord (p. 208). This is the case with cockroaches, so that probably only Schradan metabolized in the nerve cord itself can affect the insect. Three observations support this view. The nerve cords of cockroaches (*Periplaneta americana*) and squash bugs (*Anasa tristis*) can oxidize Schradan to an anti-ChE; the nerve cords of cockroaches are only inhibited *in vitro* when damaged;[63] and the fat bodies of three resistant species oxidize Schradan more rapidly than those of two susceptible species, and may therefore be supposed to destroy more Schradan before it can reach the nerve cords.[45] The importance of the last factor may be doubted, as, unlike in mammals, the capacity to oxidize Schradan is found in many organs.[62] The metabolite is not, however, equally effective at inhibiting AChE's from different insects.[54] It seems likely that insects can differ in all the following ways:

1. Capacity to oxidize Schradan (some differences shown).
2. Capacity to catalyse the hydrolysis of the metabolite.
3. Permeability of the nerve cord to Schradan.
4. Susceptibility of AChE's (some differences known).
5. Capacity to excrete Schradan.

Much further work will be needed to assess the importance of all these factors.

Heptamethylpyrophosphoramide is probably oxidized, both *in vivo* and *in vitro*. No attempt has been made to identify the metabolites.[63] As has been pointed out, the number of possible products is enormous, as the parent compound is not symmetrical, and there are several stereo-isomers and two or four optical isomers of each possible metabolite.[64]

Dimefox is oxidized enzymically *in vivo* and *in vitro* and by several oxidizing agents.[6, 65, 66] The products appear to be analogous to those from Schradan, but have not been so fully investigated.

Several other tetramethylphosphorodiamidic compounds are activated by liver slices *in vitro*. There is no evidence either for or against the theory that these compounds owe their toxic properties to activation *in vivo*.

Octaethylpyrophosphoramide, $(Et_2N)_4P_2O_3$, is only very slowly activated *in vitro*.[67] It is not known whether this is because the compound is not oxidized rapidly, or because the product is a weak anti-ChE. Either hypothesis accounts for its relatively low toxicity.[68] It is, however, observed that in each of the series of compounds: $(Me_2N)_2PO \cdot F$ (Dimefox),

$(Me_2N)(Et_2N)PO \cdot F$ and $(Et_2N)_2PO \cdot F$; and Dimefox, $(Me_2N)$(Morpholino)$PO \cdot F$ and $(Morpholino)_2PO \cdot F$ the first two are highly and about equally toxic, and the last is of very low toxicity.(55) This suggests that the enzyme system oxidizing phosphoramidic compounds oxidizes dimethylamide groups more rapidly than other alkylamide groups.

## REDUCTIVE REACTIONS

Nitrophenyl phosphate esters are reduced *in vitro* by bovine rumen fluid (Table 7.5). Parathion has been studied extensively *in vivo*. In one experiment(3) a cow was given 6·7 mg/kg of $^{32}P$-labelled Parathion. In 5 days 43% of the $^{32}P$ given was recovered in the urine, about 70% of it as amino-Parathion, $(EtO)_2PS \cdot O \cdot C_6H_4$-4-$NH_2$, identified by chromatography and i-r spectroscopy. Much of the $^{32}P$ in the blood and rumen fluid was also amino-Parathion, and a little was amino-Paraoxon. Cows do not excrete 4-nitrophenol, but do excrete an aminophenyl glucuronide,(69) and probably most of the 4-nitrophenyl groups given can be accounted for as this and amino-Parathion. The amino-derivatives are only about one-hundredth as toxic as the parent compounds when given by mouth, and their rapid formation probably explains why cows given such high doses were not poisoned, and their AChE levels were not lowered.(3) Goats are not similarly protected; in the few experiments carried out the AChE in kids feeding on the milk of a poisoned goat was lowered, whereas calves fed on the milk of a poisoned cow were unaffected.(70)

Reduction is not an important metabolic process in many species. Rats excrete little amino-Parathion; and, when given amino-Parathion, excrete almost entirely hydrolysis products.(3) Dogs(71) and monkeys,(72) unlike cows, excrete 4-nitrophenol.

Probably reduction only takes place in ruminants, and is brought about by micro-organisms in their rumens.

## BALANCE OF OXIDATIVE AND HYDROLYTIC REACTIONS

Very little of any phosphorus poison is excreted as such. Direct inhibitors are apparently hydrolysed, or react with proteins, and differences in their toxic effects can be attributed to the relative rates at which these reactions and inhibition of AChE take place. The fate of an indirect inhibitor is more complex. Oxidation converts it to a compound able to react faster with AChE. The product, however, is often also attacked rapidly by A-esterases, unlike the parent compound, and the reaction yielding it may also be in competition with other hydrolytic reactions. When the balance between these reactions is different in different species, differences in toxicity are likely to ensue. The protection conferred by SKF 525A can be explained in this way; oxidation is inhibited, but hydrolysis, excretion, etc., are not. In this section several other examples are given.

*Parathion.* Parathion is oxidized in the liver to Paraoxon, which is rapidly hydrolysed there. Injection of Parathion into the portal vein in rats leads to a greater inhibition of AChE than does injection into the vena cava, i.e. passage through the liver produces net activation.(18) Correspondingly, unlike Paraoxon, Parathion is more toxic after intraperitoneal than after intramuscular injection. As, however, Parathion is rapidly oxidized by organs other than the liver (p. 172), its toxicity to rats is not significantly reduced by hepatectomy.(28) The toxicity of Schradan is greatly reduced, because oxidation is nearly confined to the liver.

The $LD_{50}$'s of Parathion and Paraoxon to three strains of houseflies were determined,(73) with the results shown in Table 9.10. All three were equally

TABLE 9.10. $LD_{50}$'S OF PARATHION AND PARAOXON TO THREE STRAINS OF HOUSEFLIES(73)

The compounds were injected.

| Strain | $LD_{50}$'s $\mu$g/fly Parathion | $LD_{50}$'s $\mu$g/fly Paraoxon | Paraoxon content ($\mu$g/fly) |
|---|---|---|---|
| S | 0·013 | 0·010 | 0·017 |
| $U_2$ | 0·022 | 0·010 | 0·005 |
| D | 0·050 | 0·013 | 0·003 |

susceptible to Paraoxon, but of different susceptibilities to Parathion. The three strains were then injected with 0·02 $\mu$g Parathion/fly, and the quantities of Paraoxon present 5 hr later calculated from the rates at which fly homogenates inhibited AChE (it was assumed the active metabolite was Paraoxon). The results show that greater resistance was associated with lesser capacity to accumulate Paraoxon. This particular experiment, and the confirmatory ones given in the original paper, demonstrate very directly that a difference in toxicity may be associated with a change in balance between the rates of formation and disposal of active metabolite. In this case much of the metabolite is excreted unchanged, and the resistant strains excrete it more rapidly.

*Guthion and Gutoxon.* The active metabolite of Guthion, which is presumably Gutoxon, is destroyed rapidly by liver enzymes, probably by A-esterase action.(74) The rates of oxidation, measured on liver homogenates, fall in the order: rats > mice > guinea pigs, while the rates of hydrolysis fall in the order: guinea pigs > mice = rats. Guinea pigs are about 4 times more resistant to Guthion than rats or mice, as expected.(27) Mice, however, are a little more susceptible than rats, which is the wrong way round. The implicit assumption that Gutoxon would be equally toxic to all three species in the absence of liver A-esterase is, of course, only likely to be roughly correct.

*Co-ral.*[75] Like Guthion, Co-ral is oxidized to its active phosphate analogue, which is hydrolysed. In minces and slices of organs (oxygenated), both hydrolysis and oxidation can take place; in homogenates not fortified with DPNH, etc., only hydrolysis can take place; and in microsomes only oxidation. Using enzyme preparations of all three types it was shown that while the ox, rat, mouse, cattle grub (*Hypoderma bovis L.*) and housefly all oxidized Co-ral to a powerful anti-ChE, in the ox and rat the net effect* of liver enzymes was degradation, and in the mouse, housefly and cattle grub was activation. The first two are resistant to the compound—$LD_{50}$'s > 100 mg/kg—while the last three are susceptible. Thus those species in which oxidation outweighs hydrolysis *in vitro* are much the most susceptible. This also explains why a dermal application of Co-ral to cattle kills cattle grubs beneath the skin, whereas an oral dose does not. An oral dose can only reach the grubs via the blood stream, so that the compound can be degraded. A dermal application can reach the grubs unchanged by diffusion through the fat.

*Malathion.* Similar experiments show that mouse microsomes oxidize Malathion rapidly to Malaoxon, but that homogenates of most organs hydrolyse Malaoxon rapidly. The net effect is slight activation only. Cockroach mid-gut and fat body oxidize Malathion as fast as mouse liver microsomes, but hydrolyse Malaoxon very slowly, so that the net effect is considerable activation.[76] Similarly cow liver hydrolyses Malathion, but fly head does not.[77] The competition in mammals between oxidation and hydrolysis is complex. As already described (p. 221), mammals hydrolyse the carbethoxy groups of Malathion rapidly, thus reducing the quantity available for conversion to Malaoxon, and hydrolyse Malaoxon both in this way and by A-esterase action. Insects do not apparently hydrolyse carbethoxy groups, so that more Malathion is available for oxidation, and the Malaoxon produced is more stable.

*Acethion and some analogues.*[20] Mouse liver and minced cockroach degrade Acethion, while minced fly activates it. The $LD_{50}$'s are: mouse, i-p 1280 mg/kg; cockroach, i-h > 1000 mg/kg; and houseflies, topical, 9·4 mg/kg, as expected. Cockroaches both activate Malathion, and are susceptible to it; but are resistant to Acethion, which they degrade. Acetoxon is nearly as selective as Acethion, which again suggests that the major difference between susceptible and non-susceptible species is in their capacity to hydrolyse organic esters or amides. In agreement with this several

* The experimental conditions probably favoured the observation of net degradation.

compounds containing carboxylic ester or amide groups are selective, e.g. $(EtO)_2PS\cdot S\cdot CH_2\cdot CH_2\cdot CO\cdot OEt$, $(EtO)_2PS\cdot S\cdot CH_2\cdot CHMe\cdot CO\cdot OEt$, and $(EtO)_2PS\cdot S\cdot CH_2\cdot CH_2\cdot CO\cdot NH_2$; but several close analogues without such groups are not, e.g.

$$(EtO)_2PS\cdot S\cdot CH_2\cdot CO\cdot OH \quad \text{and} \quad (EtO)_2PS\cdot S\cdot CH_2\cdot CO\cdot Me.$$

Some tentative conclusions can be drawn about selectivity, though much more work will be required to show whether they are generally valid. Oil-soluble direct inhibitors are not selective. Oil-insoluble direct inhibitors kill mammals but not insects. Indirect inhibitors of low oil-solubility kill some species of insects but not others. They are about as toxic to mammals as to the most susceptible insects. Simple thionates and sulphides without methoxyphosphinyl groups are usually about equally toxic to insects and mammals. Compounds killing insects at doses much lower than those killing mammals contain either methoxyphosphinyl groups or carboxylic ester or amide groups, or combinations of them. Insects cannot, apparently, hydrolyse such groups, whereas mammals can. They are also usually indirect inhibitors.

A mathematical analysis of the effects of various factors suggests that indirect inhibitors should only be more selective than direct inhibitors when they can be degraded *before* activation by one species but not by another.

## REFERENCES


1. H. R. Krueger, J. E. Casida and R. P. Niedermeier. *J. Agric. Food Chem.* **7,** 182 (1959)
2. W. C. Dauterman, J. E. Casida, J. R. Knaak and T. Kowalczyk. *J. Agric. Food Chem.* **7,** 188 (1959)
3. M. K. Ahmed, J. E. Casida and R. E. Nichols. *J. Agric. Food Chem.* **6,** 740 (1958)
4. F. W. Plapp and J. E. Casida. *J. Agric. Food Chem.* **6,** 662 (1958)
5. F. W. Plapp and J. E. Casida. *J. econ. Ent.* **51,** 800 (1958)
6. B. W. Arthur and J. E. Casida. *J. econ. Ent.* **51,** 49 (1958)
7. W. E. Robbins, T. L. Hopkins and G. W. Eddy. *J. econ. Ent.* **49,** 801 (1956)
8. J. E. Casida, P. E. Gatterdam, J. B. Knaak, R. D. Lance and R. P. Niedermeier. *J. Agric. Food Chem.* **6,** 658 (1958)
9. B. W. Arthur and J. E. Casida. *J. econ. Ent.* **52,** 20 (1959)
10. J. S. Bowman and J. E. Casida. *J. econ. Ent.* **51,** 838 (1958)
11. B. W. Arthur and J. E. Casida. *J. Agric. Food Chem.* **5,** 186 (1957)
12. F. C. G. Hoskin. *Canad. J. Biochem.* **34,** 75 (1956)
13. F. C. G. Hoskin. *Canad. J. Biochem.* **34,** 743 (1956)
14. B. W. Arthur and J. E. Casida. *J. Agric. Food Chem.* **6,** 360 (1958)
15. P. A. Adie. *Canad. J. Biochem.* **34,** 731 (1956)
16. A. Mazur. *J. biol. Chem.* **164,** 271 (1946)
17. A. R. Main. *Canad. J. Biochem.* **34,** 197 (1956)
18. P. Holtz and E. Westerman. *Arch. exp. Path. Pharmak.* **237,** 211 (1959)
19. R. B. March, T. R. Fukuto, R. L. Metcalf and M. G. Maxon. *J. econ. Ent.* **49,** 185 (1956)
20. R. D. O'Brien, G. D. Thorn and R. W. Fisher. *J. econ. Ent.* **51,** 714 (1958)

21. H. R. KRUEGER and J. E. CASIDA. *J. econ. Ent.* **50,** 356 (1957)
22. R. D. RADELEFF and G. T. WOODARD. *J. econ. Ent.* **50,** 249 (1957)
23. R. L. METCALF, T. R. FUKUTO and R. B. MARCH. *J. econ. Ent.* **52,** 44 (1959)
24. W. N. ALDRIDGE and J. M. BARNES. *Nature, Lond.* **169,** 345 (1952)
25. J. E. GARDINER and B. A. KILBY. *Biochem. J.* **51,** 78 (1952)
26. T. J. BOMBINSKI and K. P. DUBOIS. *Arch. industr. Hlth.* **17,** 192 (1958)
27. K. P. DUBOIS, D. R. THURSH and S. D. MURPHY. *J. Pharmacol.* **119,** 208 (1957)
28. W. M. DIGGLE and J. C. GAGE. *Nature, Lond.* **168,** 998 (1951)
29. W. F. CHAMBERLIN and W. M. HOSKINS. *J. econ. Ent.* **44,** 177 (1951)
30. K. P. DUBOIS, M. MAZUR and K. W. COCHRAN. *Fed. Proc.* **9,** 269 (1950)
31. F. DUSPIVA. *Mitt. biol. Zent Anst. Berl.* **70,** 91 (1951)
32. W. N. ALDRIDGE. *Biochem. J.* **56,** 185 (1954)
33. D. K. MYERS, B. MENDEL, H. R. GERSMANN and J. A. A. KETELAAR. *Nature, Lond.* **170,** 805 (1952)
34. R. L. METCALF and R. B. MARCH. *J. econ. Ent.* **46,** 288 (1953)
35. J. C. GAGE. *Biochem. J.* **54,** 426 (1953)
36. W. E. ROBBINS, T. L. HOPKINS and G. W. EDDY. *J. Agric. Food Chem.* **5,** 509 (1957)
37. R. H. ROBERTS, R. D. RADELEFF and J. N. KAPLANIS. *J. econ. Ent.* **51,** 861 (1958)
38. S. D. MURPHY and K. P. DUBOIS. *J. Pharmacol.* **124,** 194 (1958)
39. A. N. DAVISON. *Biochem. J.* **61,** 203 (1955)
40. K. P. DUBOIS, J. DOULL, P. R. SALERNO and J. M. COON. *J. Pharmacol.* **95,** 79 (1949)
41. R. B. MARCH, R. L. METCALF, T. R. FUKUTO and M. G. MAXON. *J. econ. Ent.* **48,** 355 (1955)
42. D. F. HEATH and M. VANDEKAR. *Biochem. J.* **67,** 187 (1957)
43. K. P. DUBOIS, J. DOULL and J. M. COON. *J. Pharmacol.* **99,** 376 (1950)
44. A. J. OKINAKA, J. DOULL, J. M. COON and K. P. DUBOIS. *J. Pharmacol.* **112,** 231 (1954)
45. R. D. O'BRIEN and E. Y. SPENCER. *J. Agric. Food Chem.* **1,** 946 (1953)
46. A. W. LESSIN. *Brit. J. Pharmacol.* **14,** 251 (1959)
47. R. D. O'BRIEN and A. N. DAVISON. *Canad. J. Biochem.* **36,** 1203 (1958)
48. J. P. FRAWLEY, E. P. LAUG and O. G. FITZHUGH. *Fed. Proc.* **11,** 347 (1952)
49. K-K. CHENG. *Brit. J. exp. Path.* **32,** 444 (1951)
50. J. E. CASIDA, T. C. ALLEN and M. A. STAHMANN. *J. biol. Chem.* **210,** 607 (1954)
51. H. TSUYUKI, M. A. STAHMANN and J. E. CASIDA. *J. Agric. Food Chem.* **3,** 922 (1955)
52. E. Y. SPENCER, R. D. O'BRIEN and R. W. WHITE. *J. Agric. Food Chem.* **5,** 123 (1957)
53. D. F. HEATH. *Chem. Soc.* special publ. No. 8, 180 (1957)
54. R. D. O'BRIEN and E. Y. SPENCER. *J. Agric. Food Chem.* **3,** 56 (1955)
55. D. F. HEATH, D. W. J. LANE and P. O. PARK. *Phil. Trans. Roy. Soc.* **239B,** 191 (1955)
56. D. F. HEATH and P. CASAPIERI. *Trans. Faraday Soc.* **47,** 1093 (1951)
57. G. S. HARTLEY, D. F. HEATH, J. M. HULME, D. W. POUND and M. WHITTAKER. *J. Sci. Fd. Agric.* **2,** 303 (1951)
58. G. S. HARTLEY. Section 13, 15th internat. Congr. pure and applied Chem., N.Y. (1951)
59. D. F. HEATH and P. O. PARK, unpublished.
60. L. HENRY. *Bull. Acad. Belg. Cl. Sci.* (3) **28,** 365 (1894)
61. R. D. O'BRIEN. *Canad. J. Biochem.* **34,** 1131 (1956)

62. J. E. Casida, R. K. Chapman, M. A. Stahmann and T. C. Allen. *J. econ. Ent.* **47,** 64 (1954)
63. R. D. O'Brien and E. Y. Spencer. *Nature, Lond.* **179,** 52 (1957)
64. G. S. Hartley. *Chem. and Ind.* 529 (1954)
65. J. E. Casida, T. C. Allen and M. A. Stahmann. *Nature, Lond.* **172,** 243 (1953)
66. M. L. Fenwick. *Biochem. J.* **70,** 373 (1958)
67. R. D. O'Brien. *Nature, Lond.* **183,** 121 (1959)
68. G. Schrader. *Die Entwicklung neuer Insektizide usw.*, Verlag Chemie, Weinheim, 1952
69. J. E. Pankaskie, F. C. Fontaine and P. A. Dahm. *J. econ. Ent.* **45,** 51 (1952)
70. C. G. Wilber and R. A. Morrison. *Amer. J. vet. Res.* **16,** 308 (1956)
71. J. F. Gardocki and L. W. Hazleton. *J. Amer. pharm. Soc.* **40,** 491 (1951)
72. J. Lieben, R. K. Waldman and L. Krause. *Arch. industr. Hlth.* **6,** 491 (1952)
73. F. J. Oppenoorth. *Nature, Lond.* **181,** 425 (1958)
74. S. D. Murphy and K. P. DuBois. *J. Pharmacol.* **119,** 572 (1957)
75. R. D. O'Brien and L. S. Wolfe. *J. econ. Ent.* **52,** 692 (1959)
76. R. D. O'Brien. *J. econ. Ent.* **50,** 159 (1957)
77. D. E. Weidhaas. *J. Ass. off. agric. Chem. Wash.* **42,** 445 (1959)
78. L. Rosival, F. V. Selecky and L. Vbrovsky. *Chem. Abstr.* **52,** 16618 (1958)
79. A. D. Oliver and W. G. Eden. *J. econ. Ent.* **48,** 111 (1955)
80. J. A. Cohen and M. G. P. J. Warringa. *J. clin. Invest.* **33,** 459 (1954)

## CHAPTER X

# Metabolism in Plants and Soil

## PLANTS

PHOSPHORUS compounds are widely used as insecticides, mainly on crops. A knowledge of how they behave on and in crops is therefore of some practical importance. This behaviour is exceedingly complex, as physical, biophysical, chemical and biochemical factors operate simultaneously. Only the biochemical factors are dealt with in this book, but something must be said about the rest.

All insecticides can be regarded as interacting physically with plants in three ways. They may either remain on the surface to which they are applied (surface-acting insecticides*), penetrate the plant but remain only in the fatty parts near the point of contact (deep-acting insecticides[1]), or penetrate, and be translocated in the xylem or phloem to parts distant from the point of application (systemic insecticides). The distinction is not rigid, and some insecticides act in all three ways. Simple experiments distinguish the three types of action. A surface-acting insecticide will only kill insects on the surface to which it is applied; deep-acting insecticides will also kill insects on the opposite side of a leaf to the one treated; and systemic insecticides will kill insects in untreated leaves, including, often, the leaves of plants of which only the roots have been treated. Surface-acting insecticides do not enter the plant, and cannot, therefore, undergo biochemical reactions in them. Deep-acting insecticides do enter the plant, but it is difficult to study any biochemical reactions taking place because similar products are often formed chemically on the leaf surface. Systemic insecticides, therefore, have provided most information about biochemical reactions.

Far fewer compounds have been studied in plants than in animals. The results described are those which demonstrate unambiguously that reactions take place. The loss of insecticides from inside a plant is not evidence enough; some phosphorus compounds are guttated (excreted from the leaves), evaporated in the transpiration stream or are lost from the roots.[40]

Most experiments have been carried out with $^{32}$P-labelled compounds. Plants are macerated with a suitable solvent, often water, filtered, and the

* Such insecticides are often called contact insecticides. This is confusing, as the term is properly used to describe insecticides which are effective when applied to the surface of the *insect*, in contradistinction to those which must be ingested.

filtrate is extracted with chloroform. The parent compounds, and most metabolites except the products of hydrolysis, appear in the chloroform layer, while the hydrolysis products remain in the water layer. Unless there is evidence to the contrary, compounds which cannot be extracted by chloroform will be assumed to be hydrolysis products. These are non-toxic, whereas most of the other metabolites are potentially toxic. This simple separation, therefore, not only enables the proportion of $^{32}P$ as hydrolysis products to be determined, but also the proportion as possibly toxic residues. A distinction must be made between proportions and the absolute concentrations. The concentrations may be lowered by physical losses and the growth of plants, as well as by metabolism.

*Direct inhibitors*. Phosdrin, DDVP and Mipafox are the only direct inhibitors which have been shown to be decomposed in plants.

Both stereo-isomers of Phosdrin have been very fully studied on several plant species.[2] The roots of young pea plants were steeped for 12 hr in a 0·1% solution of the more toxic α-isomer, and washed, and the peas transplanted. The percentage of $^{32}P$ as α-Phosdrin fell rapidly and at an equal rate in every part of the plant analysed (Fig. 10.1). The rate of loss of $^{32}P$ from the plant (physical loss) and rate of decomposition depended on the plant species, and also upon which isomer was applied (Fig. 10.2). Similar results to those shown in Fig. 10.2 were confirmed on 10 other plant species. On the average the half life for the α-isomer was 0·86 days, and for the β-isomer 2·0 days. The half lives *in vitro* at the pH of the plant saps of the same species were (except for cucumber) 5–10 times greater. The breakdown in plants is, therefore, catalysed, probably enzymically, though it is just possible that general acid-base catalysis by salts in the sap could account for the results. A metabolite is present which yields acetone on hydrolysis, and which partitions about equally between water and chloroform. A product with these properties and the hydrogen phosphate also found could arise if the carbomethoxy group was the first group to be hydrolysed:

$$(MeO)_2PO\cdot O\cdot CMe{:}CH\cdot CO\cdot OMe \rightarrow \underset{\text{I}}{(MeO)_2PO\cdot O\cdot CMe{:}CH\cdot CO\cdot OH} + MeOH,$$

$$\text{I} \rightarrow (MeO)_2PO\cdot OH + MeCO\cdot Me + CO_2.$$

This is probably the route by which most of the compound is degraded, but some may be degraded either by oxidation of the olefinic group, or by direct hydrolysis to a hydrogen phosphate.

The decomposition of DDVP in pea plants is also catalysed, as is the decomposition of Dipterex and $(MeO)_2PO\cdot CH(O\cdot CO\cdot Me)\cdot CCl_3$, its O-acetyl derivative.[3] DDVP is the most stable, with a half life of about two days, and Dipterex the least, with a half life of under one day. Dipterex is, of course, an indirect inhibitor, and its acetyl derivative is probably

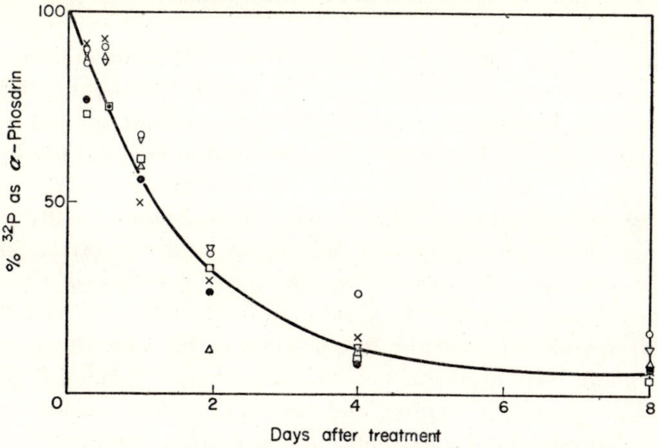


FIG. 10.1. Decomposition of α-Phosdrin in pea plants.[2]
The percentage of the $^{32}P$ as α-Phosdrin is given for roots, ×; seed, ○; stem, Δ; basal leaves, ∇; upper leaves, □; and tips, ●.

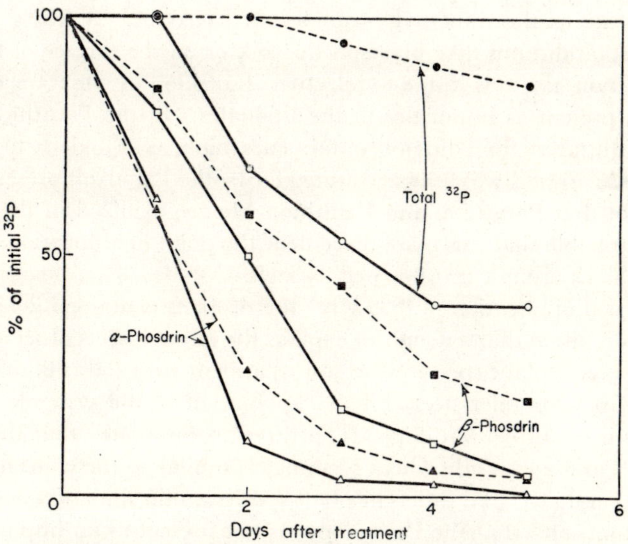


FIG. 10.2. Loss of α- and β-Phosdrin from bean and pea plants.[2]
The $^{32}P$ remaining, and the α- or β-Phosdrin are expressed as percentages of the $^{32}P$ in the plant immediately after treatment. Peas: $^{32}P$, ●; α-Phosdrin, ▲; β-Phosdrin, ■; beans: $^{32}P$, ○; α-Phosdrin, Δ; β-Phosdrin, □.

one also (p. 222). The products were not identified, so the reactions taking place are not known.

Mipafox has been studied in 5 plant species.[4] Its rate of decomposition to ionic products varies somewhat with the species, but about half is decomposed in a day. Its structure: $(Pr^iNH)_2PO \cdot F$, rules out any but hydrolytic attack. This may not be enzymic, for the hydrolysis is subject to general acid-base catalysis (pp. 75–6).

*Thionates*. Simple thionates, which do not contain other, readily oxidizable groups, are almost all deep-acting insecticides, and it is rarely possible to distinguish between reactions taking place on the surface of the leaf and biochemical reactions inside it. Oxidation inside plants of six compounds to the corresponding phosphates has, however, been demonstrated.

Wheat plants were grown in soil containing highly purified Parathion or Parathion-Me, and the guttation fluid was collected from the leaves.[5] This contained powerful inhibitors. The inhibitor from the Parathion-treated seedlings had the same hydrolysis rate as Paraoxon (assay by AChE inhibition) and rat ΨChE inhibited by the compound recovered its activity at the same rate as the diethylphosphorylated enzyme. Similarly, the inhibitor from seedlings treated with Parathion-Me was hydrolysed at the same rate as Paraoxon-Me, and rabbit AChE inhibited with it recovered its activity at the same rate as the dimethylphosphorylated enzyme. It was shown that this activation did not take place on the soil, or on the surface of the leaves after guttation, nor was it due to selective absorption by the roots of oxygen analogues present as impurities in the thionates. In the Parathion experiments the guttation fluid did not contain Parathion, as its toxicity to mosquito larvae (*Aedes aegypti*) was fully accounted for by the Paraoxon present. There is no doubt that Parathion and Parathion-Me are oxidized in these plants, but it is probable that this takes place near the point of absorption, and that the Parathions are not translocated as such.

After a soil application of 0·32 $g/m^2$ the droplets contained 20–40 p.p.m. of Paraoxon, more than enough to explain the kill of insects observed on the leaves of several plant species after soil treatment with Parathion[6] (see the original paper for references). This explanation of the systemic effects of Parathion is plausible, but depends on three assumptions: that the concentration of Paraoxon inside wheat seedlings is similar to the concentration in the guttation fluid; that different plant species attain similar concentrations of Paraoxon; and that the Paraoxon was not present as an impurity in the Parathion used. The last assumption is probably unjustified. Commercial Parathion gave higher concentrations of Paraoxon in the guttation fluid from wheat seedlings, i.e. the roots absorbed Paraoxon selectively. In most earlier experiments commercial Parathion was used. The results cannot, therefore, be held to show that other plants oxidize Parathion as rapidly as wheat seedlings do.

Dimethoate is taken up by the roots of several plant species, and a compound with the same partitioning properties and anti-ChE activity as its oxygen analogue, $(MeO)_2PO \cdot S \cdot CH_2 \cdot CO \cdot NHMe$, can be extracted from the foliage. The oxygen analogue is less stable in plants than Dimethoate itself.(7) Dimethoate is also degraded to hydrolysis products, the half life varying from two to five days with the plant species.(8)

*Cis*- and *trans*-Delnav and the related dioxene compound: O,O-diethyl S-2-(di-4-oxenyl) phosphorodithioate,

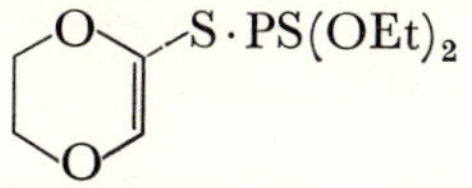

also give more powerful anti-ChE's in beans and cabbages.(9) The partitioning properties are consistent with the assumption that they are the corresponding phosphate analogues. In plants all three are decomposed to ionic products. The degradation obeys approximately first order kinetics and the half lives are as shown in Table 10.1. The rates are markedly dependent on species.

TABLE 10.1. HALF LIVES OF *cis*- AND *trans*-DELNAV AND ITS DIOXENE DERIVATIVE IN SEVERAL PLANT SPECIES(9)

| Compound | Crop | | |
|---|---|---|---|
| | Cotton | Bean | Tomato |
| *cis*-Delnav | 55 | 26 | 29 days |
| *trans*-Delnav | 50 | 25 | 55 |
| Dioxene derivative | 2·3 | 1·7 | 2·2 |

Part of Table by J. E. Casida and M. K. Ahmed. *J. econ. Ent.* **52,** 111 (1959) (with acknowledgements).

*Compounds containing sulphide groups*. Those sulphides which are not also thionates are considered first. Demeton-S and Demeton-S-Me have been studied in detail. A foliar application is rapidly absorbed and metabolized, so that within a few days only 1–5% of the original compounds remain. The foliage, however, remains toxic to phytophagous insects for several weeks.

Extraction with chloroform and then with butanol of plants treated with Demeton-S reveals the presence of three toxic products.(10) The two extracted by chloroform are the sulphoxide and sulphone,

$(EtO)_2PO \cdot S \cdot CH_2 \cdot CH_2 \cdot SO \cdot Et$ and $(EtO)_2PO \cdot S \cdot CH_2 \cdot CH_2 \cdot SO_2 \cdot Et$, as shown by the identity of their partition coefficients, hydrolysis rates and i-r spectra with those of the synthetic compounds.(10,11) The third product has only been found once,[10] in very small amounts, in the butanol extract of 30 kg of sugar-beet leaves. It was not extractable by chloroform, and was probably the sulphonium derivative,[12] $(EtO)_2PO \cdot S \cdot CH_2 \cdot CH_2 \cdot \overset{\oplus}{S}Et_2$. It was probably formed chemically on the leaf surface, and not enzymically.* In addition to the two toxic metabolites, the sulphoxide and sulphone, there

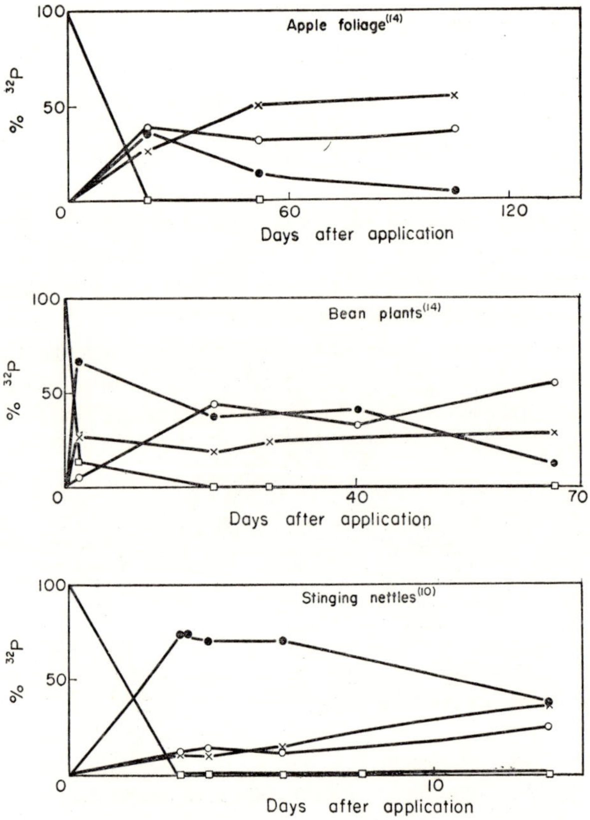


FIG. 10.3. Metabolites of Demeton-S in plants.

Amounts are expressed as percentages of the total $^{32}P$ in the plants. Demeton-S, □; its sulphoxide, ●; its sulphone, ○; and ions, ×.

are non-toxic ionic products. The percentage of $^{32}P$ remaining in three plant species as Demeton-S, sulphoxide, sulphone, and hydrolysis products at different times after treatment with Demeton-S are shown in Fig. 10.3.

* The sugar beet were sprayed with a higher concentration of Demeton-S than would be used in normal practice to obtain enough of each metabolite for identification. Under these circumstances some sulphonium derivative is likely to be produced (p. 58). A homologue has slight systemic properties.[13]

As the sulphoxide is oxidized to the sulphone,[11] the main sequence of reactions is:

Demeton-S $\rightarrow$ sulphoxide $\rightarrow$ sulphone $\rightarrow$ hydrolysis products.

The curves, however, suggest that Demeton-S, and possibly its sulphoxide, are also hydrolysed directly. The metabolism of Demeton-S is always rapid. In apple foliage and bean plants the sulphone appears to be more stable than the sulphoxide,[14] whereas in cotton plants the sulphone is the less stable.[11]

The sulphoxide and sulphone are as toxic to phytophagous insects as Demeton-S, and they persist long enough in plants to explain the prolonged toxic effects of an application of Demeton-S. Demeton-S is very sparingly soluble in water, so that its systemic effects appeared anomalous. The metabolites are miscible with water in all proportions,[10] and are expected to show systemic properties.

Demeton-S-Me also gives its sulphoxide, its sulphone and hydrolysis products in plants.[15] The sulphone was separated from plant homogenates by counter-current extraction, crystallized, and shown to be identical with the synthetic compound. The identity of the sulphoxide, separated similarly, was shown by oxidizing it to the sulphone. Metabolism of Demeton-S-Me and its products is about as fast as metabolism of Demeton-S.

Several thionates with sulphide groups have been studied.

Demeton-O-Me is simultaneously and rapidly hydrolysed, isomerized to Demeton-S-Me, and converted to two sulphonium derivatives in water (p. 58). In plants, therefore, only Demeton-S-Me is present, and this is metabolized as already described.

Demeton-O does not isomerize or hydrolyse with the same rapidity. It is absorbed slower by leaves or fruits than Demeton-S, and is metabolized more slowly after absorption, so that several per cent of the compound applied may be found unchanged in the plants 2–3 days after application.[16, 17] Even so, metabolism is very rapid. The sulphoxide,

$$(EtO)_2PS \cdot O \cdot CH_2 \cdot CH_2 \cdot SO \cdot Et,$$

has been identified by chromatography and $I_{50}$. The chromatographic separation of this compound was especially good.[18] A mixture of toxic metabolites was also found, and non-toxic hydrolysis products. The toxic metabolites were identified by i-r spectroscopy and their partition coefficients as the sulphones of Demeton-O and its oxygen analogue, $(EtO)_2PO \cdot O \cdot CH_2 \cdot CH_2 \cdot SEt$.[19] All the possible reactions are given below, and toxic compounds for which there is clear evidence are in bold type. There is some evidence that the phosphate analogue of Demeton-O is present, but it may be formed on the leaf surface.

The metabolism of Disyston,[20, 21] its dimethyl homologue,[22] and Thimet[20, 21, 23, 24] has been worked out more completely as the metabolites are more readily separated chromatographically.

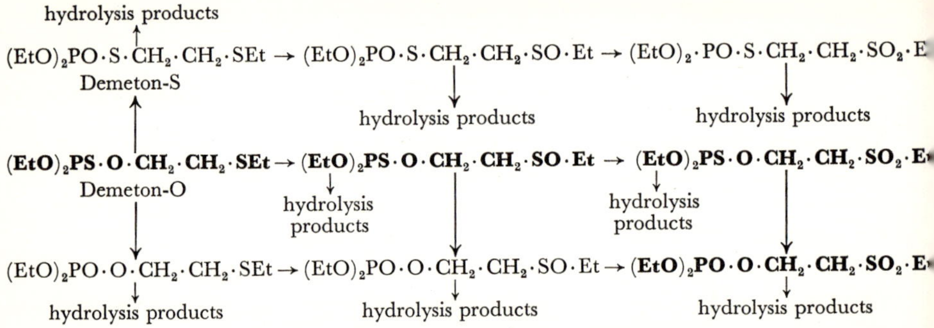


Disyston is $(EtO)_2PS \cdot S \cdot CH_2 \cdot CH_2 \cdot SEt$, O,O-diethyl S-2-ethylthioethyl phosphorodithioate and Thimet is the ethylthiomethyl analogue, $(EtO)_2PS \cdot S \cdot CH_2 \cdot SEt$. Thimet gives four metabolites, identified by chromatographic properties and infra-red spectra as the sulphoxides and sulphones of Thimet and its oxygen analogue, $(EtO)_2PO \cdot S \cdot CH_2 \cdot SEt$. According to chromatographic evidence, Disyston and its dimethyl homologue give four analogous products each. The proportions of the $^{32}P$ remaining in the leaves of cotton plants as these metabolites and hydrolysis products after stem treatment with $^{32}P$-labelled Thimet and Disyston are shown in Figs. 10.4 and 10.5. The parent compounds are fairly persistent on the surface of the leaves, but are metabolized rapidly once they have penetrated. Disyston is completely metabolized within an hour. The results obtained with Disyston can be subjected to an approximate kinetic analysis. Very rapidly indeed on entering the plant Disyston is oxidized to its sulphoxide and its oxygen analogue, and is 'hydrolysed' to ionic products,* the products being formed in the ratios: 75:10:15. The sulphoxide of Disyston is then oxidized slowly to its sulphone, and to the sulphoxide of its oxygen analogue, which is oxidized in turn to its sulphone. Probably most ionic products are formed by hydrolysis of the sulphones. The results obtained with Thimet are probably consistent with a similar sequence of reactions, except that ionic products are not formed rapidly from Thimet itself.* The metabolism of Disyston sulphoxide obeys first order kinetics with an energy of activation of 10 kcal/mole, i.e. the rate increases rapidly with temperature.[20]

The compound, O,O-diethyl O-4-methylthiophenyl phosphorothionate and its sulphoxide have also been studied in plants.[25] Chromatographic evidence suggests that the main plant metabolite is the S-ethyl isomer,

* Probably by the reaction:

$$(EtO)_2PS \cdot S \cdot CH_2 \cdot CH_2 \cdot SEt \rightarrow (EtO)_2PS_2^{\ominus} + \left.\begin{matrix} CH_2 \\ | \\ CH_2 \end{matrix}\right\rangle \overset{\oplus}{S}Et,$$

by analogy with Demeton-O-Me. Thimet cannot react like this.

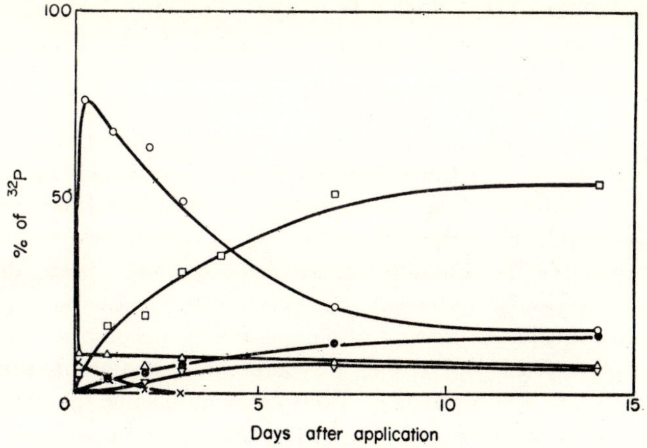


FIG. 10.4. Metabolites of Thimet in cotton leaves.[21]

The amounts of each compound are expressed as percentages of the total $^{32}P$ in the plants. $(EtO)_2PS\cdot S\cdot CH_2\cdot SEt$ (Thimet), ×; $(EtO)_2PS\cdot S\cdot CH_2\cdot SO\cdot Et$, ○; $(EtO)_2PS\cdot S\cdot CH_2\cdot SO_2\cdot Et$, ●; $(EtO)_2PO\cdot S\cdot CH_2\cdot SO\cdot Et$, Δ; $(EtO)_2PO\cdot S\cdot CH_2\cdot SO_2\cdot Et$, ∇; and ions, □.

After R. L. Metcalf, T. R. Fukuto and R. B. March. *J. econ. Ent.* **50,** 338 (1957) (with acknowledgements).

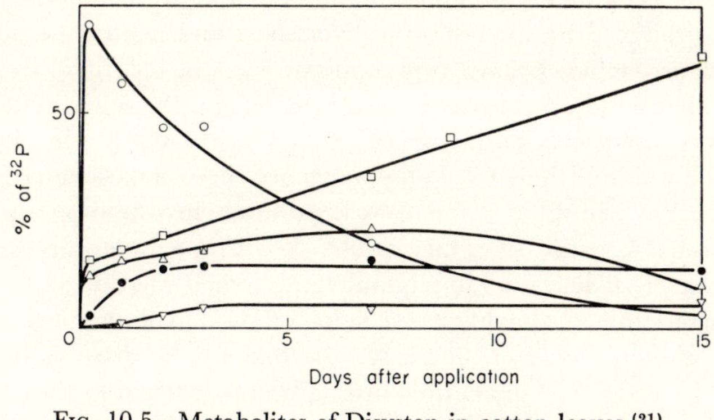


FIG. 10.5. Metabolites of Disyston in cotton leaves.[21]

$(EtO)_2PS\cdot S\cdot CH_2\cdot CH_2\cdot SO\cdot Et$, ○; $(EtO)_2PS\cdot S\cdot CH_2\cdot CH_2\cdot SO_2\cdot Et$, ●; $(EtO)_2PO\cdot S\cdot CH_2\cdot CH_2\cdot SO\cdot Et$, Δ; $(EtO)_2PO\cdot S\cdot CH_2\cdot CH_2\cdot SO_2\cdot Et$, ∇; and ions, □.

The percentage of $^{32}P$ as Disyston itself is too low to record within one hour of absorption.

After R. L. Metcalf, T. R. Fukuto and R. B. March. *J. econ. Ent.* **50,** 338 (1957) (with acknowledgements).

$(EtO)(EtS)PO \cdot O \cdot C_6H_4$-4-SMe. It is more probable, however, that the metabolite is some other compound which has, by coincidence, similar partition coefficients.

*Phosphoramidic compounds.* Schradan is the compound most investigated. In several plant species the rate of fall in concentration of Schradan is the same,[26] but varies with the season (in England) in which the plants are treated. The half lives* are about 8 days in May and June, 14 in July, 22 in August and September, and too long to determine (> 50 days) in winter. Some species, notably chrysanthemums and *Coleus*, decompose Schradan more slowly under summer conditions.[27] In most species the drop in concentration is brought about mainly by decomposition, the rate of which is increased by increasing the temperature or the intensity of light to which the plant is subjected.[27] Pre-treatment of some species with high light intensity leads to an increase in the rate of decomposition which is maintained when the light is withdrawn.

The main decomposition products are ions. From a plant treated with $^{32}$P-Schradan much of the ionic $^{32}$P can be precipitated with $Ca^{2\oplus}$. Likely ionic products are orthophosphate, $Me_2N \cdot PO_3{}^{2\ominus}$ and $(Me_2N)_2PO_2{}^{\ominus}$. The last has a soluble calcium salt, and the second is unstable in water, so probably Schradan is degraded to orthophosphoric acid.[4]

Homogenates of plants treated with Schradan inhibit AChE more rapidly than the Schradan present can.†[28, 29] Chromatographic or counter-current separation of chloroform extracts of treated plants reveals two non-ionic metabolites, heptamethylpyrophosphoramide and a metabolite identical with that produced by the action of liver slices and neutral permanganate on Schradan. Heptamethylpyrophosphoramide was identified by its partition coefficients and alkaline hydrolysis rate,[10] and the other metabolite by its partition coefficients and rates of reaction with AChE and chymotrypsin.[30] The yields of these two compounds are very small—under 7% of the $^{32}$P absorbed as Schradan is found as heptamethylpyrophosphoramide and about 0·01–0·2% as the other metabolite. For a time some workers believed that the toxicity of Schradan to phytophagous insects was due to the presence of this metabolite in the plant, in spite of the small quantities present. Schradan is not, however, increased in toxicity by passage through a plant;[31] its toxicity to susceptible insects is similar whether they take it up by feeding on a plant[32] or through the cuticle;[33] and the toxicity to mammals of homogenates of treated plants is fully accounted for by the

* The concentrations fall linearly with time (within the fairly large biological variations) so that the concentration should become zero in two half lives. This is not quite correct in practice, as low concentrations, usually under 1 p.p.m. may remain for some time after this.

† This effect cannot always be observed. Strawberries (fruits) contain an enzyme which hydrolyses ACh, and whose presence obscures any inhibition of added AChE.[4]

Schradan present.[30, 34] As it is also known that both insects and mammals can oxidize Schradan to its toxic metabolite, this theory has been dropped.

Decamethyltriphosphoramide, $(Me_2N)_2PO \cdot O \cdot PO(NMe_2)O \cdot PO(NMe_2)_2$ is decomposed in crops to ionic products at about the same rate as Schradan.[26] Nothing is known of possible non-ionic metabolites. Dimefox appears to be decomposed somewhat faster, but it is lost from foliage by evaporation in the transpiration stream, which makes analysis of the results difficult.[35] Traces

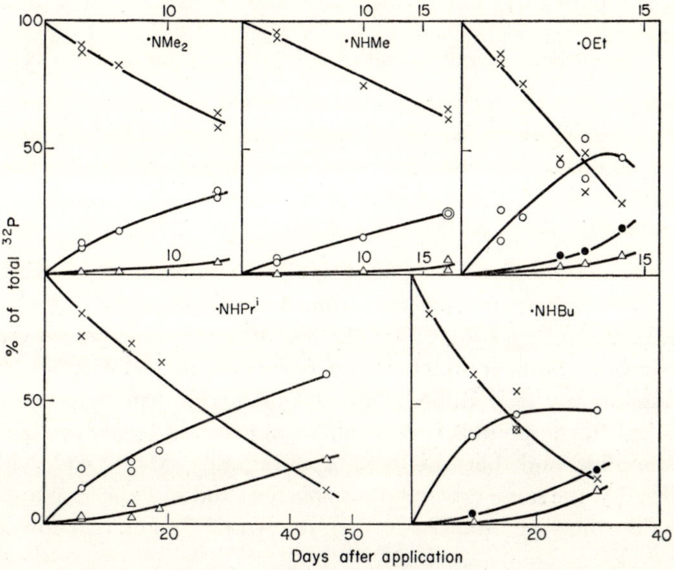


FIG. 10.6. Metabolism of some tetramethylphosphorodiamidic compounds in turnip foliage.

Parent compound, ×; first metabolite, ○; second metabolite, ●; compounds which cannot be extracted from water by chloroform, Δ. Each is given as a percentage of the total $^{32}P$, which fell little or not at all during the experiments.

After D. F. Heath, D. W. J. Lane and P. O. Park. *Phil. Trans. Roy. Soc.* **239B,** 191 (1955) (with acknowledgements).

of a compound with the partition coefficients of trimethylphosphorodiamidic fluoride have been separated by chromatography from chloroform extracts of treated plants.[36, 37]

The probable mechanism of degradation of Schradan and other phosphorodiamidic compounds can be deduced from the products obtained in plants treated with simple non-toxic phosphorodiamidic and triamidic compounds.[10] Turnips were sprayed with the following $^{32}P$-tetramethylphosphorodiamidic compounds:

$$(Me_2N)_2PO—NMe_2, —NHMe, —NHPr^i, —NHBu \text{ and } —OEt.$$

The foliage was taken up at various times after spraying and analysed by a complex partitioning process for the parent compounds and their metabolites. At least one chloroform extractable metabolite was found in each case, and also compounds which could not be extracted by chloroform. The latter will be denoted by N. The percentage of $^{32}P$ in the foliage as each compound is shown in Fig. 10.6, and the partition coefficient of each metabolite in Table 10.2. The N's are excluded from this table, as the coefficients were all less than 0·1. The results are wholly consistent with the reaction scheme:

Parent compound → metabolite I → metabolite II → N

Each step leads to the formation of a compound with a partition coefficient about 10 times more in favour of water than that of the compound from which it is formed. This change in coefficient would result from either of the reactions:

$$—NMe_2 \rightarrow —NHMe, \text{ or } —NMe_2 \rightarrow —NMe{\cdot}CH_2{\cdot}OH.$$

For reasons already given (p. 232) a hydroxymethylamide is likely to lose formaldehyde rapidly to give the primary methylamide. The changes in coefficients are not consistent with the loss of butyl or iso-propyl groups, or with hydrolysis, both of which would decrease the coefficients very much more. Apparently only dimethylamino and perhaps methylamino groups are attacked in compounds which contain no readily hydrolyseable groups. Now Schradan and hexamethylphosphoramide, $(Me_2N)_2PO{\cdot}NMe_2$, are metabolized at the same rate in both sugar beet and strawberry plants, which suggests that they are attacked in the same way.[38] The reaction scheme:

$$X—P—NMe_2 \rightarrow$$
$$X—P—NO(Me)_2 \rightarrow X—P—NMe{\cdot}CH_2{\cdot}OH \rightarrow X—P—NHMe$$
$$\downarrow H_2O$$
$$XH + \text{ionic P}$$

provides an explanation of the following observations.

(1) The metabolism of Schradan and hexamethylphosphoramide proceeds at the same rate.

(2) Schradan only gives a small yield of heptamethylpyrophosphoramide, although hexamethylphosphoramide is metabolized entirely via pentamethylphosphoramide. The amide oxide group may make the pyrophosphate bond in Schradan unstable in water, so that hydrolysis is faster than isomerization to the hydroxymethyl compound. The amide oxide group cannot have a similar effect on hexamethylphosphoramide.

(3) The ratio in the plant of the concentrations of metabolite and heptamethylpyrophosphoramide are about what one would expect from their relative stabilities if the latter is produced from the former.

TABLE 10.2. PARTITION COEFFICIENTS OF TETRAMETHYLPHOSPHORODIAMIDIC COMPOUNDS IN TURNIPS[10]

The partition coefficients determined on the compounds extracted from plants are given, and also those of the pure parent compounds at 18°.

| Compound | Solvent system | Parent compound from plant | Metabolite I | Metabolite II | Pure parent compound |
|---|---|---|---|---|---|
| $—PO(NMe_2)_2$ | | | | | |
| $Me_2N—$ | $CHCl_3/H_2O$ | 5·7–6·3 | 0·35–0·40 | — | 6·6 |
| MeNH— | $CHCl_3/H_2O$ | — | — | — | 0·43 |
| | $CHCl_3$/2N NaOH | 2·5 | 0·25 | — | 2·6 |
| $Pr^iNH—$ | $CHCl_3/H_2O$ | 4·0 | 0·5 | — | 3·6 |
| BuNH— | $CHCl_3/H_2O$ | 21 | 2·0–2·5 | 0·32 | 21·6 |
| EtO— | $C_6H_6$/N NaOH | 0·71–0·95 | 0·05–0·15 | 0 | 0·81 |
| | $CHCl_3$/2N NaOH | — | 11·0 | 1·0 | > 50 |

From Table by D. F. Heath, D. W. J. Lane and P. O. Park. *Phil. Trans. Roy. Soc.* **239B,** 191 (1955) (with acknowledgements).

## SOIL

Isolated soil bacteria metabolize Thimet and its homologues, their phosphate analogues, Parathion-Me, Chlorthion, Trolene, Schradan and Dimefox.[39] Thimet is converted to its sulphoxide, which is further oxidized to its phosphate analogue, but the main reaction in every case is hydrolysis, the sulphides hydrolysing more rapidly than the sulphoxides or sulphones. Under the conditions of the experiment about half of every compound listed was hydrolysed in eight days. Rapid oxidation of Demeton-S to its sulphoxide has been demonstrated on soil, but does not take place on sand.[14]

Dimefox and an analogue are degraded to ionic products by natural soils (clay, sand and peat). The rates of decomposition were equal to the hydrolysis rates calculated from the soil pH—i.e. hydrolysis was not catalysed enzymically.[37]

## GENERAL CONCLUSIONS AND DISCUSSION

Plant enzymes do not catalyse the hydrolysis of phosphorus compounds very greatly but they do catalyse their oxidation. The oxidation products consequently accumulate more in plants than in animals, so that it has usually been possible to identify them with reasonable certainty. They are the same as those produced by animals, as far as can be judged from the less complete identifications of animal metabolites.

Oxidation increases both the water-solubility and the anti-ChE activity. High water solubility is necessary if a compound is to be readily translocated. Oxidation therefore converts several deep-acting insecticides into systemic ones.

As insects and mammals can oxidize the parent compounds, oxidation in plants never increases the toxicity of an application very much, though of course, untreated parts of a plant may become toxic through systemic action.

Insecticides can be divided into those which are systemic in themselves, and those which are only systemic after oxidation. Thimet, Disyston and the Demetons are in the second group, but all their oxidation products and phosphoramidic compounds are in the first. Probably most thionates are oxidized to some extent, like Parathion, to the corresponding phosphates, and these may be weakly systemic.

## REFERENCES

1. E. Kramm. *Z. PflKrankh.* **60,** 20 (1953)
2. J. E. Casida, P. E. Gatterdam, L. W. Getzin, Jr. and R. K. Chapman. *J. Agric. Food Chem.* **4,** 236 (1956)
3. B. W. Arthur and J. E. Casida. *J. Agric. Food Chem.* **5,** 186 (1957)
4. D. F. Heath, unpublished.
5. W. A. L. David and W. N. Aldridge. *Ann. appl. Biol.* **45,** 332 (1957)

6. M. I. ZEID and L. K. CUTKOMP. *J. econ. Ent.* **44,** 898 (1951)
7. R. SANTI and P. DE PIETRI-TONELLI. *Nature, Lond.* **183,** 398 (1959)
8. E. D. CHILWELL and P. T. BEECHAM. *J. Sci. Fd. Agric.* **11,** 400 (1960)
9. J. E. CASIDA and M. K. AHMED. *J. econ. Ent.* **52,** 111 (1959)
10. D. F. HEATH, D. W. J. LANE and P. O. PARK. *Phil. Trans. Roy. Soc.* **239B,** 191 (1955)
11. T. R. FUKUTO, J. P. WOLF, 3rd, R. L. METCALF and R. B. MARCH. *J. econ. Ent.* **49,** 147 (1956)
12. D. F. HEATH and M. VANDEKAR. *Biochem. J.* **67,** 187 (1957)
13. R. L. METCALF, T. R. FUKUTO, R. B. MARCH and E. M. STAFFORD. *J. econ. Ent.* **49,** 738 (1956)
14. W. D. E. THOMAS, S. H. BENNETT and C. P. LLOYD-JONES. *Ann. appl. Biol.* **43,** 569 (1955)
15. R. MÜHLMANN and H. TIETZ. *Höfchenbriefe* 116 (1956)
16. R. L. METCALF, R. B. MARCH, T. R. FUKUTO and M. G. MAXON. *J. econ. Ent.* **47,** 1045 (1954)
17. R. L. METCALF, R. B. MARCH, T. R. FUKUTO and M. G. MAXON. *J. econ. Ent.* **48,** 364 (1955)
18. T. R. FUKUTO, R. L. METCALF, R. B. MARCH and M. G. MAXON. *J. econ. Ent.* **48,** 347 (1955)
19. T. R. FUKUTO, J. P. WOLF, 3rd, R. L. METCALF and R. B. MARCH. *J. econ. Ent.* **50,** 399 (1957)
20. R. L. METCALF, H. T. REYNOLDS, M. WINTON and T. R. FUKUTO. *J. econ. Ent.* **52,** 435 (1959)
21. R. L. METCALF, T. R. FUKUTO and R. B. MARCH. *J. econ. Ent.* **50,** 338 (1957)
22. O. JUCKER. *Mitt. Gebiete Lebensmitt. u. Hyg.* **49,** 299 (1958)
23. J. S. BOWMAN and J. E. CASIDA. *J. econ. Ent.* **51,** 838 (1958)
24. J. S. BOWMAN and J. E. CASIDA. *J. Agric. Food Chem.* **5,** 192 (1957)
25. E. BENJAMINI, R. L. METCALF and T. R. FUKUTO. *J. econ. Ent.* **52,** 99 (1959)
26. D. F. HEATH, D. W. J. LANE and M. LLEWELLYN. *J. Sci. Fd. Agric.* **3,** 60 (1952)
27. R. F. BATT, S. H. BENNETT and W. D. E. THOMAS. *Ann. appl. Biol.* **41,** 475, 484, 501 (1954)
28. S. A. HALL, J. W. STOHLMAN, 3rd, and M. S. SCHECHTER. *Analyt. Chem.* **23,** 1866 (1951)
29. J. E. CASIDA, R. K. CHAPMAN and T. C. ALLEN. *J. econ. Ent.* **45,** 568 (1952)
30. J. E. CASIDA, R. K. CHAPMAN, M. A. STAHMANN and T. C. ALLEN. *J. econ. Ent.* **47,** 64 (1954)
31. E. E. IVY, W. IGLINSKY and C. F. RAINWATER. *J. econ. Ent.* **43,** 620 (1950)
32. W. A. L. DAVID. *Ann. appl. Biol.* **38,** 508 (1951)
33. H. R. KRUEGER and J. E. CASIDA. *J. econ. Ent.* **50,** 356 (1957)
34. A. DAVID, G. S. HARTLEY, D. F. HEATH and F. W. POUND. *J. Sci. Fd. Agric.* **2,** 310 (1951)
35. P. O. PARK, personal communication.
36. B. W. ARTHUR and J. E. CASIDA. *J. econ. Ent.* **51,** 49 (1958)
37. D. F. HEATH and P. O. PARK, unpublished.
38. D. F. HEATH, D. W. J. LANE and M. LLEWELLYN. *J. Sci. Fd. Agric.* **3,** 69 (1952)
39. M. K. AHMED and J. E. CASIDA. *J. econ. Ent.* **51,** 59 (1958)
40. P. O. PARK and D. F. HEATH. *Phytiat Phytopharm.* Special issue, 742 (1954)

Part III

# PHARMACOLOGY IN MAMMALS

CHAPTER XI

# Introductory Chapter

THE THREE subjects, the mechanism of nerve function, the pharmacology of phosphorus compounds and the pharmacology of reversible anti-AChE's have developed together, and each has influenced the interpretation of observations made in studies of the other two. They are inseparable, and should, ideally, be treated together. This is obviously impracticable—a book longer than the present one would be needed for an elementary treatment. The following procedure has therefore been adopted. In this Chapter a brief summary is given of some aspects of nerve function, of how nerve function is influenced by various drugs, and of the main experimental set-ups used in studying the pharmacology of phosphorus compounds, with particular reference to the types of figures in which results are presented and their interpretation. Very little is said of the sensory nervous system, which is not directly influenced by anti-AChE's. It will be assumed that the reader knows the gross anatomy of the nervous system, and the neurone doctrine and its terminology. The level of exposition is intentionally varied greatly. The sections on electrophysiology and pharmacological methods are very elementary, as only a few of the finer points of these subjects are relevant in the following Chapters, and these are best described as they arise. The action of drugs on end-plates and synapses, and the ACh theory of synaptic transmission are dealt with in more detail, as here detail is relevant.

The action of reversible anti-AChE's is described rather briefly. It resembles in many ways the action of phosphorus compounds, and to summarize it separately would lead to much repetition. The close resemblance is pointed out frequently in later Chapters, and is one of the strongest reasons for believing that both groups of compounds act by inhibiting AChE, as there is no other obvious way in which they are alike. It follows that the action of phosphorus compounds is best interpreted in terms of AChE inhibition whenever possible. Isolated researches have often led to other, most plausible theories, but I think it is a mistake in such cases to ignore the general trend of results in many different studies. It has been very noticeable that results which seemed paradoxical on the anti-AChE theory when first obtained have later been shown to fit in very well with it as knowledge has advanced.

## ELECTROPHYSIOLOGY

Axons and skeletal muscle fibres are bounded by insulating sheaths, inside which the cells are at a negative potential of about − 70 mV.[2] A transient electric shock partly depolarizes the fibres (or axons) for a short time by an amount increasing with the stimulus. When depolarization exceeds a critical amount (10–50 mV depending on the fibre) an action potential (spike) is initiated (Fig. 11.1A), which is propagated along the fibre's length. At the height of the spike the cell potential is positive (+ 10–40 mV). While a spike is being propagated the fibre cannot respond to a second stimulus. For this reason the axons to voluntary muscles can only carry about 1000 stimuli/sec, post-ganglionic fibres about 100, and muscle fibres 200–1000, depending on the muscle. Action potentials are best recorded with intracellular recording electrodes. Extracellular electrodes can be used, and then in physiological media single axons give records like Fig. 11.1B (diagrammatic).[4] Recordings from nerves, i.e. bundle of axons, are more complex, as the spikes travel at different velocities in axons with different diameters. A spike in a muscle fibre sets off a wave of contraction, which lasts about 100 msec. This response to one stimulus is the twitch response. When the spikes are more frequent than 10/sec the fibre does not relax completely between each contraction. When the contractions fuse completely, as they do above about 25 spikes/sec, the response is a tetanus, which is stronger than a twitch. When fusion is incomplete the response is a clonus. Twitch and tetanic responses are shown diagrammatically in Fig. 11.2.

For the mechanism of propagation of spike potentials and the variations in sensitivity of fibres after direct stimulation the reader is referred to other sources.[4, 5]

Stimulation of a motor neurone causes spikes to be propagated in the fibres it innervates, and stimulation of pre-ganglionic fibres causes spikes to be propagated in post-ganglionic fibres. This is indirect stimulation. The reverse stimulation does not take place—stimulating a muscle fibre does not cause spikes in an axon. The junction between a nerve ending and a muscle fibre (the myoneural junction) is similar in function to that between a pre- and post-ganglionic fibre in a ganglion (synapse). There is a difference, in that each muscle fibre is innervated by one neurone only, excitation of which always stimulates the fibre, whereas each post-ganglionic fibre is innervated by many pre-ganglionic fibres, several of which have to be excited nearly simultaneously to excite it. The events in the synaptic gap, are, however, similar in the two cases.

The events at the myoneural junction have been more fully studied, and are described here. When the end of an intracellular electrode is introduced into the muscle fibre just below the end-plate, small, random depolarizations,

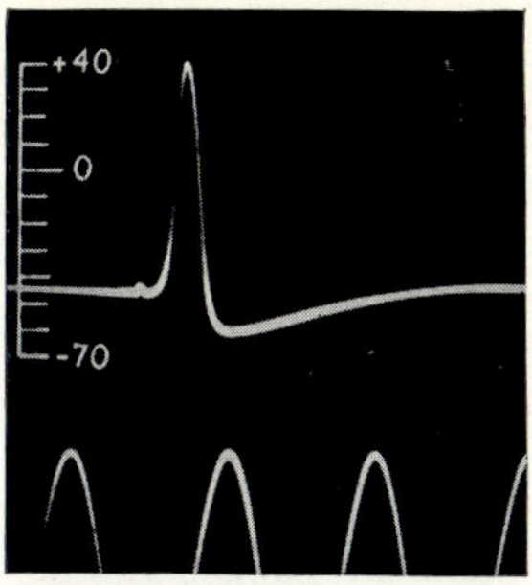


FIG. 11.1A. Action potential in the giant axon of the squid. The scale at the side shows the voltage in mV relative to the external potential. Time marker: 500 c/s.[3]

Reproduced from A. L. Hodgkin and A. F. Huxley. *J. Physiol.* **104,** 176 (1945) (with acknowledgements).

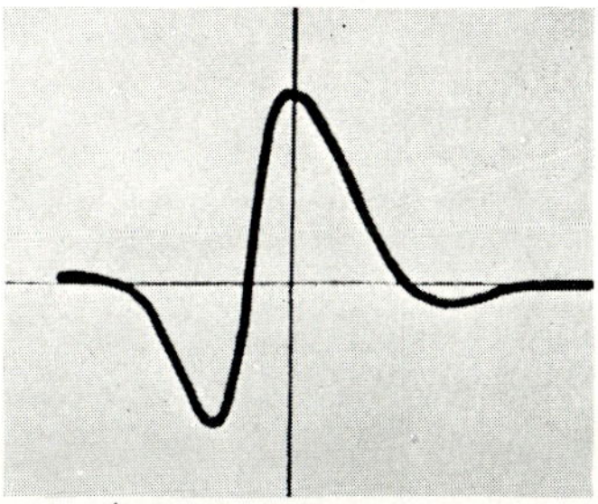

FIG. 11.1B. Record of an action potential taken on the surface of an axon in a conducting medium (diagrammatic).[4]

Reproduced from M. A. Brazier. *Electrical Activity of the Nervous System.* Pitman, London, 1953 (with acknowledgements).

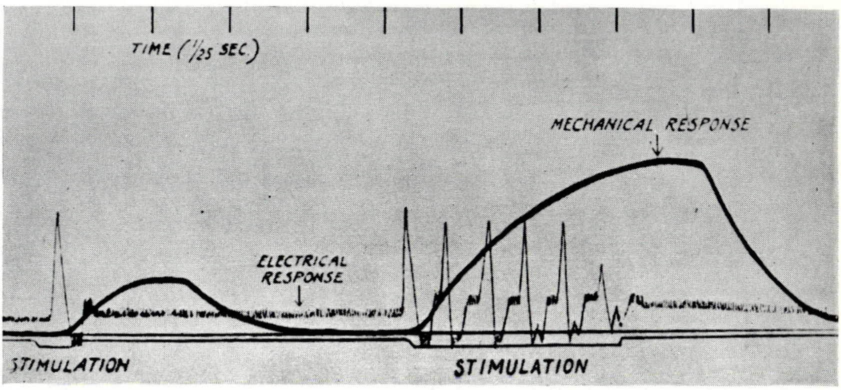


FIG. 11.2. Twitch and tetanic responses to single and rapid, repetitive indirect stimulation.[1]

Reproduced from C. Lovatt Evans. *Principles of Human Physiology*, 3rd Ed. Churchill, London, 1949 (with acknowledgements).

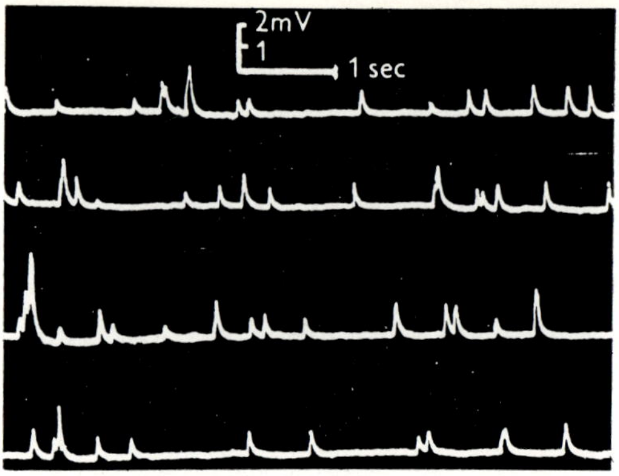


FIG. 11.3A. Miniature end-plate potentials in a frog muscle fibre. The muscle had been treated with $10^{-6}$ (w/w) prostigmine, which increased the size of the potentials.[6]

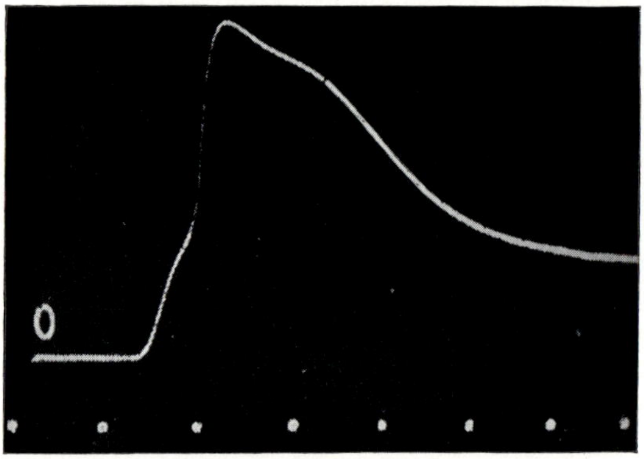


FIG. 11.3B. An action potential recorded intracellularly in a frog muscle fibre beneath the end-plate. Time marks, msec.[7]

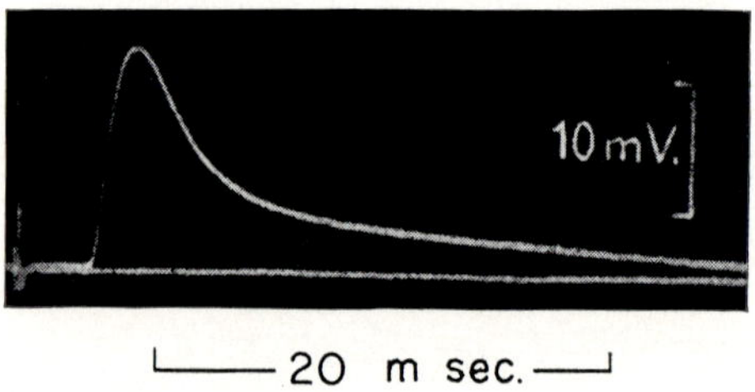


FIG. 11.3C. An end-plate potential recorded in a frog muscle fibre in a medium deficient in sodium.[7]

Reproduced from P. Fatt and B. Katz. *J. Physiol.* **117,** 109 (1952) (A) and *J. Physiol.* **115,** 320 (1951) (B and C) (with acknowledgements).

all of about 1 mV, are observed (Fig. 11.3A).[6] These are miniature end-plate potentials (miniature e.p.p.'s). They die away in under a millisecond and within a few millimetres of the end-plate. A spike in the axon causes a spike to start in the muscle from the end-plate region. Recorded under the end-plate it appears as in Fig. 11.3B, with an initial depolarization before the spike is propagated. This initial response is absent from records taken farther from the end-plate. In suitable media, which inhibit spike formation, it can be observed in isolation (Fig. 11.3C).[7] It is called the end-plate potential (e.p.p.). It dies away rapidly with time and distance from the end-plate, and is the trigger which sets off the spike. Each e.p.p. seems to consist of about 100 miniature e.p.p.'s set off simultaneously. This is the quantal response of the muscle fibre.[8, 9]

Artificial stimulation of a nerve excites all the axons in it simultaneously, so that all the fibres it serves respond almost simultaneously (not quite because some end-plates are further than others from the point of stimulation). In natural stimulation the impulses are not synchronous, so that a low rate of stimulation in any one axon is enough to maintain a state in which some fibres are always contracting, and the muscle as a whole sustains a tetanus although the fibres of individual motor units do not.

## THE FUNCTION OF ACh AND AChE

The transmission of an impulse from a nerve to a muscle fibre, or from a pre-ganglionic to a post-ganglionic fibre, with certain exceptions, takes place in the following way.[10, 11, 12] The excited nerve ending releases ACh, which migrates to the surface of the muscle or post-ganglionic neurone immediately beneath it, the post-synaptic membrane. It is there adsorbed on certain parts of the surface, called receptors. The adsorption causes transient depolarization of the post-synaptic membrane, i.e. it produces the e.p.p. and this sets off the spike. The ACh is rapidly hydrolysed by AChE. When most of it has been hydrolysed the membrane can become polarized again, and the whole process can be repeated. This theory is now generally held to be correct, although certain details are still discussed. The evidence for it is quite overwhelming.[11, 13, 14, 15, 16] Some results, which fit in well with the theory, and which only appear in the most recent reviews, are given here. Others appear in the section on drug action.

It has been shown that enough ACh is released by nerve endings to transmit impulses across the synaptic gaps. The ACh released per impulse from a nerve ending can be calculated from the rate of release of ACh during stimulation of the pre-ganglionic nerve or motor nerve. The ganglion or muscle is perfused with a dilute solution of eserine, which inhibits the AChE, and so preserves the ACh.[17, 18, 19, 20, 88] The ACh released per impulse in some experiments on rat diaphragm was about $1{\cdot}1 \times 10^{-17}$

moles,[88] an estimate which is more likely to be low than high. By introducing a micro-pipette into the gap in a myoneural junction, and extruding from it by an electrical discharge a minute amount of ACh very rapidly (ionophoretic injection) an e.p.p. was produced in the muscle fibre very like that induced by excitation of the nerve serving it. About $1{\cdot}5 \times 10^{-17}$ moles of ACh were needed,[21] a quantity in very good agreement in view of the experimental difficulties. When the ACh was released inside the muscle no e.p.p. was produced. This, with similar evidence using other drugs, shows that the receptors are not inside the muscle fibre, but must be outside it.

The appearance of myoneural junctions under the electron microscope is also consistent with pharmacological evidence. A reptilian myoneural junction is shown in Fig. 11.4. The junction in man is similar.[23] The post-synaptic membrane is folded, giving a large surface area, and the muscle and nerve membranes are only separated by a gap of about 100 Å. The axon contains many minute vesicles (possibly cross-sections of tubules) about 400 Å in diameter, and there is some evidence that these contain the ACh released. As the vesicles are all of about the same size the quantal response has an anatomical explanation, as each vesicle is likely to hold the same quantity of ACh.[24]

It is usually assumed that ACh is removed solely by AChE at synapses or myoneural junctions. At the latter the assumption is probably justified, as the synaptic gap is very narrow, so that diffusion out of it will be slow, but it has been shown that diffusion may be rapid enough in ganglia to remove the ACh, if certain assumptions are made.[20]

Two questions must now be answered. Does axonal conduction also require the mediation of ACh, i.e. is it released down the length of the axon during conduction? And what synapses (or neurones) depend on ACh as transmitter substance?

The answer to the first question is 'No'. Many results which lead to this answer have been reviewed,[15, 16] and others, which involve the use of phosphorus compounds, are given in the next Chapter.

ACh only mediates at some synapses. Those nerves which excrete ACh as the transmitter substance are called cholinergic nerves. The following nerves are cholinergic:

(1) All neurones serving voluntary muscles.

(2) All pre-ganglionic fibres.

(3) All post-ganglionic fibres in the parasympathetic system.

(4) Some post-ganglionic fibres in the sympathetic system—those to sweat glands and to some vascular beds.[32] The rest are adrenergic.

(5) Perhaps most in the central nervous system (CNS), but not all.[35] In some parts cholinergic and other neurones may alternate.[33, 34, 35] (Some internuncial fibres may be neither cholinergic nor adrenergic. They inhibit the action of the nerves they innervate,[40, 41] and may excrete an inhibitory substance.)

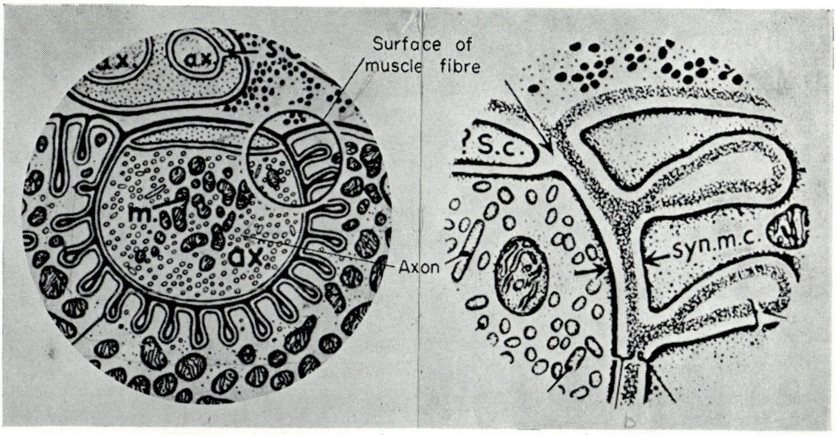


FIG. 11.4. A myoneural junction in the hind leg of the chameleon lizard (*Anolis carolinensis*). The figures are based on electron micrographs. The general appearance is shown to the left. The axon penetrates the muscle fibre, and there are many mitochondria (m). The enlargement on the right shows details of the synaptic membrane complex (syn. m. c.) and the many vesicles in the axoplasm.[22]

Reproduction of part of a figure by J. D. Robertson. *J. biophys. biochem. Cytol.* **2,** 381 (1956) (with acknowledgements).

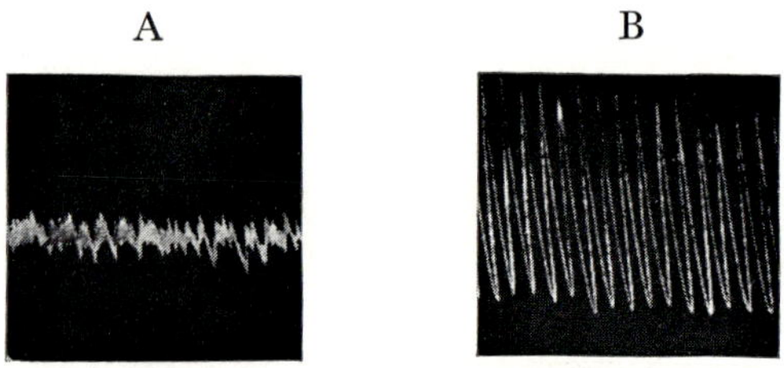


FIG. 11.6. Pendular (A) and peristaltic (B) motion of isolated rabbit ileum.[43] Reproduced from P. R. Salerno and J. M. Coon. *J. Pharmacol.* **95,** 240 (1949) (with acknowledgements).

There is only weak evidence that any sensory neurones are cholinergic.[36, 37, 38]

Cholinergic nerves possess the following properties. They contain a high concentration of AChE,* and choline-acetylase to synthesize ACh.[16, 26] The neurones or muscle fibres they innervate are excited by low doses of ACh and 'blocked' by high; and they become much more sensitive to ACh if they are denervated, i.e. if the cholinergic fibres serving them are cut and allowed to die. This sensitization is accompanied in ganglia,[28] but not always at myoneural junctions,[29] by loss of most of the AChE from the end-plate regions. Pre-ganglionic fibres carry the AChE for the end-plates they serve on their own endings, i.e. the AChE is pre-synaptic.[27, 28] The same is probably true in the CNS, but not at myoneural junctions of voluntary muscle, where the AChE is probably mainly post-synaptic, i.e. on the muscle membrane.[30] Many cholinergic nerves also contain vesicles, which probably hold ACh.[24, 31] Synapses mediated by ACh also respond in typical ways to anti-AChE's and other drugs, as described later.

## PHARMACOLOGICAL METHODS

Only the most important and least complex preparations are described. The aim is to help those who know no pharmacology.

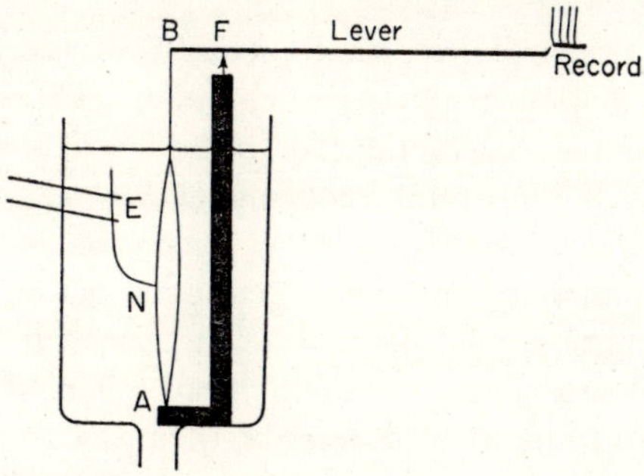


FIG. 11.5. Nerve-muscle preparation *in vitro*. The electrodes, E, are insulated from the bath fluid. For other details see text.

*Muscle preparations* in vitro. The basic set-up is shown in Fig. 11.5. The muscle is in a bath, so designed that fluids can be washed out quickly. It is

* This helps to explain why there is no close correlation between inhibition of end-plate AChE and toxic action.[27] The extracellular AChE can be inhibited by any anti-AChE, but the intracellular AChE only by inhibitors which penetrate the lipophilic axon sheath. This probably needs a combination of high oil-solubility and fair persistence. DFP has both, whereas Paraoxon is attacked by A-esterases rapidly in some animals and TEPP is of low lipoid-solubility. Thus the total inhibition of AChE by toxic doses is higher by DFP than by the others.

fixed at one end near the floor of the bath at A, and at the other to one end of a lever, L, at B. The lever rotates about its fulcrum F, and is constrained, to keep the muscle taut, either by a weight or a spring. The other end of the lever scratches a record on smoked paper at R. The paper moves past the stylus on drums at a controlled rate. Two lever systems are used, isotonic and isometric. In the isotonic system the muscle is held taut by a weight. The muscle thus contracts against constant force, and the tracing is proportional to the distance contracted. In the isometric system the muscle is attached to a strong spring, against which it can only move a small distance. The distance of movement recorded is then proportional to the force developed. This system is preferred with strong muscles. In nerve-muscle preparations the nerve, N, is attached to exciting electrodes, E, sometimes in oil.

Striated muscles are normally at rest *in vitro*, and many smooth muscles are normally in motion. The frog rectus abdominis and the leech dorsal muscle* are examples of striated muscles in which the nerves cannot be separately stimulated; the rat phrenic nerve-diaphragm, and frog sciatic nerve-gastrocnemius preparations are examples of nerve-muscle preparations. The isolated electroplax[42] is one innervated cell from the electric organ of the electric eel (*Electrophorus electricus*). It is essentially a nerve-muscle preparation, although it does not contract, and its responses are entirely electrical. The great size of the cell makes intracellular recording easy.

Smooth muscle preparations usually consist of pieces of intestine or duodenum. Their motion is extremely complex. The so-called pendular motion causes rhythmic contractions of rather variable amplitude to be recorded (Fig. 11.6A). Peristaltic movements also take place in pieces of intestine. Only strong peristalsis is clearly observable as such in records of tension variation, and then only when pendular motion is inhibited (Fig. 11.6B). 'Pendular motion' and 'peristalsis' are terms which are only strictly applicable to movements *in vivo*, but the motions observed *in vitro* are similar enough for the terms to be used to describe them.

Heart preparations are in continuous motion. Most heart preparations are not suspended in a bath of fluid, though some are perfused. The heart beat is recorded with a very light-weight lever attached either to a ventricle or the apex, or sometimes by direct visual counting.

*Muscle preparations* in vivo. These do not differ essentially from those *in vitro*. The tendon at one end of the muscle is separated from the bone, and attached to a lever. The original blood supply is retained. The bone to which the other end is attached is riveted to a firm base. The muscle is either bathed in oil to stop it drying out, or is kept perpetually moistened with a physiological medium. The nerve serving the muscle is normally

* Results on these preparations are usually regarded as relevant to studies on mammals, although the innervation and the action of drugs are in some ways different.[44]

teased away from the tissue so that it can be stimulated, and is also cut, so that central effects do not interfere. When reflexes are studied, however, the sensory nerve is often stimulated electrically, in order to provide a constant sensory impulse, and then the nerves are of course not separated from the CNS. Typical nerve-muscle preparations are the sciatic nerve-gastrocnemius muscle, the femoral nerve—sartorius muscle and superior-cervical ganglion —nictitating membrane of the cat. The nictitating membrane is a smooth muscle, which is innervated by the sympathetic system. It has no voluntary innervation.

*Close arterial injection.*(45) In this technique the side branches of the artery feeding the organ to be studied are tied off, and the needle of the injecting syringe is fixed in place in the artery so that the compound can be given in a rapid and reproducible way.

*Denervation and decentralization.* In denervation the nerve innervating an organ is cut, and the peripheral fibres are allowed to die. In decentralization the pre-ganglionic fibres are treated similarly.

The responses of these preparations to various drugs are described in the next section. More complicated preparations are described briefly at the relevant places.

## DRUG ACTION

Drugs may affect spike propagation and synaptic transmission in various ways, illustrated in Fig. 11.7. A motor neurone with one myoneural junction is shown. In terms of this discussion a pre-ganglionic fibre and its synapse with a post-ganglionic fibre are essentially similar.

Choline-acetylase is distributed throughout the neurone,(16) so that its inhibition could lead to a deficiency of ACh. In fact the axon holds enough ACh for several thousand synaptic events,(18, 46) so that the action of a choline-acetylase inhibitor would be slow. Some hemicholiniums may act in this way.(47)

Botulinum toxin appears to inhibit the release of ACh. Its action is described in more detail in the next Chapter.

The action of blocking and depolarizing agents* is best explained in terms of receptor theory. The post-synaptic membrane is believed to contain centres which are sterically and electrostatically very like the active centres of AChE, i.e. each centre consists of an anionic site and a site which attracts ester groups and is equivalent to the esteratic site of the enzyme. A molecule which 'fits' the anionic site particularly well brings about depolarization. The best groups are apparently $—\overset{\oplus}{N}Me_3$ or $>\overset{\oplus}{N}Me_2$.(48, 49) The adsorption, and hence the action, is increased if the compound is also attracted to the

* *Note added in proof:* W. D. M. Paton (*Proc. roy. Soc.* **154B,** 21 (1961)) has modified the theory of the action of such drugs in a way which eliminates most of the paradoxes from earlier theories, including that given here.

'esteratic' site, thus explaining the activity of ACh and its analogues.* A compound which is attracted to the anionic site and is adsorbed over it, but which does not fit it well, blocks access of ACh or other depolarizing drugs,

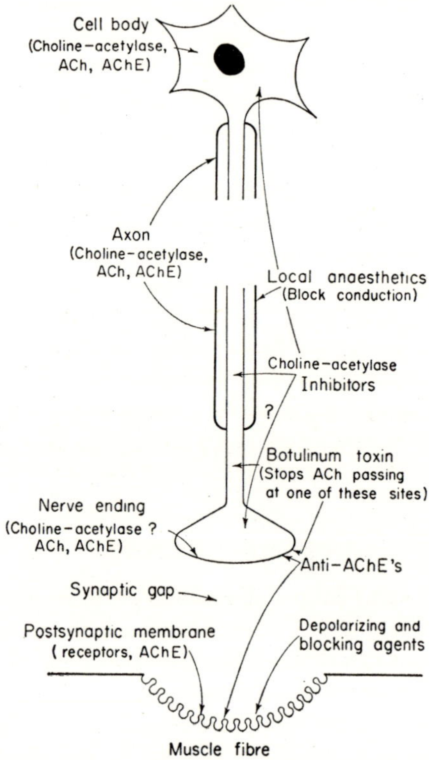


FIG. 11.7. Diagrammatic representation of a motor neurone and one myoneural junction to show the sites of drug action. The figure is not to scale. Some features are shown on the left-hand side, and the types of compound acting on the right.

without itself depolarizing. A drug can also block by being adsorbed on the 'esteratic' site.

Consider a blocking drug. Typical examples are D-tubocurarine and tetraethylammonium ion. They do not affect the electrical properties of the

* It has been suggested that the 'receptors' are AChE.[50] The action of phosphorus inhibitors is difficult to reconcile with this theory. They phosphorylate the esteratic site in AChE, and must therefore decrease the adsorption of ACh. As shown later they in fact greatly potentiate the action of ACh on the receptors. This argument has been opposed by the hypothesis that phosphorylation of the esteratic site sensitizes the anionic site enough to overcome the difference in adsorption. I shall assume that the receptors and the active centres of AChE are different, although they may be on the same protein. Other evidence seems to indicate that they are different.[51]

muscle fibre, which can therefore respond normally to direct stimulation. They block access of ACh to the receptors, so that an indirect stimulation no longer leads to contraction of the muscle, which remains relaxed. D-Tubocurarine competes with ACh for the active sites.(89) In agreement with this the e.p.p.'s are reduced in size, and also, although not so drastically, in duration. The only effect of a blocking drug is to depress transmission at the myoneural junction or synapse.(52, 53, 54) It is antagonized by ACh.

Typical depolarizing drugs acting at the myoneural junction are tetramethylammonium and decamethonium, $Me_3\overset{\oplus}{N}\cdot(CH_2)_{10}\cdot\overset{\oplus}{N}Me_3$.* Decamethonium has been studied in great detail. It acts at the end-plate, and depolarizes the fibre near it. A treated fibre therefore propagates a spike after direct stimulation as far as the end-plate, but not past it.(52, 55) Rapid depolarization of an end-plate by any means causes a fibre to propagate spikes. Depolarization by decamethonium is no exception; it causes fibres to contract, and potentiates the action of ACh until depolarization is so great that the fibre becomes inexcitable.(56) Thus a short period of muscle excitation is followed by a long period of relaxation. Some muscles, however, become adapted to depolarizing agents, and repolarize themselves, as shown in the last section of this Chapter.(57) The depolarizing agent still occupies the receptors, so that it now acts as a blocking agent. A depolarizing agent is not antagonized by ACh, which can only increase the depolarization, and inhibit transmission more completely. When D-tubocurarine is applied before a depolarizing agent the compounds are mutually antagonistic, but when it is applied after they exert an additive action, i.e. the muscle relaxation is increased. The differences in behaviour of decamethonium and D-tubocurarine at the myoneural junction are summarized in Table 11.1.

The actions of various drugs at myoneural junctions and ganglia are summarized in Table 11.2. Note that not all compounds containing $\cdot\overset{\oplus}{N}Me_3$ groups depolarize, and that decamethonium, which depolarizes myoneural junctions at low doses, blocks ganglia at high ones. Nicotine appears to act simultaneously as a blocking and as a depolarizing agent.

Atropine, which is not included in the table, blocks the receptors of muscles and glands served by the parasympathetic nervous system, probably by adsorption on the 'esteratic' site.

The action of ACh and the action of anti-AChE's are closely related, and are therefore described together.

* The methonium nomenclature is as follows. A compound:

$$Me_3\overset{\oplus}{N}\cdot(CH_2)_n\cdot\overset{\oplus}{N}Me_3$$

is called a methonium, with a latinate prefix according to the number, n, of carbons separating the two quaternary nitrogens.

Immersion of some striated muscles, notably the rectus abdominis of the frog, into a very dilute solution of ACh produces a prolonged contracture.(44) This 'slow phase' action is also observed in other amphibian muscles, and in the smooth muscles of mammals. It appears to be the result of local depolarization at the end-plates unaccompanied by propagated spike potentials.

TABLE 11.1. DECAMETHONIUM AND D-TUBOCURARINE AT MYONEURAL JUNCTIONS(55)

| | Decamethonium | D-Tubocurarine |
|---|---|---|
| 1. Resting potential | makes less negative | does not affect |
| 2. ACh | increases depolarization | antagonizes |
| 3. Anti-AChE's | increases effect | antagonizes |
| 4. D-Tubocurarine given first | antagonizes | — |
| 5. Decamethonium given first | — | acts additively |
| 6. Response of muscle to indirect tetanic stimulation | sustained contraction | no contraction |
| 7. Muscle selectivity in cats | relaxes limb more than respiratory muscles | relaxes respiratory more than limb muscles |
| 8. Species selectivity | shows great selectivity: cats and men are v. sensitive, rats under 0·01 times as sensitive | shows little selectivity |

TABLE 11.2. BLOCKING AND DEPOLARIZING AGENTS(42, 58)

| | Ganglion* | Myoneural junction |
|---|---|---|
| Blocking agents | D-tubocurarine,<br>pentamethonium,<br>hexamethonium,<br>tetraethylammonium<br>decamethonium (large doses)<br>nicotine | D-tubocurarine<br>pentamethonium (large doses)<br>hexamethonium (large doses) |
| Depolarizing agents | acetylcholine<br>nicotine<br>tetramethylammonium | acetylcholine<br>carbamoylcholine<br>nicotine<br>tetramethylammonium<br>decamethonium |

* Cat's superior cervical sympathetic ganglion.

The effect of anti-AChE's on these muscles is to sensitize them still more to ACh, and so to prolong the contracture. This is the basis of many ways of determining ACh.[46, 59, 60, 61]

Mammalian striated muscles do not contract on immersion in solutions of ACh, but they contract when ACh is given by close arterial injection. The contraction lasts under a second, and exceeds in strength the normal twitch response to a single, indirect stimulus.[45] It is accompanied by a burst of propagated action potentials in the muscle fibres, i.e. it is a brief tetanus.[62] While depolarization is at its height the muscle is less responsive than usual to indirect stimulation. The effects of anti-AChE's can be summarized under four heads.

(1) Anti-AChE's cause fibrillations in resting muscles. These appear as a rapid, unco-ordinated, weak flutter in a muscle when it is viewed as a whole. It is probable that they can be accounted for entirely by the effect of anti-AChE's on miniature e.p.p.'s, which are increased in size to such an extent that when a few happen to follow each other closely in a fibre they add up to an e.p.p., and cause it to contract.[6, 25]

(2) Anti-AChE's cause marked twitchings in 'resting' muscles, called fasciculations. A 'resting' muscle receives some random impulses from the CNS, and each impulse normally causes a very brief contraction of the fibres in the motor unit affected. Anti-AChE's increase the normal 'twitch' response to a brief tetanus, which is much more violent, and is observed as a fasciculation. The conversion of the normal response to a tetanic response can be explained in two ways. In both the primary cause is the prolongation of the e.p.p. resulting from the inhibition of AChE, as shown in Fig. 11.8A and B. On one theory prolonged e.p.p.'s automatically set up repetitive action potentials. On the other an action potential in a highly depolarized *fibre* can sometimes induce an action potential in the *nerve ending* serving it. (This cannot happen normally.) The action potential is propagated up the axon branch, and excites the whole neurone, and with it the whole motor unit of 100 fibres or more.[63] The process is called an axon reflex. On this theory a fasciculation can, of course, originate from a fibrillation as well as from an impulse from the CNS. I am not qualified to judge which of the two theories is correct. They do not seem to be mutually exclusive.

(3) Anti-AChE's greatly increase the response to low-frequency indirect stimulation. Each stimulus gives rise to a burst of action potentials in each fibre, so that a brief tetanus results (Fig. 11.9). The increase in response can be explained in the same ways as fasciculations.

(4) Anti-AChE's lower the response to high frequency tetanic stimulation. As the fibres only repolarize slowly the tails of the e.p.p.'s sum in a few stimuli, and the muscle becomes too depolarized to respond (Fig. 11.10A and B).[7, 64] The result is that the myogram shows only a single twitch.

D-Tubocurarine antagonizes the action of anti-AChE's by lowering both the voltage of e.p.p.'s and their duration. Thus a muscle paralysed by D-tubocurarine will respond to stimulation when treated with an anti-AChE; and if tetanus is inhibited by accumulation of ACh resulting from AChE inhibition D-tubocurarine will restore it. It is, however, obvious that the last effect is limited. The ACh persists longer and at a higher concentration than in the normal muscle, so the time course of the e.p.p.'s is necessarily slower than in a normal fibre.[7, 65] At very high rates of stimulation, therefore, the e.p.p.'s will still sum until the fibre can no longer respond, although a normal muscle would sustain a tetanus.

The treatment given has been very generalized. There are considerable quantitative differences from muscle to muscle, and some, for example, respond to a combination of prolonged rapid stimulation in the presence of an anti-AChE by giving a residual prolonged contracture, superposed on the initial brief tetanus.[44, 66]

Ganglia have also been studied. Perfusion of a low concentration of ACh through a ganglion releases a train of spikes in the post-ganglionic fibres.[67] Injection of ACh causes transient depolarization,[58] which is prolonged by eserine and causes a reduction in the number of post-ganglionic fibres responding to pre-ganglionic stimulation, i.e. the fibres become refractory. The evidence that anti-AChE's can also facilitate transmission at ganglia comes mainly from the study of phosphorus inhibitors, and is described in later Chapters.

The effect of ACh on the heart is inhibitory. Both the rate is slowed and the amplitude decreased. The effect is used for the estimation of ACh, the hearts of the molluscs *Venus mercenaria*[68] and *Mya arenaria*[69] being particularly sensitive. These responses are not greatly potentiated by anti-AChE's.

The main sites of drug action and the types of effect observed have now been reviewed. It is not, however, correct to assume that any one drug only affects one type of organ. Specific drugs are now listed, and their sites of action.

D-*Tubocurarine*. At normal doses this only blocks myoneural junctions. At higher doses, such as can be given when the effects are localized in some way, or are minimized by giving anti-AChE's, ganglia may also be blocked.[58] It affects the CNS when injected into it.[70]

*Atropine*. At low doses this blocks the receptors of organs served by the parasympathetic system. As anti-AChE's antagonize it, it is often given at high doses with phosphorus compounds. Then 1–2 mg/kg in cats and rabbits block sympathetic ganglia;[71, 72, 73] 3 mg/kg in rats and monkeys depress responses in the CNS;[74] and high doses in man cause central excitement.[75] 25 mg/kg in several species even block to some extent myoneural junctions. These doses are all below the $LD_{50}$'s to cats and dogs (about 40 and 100 mg/kg respectively)[76] though not below that to man (about 2 mg/kg).[75]

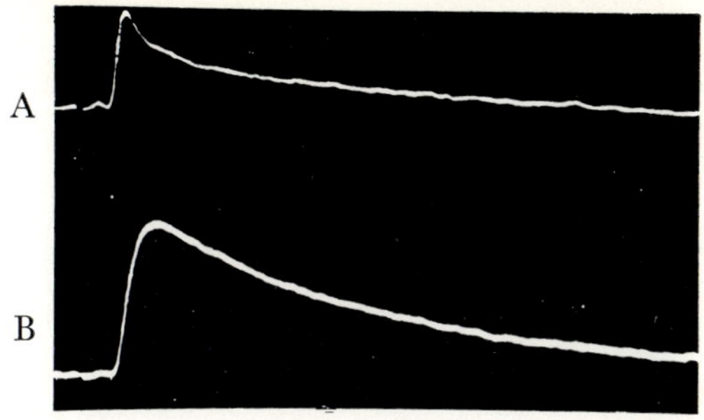


FIG. 11.8. E.p.p.'s in curarized frog sartorius muscle. A, normal; B, after treatment with 3·3 × $10^{-6}$ M prostigmine.(64)

Reproduced from J. C. Eccles and F. W. McFarlane. *J. Neurophysiol.* **12,** 59 (1949) (with acknowledgements).

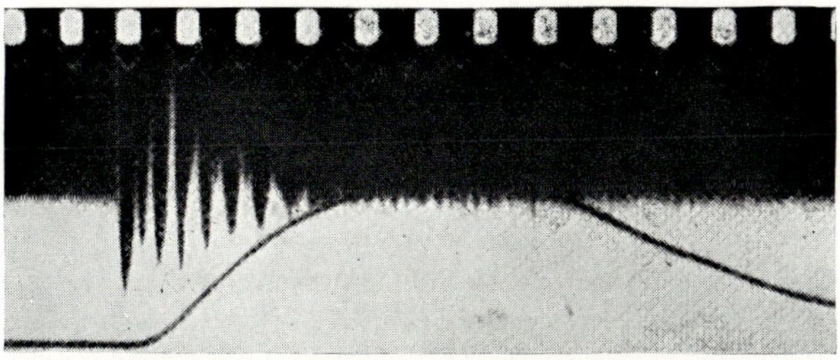

FIG. 11.9. The response of the tibialis anticus of the cat to one indirect stimulus after the cat had received 0·72 mg of eserine. The continuous line shows the tetanic response, and the spiky outline the repetitive action potentials.(62)

Reproduced from G. L. Brown. *J. Physiol.* **89,** 220 (1937) (with acknowledgements).

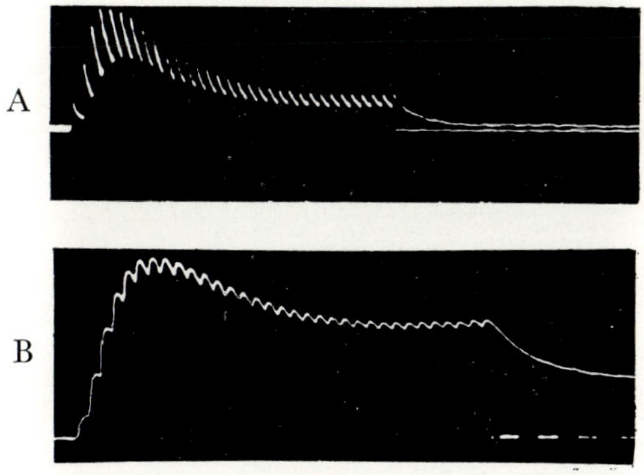


FIG. 11.10. E.p.p.'s in curarized frog sartorius muscle on stimulating the nerve at 100/sec, A—in untreated muscle; B—in muscle in 6·6 × $10^{-5}$ M prostigmine.(64)

Reproduced from J. C. Eccles and F. W. McFarlane. *J. Neurophysiol.* **12,** 59 (1949) (with acknowledgements).

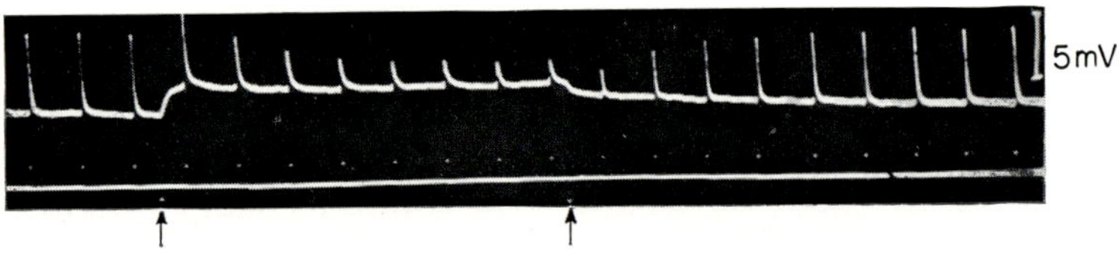


FIG. 11.11. E.p.p.'s recorded in a fibre of the serratus posterior of the rabbit. Between the arrows ACh was infused into the myoneural gap at a constant rate. Time marker: 1·5 sec; voltage marker: 5 mV.[86]

Reproduced from J. Axelsson and S. Thesleff. *Acta. physiol. scand.* **43,** 15 (1958) (with acknowledgements).

*Pentamethonium and hexamethonium.* These block both sympathetic and parasympathetic ganglia. The effective doses vary greatly with species.[77] They do not affect the CNS directly according to a recent paper,[78] but signs of interference with central nervous function are observed. This illustrates the difficulty of separating effects on ganglia from effects on the CNS. Interference with function at one level almost always leads to some effects at another, via the sensory system. Very high doses also block myoneural junctions; and lower doses antagonize decamethonium.[55]

*Decamethonium.* Besides depolarizing or blocking muscle fibres, high doses block ganglia.[58] It also inhibits AChE,[79] with complex effects. Apparent central effects[80] may be indirect, as with penta- and hexa-methonium.

*Succinylcholine.* This[81] and various analogues[82] are short-acting depolarizing agents, acting mainly on muscles.

*Anti-AChE's.* Some reversible anti-AChE's are also blocking or depolarizing agents. Eserine blocks ganglia,[58] and prostigmine depolarizes muscle fibres.[42, 83]

## ADAPTATION PHENOMENA

Some muscles and post-ganglionic fibres, perhaps all, become insensitive to depolarizing agents after prolonged exposure to them. Repolarization after treatment with decamethonium has already been mentioned.

The muscle fibre in the nerve-sartorius muscle preparation of the frog[84] was rapidly depolarized by $2 \times 10^{-5}$ g/ml of ACh, but within about 15 min the fibres became repolarized and their resting potentials returned to normal. The muscle could not, however, be excited by indirect stimuli, i.e. it was blocked. Repetitive stimulation produced e.p.p.'s like those obtained in curarized preparations. Neuromuscular transmission was restored on washing the muscle. The same effects were observed with nicotine and decamethonium as depolarizing agents. Short-term effects have been observed at the end-plates of one amphibian[85] and five mammalian nerve-muscle preparations.[86] Twin-barrelled micro-pipettes were introduced into a myoneural gap and a recording electrode into the fibre beneath it. From one barrel of the micro-pipette tiny doses of ACh were delivered in discrete amounts every second. Each caused a small e.p.p., recorded intracellularly. A conditioning dose of ACh was then delivered at a steady rate for a second or more from the second barrel at such a rate that the fibre became depolarized by about a millivolt. During this steady injection the response to individual discrete injections became less, and only recovered after it over several seconds, although the fibre became completely repolarized very rapidly (Fig. 11.11). Some muscle fibres after several experiments never regained their initial sensitivity to ACh. The state of an end-plate in a muscle poisoned with an anti-AChE is likely to be somewhere between those of the two types of experiment described. The ACh levels are likely to be nearer those in the second type of experiment, but the duration of action is similar to that in the first. It is therefore probable that a poisoned muscle will eventually repolarize, but will remain less responsive to indirect stimulation.

When the inferior mesenteric ganglion of the cat or the superior cervical ganglion of the rat was perfused continuously with ACh at a concentration which just stopped synaptic transmission the initial block was not maintained.[87] Transmission returned within about 3 min. In this case the fibre after adaptation can respond to indirect stimuli.

It is obvious that both in an animal and in isolated pharmacological preparations the degree of disfunction brought about by an anti-AChE may pass through a maximum and then subside, even though the AChE levels remain constant, or continue to fall. These observations may explain how low concentrations of ACh inhibit the twitch response to indirect stimulation of a muscle treated with an anti-AChE (p. 279), and is an alternative to the depolarization theory already put forward.

## REFERENCES

1. C. Lovatt Evans. *Principles of Human Physiology*. Churchill, London, 1949
2. A. F. Huxley and R. Stampfli. *J. Physiol.* **112,** 476 (1951)
3. A. L. Hodgkin and A. F. Huxley. *J. Physiol.* **104,** 176 (1945)
4. M. A. Brazier. *Electrical Activity of the Nervous System*. Pitman, London, 1953
5. A. L. Hodgkin. *Proc. roy. Soc.* **148B,** 1 (1958)
6. P. Fatt and B. Katz. *J. Physiol.* **117,** 109 (1952)
7. P. Fatt and B. Katz. *J. Physiol.* **115,** 320 (1951)
8. A. W. Liley. *J. Physiol.* **133,** 571 (1956)
9. J. del Castillo and B. Katz. *J. Physiol.* **124,** 560 (1954)
10. G. L. Brown. *Proc. roy. Soc.* **123B,** 406 (1937)
11. W. L. M. Perry. *Brit. med. Bull.* **13,** 220 (1957)
12. J. C. Eccles. *Annu. Rev. Physiol.* **10,** 93 (1948)
13. G. H. Acheson. *Fed. Proc.* **7,** 447 (1948)
14. P. Fatt and B. Katz. *J. Physiol.* **118,** 73 (1952)
15. B. Katz. *Brit. med. Bull.* **12,** 210 (1956)
16. C. O. Hebb. *Physiol. Rev.* **37,** 196 (1957)
17. W. Feldberg and A. Vartianen. *J. Physiol.* **83,** 103 (1934)
18. W. L. M. Perry. *J. Physiol.* **119,** 439 (1953)
19. W. Feldberg and J. H. Gaddum. *J. Physiol.* **81,** 305 (1934)
20. N. Emmelin and F. C. MacIntosh. *J. Physiol.* **131,** 477 (1956)
21. K. Krujevic and R. Miledi. *Nature, Lond.* **182,** 805 (1958)
22. J. D. Robertson. *J. biophys. biochem. Cytol.* **2,** 381 (1956)
23. E. R. Bickerstaff, J. V. Evans and A. L. Woolf. *Nature, Lond.* **184,** 1500 (1959)
24. V. P. Whittaker. *Biochem. J.* **72,** 694 (1959)
25. A. W. Liley. *J. Physiol.* **132,** 650 (1956)
26. W. Feldberg. *J. Physiol.* **101,** 432 (1943)
27. R. J. McIsaac and G. B. Koelle. *J. Pharmacol.* **126,** 9 (1959)
28. W. A. Koelle and G. B. Koelle. *J. Pharmacol.* **126,** 1 (1959)
29. V. B. Brooks and D. K. Myers. *J. Physiol.* **116,** 158 (1952)

30. R. Couteaux. *Exp. Cell Res. Suppl.* **5,** 294 (1958)
31. V. P. Whittaker. *Biochem. J.* **68,** 21P (1958)
32. E. Giacobini. *J. Neurochem.* **1,** 234 (1957)
33. W. Feldberg and M. Vogt. *J. Physiol.* **107,** 372 (1948)
34. W. Feldberg, G. W. Harris and R. C. Y. Lin. *J. Physiol.* **112,** 400 (1951)
35. C. O. Hebb and A. Silver. *J. Physiol.* **134,** 718 (1956)
36. G. B. Koelle. *J. comp. Neurol.* **100,** 211 (1954)
37. E. Giacobini. *Acta physiol. scand.* **45,** 238 (1959)
38. Symposium on Neurohumoral Transmission. *Pharm. Rev.* **6,** 3 (1954)
39. W. R. Loewenstein and D. Molins. *Science* **128,** 1284 (1958)
40. J. C. Eccles, P. Fatt and K. Koketsu. *J. Physiol.* **126,** 524 (1954)
41. J. C. Eccles, R. M. Eccles and P. Fatt. *J. Physiol.* **131,** 154 (1956)
42. E. Schoffeniels and D. Nachmansohn. *Biochem. biophys. Acta* **26,** 1 (1957)
43. P. R. Salerno and J. M. Coon. *J. Pharmacol.* **95,** 240 (1949)
44. W. F. Rikker, Jr. *Pharm. Rev.* **5,** 1 (1953)
45. G. L. Brown, H. H. Dale and W. Feldberg. *J. Physiol,* **87,** 394 (1936)
46. J. M. Barnes and J. I. Duff. *Brit. J. Pharmacol.* **9,** 153 (1954)
47. F. C. MacIntosh, R. I. Birks and P. B. Sastry. *Nature, Lond.* **178,** 1181 (1956)
48. P. Holton and H. R. Ing. *Brit. J. Pharmacol.* **4,** 190 (1949)
49. R. B. Barlow. *Introduction to Chemical Pharmacology*, Methuen, London, 1955
50. Z. Majcen and A. O. Zupancic. *Arch. internat. Pharmacodyn.* **108,** 232 (1956)
51. M. Altamirano, W. L. Schleyer, C. W. Coates and D. Nachmansohn. *Biochem. biophys. Acta* **16,** 268 (1955)
52. B. D. Burns and W. D. Paton. *J. Physiol.* **115,** 41 (1951)
53. J. del Castillo and B. Katz. *Proc. roy. Soc.* **146B,** 339 (1957)
54. J. del Castillo and B. Katz. *Proc. roy. Soc.* **146B,** 357 (1957)
55. W. D. M. Paton and E. J. Zaimis. *Brit. J. Pharmacol.* **4,** 381 (1949)
56. E. J. Zaimis. *J. Physiol.* **112,** 176 (1951)
57. D. J. Jenden, K. Kamijo and D. B. Taylor. *J. Pharmacol.* **111,** 229 (1954)
58. W. D. M. Paton and W. L. M. Perry. *J. Physiol.* **119,** 43 (1953)
59. H. Scheiner. *C.R. Soc. Biol., Paris* **130,** 748 (1939)
60. H. Scheiner. *C.R. Soc. Biol., Paris* **130,** 752 (1939)
61. J. Ruiz-Gijon, J. Tamarit T. and C. R. Murcia. *Arch. internat. Pharmacoayn.* **116,** 357 (1958)
62. G. L. Brown. *J. Physiol.* **89,** 220 (1937)
63. J. C. Eccles, B. Katz and S. W. Kuffler. *J. Neurophysiol.* **5,** 211 (1942)
64. J. C. Eccles and F. W. McFarlane. *J. Neurophysiol.* **12,** 59 (1949)
65. W. W. Douglas and W. D. M. Patton. *J. Physiol.* **124,** 325 (1954)
66. G. L. Brown and A. M. Harvey. *J. Physiol.* **99,** 379 (1941)
67. F. Brink, D. W. Bronk and M. G. Larrabee. *Ann. N.Y. Acad. Sci.* **47,** 457 (1946)
68. D. B. Tower and D. McEachern. *Canad. J. Res.* **26E,** 183 (1948)
69. B. Hughes. *Brit. J. Pharmacol.* **10,** 36 (1955)
70. S. Salama and S. Wright. *Brit. J. Pharmacol.* **5,** 49 (1950): **7,** 14 (1952)
71. L. Gyermek. *Acta physiol. Acad. Sci. Hung.* **2,** 511 (1951)
72. A. S. Marrazzi. *J. Pharmacol.* **65,** 18 (1939)
73. L. D. Fink and P. Cervoni. *J. Pharmacol.* **109,** 372 (1953)
74. W. C. Wescoe, R. E. Green, B. P. McNamara and S. Krop. *J. Pharmacol.* **92,** 63 (1948)
75. L. Goodman and A. Gilman. *The Pharmacological Basis of Therapeutics* p. 470, Macmillan, N.Y. (1940)

76. T. Koppanyi. *Proc. Soc. exp. Biol. N.Y.* **40,** 244 (1939)
77. W. D. M. Paton and E. J. Zaimis. *Brit. J. Pharmacol.* **6,** 155 (1951)
78. R. Murray, L. Beck, P. A. Rondell and D. F. Bohr. *J. Pharmacol.* **127,** 157 (1959)
79. J. del Castillo and B. Katz. *Proc. roy. Soc.* **146B,** 369 (1957)
80. R. L. Irwin and J. B. Wells. *J. Pharmacol.* **119,** 329 (1957)
81. H. Low and L-E. Tammelin. *Acta physiol. scand.* **23,** 78 (1951)
82. D. Della Bella, R. Villani and G. F. Zuanazzi. *Arch. exp. Path. Pharmak.* **229,** 536 (1956)
83. W. F. Riker, Jr. and W. C. Wescoe. *J. Pharmacol.* **88,** 58 (1946)
84. S. Thesleff. *Acta physiol. scand.* **34,** 218 (1955)
85. B. Katz and S. Thesleff. *J. Physiol.* **138,** 63 (1957)
86. J. Axelsson and S. Thesleff. *Acta physiol. scand.* **43,** 15 (1958)
87. W. A. Krivoy and J. H. Wills. *J. Pharmacol.* **116,** 220 (1956)
88. K. Krujevic and J. F. Mitchell. *Nature, Lond.* **186,** 241 (1960)
89. D. H. Jenkinson. *J. Physiol.* **152,** 309 (1960)

CHAPTER XII

# Effects on Isolated Organs *in vitro*

THE STUDY of any organ *in vivo* is complicated by secondary effects. To take a simple case, when a muscle is stimulated its contraction excites the stretch receptors, and so sets up reflexes which modify the original response. *In vitro* secondary effects are absent or less important, so that *in vitro* experiments can be of great help in interpreting responses observed *in vivo*.

The signs and symptoms of poisoning show that all the muscles and glands innervated by the voluntary and parasympathetic systems are affected by phosphorus compounds. All the pharmacological preparations commonly used *in vitro* therefore show some effects which are relevant to *in vivo* studies. Not all the effects observed are relevant, however. In deciding which are, the concentration of phosphorus compound present is often the most important factor. Consider, for example, TEPP, DFP and Paraoxon, all three popular with pharmacologists. They are toxic at about 1 mg/kg, i.e. at an initial concentration in the whole animal of about $10^{-5}$ M. As they are fat-soluble and are attacked by A-esterases, the concentration in the aqueous part of the animal soon drops to $10^{-6}$ M or less. *In vitro* the biological system under test is suspended in a large volume of aqueous medium, which may reasonably be regarded as analogous to the extracellular fluid *in vivo*. It is obvious, therefore, that results obtained with bath concentrations of these compounds above about $10^{-6}$ M can throw no light on the response of animals given marginally lethal doses. Some mammals can be made to survive for a time very large doses by giving them atropine and applying artificial respiration (Chapter XV), and experiments with bath concentrations up to $10^{-4}$ M may then have some relevance; but concentrations above this can have nothing to do with the action of the more toxic compounds *in vivo*.

As explained at the beginning of the preceding Chapter, I have assumed that the anti-AChE theory of poisoning is correct except when there is definite evidence to the contrary, and I have, in consequence, sometimes explained results in a very different way from the original workers.

The effects in various preparations are now considered in turn.

## AXONS

Immersion in 0·01–0·02 M DFP blocks conduction in both motor and sensory nerves from several species,[1, 2, 3] and in the isolated giant axon of the squid.(6) At first the block can be removed by washing the nerves, but after about an hour the block becomes irreversible. The development of the block has been studied on the sciatic nerve of the frog.(4) Block was not preceded by a period in which the nerve was more readily excited, and the fibres did not appear to be depolarized when block was established. Such a block differs from that induced by potassium chloride while resembling that induced by procaine.

Attempts have been made to determine the AChE levels in blocked axons. As DFP is very soluble in the lipoidal sheaths round the axons, correct results cannot be obtained if the axons are homogenized, as the DFP then released inhibits any AChE remaining.(5) The levels can be estimated in intact axons with the lipoid-soluble substrate N-dimethylaminoethyl acetate, and the results obtained on the leg nerves of spider crabs (*Libinia emarginata Rathbun*) suggest that conduction ceases when the levels drop to 20% of normal.(5) The giant axon of the squid can tolerate high concentrations of DFP in its axoplasm without becoming irreversibly blocked. Thus an internal concentration of $3 \times 10^{-4}$ M set up along several millimetres length by intra-axonal injection did not cause block in 40 min;(7) and during immersion of the axon in DFP solution block only became irreversible when the concentration in the axoplasm reached about $5 \times 10^{-5}$ M.(6)

These results have no possible relevance to the toxic action of DFP *in vivo*—the concentrations are far too high—but the mechanism of block is of interest because it has been suggested that conduction, like synaptic transmission, depends on an AChE–ACh system, in this case acting along the length of the axon. On this theory inhibition of AChE should lead to an accumulation of ACh, which blocks conduction by depolarizing the axonal membrane. The evidence is strongly against this theory.

(1) On this theory an axon should be unusually easy to excite while it is becoming depolarized; and when it is blocked its demarcation potential should be reduced. Neither effect is observed.

(2) Block by DFP can be reversed by washing, at least in the early stages. Washing does not remove all DFP from axons,(5) so inhibition must continue after washing. As di-isopropylphosphorylated AChE is stable *in vivo* (Chapter VIII) the AChE level must continue to fall during the restoration of conduction.

(3) Injection of high DFP concentrations into the axoplasm of the squid axon does not cause block.

(4) The maximum levels of AChE which can be supposed to remain in some conducting nerves are very low. Thus block in the squid axon took place when the DFP concentration reached $5 \times 10^{-5}$ M. As the AChE is distributed mainly on the inside of the sheath,(8) the DFP must pass near it to reach the axoplasm. As, also, the substrate, ACh, of AChE is insoluble in oils, the AChE must be in an essentially aqueous medium, like the axoplasm.

Thus the concentration of DFP near the enzyme surface must also be about $5 \times 10^{-5}$ M, and this is enough to bring about 99 per cent inhibition in a few minutes. In the experiments with N-dimethylaminoethyl acetate on the nerves of the spider crab the higher AChE levels found at block may either be due to genuine species differences, or may be artefacts. They could be artefacts if the substrate is hydrolysed by an enzyme in the nerve resistant to anti-AChE's. It should be pointed out that for any one compound some correlation will almost certainly exist between AChE levels and block, as both are likely to depend upon the concentration of the drug and how long it is in contact with the nerve.

To conclude, nerve conduction is blocked by DFP at concentrations much higher than could be found *in vivo*. The effect is not related to inhibition of AChE.

## VOLUNTARY MUSCLES

A group of seven anti-AChE's, consisting of DFP and six non-phosphorus compounds, affected the end-plate potentials (e.p.p.'s) in curarized frog's sartorius muscle in the same way.[19] They increased their voltage and lengthened their time course; and during repetitive stimulation they caused the muscle to become more rapidly and more completely depolarized, and consequently caused e.p.p.'s to become smaller sooner (Fig. 11.8 and 11.10). Sarin affected the time course of e.p.p.'s in the rat diaphragm similarly, and the e.p.p.'s were rapidly restored to normal by the AChE re-activating agent, dihydroxyimino-acetone.[26]

These results can be interpreted readily on the anti-AChE theory of poisoning, and not easily in any other way. They therefore lend some fundamental support for the explanations offered later for some of the responses of muscles treated with phosphorus compounds.

The rest of this section is concerned with the rat phrenic nerve–diaphragm preparation excited indirectly and supramaximally, except when it is stated otherwise. A normal preparation stimulated at 25/sec or faster sustains a tetanus for the duration of the stimulation, and stimulated at 1/sec or slower responds with discrete twitches.

The responses of diaphragms to low and high frequency stimulation, to D-tubocurarine and to ACh are all changed by phosphorus inhibitors in much the same way as they are changed by reversible inhibitors.

The response to low frequency stimulation is initially increased, passes through a maximum, and levels off usually somewhat above that of the diaphragm before treatment[20, 21] (Fig. 12.1). As with eserine, the increased response in the early stages of inhibition is associated with fibrillations and fasciculations. The motor units fire repetitively after a single stimulus, due to axon reflexes[22] (p. 269), so that the response is a short tetanus instead of a single twitch. With both reversible and phosphorus inhibitors the enhanced response appears when the AChE level has dropped to about 20% of normal.[20, 22] The change of response on washing out the inhibitors is different, as expected. Washing out a reversible inhibitor brings the response back to normal, whereas washing out DFP leaves the response unchanged, as the AChE level does not rise.

The relative time courses of AChE inhibition, and of the variations of strength of contraction, of depolarization of the muscle fibres and of blocking time in the fibres[23] are shown diagrammatically in Fig. 12.2. In the

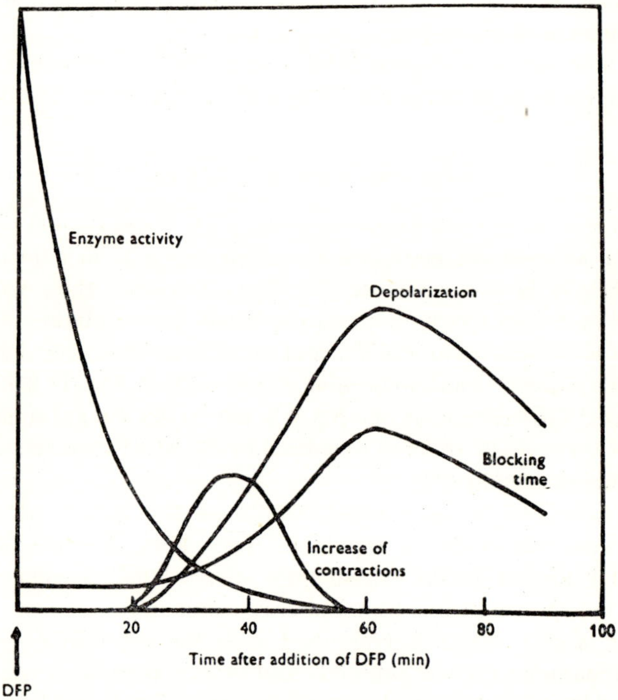


FIG. 12.2. Variation with time in the rat diaphragm preparation of AChE, response to low frequency stimulation, depolarization of the muscle fibres and blocking time after adding DFP to the bath to $2 \times 10^{-6}$ M.[23]

Reproduced from A. Meeter. *J. Physiol.* **144,** 38 (1958), with acknowledgements.

experiments summarized there the contractions declined to the original strength, and at this point axon reflexes disappeared, although fibrillatory after-discharges in individual fibres were still present. The muscle fibres finally became desensitized to ACh, as shown by the re-polarization of the muscle.

Most workers find that the contractions are not finally reduced to the original force, but remain somewhat above it, so that in their preparations axon reflexes

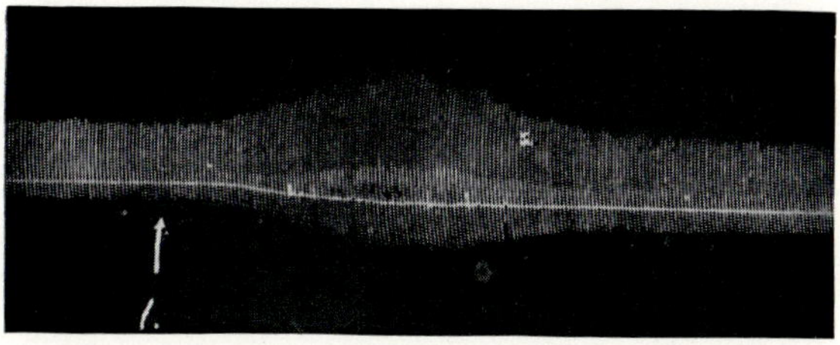

FIG. 12.1. Rat nerve–diaphragm preparation. The nerve was stimulated at 26/min, and DFP to $10^{-5}$ M added to the bath at the arrow.[21]

Reproduced from J. A. B. Barstad. *Arch. internat. Pharmacodyn.* **107,** 21 (1956) (with acknowledgements).

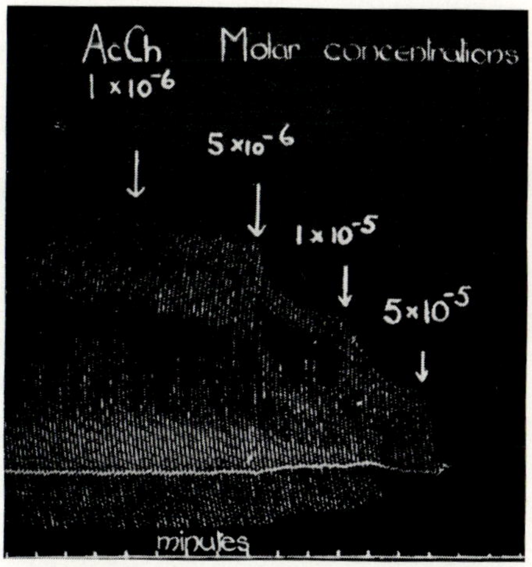


FIG. 12.3. Reduction of the response to low frequency stimulation of a diaphragm in $4 \times 10^{-7}$ M Paraoxon caused by adding various concentrations of ACh to the bath.[20]

Reproduced from J. M. Barnes and J. I. Duff. *Brit. J. Pharmacol.* **8,** 334 (1953) (with acknowledgements).

are not, presumably, completely abolished. As responses in fibres are normally reduced by prolonged exposure to ACh, the reduction in response may result in part from this. The response finally obtained when AChE is nearly completely inhibited is steady because it represents a balance between the factors tending to increase it: fibrillatory after-discharges, axon reflexes, and repolarization of the fibres; and that tending to reduce it, namely the desensitization of the fibres to ACh. Even in the absence of AChE the ACh is effectively removed from the system by diffusion into the bath, which is very large compared with the muscle. The introduction of ACh into the bath reduces the response (Fig. 12.3) as expected, but the concentrations required are very much greater than those released from the muscle into the bath even in a long experiment,[24] i.e. the endogenous ACh released is not enough to affect appreciably the diffusion of ACh from the end-plates.

When the muscle is in the final steady state the effects of washing it depend upon the rate of re-activation of the AChE. Washing after treatment with eserine or a dimethyl phosphate ester restores the muscle to normal (it does

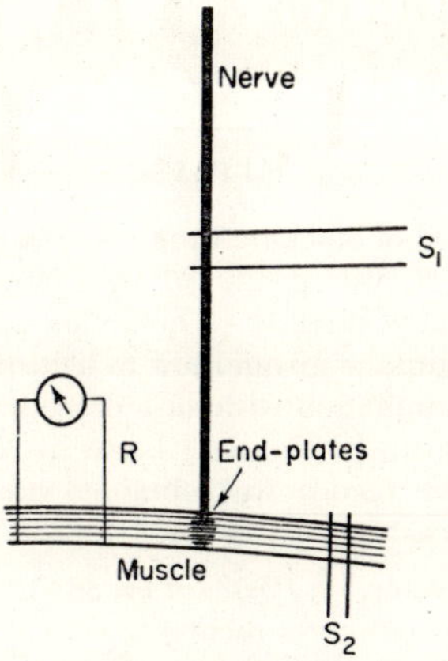


FIG. 12.4. Set-up for determining blocking times. The muscle is stimulated first by a single stimulus to the nerve by the electrodes $S_1$, and thereafter directly at about 500/sec by the electrodes $S_2$. The action potentials passing under the end-plate region are recorded at R.

not pass through a stage of increased response during washing), and on re-inhibiting the AChE the muscle goes through the same stages as before. Washing after treatment with a diethyl or a di-isopropyl ester or Sarin does not restore it,[20] but treatment with P2-AM does.[23]

These results lead to the conclusions that the compounds act only by inhibiting AChE, and that to restore the muscle to normal, enough AChE must be re-activated for the desensitization to ACh of the fibre to be lost.

The effects on capacity to hold a tetanus are linked with those on 'blocking time' (Fig. 12.2). Blocking times were determined as follows.[23] The muscle was stimulated indirectly and then stimulated directly at intervals afterwards. The action potentials produced by direct stimulation were recorded on the other side of the end-plate region to the point of stimulus (Fig. 12.4). Immediately after indirect stimulation action potentials could not pass through the depolarized regions under the end-plates. As the fibres became repolarized, more action potentials passed through. The recorded action potential was the sum of the action potentials in many individual fibres, and therefore increased. The time taken for the recorded action potential to reach half its value before indirect stimulation was taken as the blocking time (Fig. 12.5). In DFP the blocking time rose from about 4 to

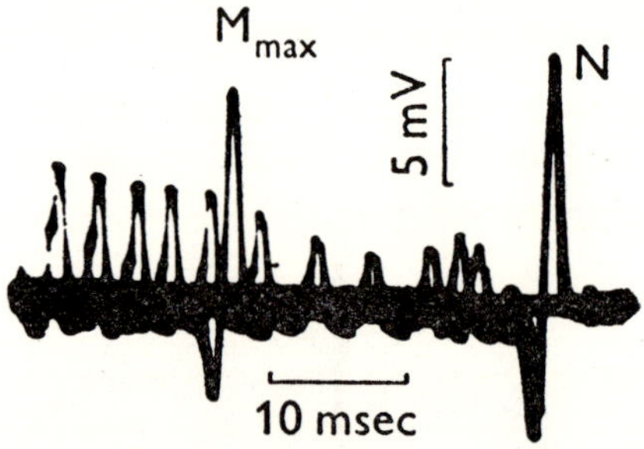


FIG. 12.5. Determination of blocking time. Action potentials in the muscle are shown. The first, N, was produced by stimulating the nerve. The muscle was then stimulated directly. The action potentials increased with time after the indirect stimulus (from right to left in the figure). $M_{max}$ was the response to direct stimulation without a preceding nerve stimulus. The blocking time was taken to be the time between N and a response equal $\frac{1}{2}M_{max}$. This particular record was obtained after a muscle had been immersed in $4 \times 10^{-6}$ M DFP for 30 min.[23]

Reproduced from E. Meeter. *J. Physiol.* **144,** 38 (1958) (with acknowledgements).

about 30 msec. It follows that unless the stimuli are much more than 30 msec apart a few stimuli will produce such depolarization that the muscle will become refractory, and a tetanus will not be fully sustained. The tetanus will be weaker the higher the frequency of stimulation, as shown in Fig. 12.6. Some data on the loss and return of the capacity to hold a tetanus when the frequency of stimulation is 50/sec are shown in Table 12.1. The tetanus is blocked when the AChE drops below about 10% of normal.* When the muscle is washed the tetanus returns in the times shown, recovery taking

* In a very recent paper[47] it was shown that when the capacity to hold a tetanus began to fail the AChE levels and the frequency of stimulation were inversely related, as expected—15% at 120 c/sec, 10% at 60 c/sec and 8% at 30 c/sec. Substantially the same results were got with DFP, $(EtO)_2PO \cdot F$, TEPP, $(MeO)_2PO \cdot F$ and Paraoxon-Me. This is a particularly convincing demonstration of the over-riding importance of the inhibition of AChE in the induction of muscular paralysis by phosphate esters.

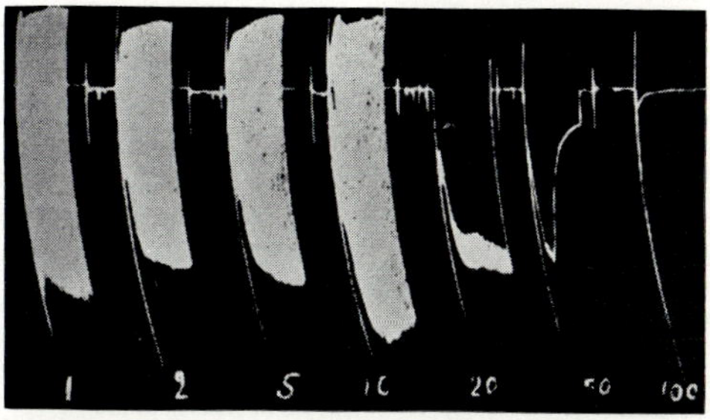


FIG. 12.6. Contractions of a diaphragm after treatment with $2 \times 10^{-5}$ M DFP. The DFP was washed out before recordings were made. The rates of stimulation in stimuli/sec are shown under the responses. At 1/sec the twitch response was not fully maintained, and up to 10/sec was abnormal. At higher frequencies the response was a tetanus, which was worse maintained the higher the frequency.[22]

Reproduced from C. van der Meer and E. Meeter. *Acta physiol. pharm. Neerl.* **4,** 454 (1956) (with acknowledgements).

TABLE 12.1. COMPOUNDS INHIBITING A TETANUS IN THE RAT NERVE DIAPHRAGM PREPARATION[20, 25]

The stimulation rate was 50/sec. The concentrations used, the times taken to remove the capacity to hold a tetanus, and the times taken after washing out the inhibitor for the capacity to return are given

| Compound | $I_{50}$ | conc. (M) | Time to block (min) | Time to recovery (min) |
|---|---|---|---|---|
| Paraoxon | $2{\cdot}1 \times 10^{-8}$ | $4 \times 10^{-8}$ | 39 | |
| | | $1 \times 10^{-7}$ | 18 | 30–40 |
| | | $4 \times 10^{-7}$ | 8 | |
| TEPP | $1{\cdot}3 \times 10^{-8}$ | $2 \times 10^{-7}$ | 10 | 30–40 |
| Diethyl ester* | $8 \times 10^{-8}$ | $10^{-4}$ | — | 50–90 |
| Paraoxon-Me | $2{\cdot}5 \times 10^{-8}$ | $10^{-6}$ | 10 | 3 |
| Dimethyl ester* | $1{\cdot}5 \times 10^{-7}$ | $5 \times 10^{-5}$ | — | 10 |
| Paraoxon-Pr$^{i}$ | $4{\cdot}7 \times 10^{-7}$ | $10^{-6}$ | 10 | > 240 |
| DFP | $4{\cdot}6 \times 10^{-6}$ | $10^{-5}$ | 10 | > 240 |
| Di-isopropyl ester* | $2{\cdot}1 \times 10^{-7}$ | $10^{-4}$ | — | > 90 |
| Di-sec-butyl ester* | $1{\cdot}6 \times 10^{-7}$ | $10^{-4}$ | — | > 90 |

* Dialkyl esters of the type: $Me_3\overset{\oplus}{N}\cdot C_6H_4\text{-3-O}\cdot PO(OR)_2$, the phosphostigmines.

longer the more stable the inhibited AChE, i.e. the rate of recovery decreases in the order:

$$\text{reversible inhibitor/AChE} > (MeO)_2PO\cdot AChE > (EtO)_2PO\cdot AChE > (Pr^iO)_2PO\cdot AChE \text{ and } (sec\text{-}BuO)_2PO\cdot AChE.$$

Recovery is accelerated by AChE re-activators,[23, 26] especially the recovery of diaphragms treated with TEPP or Sarin, but also of those treated with DFP and methylfluorophosphinylcholines.† The rates of recovery and percentages of AChE re-activated are correlated in diaphragms treated with TEPP. At $10^{-3}$ M and above, hydroxyimino-acetone and dihydroxyimino-acetone bring about recovery faster than P2-AM, although P2-AM is the most powerful re-activator, because P2-AM is itself a blocking agent. At lower concentrations P2-AM is the most effective.

† In preparations *in vitro* it must be remembered that re-activators also hydrolyse phosphorus compounds. This probably explains how P2-AM brings about the recovery of diaphragms treated with methylfluorophosphinylcholines, although it does not re-activate AChE inhibited with them. AChE's inhibited by phosphinylcholines are re-activated *in vivo* more rapidly than diethylphosphorylated AChE's, so that removal of the inhibitor should lead to recovery in under 40 min (see Table 12.1). In fact recovery takes about 15 min.[27]

The capacity to hold a tetanus returns at a far lower level of AChE than that at which it is lost (10% of normal). Thus, if inhibition is continued until the level is calculated to be below 1%, and the inhibitor is washed out, the capacity to hold a tetanus returns when only 0·5% of the AChE has been re-activated. A second treatment of a diaphragm which has recovered functionally after washing again leads to loss of tetanus, but the tetanus returns as rapidly as before although the AChE levels must have been lower throughout. The quantity of ACh released per stimulus does not lessen after many stimuli,[24, 45]* so the recovery of tetanus at low AChE levels cannot be explained by any decrease of ACh near the end-plates. It is best explained by desensitization of the muscle to ACh,[28] so that it needs less AChE to function properly. The degree of desensitization should not be exaggerated. ACh is lost from the end-plate region by diffusion at a rate equivalent to hydrolysis by about 2% of the normal AChE.† Tetanus fails at 10% AChE, and returns at a rate of removal of ACh equivalent to about 2·5%, i.e. the fibres are about one-quarter as sensitive on the average.

The interaction between D-tubocurarine and anti-AChE is complex. As expected, the addition of an anti-AChE to the bath containing a curarized preparation will restore its capacity to respond to indirect, low frequency stimulation (Fig. 12.7). When D-tubocurarine is added after the inhibitor the effects depend on the frequency of stimulation and on how long the muscle has been in contact with the anti-AChE. At low frequencies adding D-tubocurarine shortly after the response has fallen from its maximum restores it to its maximum value (Fig. 12.8A), whereas when it is added later the response is reduced (Fig. 12.8B). In the first case D-tubocurarine repolarizes the fibres, which are still sensitive to ACh, so that the maximum response is attained. In the second the fibres have become desensitized to ACh, which may then act as a blocking agent, like decamethonium under similar circumstances. D-Tubocurarine therefore increases the block, and the muscle contracts less.‡ The effects of adding D-tubocurarine to the bath after DFP on the capacity of the diaphragm to hold a tetanus at 100 stimuli/sec are shown in Fig. 12.9A, B, C and D.[29] In the first three experiments the D-tubocurarine was added shortly after the DFP. After $5 \times 10^{-7}$ M DFP D-tubocurarine improved the tetanic response for a time, but eventually

* The quantity of ACh released per stimulus decreases with the rate of stimulation in some cases, but not with the duration of stimulation.[46]

† This can be calculated either from the increase in blocking time with AChE inhibition, or from the least frequency of indirect stimulation to which the response is subnormal when the AChE is completely inhibited. Complete inhibition can be achieved by treating the preparation for several minutes with a high concentration of the di-isopropyl phosphate ester, Paraoxon-Pr$^i$.[20] The calculations are very rough, but as both methods give the same value, the estimate is probably not far out.

‡ The paper referred to in footnote, p. 265, suggests a better explanation.

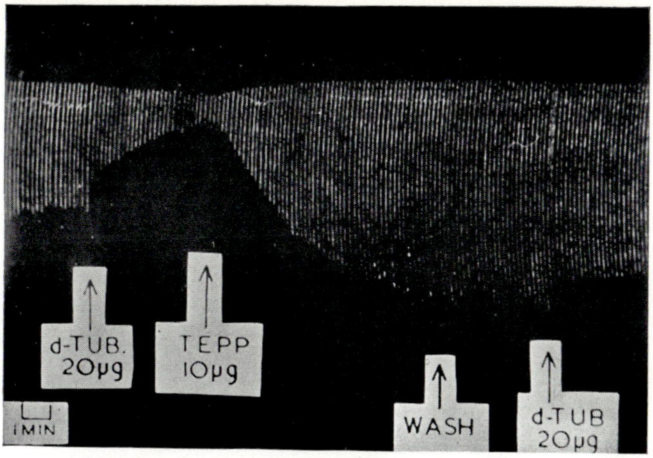


FIG. 12.7. Antagonism between TEPP and D-tubocurarine.[14]
Reproduced from A. S. V. Burgen, C. A. Keele and D. Slome. *J. Pharmacol.* **96,** 396 (1949) (with acknowledgements).

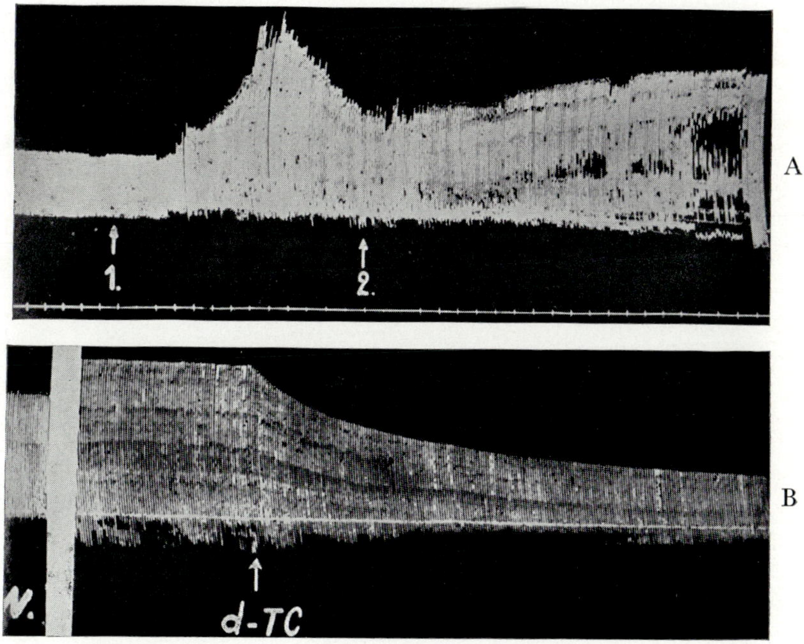


FIG. 12.8. The effects of adding D-tubocurarine to 0·15 $\mu$g/ml on the response to low frequency stimulation (26/min) of a diaphragm treated with $2 \times 10^{-5}$ M DFP. In 8A the DFP was added at 1 and the D-tubocurarine at 2. In 8B the D-tubocurarine was added 25 min after the DFP. N shows the normal response of the diaphragm before treatment.[12]
Reproduced from J. A. B. Barstad. *Arch. internat. Pharmacodyn.* **107,** 4 (1956) (with acknowledgements).

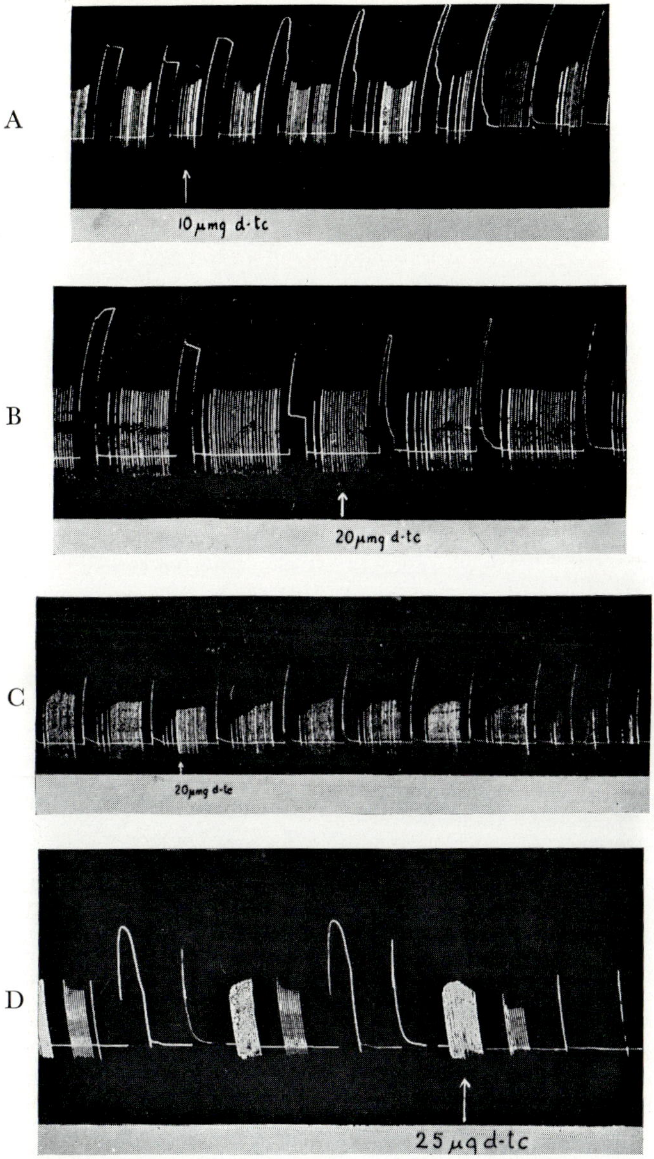


FIG. 12.9. The effects on the response to indirect high frequency stimulation of adding D-tubocurarine to the bath after DFP are shown. (Responses to low frequency stimulation can also be seen on the figures.) In A, B and C the D-tubocurarine was added shortly after the DFP, and the frequency of stimulation was 100/sec. DFP concentrations were as follows: A, $5 \times 10^{-7}$ M; B, $2{\cdot}5 \times 10^{-6}$ M; C, $6 \times 10^{-6}$ M. For D the treatment given is described in the text, and rates of stimulation were 50 and 100/sec.[29]

Reproduced from J. Axelsson, E. Gjone and K. Naess. *Acta pharmacol. tox.* **13,** 319 (1957) (with acknowledgements). The originals have been shortened.

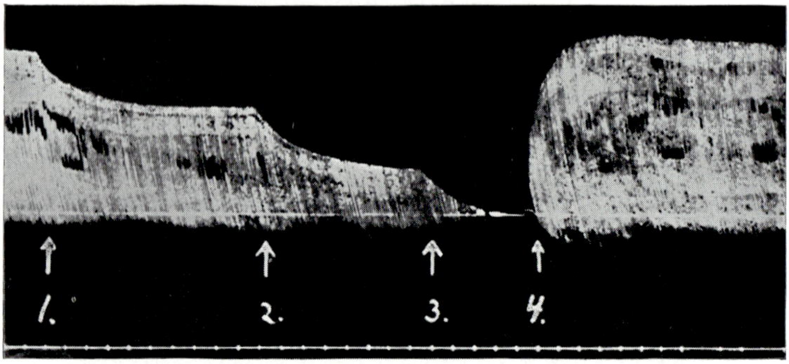


FIG. 12.10. Curarimimetic effects of DFP on response to stimulation at 26/min. The diaphragm was treated with $10^{-5}$ M DFP 5 min before 1, and at 1, 2 and 3 the DFP concentration was increased by $4 \times 10^{-4}$ M each time. At 4 the diaphragm was washed. Time marker, 30 sec.[21]

Reproduced from J. A. B. Barstad. *Arch. internat. Pharmacodyn.* **107,** 21 (1956) (with acknowledgements).

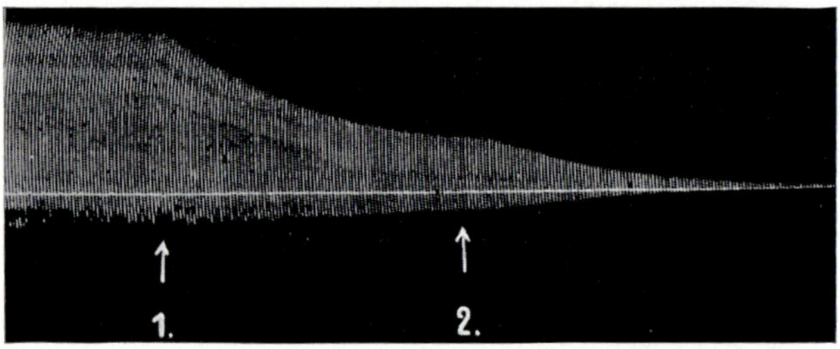


FIG. 12.11. Additive curarization by DFP and D-tubocurarine. The diaphragm was treated with eserine (1·67 μg/ml) a few minutes before the record was taken. $4 \times 10^{-4}$ M DFP added at 1 decreased the response to low frequency stimulation, and this was further decreased by D-tubocurarine added at 2 (0·25 μg/ml).[12]

Reproduced from J. A. B. Barstad. *Arch. internat. Pharmacodyn.* **107,** 4 (1956) (with acknowledgements).

it became worse than before (Fig. 12.9A); after $2 \cdot 5 \times 10^{-6}$ M DFP the improvement was very transitory (Fig. 12.9B); and after $6 \times 10^{-6}$ M DFP the only improvement in response was a slight increase in the force of the twitches to which the tetani had previously degenerated (Fig. 12.9C). In the last experiment the diaphragm was first treated with $3 \times 10^{-5}$ M DFP for 30 min, during which it had been stimulated at high frequencies. The DFP was then washed out, when the diaphragm responded to high frequency stimulation with a feeble tetanus. This was rapidly reduced to a weak twitch on adding D-tubocurarine (Fig. 12.9D). It is stated that TEPP gave similar results in experiments analogous to the four described. These results have been held to show that DFP has a direct curarimimetic action on the end-plates, and therefore synergizes or adds to the action of D-tubocurarine. This theory fails to explain the transient increase in response in the first three experiments, and does not take into account desensitization in the fourth. Alternative explanations can be put forward which are consistent with the anti-AChE theory. In the last experiment the response was feeble but tetanic in nature because the muscle was desensitized to ACh. Curarization therefore relaxed it in the usual way. In the first three experiments desensitization was still progressing when the D-tubocurarine was put in, and was probably reversible, so that the response was improved to some extent. The improvement was not maintained because the ACh concentration was rising continually as the AChE was progressively inhibited. Reversible inhibitors nearly reach equilibrium very rapidly, so that their action was antagonized much more successfully in other experiments. At such high frequencies of stimulation (100/sec) no great improvement can be expected, as curarization cannot reduce to below 10 msec the duration of the e.p.p.'s, which, therefore, sum and depolarize the muscle.

DFP at higher concentrations than those used in the preceding experiments does block receptors. Thus high concentrations inhibited single twitches(21) (Fig. 12.10), and this block was removed by washing. The response to direct stimulation was unchanged throughout, showing that DFP did not depolarize the muscle. Its action must therefore have been curarimimetic. The most convincing demonstration was carried out on eserinized diaphragms. In eserinized diaphragms the AChE is protected from the DFP, which can, therefore, only act in ways other than by inhibiting it. The twitch response was decreased by $4 \times 10^{-4}$ M DFP, and the effect was potentiated by D-tubocurarine(12) (Fig. 12.11).

Botulinum toxin inhibits the secretion of ACh from nerve endings, thus stopping all neuromuscular transmission.(31, 32, 33) DFP did not antagonize this action in rat's diaphragm.(30) It was argued that if DFP inhibits the tetanic response by increasing the persistence of ACh in the end-plate region, botulinum toxin should improve the response, at least transitorily, by lowering the ACh concentration. As it did not, DFP must inhibit tetani by blocking the receptors. This assumes that the action of the toxin is progressive at each nerve ending. If the toxin cuts out the nerve endings suddenly one by one, and there is some suggestion that this is more nearly its action,(32, 33) anti-AChE's will not antagonize it.

Choline restored to some extent the capacity of a diaphragm to hold a tetanus after DFP treatment, although it is itself a depolarizing agent.(34) It has therefore been suggested that the block, far from being due to an excess of ACh, is due to a deficiency. The ACh is not hydrolysed, but diffuses into the bath, so that there is not enough choline for resynthesis. The argument seems to imply both that much of the choline-acetylase is in the nerve endings and that the nerve holds very inadequate stocks of ACh. The first assumption is doubtful, and the second probably incorrect. The concentrations of choline were very high (about $10^{-3}$ M), so that the results are unlikely to have much relevance to *in vivo* studies.

We now consider briefly the effects of very high concentrations of some inhibitors, and the behaviour of another muscle preparation, the frog's rectus abdominis.

High concentrations of phosphorus compounds block conduction in muscle, and the block is associated with depolarization. Block can sometimes be reversed by washing if the muscle and compound have only been in contact a short time. Electroplax is blocked by 0·013 M DFP and 0·012 M Tabun,(9) and frog sartorius muscle by $10^{-4}$ M TEPP,*(10) but rat diaphragm is not blocked by $10^{-3}$ M DFP.(11, 12) These concentrations are far higher than can be tolerated *in vivo*.

The only direct response of the frog's rectus abdominis muscle is a slow contraction in 0·01 M TEPP or DFP.(13) Much lower concentrations greatly potentiate the contraction induced by low concentrations of ACh.(14) The potentiation is only caused by direct inhibitors, is progressive with time, and is annulled by P2-AM.(15, 16) Antagonism by P2-AM is correlated with the re-activation of AChE at the myoneural junctions. The degree of potentiation increases with concentration of inhibitor up to a maximum, which with TEPP is attained at about $10^{-5}$ M.(13, 14) Qualitatively similar effects are produced by reversible inhibitors, and the differences are such as one expects if both reversible inhibitors and phosphorus compounds act solely by inhibiting AChE. Inhibition by the phosphorus compounds, unlike that by reversible inhibitors, is progressive with time and reversed by P2-AM. BuCh also produces a contraction in the normal muscle, and this is potentiated by $10^{-7}$ M DFP, a concentration which does not inhibit AChE, but which inhibits ΨChE, for which BuCh is a substrate.(17)

In this preparation also phosphorus compounds may act as curarizing

* The author used 'hexaethyl tetraphosphate' (HETP). It is assumed in this and similar cases throughout this part of the book that the active ingredient was TEPP, at one-fifth the concentration given in the original paper.

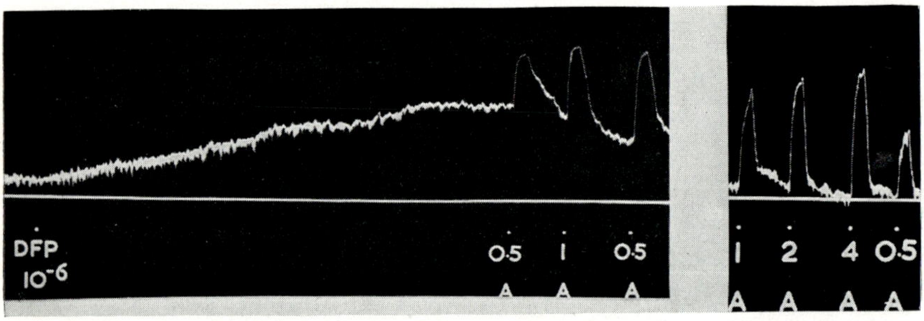


Fig. 12.12. Increased tonus and response to ACh brought about in the longitudinal muscle layer of the rabbit duodenum by treatment with DFP. The normal response of a similar muscle is shown on the right. The doses of ACh are shown in $\mu$g under the responses.[36]

After: H. Shelley. *Brit. J. Pharmacol.* **10,** 26 (1955) (with acknowledgements).

agents,* i.e. they may block access of ACh to the receptors. This is shown by the increase in potentiation which is observed if DFP or TEPP is washed out of the preparation before it is treated with ACh.[13, 14]

## GASTRO-INTESTINAL TRACT

Intestinal preparations exert a fairly constant force (tonus), superposed on which are somewhat irregular contractions (pendular motion). Segments of intestine, but not, of course, strips, may also show peristalsis.

Low concentrations of ChE inhibitors increase the tonus (Fig. 12.12), the rapid contraction caused by ACh, and, to a lesser extent, the pendular motion, which may be replaced by slow peristalsis.[13, 14, 35, 36, 37] Normal behaviour is restored by $10^{-8}$–$10^{-7}$ M atropine.[13, 35] The effects of diethyl and di-isopropyl phosphate inhibitors are only partially reversed by prolonged washing, but otherwise resemble closely those of eserine and other reversible AChE inhibitors. There is no doubt that the effects observed result from inhibition of AChE:

(a) Potentiation of the effects of ACh increases with AChE inhibition.[36]

(b) The effects of TEPP or Paraoxon are all completely reversed by P2-AM.[38]

(c) Inhibition of choline-acetylase in the preparations greatly reduces the effects, without reducing the direct response to ACh.[39] ACh is released continuously at a high rate from intestines,[40] and anti-AChE's enable it to act unhindered on the muscles. A fall in ACh production therefore reduces the response.

Inhibition of ΨChE as well as of AChE appears to be important. In the cat ileum preparation the increase in tonus was more closely related to inhibition of ΨChE than to the inhibition of AChE;[41] and in strips of longitudinal muscle from rabbit duodenum the first response was obtained when only 10% of the AChE was inhibited by DFP.[36] In the latter experiments the response was closely correlated with the inhibition of AChE, but it is likely that complete inhibition of ΨChE, which took place first, had increased the sensitivity of the muscle—no other organ is known in which the inhibition of only a few per cent of the AChE present leads to a definite change in behaviour.

The response passed through a maximum as the concentrations of direct phosphorus inhibitors were raised, in DFP[13, 36] at about $10^{-4}$ M, and in Paraoxon[38] at about $5 \times 10^{-7}$ M. At still higher concentrations both direct inhibitors (DFP and Paraoxon) and indirect ones (Parathion and 'Systox') relaxed the muscle. The relaxant effect was not antagonized by atropine,

* DFP potentiates slightly the action of several depolarizing drugs, e.g. decamethonium.[18] It has been suggested that it does so by adsorbing on the 'esteratic' (B) sites of the *receptors*, thus sensitizing the anionic sites to the action of depolarizing agents. Most of the evidence for this theory can be explained in other ways.

but was reversed by washing out the preparation. The relaxant effects of Paraoxon were demonstrated on eserinized rabbit's intestine—the contraction caused by the eserine disappeared as the Paraoxon concentration approached $3 \times 10^{-4}$ M.[38]

## HEART

Three types of preparation have been studied; isolated hearts, isolated auricles and isolated aortic strips.

Hearts do not respond to low concentrations of phosphorus compounds, but $10^{-4}$ M DFP,[35] TEPP[14] or Paraoxon[42] slow the heart beat and reduce its amplitude. These responses are not influenced by atropine, and cease on washing the heart, i.e. they result from the usual blocking action of phosphorus compounds at high concentrations. The heart is, however, greatly sensitized to ACh. The sensitization is not related to the rates at which the compounds inhibit AChE. This lack of correlation is not unexpected, as it can be assumed that the AChE was substantially completely inhibited by all the inhibitors at the concentrations used in the experiments described.

The movement of isolated auricles from guinea pigs is slowed by about $4 \times 10^{-6}$ M Paraoxon, and the normal rate is restored by atropine or P2-AM, but not by washing.[42]

Aortic strips do not contract spontaneously, but are made to do so by ACh. DFP, etc., up to $10^{-5}$ M potentiates the action of ACh, but higher doses antagonize it, i.e. as usual excitation gives way to block at high concentrations.[11]

All the effects follow the familiar pattern: low concentrations of direct inhibitors produce the responses expected on the anti-AChE theory of poisoning, but high doses relax the muscles.

## OTHER PREPARATIONS

*Isolated superior cervical ganglion of the rat.*[43] The ganglia were stored in Krebs–Ringer bicarbonate solution, but were tested in oil. Pre-treatment with $3 \times 10^{-6}$ M DFP caused an increase in the response of the post-ganglionic fibres to submaximal pre-ganglionic stimulation, especially at low frequencies. $10^{-5}$ M DFP slightly decreased the post-ganglionic response, but caused some post-ganglionic fibres to continue firing for several seconds after stimulation had ceased (after-discharge). The degree of facilitation was correlated with inhibition of AChE; but the depression of response took place when the AChE was apparently completely inhibited.

*Isolated guinea pig lungs.*[44] The amplitude of expansion and contraction was greatly decreased by Paraoxon and DFP at low doses. (The exact doses cannot be calculated from the experimental results. They were probably in the range $5 \times 10^{-5}$—$5 \times 10^{-6}$ M.) The amplitude was not restored by atropine, but was by adrenaline. The mechanism of the response is not known.

## SUMMARY AND EXTENSION

There are three types of effect, depending upon the concentration of phosphorus compound.

Very high concentrations block axons and depolarize some muscles. These actions cannot be reversed by washing the preparations after exposure has been maintained for more than a few minutes. It is probable that any oil-soluble phosphorus compound, whether an inhibitor or not, can act in this way, and that the effect is, therefore, a physical one on the cell membranes.

At concentrations around $10^{-4}$ M many compounds, including indirect inhibitors, block the receptors on muscles, which are not depolarized. This action is rapidly reversed by washing. Slow reversal of any action after washing could, perhaps, be explained by the capacity of muscles to adapt. The rapid reversal in these cases is strong evidence for a curarimimetic mechanism. Similar, but less complete blocking, must be caused by lower concentrations, but is swamped, in the case of direct inhibitors, by the third type of action.

The third type of action depends upon the inhibition of AChE, and, in intestinal preparations, of ΨChE. The results are those expected to follow accumulation at the end-plates of ACh released from the nerve endings. Atropine antagonizes the effects on end-plates served by the parasympathetic system, and D-tubocurarine the effects on end-plates innervated by the voluntary motor system. Curare antagonism is most effective when the frequency of stimulation is fairly low, and is therefore likely to be more effective *in vivo* than in many of the experiments described.

At toxic doses *in vivo* only the third type of effect could normally be observed, but in animals protected by atropine curarimimetic effects could also be important.

## REFERENCES

1. T. H. Bullock, H. Grundfest, D. Nachmansohn, M. A. Rothenberg and K. Sterling. *J. Neurophysiol.* **9,** 253 (1946)
2. F. Crescitelli, G. B. Koelle and A. Gilman. *J. Neurophysiol.* **9,** 241 (1946)
3. T. H. Bullock, H. Grundfest, D. Nachmansohn and M. A. Rothenberg. *J. Neurophysiol.* **10,** 11 (1947)
4. J. E. P. Toman, J. W. Woodbury and L. A. Woodbury. *J. Neurophysiol.* **10,** 429 (1947)
5. I. B. Wilson and M. Cohen. *Biochim. biophys. Acta* **11,** 147 (1953)
6. E. A. Feld, H. Grundfest, D. Nachmansohn and M. A. Rothenberg. *J. Neurophysiol.* **11,** 125 (1948)
7. R. O. Brady, C. S. Spyrolpoulos and I. Tasaki. *Amer. J. Physiol.* **194,** 207 (1958)
8. E. J. Boell and D. Nachmansohn. *Science* **92,** 513 (1940)

9. E. Schoffeniels and D. Nachmansohn. *Biochim, biophys. Acta* **26,** 1 (1957)
10. W. G. van der Kloot. *J. gen. Physiol.* **41,** 879 (1958)
11. K. Naess. *Acta pharmacol. tox.* **12,** 154 (1956)
12. J. A. B. Barstad. *Arch. internat. Pharmacodyn.* **107,** 4 (1956)
13. A. S. V. Burgen, C. A. Keele, M. Chennells, J. del Castillo, W. F. Floyd, D. Slome and S. Wright. *Nature, Lond.* **160,** 760 (1947)
14. A. S. V. Burgen, C. A. Keele and D. Slome. *J. Pharmacol.* **96,** 396 (1949)
15. J. H. Fleisher, J. P. Corrigan and J. W. Howard. *Brit. J. Pharmacol.* **13,** 291 (1958)
16. J. H. Fleisher, J. W. Howard and J. P. Corrigan. *Brit. J. Pharmacol.* **13,** 288 (1958)
17. J. Jacob and M. Dechavassine. *Experientia* **12,** 434 (1956)
18. J. A. Cohen and C. H. Posthumus. *Acta physiol. pharm. Neerl.* **4,** 17 (1955)
19. J. Eccles and F. W. McFarlane. *J. Neurophysiol.* **12,** 59 (1949)
20. J. M. Barnes and J. I. Duff. *Brit. J. Pharmacol.* **8,** 334 (1953)
21. J. A. B. Barstad. *Arch. internat. Pharmacodyn.* **107,** 21 (1956)
22. C. van der Meer and E. Meeter. *Acta physiol. pharm. Neerl.* **4,** 454 (1956)
23. E. Meeter. *J. Physiol.* **144,** 38 (1958)
24. J. M. Barnes and J. I. Duff. *Brit. J. Pharmacol.* **9,** 153 (1954)
25. A. S. V. Burgen and F. Hobbiger. *Brit. J. Pharmacol.* **6,** 593 (1951)
26. R. Holmes and E. L. Robins. *Brit. J. Pharmacol.* **10,** 490 (1955)
27. T. Fredriksson and G. Tibbling. *Biochem. Pharm.* **2,** 63 (1959)
28. J. Axlesson and S. Thesleff. *Acta physiol. scand.* **43,** 15 (1958)
29. J. Axelsson, E. Gjone and K. Naess. *Acta pharmacol. tox.* **13,** 319 (1957)
30. K. Naess and E. Sognen. *Acta pharmacol. tox.* **14,** 333 (1958)
31. A. S. V. Burgen, F. Dickens and L. J. Zatman. *J. Physiol.* **109,** 10 (1949)
32. V. B. Brooks. *J. Physiol.* **123,** 501 (1954)
33. V. B. Brooks. *J. Physiol.* **134,** 264 (1956)
34. J. Stovner. *Acta pharmacol. tox.* **12,** 175 (1956)
35. P. R. Salerno and J. M. Coon. *J. Pharmacol.* **95,** 240 (1949)
36. H. Shelley. *Brit. J. Pharmacol.* **10,** 26 (1955)
37. E. D. Adrian, W. Feldberg and B. A. Kilby. *Brit. J. Pharmacol.* **2,** 56 (1947)
38. W. D. Erdmann and D. Heye. *Arch. exp. Path. Pharmak.* **232,** 507 (1957)
39. A. Gioia and C. Morpurgo. *Brit. J. Pharmacol.* **13,** 467 (1958)
40. W. Feldberg and R. C. Y. Lin. *J. Physiol.* **109,** 475 (1949)
41. G. B. Koelle, E. S. Koelle and J. S. Friedenwald. *J. Pharmacol.* **100,** 180 (1950)
42. K. Bethe, W. D. Erdmann, L. Lendle and G. Schmidt. *Arch. exp. Path. Pharmak.* **231,** 3 (1957)
43. D. A. Holaday, K. Kamijo and G. B. Koelle. *J. Pharmacol.* **111,** 241 (1954)
44. T. O. King, E. Poulsen and L. A. Rovati. *Arch. internat. Pharmacodyn.* **110,** 77 (1957)
45. K. Krujevic and R. Miledi. *J. Physiol.* **140,** 440 (1958).
46. D. W. Straughan. *Brit. J. Pharmacol.* **15,** 417 (1960)
47. J. A. B. Barstad. *Arch. internat. Pharmacodyn.* **128,** 143 (1960)

## CHAPTER XIII

# Effects on Partly Isolated Organs *in vivo*

THIS and the following Chapter are on the effects most commonly observed. Unusual effects, caused by a few compounds only, are described in Chapter XVI. The present Chapter is devoted to experiments in which the action of phosphorus compounds is mainly confined to the particular part of the body into which they are injected. Under ideal circumstances such experiments show the effect on the animal of poisoning one part of it only, and can be interpreted more simply and with more certainty than experiments in which the whole animal is poisoned. In practice the compounds are not confined solely to one part of the body, though they can often be confined sufficiently for it to be reasonably certain that the effects observed are caused only by their action on the part treated. Because, however, the compounds do become generally distributed to some extent, it is very difficult to assess the doses actually operating on the treated part, and hence to compare the effects quantitatively with those in a generally poisoned animal. Further uncertainty is introduced by the drastic surgery needed to isolate some organs, the effects of which are hard to assess.

The effects now described are grouped according to the part isolated.

## VOLUNTARY MUSCLE

The only 'isolation' of voluntary muscle which is practicable is that involved in close arterial injection. This ensures that the muscle is the first organ to receive the compound, but signs of poisoning usually appear elsewhere before the response of the muscle is at a maximum, as phosphorus inhibitors act progressively, and a good circulation through the muscle has to be maintained during the experiment if reproducible results are to be got.

Very high doses of some compounds elicit an immediate response. Thus 1–10 mg/kg of DFP or Sarin produce an immediate contraction, lasting about a minute, in the denervated gastrocnemius muscle of the cat.[1] Concentrations greater than 0·1% must be used. The mechanism of action is unknown. The response does not depend on the inhibition of AChE, for injection of the salts of several hydrogen phosphates at similar concentrations cause similar contractions. Results at these concentrations are not, of course, relevant to results obtained in generally poisoned animals.

Lower concentrations do not elicit this response, and their effects may be attributed entirely to inhibition of AChE. In the cat close arterial injection of sublethal doses of TEPP, DFP or Sarin into the tibialis anterior or gastrocnemius muscles have the following effects:

(a) The capacity to hold a tetanus is lost.[2, 3]

(b) Low frequency stimulation of the motor nerve causes repetitive firing in the muscle fibres, which is accompanied by an increase in response (Fig. 13.1). The response passes through a maximum like that of the rat nerve-diaphragm preparation.[3]

(c) The twitch response of the curarized muscle is restored[4] (Fig. 13.2).

(d) The injection of ACh, or tetanic stimulation of the nerve, leads to a depression in the later response of the muscle to low frequency stimulation, either because the muscle becomes depolarized, or because it becomes adapted to ACh.

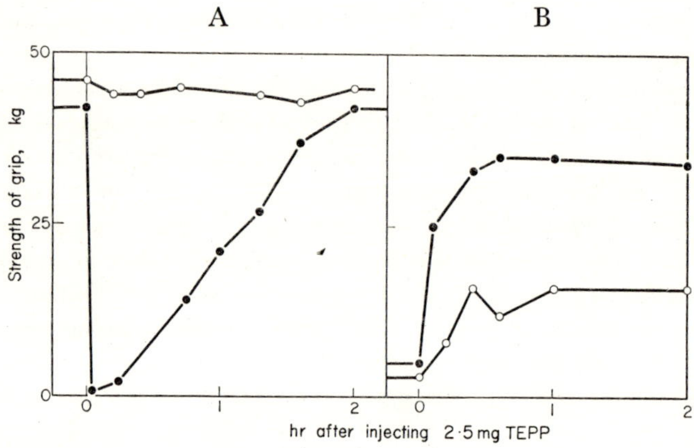


FIG. 13.3. Effect on the grip strength of injecting TEPP into the right brachial artery of a normal man and a myasthenic. Dose: 2·5 mg. 3A, normal; 3B, myasthenic; right hand, ●; left hand ○.[5]

After D. Grob and A. M. Harvey. *Bull. Johns Hopkins Hosp.* **84,** 532 (1949) (with acknowledgements).

These effects are very similar to those caused by eserine and other reversible inhibitors, except that they are somewhat slower in onset.

In man effects can be produced with less interference from poisoning outside the region of the injection. The subjects in the following experiments were controls and sufferers from myasthenia gravis.[5] This condition is characterized by great muscular fatiguability due to the failure of neuromuscular transmission. Whether transmission fails because less ACh than usual is excreted from the endings of excited nerves, or because the end-plates are less sensitive to ACh, is not known.[6, 7, 8, 9] Recent evidence favours the second theory.[10] Anti-AChE's restore the function, as they

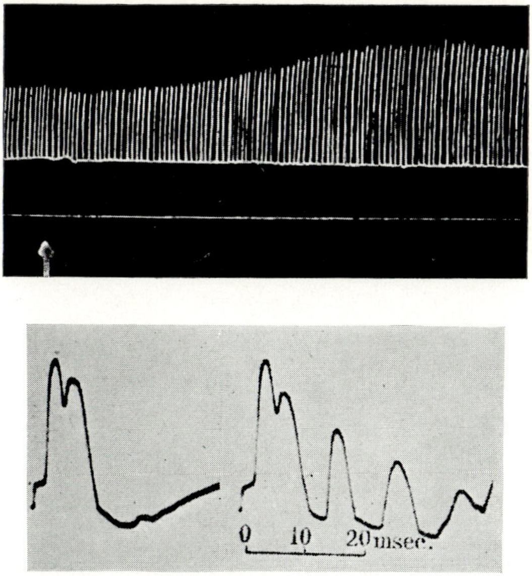


Fig. 13.1. Increased contractile response of the cat's tibialis anterior muscle to low frequency, supramaximal indirect stimulation after injecting 100 $\mu$g of DFP i-a at the arrow. The single action potential set up in the normal muscle fibre is shown below to the left, and the burst of potentials set up by a stimulus after DFP on the right.[3]

Reproduced from G. L. Brown, B. D. Burns and W. Feldberg. *J. Physiol.* **107,** 346 (1948) (with acknowledgements).

Fig. 13.2. D-Tubocurarine–DFP antagonism in the cat's gastrocnemius muscle. D-Tubocurarine (0·1 mg/kg) was injected at 11·06 and DFP (0·2 mg/kg) at 11·12.[4]

Reproduced from C. C. Hunt. *J. Pharmacol.* **91,** 77 (1947) (with acknowledgements).

should on either theory.* Both controls and patients were given an intra-arterial injection of TEPP into the right brachial artery, and the changes in the grip strengths of the right and left hands recorded. In the controls the strength of the right, injected hand decreased, and remained low for 2 hr, i.e. the muscles could not sustain a tetanus (Fig. 13.3A), while the left hand was unaffected. In the patients the right hand gained strength rapidly, and the left hand more slowly (Fig. 13.3B). d-Tubocurarine increased the rate of recovery of the controls, and caused the hands of the myasthenics to lose strength again.

## EYES

Low doses of direct AChE inhibitors, whether phosphorus compounds or not, cause myosis (constriction of the iris, pin-point pupils) when placed in the eye. Only the action of phosphorus compounds is described here, but it does not differ except in duration and reversibility from that of reversible inhibitors. The least concentration causing myosis depends upon $I_{50}$ and species, and to a minor extent upon the volume applied. Ionic compounds are less effective than oil-soluble ones of the same $I_{50}$, because they are less readily absorbed.[13] In the guinea pig the least concentrations of oil-soluble compounds causing maximal myosis are proportional to their AChE $I_{50}$'s,[14] and other, scattered observations suggest that the correlation holds for other species also. Indirect inhibitors do not cause myosis. The quantity of a given compound required per eye varies greatly with species, but if the dose is expressed in mg/kg total body weight the differences are far less, presumably because, in the animals studied, the smaller species have smaller eyes. On this basis, guinea pigs and rabbits[14] are more sensitive than dogs[15] and men.[16] The doses needed per eye are very small—that per eye of Sarin is about $2 \times 10^{-5}$ i-v $LD_{50}$'s in the guinea pig and $2 \times 10^{-4}$ $LD_{50}$'s in man. Such doses cause strictly localized effects—they do not cause signs of poisoning elsewhere, or even inhibit blood AChE to a significant extent.[17] The duration of myosis does not vary much with dose, providing it is great enough to cause the maximal effect. Myosis lasts longer the more stable the inhibited AChE. Thus in the guinea pig[14] myosis from dimethyl phosphate esters last only about 2 hr, whereas that from diethyl phosphate esters lasts for over 10; and in man myosis from Sarin and diethyl phosphate esters[16] lasts about 3 days, whereas that from DFP lasts over a week.[18] Doses just inadequate to cause myosis reduce the capacity of the pupil to accommodate to sudden changes in light intensity, a sign also found during recovery from myosis. Myosis and poor accommodation are antagonized by atropine,[19, 20]

* The illness is in fact treated with anti-AChE's. Both phosphorus and reversible inhibitors have been used successfully.[9, 11, 12] The former give more prolonged relief from a single dose, but require more skill in management, as the effects of an overdose are more difficult to control.

which, however, acts much more slowly against some phosphorus compounds than against reversible inhibitors such as eserine.*

Myosis is caused by inhibition of AChE in the sphincter pupillae muscle. This is parasympathetically innervated, and nervous excitation causes contraction of the iris. Effects first appear when the AChE drops below about 15% of normal.[21] Complete inhibition does not cause the iris to relax in the way a voluntary muscle does, i.e. mydriasis is not caused.

Higher concentrations than are needed to cause myosis cause lacrimation by inhibiting the AChE in the lacrimal gland, so that it is stimulated by the ACh released.[19] Sarin, and presumably other inhibitors, also raise the absolute scotopic threshold, i.e. the light intensity which is just visible. This effect, unlike those described before, seems to be central in origin, as treatment of one eye affects both, and the threshold is also raised when the inhibitor is inhaled.[23]

Still higher doses lead to the general poisoning of the animal. The eye rapidly absorbs oil-soluble compounds, and the ocular $LD_{50}$ may be not much greater than the intravenous. In rabbits for example the $LD_{50}$ of Sarin is only about twice times greater.[24]

## SALIVARY GLANDS

Both the submaxillary and parotid glands can be studied in isolation. Close arterial injection into the gland on one side barely affects its twin on the other, which is therefore used as a control. It is unlikely that the slight effects observed in other parts of the body, especially the facial muscles, affect observations on the glands.[25] For experiments the glands are separated from the central nervous system by cutting the chorda tympani (to the submaxillary) or Jacobsen's anastomosis (to the parotid), and the nerve stumps are prepared for stimulation. Stimulation causes the glands to excrete saliva, which stops flowing when stimulation is stopped. Secretion begins at a definite frequency of supramaximal stimulation, and increases in rate with the frequency up to a maximum. The glands also excrete after a close arterial injection of ACh, and themselves excrete ACh, slowly when not stimulated,[60] and faster when stimulated.[26]

The effects on the submaxillary gland of several direct reversible and irreversible inhibitors have been studied. The results are wholly consistent with the anti-AChE theory of poisoning, and so will be described in terms of it.

In the cat inhibition of up to 50% of the AChE had no effect. Greater inhibition caused the threshold frequency to decrease; the rate of flow of

* Myosis is accompanied by a lowering of intra-ocular pressure. Certain phosphorus compounds therefore find a use in the treatment of glaucoma.[22] It is outside the scope of this book to review the extensive literature published on this subject, especially in Russia.

saliva at any given frequency to increase; the saliva to continue flowing after stimulation had ceased (after-flow); and, when AChE inhibition reached about 90%, caused a spontaneous excretion to begin without the nerve being stimulated. The magnitude of all these effects was closely correlated with the percentage of AChE inhibited.[27] Experiments with TIPA and Mipafox as selective inhibitors of ΨChE in the cat and the dog showed that the effects did not follow the inhibition of most of the ΨChE, but did follow inhibition of AChE.[25] It seemed, however, that inhibition of ΨChE sensitized the gland to some extent, as in intestines. Low doses of an inhibitor potentiated the action of injected ACh, but high doses caused injected ACh to stop spontaneous flow for a short time. As there was a maximum attainable flow after AChE inhibition, which was not increased by further inhibition, it is likely that ACh can cause a depolarizing block, but that endogenous ACh is just inadequate to do so. The parasympathetic ganglia in the submaxillary cannot be extirpated surgically, so that it is not quite certain whether the effects are caused by inhibition in the ganglia or at the nerve–gland junctions. As hexamethonium only slowed excretion slightly, inhibition in the ganglia is probably unimportant. Excretion was stopped rapidly by a small dose of atropine, and was then not easily re-started, even by high doses of anti-AChE's.

The parotid gland in the cat can be either decentralized or denervated, by cutting Jacobsen's anastomosis or the auriculo-temporalis nerves respectively, and allowing the nerves to degenerate. The response of the normal gland was like that of the submaxillary. Decentralization increased the spontaneous flow induced by both ACh and anti-AChE's, whereas denervation increased the sensitivity to ACh but rendered anti-AChE's almost without effect. These experiments establish that the ACh which causes the spontaneous flow when the AChE is inhibited comes from the peripheral nerve endings, and is secreted there whether the nerves are stimulated or not.[28]

The results described suggest that the effects of anti-AChE's on salivary glands will be potentiated in generally poisoned animals by any general increase in the ACh concentration in the blood.

## GANGLIA

Few attempts have been made to confine the action of phosphorus inhibitors to ganglia. One example is given. 3-(di-isopropoxyphosphinyl-oxy)-N-methylpyridinium methosulphate,[29]

$$Me\overset{\oplus}{N}C_5H_4\text{-3-O}\cdot PO(OPr^i)_2\ MeSO_4^{\ominus},$$

causes similar effects to eserine[30] in the perfused superior cervical ganglion of the cat. The pre-ganglionic fibres were stimulated supramaximally at 15/sec, and the response of the nictitating membrane was recorded. When no anti-AChE had been given the response was a sustained contraction (Fig. 13.4A), which changed on dosing to that shown in Fig. 13.4B: a rapid rise was followed by a transient fall, and then by a further slow rise to a second maximum. Injection of ACh also caused a transient fall. The reduction in response in the inhibited ganglion is plainly caused by accumulation of ACh. The recovery takes place partly because the quantity of ACh released in the ganglion per stimulus falls off with time when stimulation is continuous,[30] and partly because the post-ganglionic fibres adapt to ACh[31] (p. 272).

The results show that it is possible to block synaptic transmission in ganglia with anti-AChE's, but not whether this is likely to take place in a generally poisoned animal. Probably any block will be transient.

## THE CENTRAL NERVOUS SYSTEM

Compounds can be administered so that their effects are mainly confined to the central nervous system (CNS) by injecting them into the mass of the brain cells (intracerebrally), or into the cerebrospinal fluid (by intracisternal, intraventricular or intrathecal injection), or into an artery, usually the carotid, which serves the CNS. Given by any of these routes some poison enters the general circulation, and that remaining in the CNS always appears to be unevenly distributed, as shown by uneven inhibition of AChE. (As shown later, uneven distribution itself has interesting results.) Intra-carotid injection of ionic inhibitors causes little inhibition in the CNS, as is to be expected from results already given (Chapter VIII). Even when injected intraventricularly such compounds cause much less inhibition than might be expected. For example, 0·5 i-v $LD_{50}$'s of 3-(di-isopropoxyphosphinyloxy)-N-methylpyridinium methosulphate injected into one lateral ventricle of the dog mainly inhibited AChE near the site of the injection, and that incompletely;[29] and 2-(diethoxyphosphinylthio)-ethyltrimethylammonium iodide, $(EtO)_2PO \cdot S \cdot CH_2 \cdot CH_2 \cdot \overset{\oplus}{N}Me_3, I^{\ominus}$, injected similarly into the rabbit caused partial inhibition only, although appreciable concentrations persisted in the brain for up to 4 days.[33] The CNS obviously has barriers which protect its AChE from ionic inhibitors, however they are given. Even an oil-soluble compound given intraventricularly may have only local effects, as compounds do not move rapidly from the lateral ventricles into other parts of the CNS.[34] The distribution of a compound after injection into the CNS is thus likely to be different from its distribution when the CNS receives it in the normal circulation, as after intravenous injection, and the effects may be different also.

Central injection of oil-soluble compounds changes reflex behaviour, causes convulsions, and, when confined to one side, causes one-sided responses.

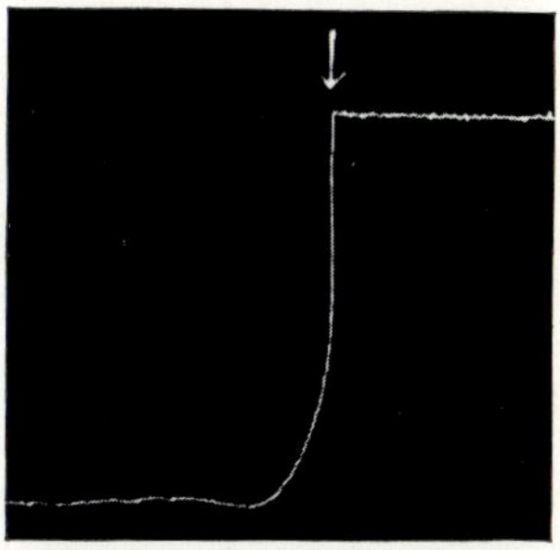

A

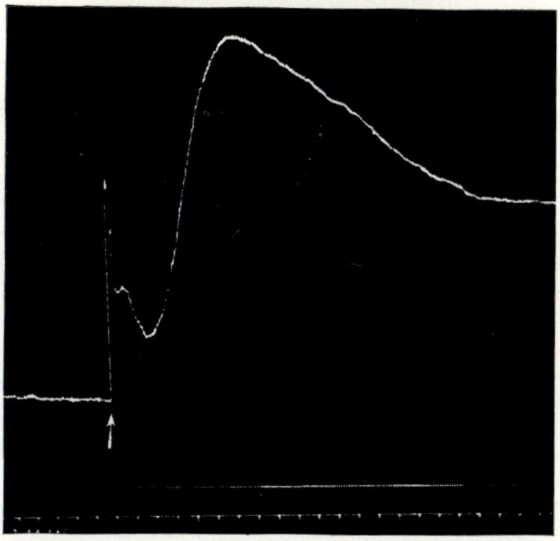

B

FIG. 13.4. Contraction of the nictitating membrane of the cat caused by pre-ganglionic, supramaximal stimulation at 12/sec, started at the arrow. 4A, recording from a normal cat; 4B, recording from a cat in which the superior cervical ganglion was perfused with $2 \times 10^{-6}$ eserine (g/g).[30] Reproduced from G. L. Brown and W. Feldberg. *J. Physiol.* **88,** 265 (1936) (with acknowledgements).

Spinal reflexes may be increased, decreased or unchanged by anti-AChE's. Injection by lumbar punction of 0·12–0·4 mg/kg of TEPP[35] or 0·7 mg/kg of DFP[36] into chloralosed cats greatly increased the knee jerk and flexor reflexes, and initiated the crossed extensor reflex, which is normally absent from chloralosed cats. Higher doses of DFP suppressed all three. Tabun was injected into decerebrate or deep spinal cats in such a way that it was nearly confined to a lumbar artery, and hence to the spinal cord.[37] 5–40 $\mu$g increased multisynaptic flexor reflexes greatly, and higher doses decreased

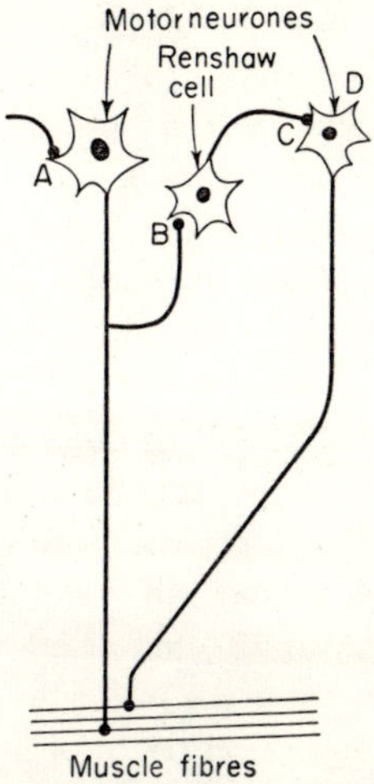


FIG. 13.5. Diagrammatic representation of an inhibitory neurone.

them. Facilitation was accompanied by repetitive firing in the motor-neurones. The monosynaptic flexor reflex was, however, only slightly increased, and the monosynaptic extensor reflex was decreased by all doses. Tabun increased monosynaptic extensor reflexes when injected into the lumbar artery via a donor animal, i.e. an artery of a donor cat was joined to the bottom of the aorta of the recipient. When the donor was given Tabun intravenously the recipient received an intra-arterial dose, which had been pre-mixed with a large volume of blood in the donor.[38] This may well have caused a different distribution of Tabun in the recipient's spinal cord, and so a different response.

Such varied results can be explained in principle by later discoveries. The ventral horn contains neurones (Renshaw cells) which are stimulated by cholinergic motor neurones, and which inhibit the responses of the motor neurones with which they synapse.[39] The basic 'circuit' is illustrated in Fig. 13.5. Stimulation of a motor neurone by a sensory or internuncial fibre at A causes the axon to fire. The axon branches, so the collateral axon also fires, and excites a Renshaw cell at B. This in turn synapses with a second motor neurone at C, and makes it refractory, so that it is less readily excited by a sensory or internuncial fibre at D. The synapse, B, is cholinergic, and injection of DFP, TEPP or eserine into the lumbar artery causes the Renshaw cells to fire repetitively after a single excitation of the motor axon.[39, 40] The conclusions reached have been fully confirmed by ionophoretic injection of ACh and other compounds into the synaptic gap at B. The cells are surrounded by complex barriers to diffusion.[41] Not enough is known of the pattern of synapses involving Renshaw cells to say that any given effect on a reflex can be explained anatomically. It is likely, however, that such cells would be linked differently to different groups of neurones, so that some reflexes may be more inhibited than others; and it is known that they are protected by diffusion barriers, so that they will not be equally readily affected by anti-AChE's differing in molecular size and oil-solubility. The discovery of these cells removes one of the major objections to the theory that poisoning in the CNS results from inhibition of AChE, namely that some compounds inhibit while others potentiate some types of central nervous activity.

Intracisternal injections of TEPP into dogs at low doses potentiated the hyperpnoea (abnormally deep and rapid breathing) caused by stimulation of Hering's nerve, and at higher doses inhibited it. The response was closely correlated with the AChE levels near the site of injection. It was maximally potentiated at 85% of normal, unaffected at 40%, and completely inhibited at 10%. The level fell to 10% after the injection of 0·116 mg/kg of TEPP.[42]

Injection of DFP into a lateral ventricle of an unanaesthetized cat gave rise to a complex pattern of behaviour.[43] 100 $\mu$g caused the cat to act as though it itched severely, to make general washing and wiping movements, and to move its jaws in a distinctive fashion. After a time the cat showed difficulty in breathing, changes in stance and gait, quivering in its head, shoulders and ears, and finally went into a state of reduced awareness resembling catatonia. 200 $\mu$g caused agitation and seizure-like actions; and 1 mg caused convulsions in a few minutes, followed by a deep stupor. Eserine and ACh given by the same route caused many similar effects, suggesting that DFP acted by AChE inhibition, leading to accumulation of ACh. The enormous doses are noteworthy. The top dose given was about 0·3 i-v $LD_{50}$'s,[44] yet the cat survived. It must be concluded that action of phosphorus compounds on the CNS of cats is not likely to be the major cause of death in most experiments.

Injection of oil-soluble anti-AChE's into one side of the brain tends to cause the animal to circle towards the other side (aversive circling).[45, 47, 48, 49, 50] Compounds which cannot be extracted from water by oils and fats do not cause this effect.[32] Most experiments have been carried out with DFP on rabbits. Injection of from 0·01 to 0·32 mg of DFP into one caudate nucleus caused prolonged aversive circling.[51] Up to 0·16 mg only the AChE in the caudate nucleus treated was inhibited, i.e. the DFP was confined there.

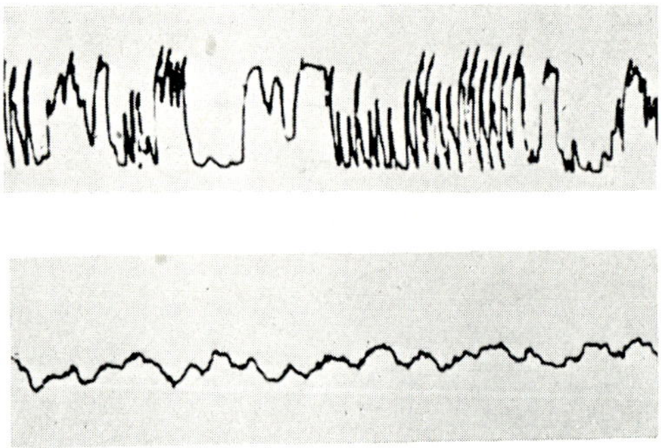

FIG. 13.6. Electroencephalographic records from the right cortex of a rabbit. The top line shows the 'grand mal' pattern obtained on injecting 1–1·5 mg/kg of DFP intra-carotidally. The bottom line shows the substantially normal pattern obtained on giving the treated rabbits 2–4 mg/kg of parpanit. Atropine gave similar results.[58]

Reproduced from C. F. Essig, J. L. Hampson, P. D. Bales, A. Willis and H. E. Himwich. *Science* **111,** 38 (1950) (with acknowledgements).

Somewhat higher doses caused aversive circling for a few minutes only, and the DFP spread to other tissues, including the other caudate nucleus. Doses above 2 mg killed the rabbit. The results are wholly consistent with the view that aversive circling is caused by selective inhibition of about 60% of the AChE in one caudate nucleus. Doses above 0·16 mg increased the inhibition on the side injected, but also inhibited AChE on the other side, so that the difference between the two sides rapidly became less and the syndrome was not then maintained. The rabbit died if the cortex AChE fell below about 13% of normal.* Aversive circling is also caused, though not reliably, by intra-carotid injection of DFP. In rabbits treated by this route circling was rarely stopped by removing the cortex, but was always stopped by removal of the caudate nucleus of the side affected; and aversive circling could develop when the cortex was removed before injection.[(52)] As, in addition, multiple, unilateral injections into the cortex did not cause circling,[(53)] it is clear that in the rabbit aversive circling is caused by action on the caudate nucleus and not by action on the cortex. In man[(54)] and monkey[(55)] action on the cortex may be more important, as circling can be caused either by removing portions of one side of the cortex, or by stimulating one side of it electrically. In rabbits intracarotid DFP sometimes did not cause circling, and sometimes caused circling towards the injected side (ipsilateral circling). The direction (aversive or ipsilateral) and rate of rotation were closely correlated with the difference between the concentrations of ACh in the two sides of the *cortex*, suggesting that the cortex controls circling.[(56, 57)] This theory is probably wrong. Action on the caudate nucleus is enough to explain aversive circling. Ipsilateral circling can be induced when the cortex and the caudate nuclei are all removed.[(52)] The correlation may be the result of action at a lower level in the CNS, i.e. unilateral action at a low level may induce unilateral accumulation of ACh at a higher. Both circling syndromes are abolished by high doses of atropine —in rabbits by 2–4 mg/kg. It is generally assumed that they are caused by AChE inhibition, and that atropine acts as an ACh antagonist in the CNS[(46)] as it does in ganglia at similar doses.

The injection of high doses of DFP by the carotid artery causes convulsions,[(46)] and the electroencephalograph records waves similar to those obtained during epileptic seizures (the 'grand mal' pattern, Fig. 13.6). The pattern is suppressed by 2–4 mg/kg of atropine or parpanit. Such patterns have often been recorded in rabbits given several i-v $LD_{50}$'s, the peripheral effects being controlled with low doses of atropine. It follows that action in the brain cannot be a major cause of death in rabbits poisoned with DFP.

* As DFP tends to be stored in brain tissue, and is released on homogenizing, the levels stated are probably too low. They are, however, a reliable indication of the quantity of DFP present in each part of the brain, whether inhibition by it took place *in vivo*, or, later *in vitro*.

The brain of one dog can be isolated completely from its body except for the spinal cord, and a blood supply for it provided by a donor dog. In a series of such experiments[59] Sarin was injected into the carotid of the recipient. Two to three $LD_{50}$'s barely affected the heart rate of the donor dog, showing that the brain of the recipient had received much of the dose. At higher doses the donor was affected, and the effects were controlled with atropine. In the recipient 2–3 $LD_{50}$'s first greatly increased the respiration frequency and slightly increased the depth of breathing. The depth then fell over 15 min to normal, while the frequency remained high. A further 2 $LD_{50}$'s caused a transient increase in frequency, followed by reduction in both frequency and depth, and the arterial pressure rose from 150 to 210 mm Hg. Thus these enormous doses stimulated the vasomotor centre and partially blocked the respiratory centre. Atropine injected into the carotid artery antagonized both effects. The results show that the normal lethal action of Sarin in the dog cannot be due to direct inhibition of the respiratory centre, which must receive very much less Sarin than in these experiments.

## REFERENCES


1. G. E. Groblewski, B. P. McNamara and J. H. Wills. *J. Pharmacol.* **118,** 116 (1956)
2. R. Holmes and E. L. Robins. *Brit. J. Pharmacol.* **10,** 490 (1955)
3. G. L. Brown, B. D. Burns and W. Feldberg. *J. Physiol.* **107,** 346 (1948)
4. C. C. Hunt. *J. Pharmacol.* **91,** 77 (1947)
5. D. Grob and A. M. Harvey. *Bull. Johns Hopkins Hosp.* **84,** 532 (1949)
6. J. E. Desmedt. *Nature, Lond.* **182,** 1673 (1958)
7. A. Lanari. *C.R. Soc. Biol. Paris* **126,** 827 (1937)
8. A. M. Harvey and J. L. Lilienthal, Jr. *Bull. Johns Hopkins Hosp.* **69,** 566 (1941)
9. N. P. Bergh. *Scand. J. lab. clin. Invest.* **5,** Suppl. 5–7 (1953)
10. D. Grob, R. J. Johns and A. M. Harvey. *Bull. Johns Hopkins Hosp.* **99,** 115 *et seq.* (1956)
11. A. S. V. Burgen, C. A. Keele and D. McAlpine. *Lancet* **254,** 519 (1948)
12. H. Aranow, Jr., P. F. Hoefer and L. P. Rowland. *J. chronic Diseases* **6,** 457 (1957)
13. T. Fredriksson. *Arch. internat. Pharmacodyn.* **115,** 474 (1958)
14. D. M. Sanderson. *J. Pharm. Lond.* **9,** 600 (1957)
15. C. Dayrit, C. H. Manry and M. H. Seevers. *J. Pharmacol.* **92,** 173 (1948)
16. D. Grob and J. C. Harvey. *J. clin. Invest.* **37,** 350 (1958)
17. D. Grob. *Arch. int. Med.* **98,** 221 (1956)
18. E. D. Adrian, W. Feldberg and B. A. Kilby. *Brit. J. Pharmacol.* **2,** 56 (1947)
19. R. O. Scholz. *J. Pharmacol.* **88,** 23 (1946)
20. W. Wirth. *Deutsch. med. Wochsch.* **74,** 1243 (1949)
21. A. da Roetth, Jr. *J. Neurophysiol.* **14,** 55 (1951)
22. W. M. Grant. *Pharmacol. Rev.* **7,** 143 (1955)
23. L. S. Rubin and M. N. Goldberg. *J. appl. Physiol.* **12,** 305 (1958)
24. A. A. Kondritzer, W. H. Mayer and P. Zvirblis. *Arch. industr. Hlth.* **20,** 50 (1959)
25. P. Dirnhuber and C. Lovatt Evans. *Brit. J. Pharmacol.* **9,** 441 (1954)

26. N. Emmelin and A. Muren. *Acta physiol. scand.* **20,** 13 (1950)
27. W. F. Riker and W. C. Wescoe. *J. Pharmacol.* **95,** 515 (1949)
28. N. Emmelin and B. C. R. Strömblad. *Brit. J. Pharmacol.* **13,** 193 (1958)
29. A. S. V. Burgen and L. M. Chipman. *Quart. J. exp. Physiol.* **37,** 61 (1952)
30. G. L. Brown and W. Feldberg. *J. Physiol.* **88,** 265 (1936)
31. W. A. Krivoy and J. H. Wills. *J. Pharmacol.* **116,** 220 (1956)
32. T. Fredriksson. *Arch. internat. Pharmacodyn.* **113,** 101 (1957)
33. G. B. Koelle and E. C. Steiner. *J. Pharmacol.* **118,** 420 (1956)
34. W. B. Bhawe. *J. Physiol.* **140,** 169 (1958)
35. M. Chennells, W. F. Floyd and S. Wright. *J. Physiol.* **108,** 375 (1949)
36. M. Chennells, W. F. Floyd and S. Wright. *J. Physiol.* **114,** 107 (1951)
37. B. Holmstedt and C. R. Skoglund. *Acta physiol. scand.* **29,** 410 (1953)
38. J. Haase, D. Lücke, F. Scheler, R. Schütz, B. Mühlberg and W. Koll. *Arch. exp. Path. Pharmak.* **232,** 274 (1957)
39. J. C. Eccles, P. Fatt and K. Koketsu. *J. Physiol.* **126,** 524 (1954)
40. J. C. Eccles, R. M. Eccles and P. Fatt. *J. Physiol.* **131,** 154 (1956)
41. D. R. Curtis and R. M. Eccles. *J. Physiol.* **141,** 446 (1958)
42. B. Metz. *Amer. J. Physiol.* **192,** 101 (1958)
43. W. Feldberg and S. L. Sherwood. *J. Physiol.* **125,** 488 (1954)
44. R. G. Horton, G. B. Koelle, B. P. McNamara and H. J. Pratt. *J. Pharmacol.* **87,** 414 (1946)
45. A. M. Freedman and H. E. Himwich. *Amer. J. Physiol.* **156,** 125 (1949)
46. A. M. Freedman, P. D. Bales, A. Willis and H. E. Himwich. *Amer. J. Physiol.* **153,** 117 (1949)
47. P. Nathan and M. H. Aprison. *Biochim. biophys. Acta* **18,** 150 (1955)
48. M. H. Aprison, P. Nathan and H. E. Himwich. *Science* **119,** 158 (1954)
49. C. T. Harwood. *Amer. J. Physiol.* **177,** 171 (1954)
50. M. H. Aprison, P. Nathan and H. E. Himwich. *Amer. J. Physiol.* **177,** 175 (1954)
51. R. P. White. *Proc. Soc. exp. Biol. N.Y.* **93,** 113 (1956)
52. R. P. White and H. E. Himwich. *Amer. J. Physiol.* **189,** 513 (1957)
53. R. P. White and H. E. Himwich. *J. Neurophysiol.* **20,** 81 (1957)
54. W. Penfield and T. Rasmussen. *The Cerebral Cortex in Man* N.Y., Macmillan, (1952)
55. M. A. Kennard and L. Ectors. *J. Neurophysiol.* **1,** 45 (1938)
56. M. H. Aprison. *J. Neurochem.* **2,** 197 (1958)
57. M. H. Aprison and P. Nathan. *Amer. J. Physiol.* **189,** 389 (1957)
58. C. F. Essig, J. L. Hampson, P. D. Bales, A. Willis and H. E. Himwich. *Science* **111,** 38 (1950)
59. H. Polet and A. F. de Schaepdryver. *Arch. internat. Pharmacodyn.* **118,** 231 (1959)
60. N. Emmelin. *Nature, Lond.* **185,** 297 (1960)

CHAPTER XIV

# Effects on Mammals

THE MAIN signs of poisoning in mammals and their causes are dealt with in this Chapter. Symptoms in man, along with methods of therapy in general, are described in the next Chapter; and unusual signs, only caused by a few compounds, in the one following.

The signs of poisoning are listed in Table 14.1. Almost every sign can be observed in every mammal, whatever phosphorus compound it is poisoned with, but the intensity and relative importance of each effect varies with compound and with species, and to a lesser extent from animal to animal within a species.

The signs of poisoning can be divided according to the part of the nervous system mainly affected, in the way shown in Table 14.1. They are first related to their causes under these heads. Major variations with type of compound and species are described afterwards.

## SIGNS OF POISONING

*Organs innervated by the parasympathetic system.* Signs in these organs are caused by the potentiation of the responses of glands and muscles to ACh by the mechanisms described in the preceding two Chapters.

Salivation and lacrimation are conspicuous signs in poisoned animals. Their causes have been described sufficiently already.

Bronchosecretion is usually only slightly increased,[9] and myosis is rarely severe, except when the eye is treated directly. Signs in which the gastro-intestinal tract are involved are those expected from the violent contractions observed *in vitro*. These effects will not be discussed further.

Chromodacryorrhea results in normal rats from stimulation of the Harderian gland, or from the action of ACh injected intravenously. It takes place spontaneously in seriously poisoned rats. Of most interest, however, is the extent to which less toxic doses of phosphorus compounds lower the threshold dose of ACh required to elicit it. The threshold is lowered by a factor which is very closely correlated with the degree of AChE inhibition, as shown by the following results:

(1) The logarithm of the factor increases linearly with the dose of anti-AChE.[10]

TABLE 14.1. SIGNS OF POISONING[1, 2, 3, 4, 5]

| Innervation | Signs |
|---|---|
| Parasympathetic | Excessive salivation, retching, vomiting, borborygmus (belly rumblings), diarrhoea, tenesmus (anal spasms), lacrimation, in rats chromodacryorrhea (weeping of red tears), myosis, bronchosecretion, cardiac slowing (bradycardia), and frequent micturition with urinary incontinence. |
| Voluntary | Fibrillations, fasciculations, muscular weakness, ataxia and paralysis. |
| Complex | Transient hyperpnoea (rapid and deep breathing), followed by gasping and respiratory failure; changes in blood pressure, failure of circulation, convulsions, hyperexcitability to sensory stimuli,[6] acidosis[7] and hyperglycaemia.[8] |

(2) Maximum potentiation is reached very soon after injecting eserine, but more slowly after injecting phosphorus inhibitors, which inhibit progressively.[10]

(3) The potentiation drops as the AChE is re-activated: rapidly after eserine, over a few days after TEPP, and over several days after di-isopropyl phosphate esters.[11, 12]

(4) Recovery takes place at an equal rate in rats given different phosphorus compounds with the same basic (alkyl) groups.[11, 12]

Micturition results because the bladder empties itself reflexly when containing lower volumes than usual.[13] The potentiation of micturition is probably not only an effect of ACh accumulation on the smooth muscle of the bladder. The micturition reflex is complex,[14] and effects on the CNS and ganglia may also contribute.

Bronchoconstriction is caused by parasympathetic stimulation. It is dealt with in more detail later, when syndromes are described, as it is often an important contributory cause of death.

Phosphorus compounds cause bradycardia (heart-slowing) at an early stage of poisoning.[15] Bradycardia develops into A–V block, and sometimes the heart stops beating. For more details the reader is referred elsewhere.[1, 4, 15] Bradycardia is not centrally determined.[16] In dogs, for example, Sarin caused bradycardia after vagotomy and in a decapitated trunk, but the heart did not slow in dogs in which the circulation for the head was provided by a donor-dog treated with DFP. The bradycardia caused by intravenous injection of ACh or by vagal stimulation is greatly potentiated by anti-AChE's. Atropine abolishes bradycardia, and causes

ACh to accelerate the heart.[13, 17, 18, 19] The main effect of phosphorus compounds is therefore on the parasympathetic peripheral synapses, and can be explained by accumulation of ACh. Atropine in small doses antagonizes ACh at these sites but not at the autonomic ganglia. In most species the heart is not much affected when the poison is given after atropine, which suggests that the sympathetic and parasympathetic systems are restored to a rough balance. The sympathetic ganglia and/or the adrenal medulla are, however, somewhat excited, as injection of ACh after atropine accelerates the heart. Bradycardia and other effects on the heart are rarely the main causes of death, but are often contributory factors.

*Signs in voluntary muscles.* Fibrillations and fasciculations are a marked feature of poisoning, and often involve most of the superficial musculature. The probable mechanisms of these responses have been discussed already (p. 269). In the whole animal fasciculations may also be caused by monosynaptic reflexes, i.e. fibrillations stimulate the stretch receptors, which in turn set up a reflex.

The effects of phosphorus compounds on nerve-muscle systems have been investigated in great detail. Most of the observations can be described under seven headings.

(1) Twitch responses to single stimuli give way to brief tetani. The force of contraction often passes through a maximum.

(2) As AChE inhibition proceeds the twitch response is reduced, and is finally *suppressed completely*. Often one dose will increase the response, and another, only a little higher, will suppress it. The doses which suppress responses give concentrations *in vivo* far lower than those which exert curarimimetic effects *in vitro*.

(3) D-Tubocurarine antagonizes both of these effects.

(4) Succinylcholine or decamethonium causes the response to fall more rapidly, and to recover more slowly.

(5) The capacity to hold a tetanus is lost, especially at high frequencies of stimulation.

(6) D-Tubocurarine restores the capacity to hold a tetanus, especially in the early stages of inhibition.

(7) Re-activators of phosphorylated AChE's such as P2-AM and hydroxyimino-acetone restore the response to both low and high frequency stimulation.

Many of the published results are summarized in Table 14.2. The numbers under 'Effects' in the table refer to those in the text above. The responses were only elicited in many instances at doses lethal in the absence of therapy, and were therefore observed in animals given atropine and artificial respiration. In every animal different muscles are affected unequally, and the distribution of the effects varies with species and compound.[2, 33, 37]

All but the second effect described so far are obviously similar to those described in the preceding two Chapters, and are in agreement with the theory that phosphorus compounds act entirely as anti-AChE's at myoneural junctions. The complete suppression of the response to low frequency stimulation has no analogy in experiments *in vitro*. Three theories of this effect will be discussed.

TABLE 14.2. EFFECTS ON NEUROMUSCULAR SYSTEMS *in vivo*

The numbers under 'Effects' refer to the list of effects in the main text.

| Species | Muscle | Compound | Dose mg/kg | Effects | Refs. |
|---|---|---|---|---|---|
| Cat | diaphragm | TEPP | 0·45* | 2 | (22) |
| | | Sarin | 0·2* | 2, 5 | (23) |
| | gastrocnemius | DFP | 15* | 5 | (24) |
| | | Sarin | 0·20* | 2, 3, 4, 7 | (25) |
| | | Tabun | 0·80* | 2, 5, 7 | (23) |
| | gracilis | TEPP | < 10* | 3 | (20) |
| | quadriceps | TEPP | 0·1 | 1, 3 | (21) |
| | | | 0·2 | 2, 5 | (21) |
| | tibialis anterior | MePFCh's‡ | about 0·02* | 1, 3, 4 | (26, 27) |
| | | | about 0·03* | 2, 3, 4 | (26, 27) |
| | | Sarin | 0·3* | 5, 7 | (28) |
| | | Tabun | 0·01 | 1, 3, 4 | (15) |
| | | | 0·02 | 2, 3, 4, 5, 6 | (15) |
| | | TEPP | 1·0* | 5, 7 | (28) |
| Rabbit | neck muscles (head drop test) | DFP | various | 3 | (29) |
| | | TEPP | various | 4 | (30) |
| | diaphragm | Sarin | 0·03* | 1, 5 | (31, 32) |
| | | DFP | 1–2 | 5 | (33) |
| | gastrocnemius | DFP | 1–2 | 1, 5 | (33) |
| | | Parathion | 7 | 5, 7 | (34) |
| | | Endothion § | 12 | 5, 7 | (34) |
| | tibialis anticus | DFP | 1–2 | 1, 5 | (33) |
| Rat | diaphragm | 217-MI† | 0·05 | 1 | (35) |
| | | | 0·10* | 2 | (35) |
| | | 217-AO† | 0·5 | 1, 2 | (35) |
| | | Paraoxon | 0·1 | 5 | (36) |
| | masseter | Sarin | 0·03 | 1, 3 | (16) |
| | | | 0·09* | 2, 3 | (16) |

* Doses which are lethal unless the animals are given therapy.

† 217-MI = $(EtO)_2PO \cdot S \cdot CH_2 \cdot CH_2 \cdot \overset{\oplus}{N}Me_3$; 217-AO = $(EtO)_2PO \cdot S \cdot CH_2 \cdot CH_2 \cdot NMe_2$

‡ Methylfluorophosphinylcholines, various.

§ Endothion = 2-(dimethoxyphinylthiomethyl)-5-methoxy-4-pyrone.

The response could be inhibited if the phosphorus compounds block the muscle receptors. This explanation is inconsistent with the antagonistic action of D-tubocurarine, which, on this theory, should increase the block. The concentrations of phosphorus compounds present are also very much smaller, sometimes over 1000 times smaller, than concentrations which block *in vitro*. The theory cannot hold generally. It may be correct in some individual cases, or with compounds less toxic than those used in the experiments summarized.

Inhibition could result from block of nerve conduction. There is some evidence in favour of this theory. Cats were given about 10 i-v $LD_{50}$'s of DFP or TEPP, and the sciatic nerves stimulated.[24, 38] Nerve conduction and the responses of the gastrocnemius were recorded. Both fell simultaneously, and the return of conduction was accompanied by the return of the twitch response to low frequency stimulation. (The capacity to hold a tetanus did not return until much later.) It is more often found, however, that the muscle response fails while the nerve still conducts.[20, 32, 35, 36] Block of nerve conduction is a rare phenomenon. The block in cat's sciatic nerve was caused by concentrations under a thousandth of those which block nerve conduction *in vitro*, and may have originated from the nerve endings, in some unknown way.

The effect can be explained more plausibly by assuming that ACh accumulates near the end-plates and depolarizes or blocks them. A detailed study of the gracilis muscle in cats poisoned with TEPP gave results which accord with this explanation. The muscle depolarized rapidly after 10 mg/kg of TEPP, and the muscle then returned to normal over about 40 min[20] (Fig. 14.1). The denervated gracilis depolarized somewhat more slowly, and the degree of depolarization followed closely the ACh concentration in the blood (Fig. 14.1). As direct action by TEPP was ruled out, the depolarization in the second case can reasonably be attributed to blood ACh; and the initial sharp depolarization in the first to ACh released from the nerve endings as a result of the random impulses normally reaching a resting muscle, perhaps reinforced by axon reflexes. The block in transmission may therefore be due to fibrillations and fasciculations, and therefore takes place when poisoning in the animal reaches a certain severity. The effect should not depend greatly upon the physical properties of the compound, and should take place when a given fraction of the $LD_{50}$ has been allowed to act. Both these conclusions are in accordance with the facts. The difficulty encountered by the theory is that the concentration of ACh in the blood of poisoned animals is only about 50 m$\mu$g/ml,[20, 39, 40] whereas 1 $\mu$g/ml is needed to inhibit completely the twitch response in rat's diaphragm *in vitro*.[41] Some difference is to be expected, and even a twentyfold difference may not be unreasonable. In the *in vitro* experiments AChE inhibition was not complete, so that as the synaptic gaps are very narrow there may well have been a big difference between the concentration in the bath and the concentration near the receptors. *In vivo* for the same reason the concentration near the receptors may have been much higher than the concentration in the blood. Whether such an explanation is correct is, of course, an open question. These responses still need to be investigated further.

The above theory cannot explain two sets of observations. In rabbits D-tubocurarine when given after TEPP or DFP always decreased the

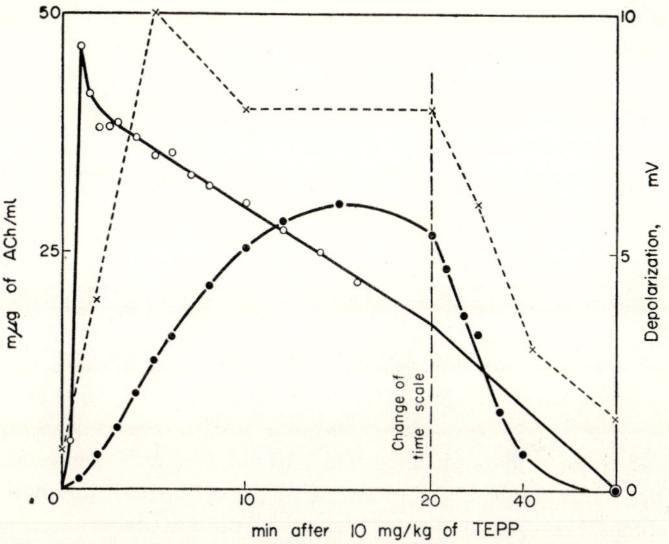


FIG. 14.1. End-plate depolarization of cat's gracilis muscle and ACh concentrations in the blood after TEPP.

Cats were given 10 mg/kg of TEPP i-v. The time courses of depolarization in the denervated (–●–) and innervated (–○–) muscles were recorded, and the blood ACh determined (–×–). The 3 experiments were carried out on different cats. The time scale is changed at 20 min in the figure.[20]

response of the longus digitorum muscle to tetanic stimulation[42] and Paraoxon always decreased the twitch response of the tibialis anterior and soleus muscles, and never increased it.[3] These results could be explained by postulating that the phosphorus compounds blocked the muscle receptors. An alternative hypothesis is that the muscles concerned (all in the same species, rabbits) were rapidly desensitized to ACh by contact with it. This is consistent with the observation that in the rabbits given Paraoxon the muscles lost their sensitivity to ACh after the first dose.[3]

Muscles in poisoned animals normally recover before any appreciable amount of AChE has been re-activated.[24] Three factors probably aid recovery: the ACh level in the blood falls;[20, 39] the nerve endings, which have discharged frequently and heavily, may excrete less ACh per stimulus;[43] and the muscles are desensitized to ACh.[20] There is no evidence on which to assess the relative importance of these factors.

Paralysing the musculature, especially that used in breathing, can obviously cause death, and in some cases does so, as shown later.

*Complex signs.* The signs listed after 'Complex' in Table 14.1 all involve the central nervous system. It is often uncertain, however, whether a phosphorus compound acts on the CNS directly or whether its peripheral action elicits central nervous responses. The signs are all to some extent interdependent, so that a discussion starting with any one will bring in the rest.

The immediate cause of death is asphyxia in almost all cases, as shown by the following observations.

(1) Respiration fails before the blood pressure falls dangerously and the heart does not stop until a minute or more later.[15]

(2) Artificial respiration will keep the heart of an animal beating after an otherwise lethal dose, unless the dose is very great.

(3) If artificial respiration establishes normal breathing the animal survives.

Autopsies carried out immediately after death on animals which died rapidly after treatment show that circulatory failure may also be important.[44, 45, 46] The splanchnic vascular bed is engorged with venous blood, and there is marked vasodilatation and passive hyperaemia. Death takes place by the asphyxia of an animal with an impaired circulation. The improvement brought about by forced ventilation shows, however, that respiratory failure is the controlling factor. Effects on respiration are therefore considered first.

In the early stages of poisoning, lung movements normally increase in frequency for a time. This transient effect may be associated with an increase or a decrease in the air intake, i.e. the breathing may become much shallower. This is the case with dogs treated with Sarin,[16] and cats treated with Tabun.[15] In cats treated with TEPP,[22] or rabbits[15] or dogs[16] treated with Tabun the tidal volume is increased transiently. Where the tidal volume falls the increase in rate is probably the normal reflex response to partial asphyxia. The mechanism may be the same even when the tidal air is increased, as the violent contractions of the gastro-intestinal tract, fasciculations and other peripheral effects presumably increase the demand for oxygen, while the cardiac output appears to fall somewhat at an early stage of intoxication.[15] With some compounds in some animals it is certain that the CNS is not directly involved in this effect. The respiration rate was increased considerably in dogs by Dimefox[47] and Schradan,[5] neither of which penetrate the CNS. In rats given

$$217\text{-MI}, [(EtO)_2PO \cdot S \cdot CH_2 \cdot CH_2 \cdot \overset{\oplus}{N}Me_3\ I^{\ominus}],$$

of which the same is true, phrenic discharges were higher than normal when the musculature was paralysed;[35] and in rabbits given TEPP or Sarin the phrenic nerve traffic increased as the tidal volume fell, an effect which was abolished by bilateral vagotomy, i.e. by treatment which destroyed the *sensory* part of the reflex.[31] It is possible that in some species direct action of the anti-AChE on the CNS increases the rate, perhaps by increasing the number of neurones excited by each sensory impulse, e.g. hyperpnoea has been caused in dogs by an anti-AChE confined to the CNS, as described in the last Chapter (pp. 296, 298).

Asphyxia may result from bronchoconstriction, paralysis of the respiratory musculature, or paralysis of the respiratory centre or a combination of all three. The balance varies with species and compound, as shown later.

Bronchoconstriction is primarily caused by stimulation of the parasympathetic system, as already described. It is also caused in the dog by decreasing the oxygen tension or increasing the carbon dioxide tension of the blood perfusing the brain.[48] Caused in this way it is central in origin, is mediated through the vagus, and is increased by eserine. It follows that partial asphyxia, however caused, may be worsened by this central reflex, and that some degree of bronchoconstriction is to be expected in any animal, whether the anti-AChE acts directly on the bronchi or not.

Paralysis of the respiratory muscles was described in the preceding section. Paralysis is likely to be made more complete by convulsions. Convulsions, by their nature, imply a very high rate of discharge of the motor neurones, and hence an unusually rapid release of ACh from the nerve endings, with consequent inhibition of neuromuscular function. This may account for the general observation that in some mammalian species, notably rats and mice, respiration soon fails in those animals which convulse, but is maintained in those animals receiving the same dose which do not.

Failure of the respiratory centre has been demonstrated in several species.[3, 23, 32, 36] An interesting record, obtained on rats treated with Paraoxon, is shown in Fig. 14.2. The functioning of the respiratory centre was followed by recording the electrical activity in the phrenic nerve, and the functioning of the myoneural junctions by recording the action potentials in the diaphragm muscle. Thorax movements were also recorded.[36] Paraoxon (0·1 mg/kg, i-v) did not alter the activity of the phrenic nerve, but abolished transmission across the myoneural junction (second line, Fig. 14.2B), and reduced the thoracic movements. P2-AM restored the transmission to normal (Fig. 14.2C). Slow infusion over 1 hr of 1·14 mg/kg, however, paralysed the respiratory centre completely (top line, Fig. 14.2D), and respiration could not be restored by P2-AM, which, being ionic, does not penetrate the CNS. In this and other instances the myoneural junctions cease to transmit before the respiratory centre is affected.[32, 35] In other instances the converse is true.[23, 32] In cats treated with DFP, for example, action potentials ceased simultaneously in the phrenic nerves and the intercostal muscles.(49) Sometimes, however, central failure follows closely on paralysis of the respiratory musculature, and the question then arises as to whether the latter causes the former. In this context the effects of paralysis must be measured by the decrease in the tidal volume. Paralysis need not overtake all muscles at the same time. Thus in cats treated with TEPP the diaphragm was usually paralysed before the intercostal muscles.(32) As an untreated cat could breathe adequately when its diaphragm was made completely flaccid by cutting both phrenic nerves, paralysis of the diaphragm could only be a subsidiary cause of asphyxia. The relationship between

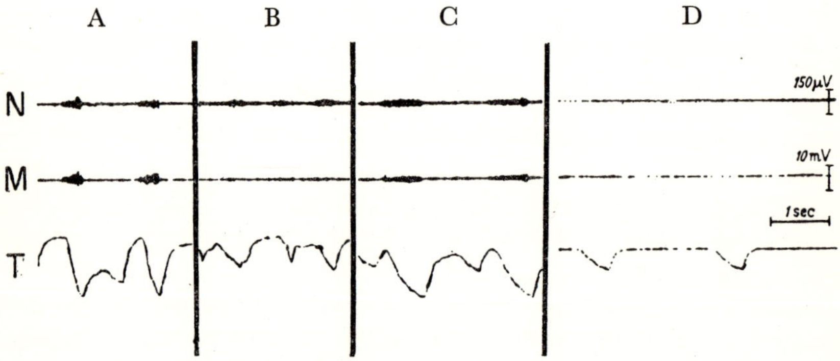


FIG. 14.2. Functioning of the phrenic nerve, diaphragm and thorax in the rat.

N, nerve traffic; M, action potentials in the diaphragm muscle, T, thoracic movements. *A*, before treatment; *B*, 3 min after 0·1 mg/kg of Paraoxon had been given intravenously; *C*, 9 min after 15 mg/kg of P2-AM had also been given, showing restoration of action potentials in the diaphragm and of thoracic movements; *D*, after infusion over 1 hr of 1·14 mg/kg of Paraoxon, showing paralysis of the respiratory centre.(36)

After F. Sakai, H. dalRi, W. D. Erdmann and G. Schmidt. *Arch. exp. Path. Pharmak.* **234,** 210 (1958) (with acknowledgements).

peripheral and central paralysis was studied in detail on rabbits poisoned with TEPP or Sarin,(31) both of which paralysed the diaphragm and thoracic muscles simultaneously. The phrenic nerve traffic increased as the tidal air fell, and then dropped rapidly to zero shortly after breathing ceased. The block in phrenic discharges was not, however, caused by asphyxia, or by a fall in blood pressure, as it still took place in rabbits artificially ventilated, and in which the blood-pressure was nearly normal. It was not completely independent of asphyxia, as erratic discharges could be maintained in some rabbits by artificial respiration. The picture in such cases is obviously complex. Partial paralysis of the musculature reduces the tidal volume, and increases the phrenic discharge, thus increasing the degree of paralysis.

When the paralysis reaches a certain stage the blood pressure falls, and the respiratory centre is starved of oxygen, thereby reducing its activity. At the same time the activity of the respiratory centre is directly reduced by the central action of the anti-AChE.

Such results may tend to exaggerate the importance of direct action on the CNS.[4, 33] In the poisoned animal there is always vasoconstriction (see later) and bradycardia, so that there is likely to be some hypoxia. Even very partial paralysis of the respiratory musculature may increase this enough to lead to bronchoconstriction and paralysis of the respiratory centre by local

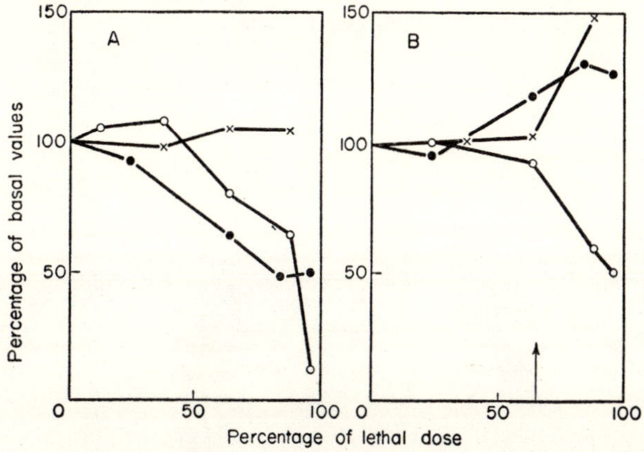


FIG. 14.3. The effects of infusing Tabun slowly on the cardiac output, oxygen consumption and blood pressure in the rabbit.

Cardiac output, ○; $O_2$ consumption, ●; blood pressure, ×. In *A* the results obtained on rabbits receiving Tabun only are shown, and in *B* those on rabbits pre-treated with 10 mg/kg of atropine. The arrow in *B* indicates the dose lethal to an unatropinized rabbit.[15]

high $pCO_2$ and low $pO_2$ in the CNS. The failure to produce central respiratory failure in dogs and rabbits by large doses of anti-AChE's confined to the CNS (previous Chapter) suggests that this explanation may be the correct one. On the other hand there is no rapid change in $pO_2$ or $pCO_2$ in the general circulation of unatropinized rabbits until near death, so that for the 'peripheral' theory to hold it must be assumed that the blood-vessels serving the respiratory centre are constricted more than most.

The action of the respiratory centre is restored by atropine.[22, 31, 32, 35, 49] The doses needed are adequate to antagonize other central effects, such as the aversive syndrome in rabbits. The effects of atropine have been analysed in rabbits infused slowly with Tabun.[15] In the unatropinized rabbit the heart rate and cardiac output fell continuously. The oxygen consumption first rose, but fell after about 0·3 $LD_{50}$'s had been injected. A final catastrophic fall took place as the lethal dose was approached (Fig. 14.3A). In rabbits given 10 mg/kg of atropine the heart beat, cardiac output and oxygen consumption were all maintained up to a dose which was lethal to an unatropinized rabbit. At higher doses the oxygen consumption fell, but the heart beat and blood pressure were maintained almost until death (Fig. 14.3B). In the atropinized rabbit the respiratory failure was accompanied by a large increase in arterial $CO_2$ tension and a large decrease in arterial $O_2$ tension. It is reasonable to suppose that in these animals respiration failed because of peripheral paralysis and that failure was accelerated by the action of high $CO_2$ and low $O_2$ tension in the blood on the respiratory centre. In the non-atropinized rabbits the rise of $CO_2$ tension and fall in $O_2$ tension were small. In neither case was circulatory failure apparently important, in

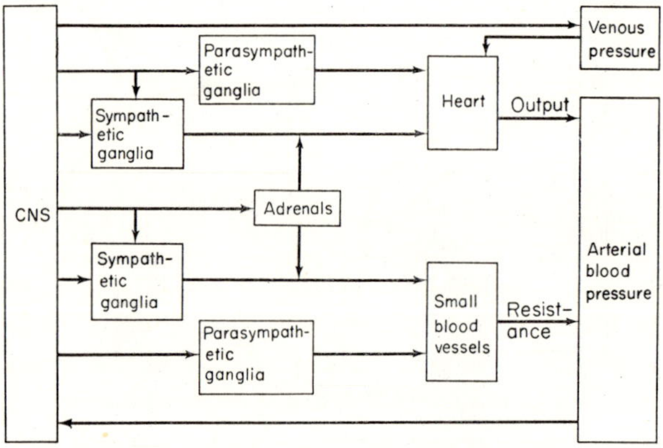


FIG. 14.4. Factors affecting circulation.[50]

atropinized rabbits because circulation was maintained and in the others because the $O_2$ and $CO_2$ tensions were not grossly affected.

The effects on the circulation are now considered in more detail. The factors which affect circulation[50] are shown diagrammatically in Fig. 14.4. Sympathetic stimulation increases the heart rate, and causes vasoconstriction, thereby increasing the arterial pressure. Stimulation of the adrenals, which are cholinergically innervated, has the same effect. Parasympathetic stimulation slows the heart and causes some vasodilatation. The arterial pressure is largely self-regulating via the carotid sinus, which acts as a baroreceptor.

The effects of phosphorus compounds on blood pressure are variable, and, in view of the bradycardia and reduced cardiac output, surprising. The most common sequence of events caused by a lethal dose is a transient rise followed by a slow fall, which only becomes catastrophic when the animal is severely cyanosed. This sequence has been observed after reversible inhibitors (neostigmine and eserine in cats and dogs[51]), direct phosphorus inhibitors (TEPP,[51] Sarin and Tabun[4] in cats and dogs[52]), and indirect phosphorus inhibitors (Schradan[5] and Dimefox[47] in dogs). In other experiments no transient rise was found (DFP, TEPP, Paraoxon and eserine in cats,[32] Sarin and Tabun in cats and dogs,[15] and DFP[33] and Tabun[15] in rabbits). The rate of fall was never great. In rats, on the other hand, Sarin, TEPP, DFP, Paraoxon and eserine all increased the blood pressure rapidly.[53] Bradycardia must tend to lower the blood pressure, and it is obvious that this tendency is counteracted, and sometimes outweighed, by compensatory mechanisms.

Injection of ACh into a normal animal decreases its blood pressure, and into an atropinized one increases it. These depressor and pressor actions are both greatly potentiated by anti-AChE's. Inhibition of ΨChE contributes to the potentiation of the depressor action, as demonstrated on dogs treated with enough DFP to inhibit most of their ΨChE but little of their AChE.[18] Some of the potentiation can therefore be attributed to the greater concentration of ACh reaching the receptor sites. The main effect must, however, arise from the action of the anti-AChE's in the receptor region. The potentiation after a near-lethal dose varies with the compound, but is usually about one-hundredfold.[5, 47, 52] The pressor effect in cats and dogs is similarly potentiated.[1, 17, 54] Both types of potentiation are sensitive tests of AChE inhibition. Thus the pressor effects in dogs lasted only a short time after they had been given reversible inhibitors, many hours after TEPP and several days after DFP;[54] and the potentiation of the depressor effect in dogs increased rapidly to a maximum after direct inhibitors had been given,[52] but only slowly after Dimefox[47] and Schradan,[5] which are metabolized slowly to inhibitors.

These effects are not fully understood. They are the same with both reversible and phosphorus anti-AChE's, so that results obtained with either may be used to elucidate the mechanism. It is obvious that a pressor response opposes the effects of bradycardia. Correspondingly ACh lowers the blood pressure drastically in the unatropinized animal, and raises it in the atropinized one, where bradycardia is abolished. The pressor response could be a result of stimulation of the carotid sinus, of an action of ACh on the adrenals or other organs liberating adrenergic hormones, or the result of central or ganglionic stimulation. Most results indicate that stimulation of the sympathetic ganglia is the most important cause.

(1) The increase in the blood pressure of rats given Sarin, etc., was greatly reduced by 10 mg/kg of hexamethonium, a ganglionic blocking agent.[53] At this dose it is also a weak curarizing agent at myoneural junctions, so some of the depression may have been caused by general improvement in the rats' condition.

(2) In atropinized cats and dogs the potentiation of the pressor response to ACh was not abolished by vagotomy, or by extirpation of the carotid sinuses, adrenals, or liver, or by decapitation. It was abolished by removal of

the sympathetic chain, providing the aorta was clamped at a level which stopped the influx of adrenergic hormones.[55]

(3) The hind limbs of dogs were prepared in such a way that the perfusion pressure across the limbs was held constant by pumps. For one set of experiments the blood supply of the dog whose limb was prepared was used (self-perfused dogs); for the rest the blood supply was provided by donor dogs (cross-perfused). The hind limbs were always normally innervated. In self-perfused dogs TEPP or Sarin increased the vascular resistance, and the effect was reduced by hexamethonium or section of the sympathetic system. The vascular system in skin and muscles was affected similarly. In cross-perfused systems poisoning the recipient still increased the resistance, showing that the effects originated in the nervous system, and not at the blood vessels themselves. An increase was also observed when the donor was poisoned, unless it was adrenalectomized. Thus vasoconstriction was partly due to the release of adrenergic hormones into the blood stream. In spite of vasoconstriction, the limb volume increased. This could be explained by an increase in venous pressure, so that the vessels distended passively.[56]

There is other evidence that the release of sympathetic hormones from the adrenals may contribute. The pressor response to ACh was still potentiated in cats and dogs in which the sympathetic ganglia had been blocked by nicotine, but it was abolished by dibenamine, which is a peripheral antagonist to sympathetic hormones.[51]

A combination of stimulation of sympathetic ganglia and of release of adrenal hormones cannot explain all the results. Increased resistance to blood flow has been demonstrated in the pulmonary system of dogs given TEPP or Sarin, and some of the effects appeared to have a post-ganglionic origin.[57, 58] The carotid body was sensitized to ACh and to low oxygen tension by some anti-AChE's, so that action at the carotid body may also contribute.[73] There are numerous possible reflexes initiated either by the blood vessels[18] or oxygen-lack,[56] and their relative importance cannot yet be assessed. The fact that vasoconstriction still takes place when these reflexes are abolished by nerve section does not prove that they are unimportant in the intact animal. Increase in blood pressure can be partly central in origin, as shown by experiments in which the compound was injected into the brain.[16, 59] The doses given were large on a mg/kg basis, and it is likely that direct action on the CNS is a subsidiary factor in a generally poisoned animal.

Methylfluorophosphinylcholines, e.g. $(Me)(F)PO \cdot O \cdot CH_2 \cdot CH_2 \cdot \overset{\oplus}{N}Me_3$, are interesting in that they cause an immediate but very short-lived decrease in the blood pressure of cats on injection at 0·2 $\mu$g/kg or more, and a similar increase in atropinized cats.[27] The pressor response is less after adrenalectomy. The compounds also potentiate the response to ACh like other anti-AChE's, but it is probable that their initial, rapid action is a direct ACh-like one on the receptors. This is plausible in view of the close resemblance in structure between ACh and the phosphinylcholines.

Convulsions are a very noticeable component of the syndrome in almost all animals poisoned by any phosphorus compound. Convulsions are not, in themselves, evidence that the compounds act in the CNS. Both ionic compounds, such as the methylfluorophosphinylcholines[27] and the sulphonium derivatives of the Demetons,[60] and very water-soluble compounds, such as Dimefox[47] and Schradan,[61] also cause them, although the brain AChE is not inhibited at death. In these cases convulsions are usually confined to the agonal stages, and it is often observed that animals which do not convulse survive. These convulsions are obviously due to asphyxia. Oil-soluble compounds sometimes cause convulsions at an early stage of poisoning, e.g. TEPP in cats[19] and 217-AO in rabbits.[6] These convulsions are abolished by doses of atropine similar to those which restore the activity of the respiratory centre,[22, 62] and are reduced by pentobarbitone,[63] unless the dose of phosphorus compound is very much greater than the $LD_{50}$. This establishes that the convulsions are in part caused by the action of the phosphorus compounds in the CNS. There is much evidence, however, that action in the CNS is not more than a predisposing factor.

(1) Convulsions are not produced in animals which do not receive near-lethal doses, whatever compound is given, although the extent to which compounds penetrate the CNS depends upon their solubility properties (Chapter VIII) and these are not related to their toxicities. Thus the onset of convulsions is more closely related to the general degree of poisoning in the animal than to specific effects on the CNS.

(2) The doses of phosphorus compounds which must be injected into the brain via the carotid arteries or intracisternally to cause convulsions are very high (Chapter XIII)—far above $LD_{50}$ doses on a w/w basis.

(3) The most toxic compounds reported are all ionic, and do not enter the CNS.

(4) Either high blood $CO_2$ or a sudden fall from high blood $CO_2$ causes violent seizures, accompanied by typical convulsive patterns on the electroencephalograph.[64, 65] The changes in $CO_2$ tension required to cause seizures in animals not otherwise affected are, however, greater than those observed in animals poisoned with phosphorus compounds.

More work is required to establish the mechanism of convulsions. In particular, data are required on the AChE levels in various parts of the brain at the times convulsions begin, obtained with due precautions against further inhibition taking place during homogenization. There is no reason to believe that the normal level of convulsive activity observed is, in itself, harmful, except insofar as it increases neuromuscular block.

*Muscle reflexes*. There is little to add to the discussion in Chapter XIII. Near-lethal doses of DFP, Paraoxon, Tabun, Parathion or eserine markedly potentiate monosynaptic extensor and polysynaptic flexor reflexes in cats, while leaving other reflexes little affected.[66] High doses inhibit reflexes.[21, 32, 66] Both potentiation and inhibition can be abolished by high doses of atropine:[66] and there is no correlation between the effects of phosphorus compounds on reflexes and on myoneural junctions.[21] Therefore these effects are central in origin.

*Acidosis.*[7] In dogs poisoned with Sarin or Parathion marked acidosis was present, due to an accumulation of $CO_2$ and lactic acid in the blood. High $CO_2$ is caused by respiratory failure, and high lactic acid by oxygen lack, aggravated by the increase in demand for oxygen resulting from cholinergic stimulation.

*Hyperglycaemia.* A wide variety of phosphorus compounds increased the blood sugar levels in rats. The levels were restored to normal by the sympathetic blocking agent, dihydroergocornine.[8, 67] Hyperglycaemia is a common stress reaction. In these cases it may arise either from stress, or by inhibition of AChE in the adrenal medulla.

*Effects on ganglia.* It has been necessary to postulate potentiation at ganglia several times in the preceding discussion. Potentiation of synaptic transmission has been demonstrated, mainly in the superior cervical ganglion of the cat. In atropinized cats the response of the nictitating membrane to pre-ganglionic stimulation was increased by TEPP,[13] Tabun,[15] 217-MI and 217-AO (p. 293).[68] Post-ganglionic action potentials also were increased by DFP at lethal and near lethal doses. The increase was small unless the pre-ganglionic trunk was partially resected between the ganglion and the point of stimulation, or the stimulation was submaximal, i.e. the increase was mainly the result of the recruitment of more fibres.[69] This reconciles the results with others, in which it was found that eserine did not much alter the form or duration of action potentials proceeding from the curarized stellate ganglion of the cat in response to pre-ganglionic stimulation.[70, 71] The type of response is not altered, but more fibres are excited per stimulus. The potentiation is likely to be greater when the ganglionic excitation consists of natural impulses instead of artificial, synchronous stimuli. In the former probably few pre-ganglionic fibres carry action potentials simultaneously, so that potentiation by recruitment should be particularly effective.

High doses of DFP (4–8 mg/kg) suppressed synaptic transmission in the superior cervical ganglion.[69]

The effects of DFP were antagonized by atropine, which itself blocked transmission at 5 mg/kg.[69]

## DIFFERENCES BETWEEN COMPOUNDS

The way signs vary with compounds depends upon the species of animal. There are, however, certain differences which are commonly observed.

Compounds which act rapidly but which do not penetrate the CNS do not cause convulsions except in the terminal stages. Otherwise they are similar to rapidly-acting compounds which do.

Some compounds given in doses near the $LD_{50}$ kill some species very slowly. This is especially true of some indirect inhibitors such as Schradan, but is also true in some species of DFP and of other direct inhibitors which phosphorylate AChE irreversibly. In these cases the animal often shows no well-defined convulsions in the agonal stages. The final cause of death is also more obscure because more general. The functioning of the respiratory system, of the heart, and of the regulatory mechanisms controlling the blood pressure are all impaired. As nothing happens quickly the various functions are in a state of near equilibrium throughout poisoning, and it is a matter of chance which fails first.

## DIFFERENCES BETWEEN ANIMALS

Several differences have already been mentioned. Those described here all relate to the prime causes of death. Attention is confined to those anti-AChE's which killed in under an hour at the doses given. Death can result from cardiac failure, failure of the respiratory centre, paralysis of the respiratory musculature, or bronchoconstriction. Cardiac failure can be regarded as the major cause only when it takes place before respiration stops. Failure of the respiratory centre and paralysis of the respiratory musculature can only be distinguished when nerve recordings are taken. Bronchoconstriction is demonstrated when co-ordinated lung movements do not result in ventilation.

Some results reported are summarized in Table 14.3. When the evidence is inadequate to show whether the respiratory centre failed before the musculature, 'respiration' is entered. It is assumed that compounds which do not inhibit AChE in the CNS do not enter it, and cannot, therefore, cause death by failure of the respiratory centre.

Three conclusions can be drawn from the table: cardiac failure rarely takes place before respiration ceases; bronchoconstriction is often the most important factor in cats and dogs, but is rarely important in monkeys, rabbits or rats; and there is considerable disagreement between different workers. The last is not surprising. It reflects the difficulty in distinguishing causes when so much is not functioning properly, and also genuine differences between the responses of different animals of the same species. Some workers also have studied anaesthetized animals, others animals unanaesthetized. The order in which symptoms appear can be quite different in the two cases, as shown by some experiments on dogs given Sarin. In the unanaesthetized dogs effects on circulation were observed before effects on respiration; in anaesthetized dogs the reverse.[(74)] Finally some workers have denied that

TABLE 14.3. THE IMMEDIATE CAUSE OF DEATH IN MAMMALS OF DIFFERENT SPECIES

The immediate cause of death is denoted by: heart failure, h; paralysis of the respiratory centre, r.c; paralysis of the respiratory musculature, r.m; and bronchoconstriction, b. Where there is no basis on which to decide whether the respiratory centre or musculature failed first the cause is denoted by 'resp'. The different causes given by different workers are noteworthy.

| Species | Compounds | Causes |
|---|---|---|
| Cat | DFP | b[32] |
| | Paraoxon | b[32] |
| | TEPP | b;[32] b rarely, usually r.c + r.m or r.c only[22] |
| | Sarin | b;[32] h + b;[4] r.c[23] |
| | Tabun | b;[32] h + b;[4] r.c[23] |
| | Methylfluorophosphinylcholines | b + r.m (sometimes h)[27] |
| Dog | Parathion | resp + b[72] |
| | Sarin | resp + b;[72] resp + b, or h;[16] b;[4] r.c[23] |
| | Tabun | resp + b, or h;[16] r.c[23] |
| | Schradan | r.m + h[5] |
| | TEPP | resp + b[1] |
| | $(Me_2N)_2PO \cdot O \cdot PO(OEt)_2$ | r.m[52] |
| | $[(EtO)(Me_2N)PO]_2O$ | r.m[52] |
| Monkey | DFP | r.c[32] |
| | Paraoxon | r.c[32] |
| | Sarin | r.c[32] |
| | TEPP | r.c + r.m[32] |
| | Tabun | r.c[32] |
| | Parathion | resp[72] |
| Rabbit | Paraoxon | r.m[32] |
| | TEPP | r.m[32] |
| | Sarin | r.c + r.m[32] |
| | DFP | r.c;[32] r.m[33] |
| | Tabun | r.m;[32] b + h[4] |
| Rat | 217-MI | r.m[35] |
| | 217-AO | r.c[35] |

failure of the respiratory centre is caused directly, and have attributed failure to hypoxia,[4, 33] as discussed earlier.

## SUMMARY

The main signs of poisoning, summarized in Table 14.1, are observed in nearly all mammals. With very few exceptions they are consistent with the theory that the only biochemical or biophysical lesion is inhibition of AChE. There are, however, numerous cases in which there is no positive evidence that the anti-AChE theory is correct, and in which other theories are tenable.

Animals usually die of asphyxia, occasionally of cardiac failure. The importance of the central component in asphyxia is debatable. Some authors regard it as the controlling factor in most cases, while others regard the central effects solely as reflex responses to peripheral action. Variations with animal and compound probably cover both extremes.

## REFERENCES


1. C. Dayrit, C. H. Manry and M. H. Seevers. *J. Pharmacol.* **92,** 173 (1948)
2. W. Modell, S. Krop, P. Hitchcock and W. F. Riker, Jr. *J. Pharmacol.* **87,** 400 (1946)
3. J. M. Barnes. *Brit. J. Pharmacol.* **8,** 208 (1953)
4. S. Krop and A. M. Kunkel. *Proc. Soc. exp. Biol. N.Y.* **86,** 530 (1954)
5. K. P. DuBois, J. Doull and J. M. Coon. *J. Pharmacol.* **99,** 376 (1950)
6. G. B. Koelle and E. C. Steiner. *J. Pharmacol.* **118,** 420 (1956)
7. A. J. Gold, J. M. Weller and G. Freeman. *Amer. J. Physiol.* **188,** 321 (1957)
8. C. Klotzsche. *Arzneimittelforsch.* **5,** 436 (1955)
9. H. Cullumbine and P. Dirnhuber. *J. Pharm. Lond.* **7,** 580 (1955)
10. A. S. V. Burgen. *Brit. J. Pharmacol.* **4,** 185 (1949)
11. F. Hobbiger. *Brit. J. Pharmacol.* **6,** 21 (1951)
12. A. S. V. Burgen and F. Hobbiger. *Brit. J. Pharmacol.* **6,** 593 (1951)
13. A. S. V. Burgen, C. A. Keele and D. Slome. *J. Pharmacol.* **96,** 396 (1949)
14. C. Lovatt Evans. *Principles of Human Physiology,* p. 1023 *et seq.,* Churchill, London, 1949.
15. B. Holmstedt. *Acta physiol. scand.* **25,** Suppl. 90 (1951)
16. C. Heymans, A. Pochet and H. van Houtte. *Arch. internat. Pharmacodyn.* **104,** 293 (1956)
17. R. Verbeke. *Arch. internat. Pharmacodyn.* **79,** 1 (1948)
18. R. Verbeke. *Arch. internat. Pharmacodyn.* **80,** 1 (1949)
19. A. S. V. Burgen, C. A. Keele, M. Chennells, J. del Castillo, W. F. Floyd, D. Slome and S. Wright. *Nature, Lond.* **160,** 760 (1947)
20. W. W. Douglas and W. D. M. Paton. *J. Physiol.* **124,** 325 (1954)
21. M. Chennells, W. F. Floyd and S. Wright. *J. Physiol.* **108,** 375 (1949)
22. W. W. Douglas and P. B. C. Matthews. *J. Physiol.* **116,** 202 (1952)
23. R. V. Brown, A. M. Kunkel, L. M. Somers and J. H. Wills. *J. Pharmacol.* **120,** 276 (1957)
24. B. P. McNamara, E. F. Murtha, A. D. Bergner, E. M. Robinson, C. W. Bender and J. H. Wills. *J. Pharmacol.* **110,** 232 (1954)

25. A. M. KUNKEL, J. H. WILLS and J. S. MONIER. *Proc. Soc. exp. Biol. N.Y.* **92,** 529 (1956)
26. T. FREDRIKSSON. *Arch. internat. Pharmacodyn.* **115,** 474 (1958)
27. T. FREDRIKSSON. *Arch. internat. Pharmacodyn.* **113,** 101 (1957)
28. R. HOLMES and E. L. ROBINS. *Brit. J. Pharmacol.* **10,** 490 (1955)
29. H. F. CHASE, J. L. SCHMIDT and B. K. BHATTACHARYA. *J. Pharmacol.* **91,** 236 (1947)
30. H. LOW and L-E. TAMMELIN. *Acta physiol. scand.* **23,** 78 (1951)
31. P. G. WRIGHT. *J. Physiol.* **126,** 52 (1954)
32. C. A. DE CANDOLE, W. W. DOUGLAS, C. LOVATT EVANS, R. HOLMES, K. E. V. SPENCER, R. W. TORRANCE and K. M. WILSON. *Brit. J. Pharmacol.* **8,** 466 (1953)
33. L. LUNDHOLM. *Acta physiol. scand.* **16,** 345 (1949)
34. J. FOURNEL. *C.R. Soc. Biol. Paris.* **151,** 1373 (1957)
35. W. SCHAUMANN and C. JOB. *J. Pharmacol.* **123,** 114 (1958)
36. F. SAKAI, H. D. RI, W. D. ERDMANN and G. SCHMIDT. *Arch. exp. Path. Pharmak.* **234,** 210 (1958)
37. B. HOLMSTEDT, S. J. LUNDIN and C. R. SKOGLUND. *Acta physiol. scand.* **29,** Suppl. 106, 428 (1953)
38. E. F. MURTHA, B. P. MCNAMARA, L. J. EDBERG, A. D. BERGNER and J. H. WILLS. *J. Pharmacol.* **115,** 291 (1955)
39. J. M. BARNES and J. I. DUFF. *Brit. J. Pharmacol.* **9,** 153 (1954)
40. W. C. STEWART. *Brit. J. Pharmacol.* **7,** 270 (1952)
41. J. M. BARNES and J. I. DUFF. *Brit. J. Pharmacol.* **8,** 334 (1953)
42. J. AXELSSON, E. GJONE and K. NAESS. *Acta pharmacol. tox.* **13,** 319 (1957)
43. N. EMMELIN and F. C. MACINTOSH. *J. Physiol.* **131,** 477 (1956)
44. B. HOLMSTEDT, L. KROOK and J. R. ROONEY. *Acta pharmacol. tox.* **13,** 337 (1957)
45. H. SIEDEK and H. THALER. *Arch. internat. Pharmacodyn.* **91,** 194 (1952)
46. F. A. DENZ. *J. Path. Bact.* **63,** 81 (1951)
47. A. J. OKINAKA, J. DOULL, J. M. COON and K. P. DUBOIS. *J. Pharmacol.* **112,** 231 (1954)
48. M. DE BURGH DALY, C. J. LAMBERTSEN and A. SCHWEITZER. *J. Physiol.* **119,** 292 (1953)
49. W. A. KRIVOY, E. R. HART and A. S. MARRAZZI. *J. Pharmacol.* **103,** 351 (1951)
50. J. M. VAN ROSSUM and E. J. ARIENS. *Arch. internat. Pharmacodyn.* **118,** 447 (1959)
51. P. R. SALERNO and J. M. COON. *J. Pharmacol.* **95,** 240 (1949)
52. K. P. DUBOIS, J. DOULL, A. J. OKINAKA and J. M. COON. *J. Pharmacol.* **107,** 464 (1953)
53. P. DIRNHUBER and H. CULLUMBINE. *Brit. J. Pharmacol.* **10,** 12 (1955)
54. T. KOPPANYI and A. G. KARCZMAR. *J. Pharmacol.* **101,** 327 (1951)
55. T. KOPPANYI, C. R. LINEGAR and R. P. HERWICK. *Amer. J. Physiol.* **130,** 346 (1940)
56. M. DE BURGH DALY and P. G. WRIGHT. *J. Physiol.* **133,** 475 (1956)
57. M. DE BURGH DALY. *J. Physiol.* **139,** 250 (1957)
58. M. DE BURGH DALY and P. G. WRIGHT. *J. Physiol.* **139,** 273 (1957)
59. H. POLET and A. F. DE SCHAEPDRYVER. *Arch. internat. Pharmacodyn.* **118,** 231 (1959)
60. D. F. HEATH, unpublished.
61. J. M. BARNES, personal communication.
62. W. C. WESCOE, R. E. GREEN, B. P. MCNAMARA and S. KROP. *J. Pharmacol.* **92,** 63 (1948)

63. B. P. McNamara, G. B. Koelle and A. Gilman. *J. Pharmacol.* **88,** 27 (1946)
64. D. M. Woodbury, L. T. Rollins, M. D. Gardner, W. L. Hirschi, J. R. Hogan, M. L. Rallison, G. S. Tanner and D. A. Brodie. *Amer. J. Physiol.* **192,** 79 (1958)
65. D. A. Brodie and D. M. Woodbury. *Amer. J. Physiol.* **192,** 91 (1958)
66. J. Haase, D. Lücke, F. Scheler, R. Schütz, B. Mühlberg and W. Koll. *Arch. exp. Path. Pharmak.* **232,** 274 (1957)
67. O. Dybing and E. Sognen. *Acta pharmacol. tox.* **14,** 231 (1958)
68. R. J. McIsaac and G. B. Koelle. *J. Pharmacol.* **126,** 9 (1959)
69. D. A. Holaday, K. Kamijo and G. B. Koelle. *J. Pharmacol.* **111,** 241 (1954)
70. J. C. Eccles. *J. Physiol.* **101,** 465 (1943)
71. J. C. Eccles. *J. Physiol.* **103,** 27 (1944)
72. R. P. Johnson, A. J. Gold and G. Freeman. *Amer. J. Physiol.* **192,** 581 (1958)
73. C. Heymans. *Pharmacol. Rev.* **7,** 119 (1955)
74. T. Fredriksson, C-H. Hanson and B. Holmstedt. *Arch. internat. Pharmacodyn.* **126,** 288 (1960)

CHAPTER XV

# Therapeutic Methods; and Symptoms, Diagnosis and Therapy in Man

## THERAPEUTIC METHODS

THE PRINCIPAL methods of therapy have already been described. To summarize, they are:

(1) Artificial respiration, to overcome respiratory paralysis in the acute stages of poisoning.

(2) Atropinization, to block ACh action at the parasympathetic nerve endings and to relieve central depression.

(3) Curarization, to block ACh action at the end-plates of striated muscles.

(4) Treatment with penta- or hexa-methonium, to antagonize ACh at ganglia.

(5) Treatment with re-activators of phosphorylated AChE's.

Only the last method reverses the fundamental biochemical lesion. In this Chapter the effectiveness of these methods is discussed.

Efficacy can be measured quantitatively in terms of the $LD_{50}$ in the untreated animal by the therapeutic index described on p. 198. For any given treatment the index depends upon how soon the treatment is applied after the poison. For some of the more rapidly acting direct inhibitors the index is unlikely to exceed about 9, as doses above 10 $LD_{50}$'s kill before any treatment can be started. Some treatments prolong life without greatly increasing the $LD_{50}$. This gives time for other measures to be taken.

Therapy in this Chapter is considered strictly in terms of saving life. Experiments in which, for example, anaesthetized cats are kept alive by forced ventilation and atropinization for several hours after receiving 100 $LD_{50}$'s of TEPP or DFP are not considered as the animals die when artificial respiration is stopped.

The efficacy of each method is described in turn first, and then the efficacies of multiple treatments.

*Artificial respiration.* This, in itself, both prolongs life and saves it. Cats and dogs have been kept alive for 20–30 min after receiving 8–10 $LD_{50}$'s of Sarin and Tabun, and then saved by atropinization.(1) Guinea pigs infused slowly with Parathion withstood about 8 $LD_{50}$'s under artificial respiration alone. It did not, however, much reduce their susceptibility to DFP.(2) Artificial respiration is very effective against Paraoxon. Paraoxon was given in divided doses over a few hours to rabbits,[3] cats and dogs,[4] mostly tracheotomized. Rabbits survived about 10 $LD_{50}$'s, dogs up to 50 $LD_{50}$'s, and cats about 5 $LD_{50}$'s. The first injection caused fasciculations and the respiration to fail rapidly. After a few $LD_{50}$'s the response to a further dose changed. There were no fasciculations and the respiration did not collapse suddenly. Breathing became shallower and less effective. It sometimes failed gradually, but could be restored by artificial respiration. These experiments suggest that artificial respiration worked by giving the animals time to adapt to high ACh concentrations.

Respiratory stimulants such as nikethamide are not generally effective. With atropine nikethamide increased the $LD_{50}$ of EPN in rats,[5] but was without effect on mice poisoned with any of a wide range of compounds,[6] and had an adverse effect on rats poisoned with Parathion, Parathion-Me or 'Systox'.[7]

*Atropine.* Some therapeutic indices are shown in Table 15.1. Atropine always decreases mortality, but it rarely increases the $LD_{50}$ by a factor of

TABLE 15.1. EFFICACY OF ATROPINE THERAPY

Species, poison, dose of atropine, time after the poison the atropine was given, therapeutic index and references are given. The therapeutic index is the factor by which the $LD_{50}$ is raised minus 1. Ro3-0412 is $(MeO)_2PO \cdot O \cdot C_6H_4\text{-3-}\overset{\oplus}{N}Me_3$, and Ro3-0422 is its diethoxy quinolinium analogue.

| Species | Poison | Atropine (mg/kg) | When given | Therapeutic index | Refs. |
|---|---|---|---|---|---|
| Cat | DFP | 1 | 5 min | 2 | (8) |
| | TEPP | 1 | 10 min | < 1 | (9) |
| Dog | Schradan | 2 | 20 min | 4 | (10) |
| | Sarin | 1 | few sec | < 1 | (11) |
| | Tabun | 1 | few sec | < 1 | (11) |
| | TEPP | 1 | few sec | 3 | (12) |
| Guinea pig | Sarin | 17·4 | 30 sec | < 0·5 | (13) |
| Monkey | Sarin | 0·029 | 1 min | 2 | (13) |
| Mouse | Parathion | 5–50 | few min | 0·3 | (14, 15) |
| | | 100 | few min | 0·8 | (16) |
| | Paraoxon | 1 | at same time | 0·2 | (17) |
| | | 10 | at same time | 0·8 | (17) |
| | Ro3-0412 | 10 | at same time | 2·5 | (18) |
| | Ro3-0422 | 10 | at same time | 1·2 | (18) |
| | TEPP | 10 | at same time | 0·8 | (18) |
| | Sarin | 17·4 | 30 sec | 0·5 | (13) |
| Rat | EPN | 90 | few sec | 1 | (5) |
| | Sarin | 17·4 | 30 sec | 0·5 | (13) |

more than two. The dose of atropine tolerated varies with the species, but there is general agreement that the therapeutic effects of atropine are not much increased by increasing the dose given above about 5 mg/kg. As described in earlier Chapters, at low doses atropine acts only at parasympathetic receptors, but at doses above 1 mg/kg it blocks ganglia, and antagonizes the action of anti-AChE's in the CNS. It is possible that in many of the experiments reported on cats and dogs the doses given were below the optima.

*Curarizing agents.* In mice 0·1 mg/kg of D-tubocurarine (a near-lethal dose) increased the $LD_{50}$ of TEPP over 3 times, of Ro3-0412 about 15 times and of Ro3-0422 about 20 times. The phosphorus compound and the D-tubocurarine were given at the same time.[18] D-Tubocurarine reduced the toxicity of the two ionic compounds, which do not penetrate the CNS, more than it reduced the toxicity of TEPP, which does. The therapeutic effects of giving D-tubocurarine some time after a phosphorus compound have not been investigated. From results already described in earlier Chapters such treatment is likely to be less effective, and may even be harmful, as D-tubocurarine may increase paralysis of muscles desensitized to ACh.

Magnesium sulphate, given intramuscularly in high doses, had a slight therapeutic effect on rabbits treated with DFP, but none on cats.[8, 19] Whether magnesium ions block the receptors curarimimetically or accelerate the re-activation of inhibited AChE[20] is not known.

*Hexamethonium.* Injected into mice at the same time as the poison, this, at 40 mg/kg, increased the $LD_{50}$ of TEPP about 3 times, that of Ro3-0412 about 6 times, and that of Ro3-0422 about 4 times.[18] The dose given also blocks receptors on striated muscles. It is therefore doubtful how much of its effect should be attributed to its action at ganglia.

Buscopan (scopolamine butyl bromide) was about as effective as atropine against Parathion, Parathion-Me and 'Systox' in rats.[7] Again, the doses were very high, so the mechanism of action is doubtful.

*Re-activating agents.* The therapeutic effects of these compounds have been described already in Chapter VIII, where some therapeutic indices are given in Tables 8.5 and 8.6. Only a few additional observations are recorded here, and the rest are summarized briefly.

P2-AM raised the $LD_{50}$ of Parathion 2–5 times in rabbits,[21] guinea pigs,[2] mice and rats,[22] and saved horses given an otherwise lethal dose.[23] For effective therapy doses greater than 10 mg/kg had to be given repeatedly, because P2-AM is decomposed in about 2 hr *in vivo*, while Parathion is converted to Paraoxon over several hours. Diacetyl monoxime had a paradoxical effect in rats.[22] It *increased* the toxicity of Parathion, and the AChE inhibition it caused, although it decreased the toxicity of Paraoxon. There are several possible explanations. Diacetyl monoxime may catalyse the conversion of Parathion to Paraoxon, inhibit the A-esterase which hydrolyses Paraoxon, or react with Parathion to form a powerful inhibitor as Sarin does with P4-AM (p. 141).

P2-AM is, in general, usually found to be at least as effective therapeutically as atropine against all but those compounds which form very stable phosphorylated AChE's, such as Tabun,* Dimefox, Schradan, and probably Mipafox. Perhaps poisoning by phosphoramidic compounds such as the four listed can never be treated successfully with re-activating agents. Against other compounds some bis-quaternary re-activators are more effective than P2-AM, as are hydroxyimino-acetone and diacetyl monoxime against a few compounds of marked central action, such as Sarin.

*Other treatments*. Pentobarbital and other hypnotics reduce convulsions, but do not increase $LD_{50}$'s significantly.[14, 19]

Compounds such as SKF 525A, which inhibit the oxidation of indirect inhibitors, have some therapeutic effects, as described in Chapter IX.

Injections of blood decreased the toxicity of DFP to rats.[24] Possibly blood transfusions make up some of the fluid lost by excretions, and also counteract shock, which appears to be present (p. 314). If this explanation is correct, plasma transfusions should be more effective.

*Multiple treatments*. No cases have been reported in which the combination of two or more treatments is not at least as efficacious as any of the component treatments. This is a significant finding, as in many treatments the therapeutic compounds were given in doses which were not much less than their $LD_{50}$'s. It suggests that the therapeutic compounds were acting in different ways, and that they and the phosphorus compounds were to some extent mutually antagonistic.

The advantage gained by using a combination of treatments is very variable. Some experiments are summarized below.

(1) Mice were given phosphorus inhibitors and therapeutic agents simultaneously.[18] With TEPP, atropine (10 mg/kg), D-tubocurarine (0·1 mg/kg) or hexamethonium (40 mg/kg) increased the $LD_{50}$ about 3 times. Either atropine + D-tubocurarine, or atropine + D-tubocurarine + hexamethonium at the same doses raised the $LD_{50}$ about 5 times. Atropine + hexamethonium was a little less effective. In similar experiments with Ro3-0412 and Ro3-0422 atropine was much less effective than either D-tubocurarine or hexamethonium, but increased the effects of either or of both combined. The increases in $LD_{50}$'s recorded are such as to suggest that the sites of action of the three compounds overlapped considerably, as is to be expected from the very high doses given.

(2) Pentamethonium about doubled the effect of atropine against Paraoxon in mice, and increased it in cats and rabbits.[17]

* Compound 16 re-activated rapidly electric eel AChE inhibited by Tabun.[51] Possibly the AChE from electric eels is particularly readily re-activated.

(3) Magnesium sulphate at 800 mg/kg with atropine was more effective than atropine alone against DFP in rabbits.[8, 19]

(4) Dogs survived 8–10 $LD_{50}$'s of Tabun or Sarin under artificial respiration only as long as it was maintained. Normal respiration was restored rapidly by atropine (0·5 mg/kg). P2-AM (25 mg/kg) also restored breathing in two out of the three dogs treated with Sarin.[1]

(5) Guinea pigs were infused with Paraoxon or DFP at a steady rate. A combination of artificial ventilation, atropine (20 mg/kg) and P2-AM (300 mg/kg in divided doses) enabled them to survive up to 160 $LD_{50}$'s of Paraoxon or 20 of DFP.[2]

(6) In monkeys 0·029 mg/kg of atropine increased the $LD_{50}$ of Sarin 3 times. With 20 mg/kg of hydroxyimino-acetone it raised the $LD_{50}$ about 14 times.[13]

(7) Hydroxyimino-acetone (35 mg/kg) or P2-AM (100 mg/kg) was markedly more effective with atropine (17·4 mg/kg) than any one of them given alone against Sarin in mice, rats and guinea pigs. The $LD_{50}$'s were raised 2–5 times.[13]

(8) In mice, rats and rabbits P2-AM and atropine combined were markedly more effective than either alone against Parathion poisoning.[23]

The combination of atropine with a re-activating agent is often very much more effective than either alone.[2, 13, 25, 26] There are three reasons for this. Firstly, atropine has no effect on the receptors of striated muscle at the doses usually given. Consequently an animal can die from paralysis of the respiratory musculature even when the central and parasympathetic effects have been overcome successfully. Re-activators antagonize the most important lethal action which remains in an atropinized animal. Secondly, atropine is not a perfect antagonist to ACh even at parasympathetic receptors. If enough AChE is inhibited, some sites will still be over-affected by ACh. Even slight re-activation may make the atropine much more effective.[26] Thirdly, any treatment which gives the animal more time to adapt to ACh increases the chances of survival. Increase in survival time is not, however, of unlimited value. In some cases, e.g. rats, mice and rabbits poisoned with massive doses of Parathion, P2-AM + atropine therapy greatly prolongs life even in those animals which eventually die of Parathion poisoning. The animals sink into a coma, which may last for days.[23] The mechanism of this effect has not been studied.

## SYMPTOMS, DIAGNOSIS AND THERAPY IN MAN

Certain groups of people are at risk from phosphorus compounds.

(1) Those who manufacture and spray phosphorus insecticides. In California in 1954, for example, there were 121 cases of disablement from

phosphorus insecticides among those engaged in agriculture.* Mild cases are included in this total. The number of deaths is not stated. Most casualties received the poison through the skin. In Japan there were about 6000 cases of Parathion poisoning in the years 1953–59.[21] It seems to be always surprising to the scientifically uneducated that a compound can penetrate the skin without damaging it, and that it may be at least as toxic by that route as when taken orally. The two main safety precautions are undoubtedly education and washing. Without these protective clothing loses much of its effect.

(2) Children of users. Several fatal cases have been reported.[27]

(3) Suicides.[27, 28] Control of sales greatly reduced the use of Parathion by suicides in Germany. At one time the compound seemed likely to supersede barbiturates and coal-gas,[28] in spite of its symptomatology.

(4) Research workers. Several cases, some severe, are known to me.

No cases have been established amongst consumers of sprayed crops. Hazards to consumers appear to be very slight indeed.

The hazard from a phosphorus compound depends in an obvious way primarily on a few factors.

(1) The acute toxicity. By mouth the $LD_{50}$'s to man of a few compounds are about as follows: TEPP, 1·7 mg/kg; Sarin, 0·28 mg/kg; DFP, 2·1 mg/kg;[29] and Parathion, about 5 mg/kg.[52] Sarin splashed in the eye may be lethal at about 0·05 mg/kg.[46] Malathion, on the other hand, is much less toxic. A child who drank some accidentally showed abnormal symptoms, which suggested that the paraffinic solvent may have contributed as much to the toxic effects as Malathion itself.[27] Undoubtedly the use as insecticides of compounds of low toxicity, such as Malathion and Dimethoate, will reduce hazards greatly. It is, however, inevitable that poor countries threatened with food shortages will use the cheapest available insecticide on the largest possible scale without much regard for acute toxicity.

(2) The rate of onset of symptoms. Indirect inhibitors are likely to prove rather safer than direct ones of the same acute $LD_{50}$.

(3) The rate at which the phosphorylated AChE is re-activated.

(4) The rate at which the compound is absorbed through the skin. Most non-ionic inhibitors are absorbed very fast, but the rate of absorption of concentrates can be greatly reduced by various additives.[47, 48]

(5) Volatility. Even the vapour of Parathion, impregnated on gauze strips for fly control, lowered the ChE levels in guinea pigs' blood under conditions which resembled those of practical use.[49] In practice, however,

* This is a very small proportion of total agricultural accidents.[27] In California in 1953 there were 14 188 such accidents of all kinds. Those with ladders accounted for 1371. In California exceptionally large quantities of insecticides are used, and frequently high temperatures make the wearing of protective clothing irksome.

only the vapours of volatile compounds designed for gas warfare seem to be hazardous to man.

The toxicity to man of phosphorus insecticides has been reviewed.[50]

*Symptoms*

Symptomatology has been studied both on human volunteers, mainly as part of chemical warfare programmes, and on casualties.

The symptoms in man given Sarin have been studied in great detail.[29] A brief summary only is given here. The Sarin was given in very dilute solution orally over 3 days. The estimated $LD_{50}$ was 0·28 mg/kg, and the doses below are expressed as fractions of this dose. 0·025 $LD_{50}$'s caused no symptoms. 0·16 $LD_{50}$'s caused mild parasympathetic symptoms: anorexia, mild abdominal cramps and excessive sweating in 1/2 subjects, and central effects: excessive dreaming, in the other subject. 0·33 $LD_{50}$'s caused marked parasympathetic symptoms in all three subjects: anorexia, nausea, abdominal cramps, sweating, epigastrial and substernal tightness, and tightness in the chest. One subject also showed bradycardia, and another urinary frequency and subjective dyspnoea. All three were easily fatigued, and one showed muscle twitching, due to interference with muscle receptors. Two showed central symptoms: giddiness, tension, anxiety, excessive dreaming with nightmares, and, in one case, depression, withdrawal, restlessness, emotional lability and tremors. Similar symptoms in more severe form were caused by 0·39 $LD_{50}$'s, and, in one of the two subjects, vomiting, tenesmus, urgent loose stools, fasciculations and pallor. The expiratory movement in breathing tended to be prolonged in the more severe cases. Interference with respiration was particularly marked in one subject with a history of infections of the upper respiratory tract. More severe symptoms have been observed in cases of accidental exposure to Sarin.[30] Bronchial secretion and salivation were severe, with appreciable bronchoconstriction. During moderate symptoms the electroencephalograph recorded epileptic-type patterns. In fatal and near-fatal cases the respiratory musculature became paralysed, and the patient became cyanosed, and convulsed. The respiratory centre did not fail except when the dose was well above the $LD_{50}$.

The order in which the symptoms appear depends somewhat on the route of administration. After inhalation, bronchoconstriction and respiratory distress appeared before symptoms involving the gastrointestinal tract became severe. Application to the eye caused complete pupillary constriction at low doses, although myosis was slight in cases who had received the compound orally.

The symptomatology does not vary much with the compound; substantially the same picture has been observed with DFP,[31, 37] TEPP,[32, 33] Parathion,[21, 28, 32, 34, 35] Schradan,[30] and 'Systox'[36] as with Sarin. Parathion poisoning in man has been reviewed in detail.[52, 53] The onset may, however, be delayed several hours with indirect inhibitors. Even with direct inhibitors death does not come so rapidly as in the smaller mammals. Sarin given orally only kills within an hour after doses well above the $LD_{50}$.[30]

The only overt sign in man which does not appear in animals is coma, which is observed at a late stage in severe poisoning. Man can describe many symptoms, such as nausea, colic, and anxiety which obviously cannot be observed in animals, and which may give fairly early warning of poisoning. With these exceptions signs are generally similar to those in animals. The syndrome has, however, some fairly distinctive features. Bronchoconstriction is not an important factor in causing death, providing the additional effects of obstruction of the bronchi by the heavy secretions of saliva and mucus are avoided by postural drainage.[29, 20, 32] The blood pressure is raised moderately,[35] and bradycardia is rarely as severe as in many species.[29, 32] The central nervous system is affected before the voluntary musculature.[29, 38] In spite of this it appears from the success of P2-AM therapy of Parathion poisoning[21] that death is caused by paralysis of the musculature and not of the respiratory centre. The apparent greater importance of central symptoms in man may only reflect his unique capacity to describe them.

*Diagnosis*

Diagnosis is of three kinds: diagnosis of incipient cases, showing no overt symptoms; diagnosis in casualties; and diagnosis post mortem.

Diagnosis of incipient cases is carried out by estimating the AChE activity of blood samples or the ΨChE activity of plasma samples. Such tests are carried out routinely on many people engaged in spraying or manufacturing phosphorus insecticides. They show whether a worker has absorbed a phosphorus compound or not, and consequently whether he is likely to be unusually sensitive to a further dose. Absorption into the general system always lowers the blood ChE's. Estimations of AChE are preferable to estimations of ΨChE, as the blood AChE is a fairly reliable indication of the general AChE level in the body. Tests on blood ΨChE can, however, be used. ΨChE is lowered faster than AChE by most phosphorus poisons, and there are no cases reported in which it is not lowered as fast initially; but ΨChE may recover faster than AChE, so that the ΨChE level may lead to an underestimate of the residual effects of a dose received some time earlier. Various methods of estimation have been standardized on fairly large numbers of people.[39, 40, 41] The variation from person to person is such that levels below 70% of the average are likely to be significant. In man it has been calculated that the oral dose of Sarin, TEPP, or DFP which causes 50% inhibition of blood AChE is about 0·1 $LD_{50}$'s.[29] Two have, however, died of Parathion poisoning and their blood AChE levels have been found to be

50% and 65% post mortem.[53] The times elapsed between death and the AChE determinations were not stated, and some re-activation may have taken place. As was pointed out in Chapter VIII man may show no symptoms when his AChE levels are lowered slowly by spaced successive doses to about 5% of normal levels, although a single dose which causes acute symptoms may have considerably less effect on his AChE. The effects of small multiple doses are more relevant to the use of ChE determinations as a safeguard against poisoning. Providing poisoning takes this form, ChE determinations are a reliable and sensitive method of diagnosing its early stages.

Diagnosis in casualties depends mainly upon recognizing the very general nervous symptoms. It may be confirmed by giving about 2 mg of atropine. In a normal subject this causes marked atropinization, but in a case of phosphorus poisoning relieves the symptoms, without atropinizing.[30, 42] Estimations of ChE are confirmatory.

Diagnosis at autopsy may be of financial importance to dependants of the deceased. There are no gross pathological changes, and diagnosis depends mainly upon determinations of AChE either in erythrocytes or histochemically at myoneural junctions (Chapter VIII).[36, 43, 44] Although tissues can be stored for a week or more in the cold without losing significant AChE activity (AChE in spots of blood dried on filter papers are particularly stable,[40, 45]) the chances of a successful diagnosis depend upon the rate at which the phosphorylated AChE is re-activated and not upon the stability of the tissue. For this reason tests should be carried out immediately after death. In cases where death is caused by a single dose of a dimethyl phosphate ester no inhibition may be observed even then, as re-activation is very fast. After a diethyl phosphate ester or Sarin the AChE levels should be markedly below normal for at least a day; and after di-isopropyl phosphate esters or phosphoramides the AChE remains inhibited almost indefinitely. This is also the case after death from successive small doses of a dimethyl phosphate ester, as dimethylphosphorylated AChE goes to its stable form in this type of poisoning.

### *Therapy*

Therapy has consisted of artificial respiration, with oxygen in some cases, atropinization, dosing with P2-AM, giving large volumes of fluids, and various ancillary measures, such as arranging the body for easy drainage of bronchial secretions, and giving small doses of barbiturates to lessen anxiety. Patients of course, have been washed thoroughly and their clothing removed when skin contamination is suspected. Opiates have not been used, as they inhibit respiration.

Artificial respiration should take a form in which inspiration is brought about positively, and not one in which it depends on the natural elasticity of the lungs. Thus the Holger Nielson method is better than the Schaeffer prone method.[30]

Atropine in a normal subject causes marked and unpleasant symptoms at about 0·03 mg/kg. It is generally agreed that this dose is inadequate in treatment of phosphorus poisoning.[28, 30, 35] Most workers recommend giving individual doses of 0·03–0·06 mg/kg intravenously or intramuscularly, repeating them at intervals of a few minutes until symptoms are markedly reduced, and then whenever symptoms regain force. Total doses of 50 mg (per person, not per kilogramme) have been given.[30] The organophosphorus poison antagonizes the action of atropine, so that signs of atropinization may not appear until several milligrammes have been given. Unless man behaves very unlike the cat and the dog, doses of under 50 mg cannot be expected to give maximal protection against central effects. Such doses are said to be unpleasant, but not dangerous.[30, 38, 42]

P2-AM has been used against Parathion poisoning with great success. 1 g was given intravenously, and the dose repeated in some cases.[21] P2-AM cannot, of course, be an effective agent against Schradan or other compounds which form very stable phosphorylated AChE's. The use of re-activators after massive doses of atropine will increase the signs of over-atropinization.

Curarizing and ganglion blocking agents have not been tried. In view of the differences in the effects of D-tubocurarine according to species and the stage of poisoning, curarization in man seems to be a risky measure.

There are no reports of plasma transfusions being given, even though liquid losses from man are great, and in some animals the condition appears to be partly one of shock.

Prognosis is markedly more favourable the sooner the treatment is begun and the slower the onset of symptoms. Thus of 5 severe cases of TEPP poisoning, none survived, whereas of 41 severe cases of Parathion poisoning 23 survived.[32] The failure to save the lives of these TEPP victims was largely attributed to the rapid onset of symptoms, so that the patients were nearing the agonal stages by the time treatment was begun. Once the crisis is past, recovery is normally complete. A few compounds, described in the next Chapter, cause permanent lesions. These are not used as insecticides.

## REFERENCES


1. R. V. Brown, A. M. Kunkel, L. M. Somers and J. H. Wills. *J. Pharmacol.* **120,** 276 (1957)
2. K. Bethe, W. D. Erdmann, L. Lendle and G. Schmidt. *Arch. exp. Path. Pharmak.* **231,** 3 (1957)
3. J. M. Barnes. *Brit. J. Pharmacol.* **8,** 208 (1953)
4. J. M. Barnes and J. I. Duff. *Brit. J. Pharmacol.* **9,** 153 (1954)
5. V. DiStefano, L. Hurwitz, W. F. Neuman and H. C. Hodge. *Proc. Soc. exp. Biol. N.Y.* **78,** 712 (1951)
6. F. Hobbiger. *Brit. J. Pharmacol.* **12,** 438 (1957)

7. W. B. Deichmann and R. Rakoczy. *Arch. industr. Hlth.* **7,** 152 (1953)
8. W. Modell and S. Krop. *J. Pharmacol.* **88,** 34 (1946)
9. W. W. Douglas and P. B. C. Matthews. *J. Physiol.* **116,** 202 (1952)
10. K. P. DuBois, J. Doull and J. M. Coon. *J. Pharmacol.* **99,** 376 (1950)
11. S. Krop and A. M. Kunkel. *Proc. Soc. exp. Biol. N.Y.* **86,** 530 (1954)
12. C. Dayrit, C. H. Manry and M. H. Seevers. *J. Pharmacol.* **92,** 173 (1948)
13. B. M. Askew. *Brit. J. Pharmacol.* **12,** 340 (1957)
14. G. Wilhelmi and R. Domenjoz. *Arch. internat. Pharmacodyn.* **86,** 321 (1951)
15. J. Fournel. *C.R. Soc. Biol. Paris.* **151,** 1373 (1957)
16. B. E. Conley. *Arch. internat. Pharmacodyn.* **116,** 375 (1958)
17. C. A. de Candole and M. K. McPhail. *Nature, Lond.* **174,** 552 (1954)
18. M. W. Parkes and P. Sacra. *Brit. J. Pharmacol.* **9,** 299 (1954)
19. B. P. McNamara, G. B. Koelle and A. Gilman. *J. Pharmacol.* **88,** 27 (1946)
20. F. Hobbiger. *Brit. J. Pharmacol.* **6,** 21 (1951)
21. T. Namba and K. Hiraki. *J. Amer. med. Ass.* **166,** 1834 (1958)
22. E. Reiner, B. Svetlicic and M. Vandekar. Proc. XIIth internat. *Congress of occup. Health* **3,** 233 (1957)
23. B. Svetlicic and M. Vandekar. *J. comp. Path. and Therap.* **70,** 257 (1960)
24. R. F. Borgman, M. D. Rumbaugh and G. L. Campbell. *Amer. J. vet. Research.* **17,** 781 (1956)
25. D. R. Davies and A. L. Green. *Brit. J. industr. Med.* **16,** 128 (1959)
26. A. L. Green. *Biochem. Pharmacol.* **1,** 115 (1958)
27. R. L. Metcalf. *Arch. industr. Hlth.* **16,** 337 (1957)
28. G. Jantzen. *Deutsche medz. Wochschr.* **76,** 1601 (1951)
29. D. Grob and J. C. Harvey. *J. clin. Invest.* **37,** 350 (1958)
30. D. Grob. *Arch. int. Med.* **98,** 221 (1956)
31. B. A. Kilby and M. Kilby. *Brit. J. Pharmacol.* **2,** 234 (1947)
32. G. Freeman and M. A. Epstein. *New England J. Med.* **253,** 266 (1955)
33. D. Grob and A. M. Harvey. *Bull. Johns Hopkins Hosp.* **84,** 532 (1949)
34. A. Harrestrap-Anderson and T. Jersild. *Acta pharmacol. tox.* **5,** 129 (1949)
35. D. Grob, W. L. Garlick and A. M. Harvey. *Bull. Johns Hopkins Hosp.* **87,** 106 (1950)
36. C. S. Petty. *Arch. Path.* **66,** 458 (1958)
37. D. Grob, J. L. Lilienthal, Jr., A. M. Harvey, B. F. Jones, O. R. Langworthy and S. A. Talbot. *Bull. Johns Hopkins Hosp.* **81,** 217 (1947)
38. W. Wirth. *Deutsche med. Wochschr.* **79,** 1205 (1954)
39. J. A. Rider, J. L. Hodges, J. Swader and A. D. Wiggins. *J. Lab. clin. Med.* **50,** 376 (1957)
40. K-B. Augustinsson and G. Heimburger. *Acta physiol. scand.* **30,** 45 (1953)
41. J. H. Wolfsie and G. D. Winter. *Arch. industr. Hlth.* **6,** 43 (1952)
42. H. H. Golz. *Arch. industr. Hlth.* **16,** 330 (1957)
43. A. D. Bergner and M. W. Bayliss. *U.S. Armed Forces med. J.* **3,** 1637 (1952)
44. A. D. Bergner and S. H. Durlacher. *Amer. J. Path.* **27,** 1011 (1951)
45. E. Heilbronn. *Scand. J. clin. Lab. Invest.* **5,** 308 (1953)
46. A. A. Kondritzer, W. H. Mayer and P. Zvirblis. *Arch. industr. Hlth.* **20,** 50 (1959)
47. W. B. Deichmann, P. Brown and C. Downing. *Science* **116,** 221 (1952)
48. L. Rosival, F. V. Selecky and L. Vbrovsky. *Chem. Abstract.* **52,** 16618a (1958)
49. J. Glomme and A. Swensson. *Brit. J. industr. Med.* **15,** 62 (1958)

50. *Bull. World Health Organization* **16,** 1219 (1957)
51. J. H. FLEISHER, H. O. MICHEL, L. YATES and C. S. HARRISON. *J. Pharmacol.* **129,** 31 (1960)
52. W. D. ERDMANN and L. LENDLE. *Ergebnisse der inneren Medizin u. Kinderheilkunde* **10,** 104 (1958)
53. O. PRIBILLA. *Arch. Toxikol.* **15,** 210 (1955)

CHAPTER XVI

# Abnormal Effects

A FAIRLY small number of compounds anaesthetize mammals, or cause in some species long-lasting paralysis of the limbs. These signs of poisoning appear independently of any signs which can be attributed to inhibition of AChE; and some compounds cause paralysis or narcotic effects which do not cause any cholinergic symptoms.

## PARALYTIC EFFECTS

The present (April, 1960) is an inopportune time for writing on paralytic effects. There is as yet no fundamental explanation of the genesis of paralysis, and several groups of workers are actively engaged on the problem. No attempt is therefore made to describe all the results—this would be confusing rather than informative in the absence of a general theory of action.

*Historical background.* Many thousands of cases of paralysis in man have been reported, and the reports reviewed.[1] In all but a very few paralysis was caused by the absorption of tri-tolyl phosphates,* i.e. compounds of the general structure:

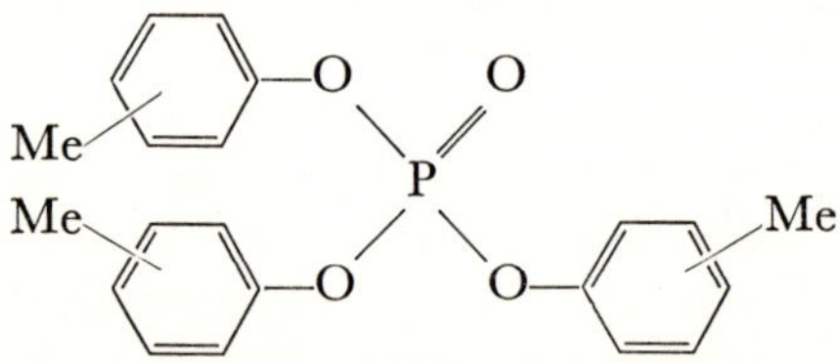


(As shown later, some tri-tolyl phosphates do not cause paralysis, and some homologues do.) In many cases the compounds were consumed, either in drinks (they were added as substitutes for ginger extracts to 'ginger jake' in the U.S.A. during prohibition) or in substitutes for cooking oils. Some victims absorbed the compounds through the skin in the processes involved in the manufacture of plastics or hydraulic oils.

* On older systems of nomenclature these compounds were called tri-cresyl phosphates. Thus tri-2-tolyl phosphate was called tri-orthocresyl phosphate, abbreviated to TOCP.

A different type of compound, Mipafox, $(Pr^iNH)_2PO \cdot F$, has also been the cause of two cases of paralysis in man.

*Typical syndrome.* Absorption of most paralysing compounds causes the same effects in the first few months. In man the first signs of paralysis, weakness in the feet, appear about a fortnight after the dose is received. Paralysis is of the flaccid type, is mainly confined to the arms and legs, and starts at the extremities. The feet are affected before the hands. Flaccid paralysis normally persists for several months. Sequelae vary with the compound, and are described later. A precise description has been given of the development of poisoning in two cases in which the time the poison (Mipafox) was received is accurately known.[(2)]

*Species susceptibility.* Compounds which cause paralysis in man do not cause paralysis in most species. There is some disagreement as to which species are affected, but it is generally agreed that only the chicken, cat and calf exhibit syndromes like those in man.[(1, 3)] Other species (rabbit, monkey and dog) show some weakness in the hind limbs after treatment with tri-2-tolyl phosphates. No symptoms of a paralytic nature can be induced in the rat.

Most work has been done on the hen. The effectiveness of a compound may be measured either by the least dose causing hind-limb weakness (ataxia) from 9 to 13 days after dosing or by that causing paralysis. The difference between the dose causing ataxia and that causing paralysis varies with the compound. In most experiments the dose causing ataxia has been regarded as the effective dose. Chicks are not susceptible until about 80 days after hatching, and full susceptibility takes about 6 months to develop.[(4, 5)] The hen is paralysed by DFP, and in this probably differs from man. Patients suffering from myasthenia gravis have been treated for long periods with high doses of DFP, but no cases of paralysis have been reported.

*Pathology.* Histological examination of the nervous system of hens poisoned with tri-2-tolyl phosphate, Mipafox or DFP reveals de-myelination of axon sheaths and degeneration of the axons in the sciatic nerve and spinal cord.[(4, 5, 6)] Both processes take place simultaneously, as in Wallerian degeneration, and the onset of histological changes is synchronous with the onset of paralytic symptoms.[(5, 7)] Few nerve cells are affected. It is therefore generally assumed that the primary lesion is de-myelination, and that axon degeneration is a consequence of it. For this reason effective compounds are often called de-myelinating agents. In hens some de-myelination may be found without paralysis after mild poisoning, and in rats treated with Mipafox de-myelination appears quite extensive, although the rats only show some muscular weakness.[(4)] The severity of the signs obviously depends upon how many and which tracts are affected.

A similar distribution of de-myelinated tracts is found in the hen whatever de-myelinating agent is given. The spinal cord is more damaged than the sciatic nerve at all stages of poisoning, and is the only part damaged in mild cases.[4, 8] Damaged tracts in the CNS are never found above the cerebellar peduncles. The distribution and type of lesion is similar to that observed in hens fed on a diet devoid of thiamine.[4] For more details the original references should be consulted.

### *The Dependence of the Syndrome on Chemical Structure*

The compounds which cause paralysis can be divided into two groups: fluoridates; and tri-tolyl phosphates and their homologues. The fluoridates and tri-tolyl phosphates differ in two ways. Firstly, the fluoridates cause paralysis at very much lower doses than the tri-tolyl phosphates; and secondly the fluoridates, unlike the phosphates, are of very high acute toxicity. The acute toxic effects are of the usual type, and result from inhibition of AChE. Doses of many fluoridates sufficient to cause paralysis can only be given to animals treated with atropine and a cholinesterase re-activator,[9] or pre-treated with eserine,[10] to overcome the acute effects, and even then it may be necessary to give divided doses over several days. Some tri-tolyl phosphates may cause mild signs of anti-AChE action at paralytic doses in some species. The type of paralysis caused by the two groups is the same.

*Fluoridates*. A wide range of compounds containing the P—F group causes paralysis in hens. Type compounds which are effective are listed in Table 16.1, and compounds of similar acute toxicity which are not in Table 16.2.

TABLE 16.1. FLUORIDATES WHICH CAUSE PARALYSIS IN HENS

The effective doses are given for the specific examples.

| Type | Number tested | Examples | Effective dose (mg/kg) | Refs |
|---|---|---|---|---|
| Phosphorodiamidic fluorides | 1 | Mipafox, $(Pr^iNH)_2PO \cdot F$ | 40 | (4) |
| Phosphorofluoridates | 11 | DFP, $(Pr^iO)_2PO \cdot F$ | 0·25 | (9) |
| | | $(MeO)_2PO \cdot F$ | 30 | (9) |
| Phosphonofluoridates | 5 | Sarin, $(Pr^iO)(Me)PO \cdot F$ | 1·0 | (9) |

Only one phosphoramidic fluoride, Mipafox, has been tested. All the phosphoro- and phosphono-fluoridates caused paralysis, but none of the phosphinofluoridates. It has been suggested that dimethyl phosphorofluoridate was less effective than other phosphorofluoridates because it is more readily hydrolysed. Two of the phosphinofluoridates are still less stable and none is much more stable. They are all very toxic, and so could only be given in small doses. These factors may explain their inefficacy. Many other typical anti-AChE's have been tested (Table 16.2), and are all without paralytic effects. The chloridates may well be too unstable, and some of the other compounds have not been tested at high doses. Two general conclusions, however, appear to be valid: compounds causing paralysis contain P—F groups; and the effects do not depend to any great extent on the basic groups in the molecule. This suggests that paralysis results from the release of hydrofluoric acid in the nervous system.

TABLE 16.2. FLUORIDATES AND COMPOUNDS OF SIMILAR ACUTE TOXICITY WHICH DO NOT CAUSE PARALYSIS IN HENS

The maximum dose given is shown under 'Max dose'.

| Type | Number tested | Examples | Max dose (mg/kg) | |
|---|---|---|---|---|
| Phosphinofluoridates | 4 | $Et_2PO \cdot F$ | 5 | (9) |
| Pyrophosphates | 1 | TEPP, $(EtO)_4P_2O_3$ | 100 | (9) |
| Phosphoric-phosphonic anhydrides | 5 | $(EtO)(MeO)PO \cdot O \cdot PO(Me)(OPr)$ | 10 | (9) |
| Phosphorochloridates | 3 | $(EtO)_2PO \cdot Cl$ | 100 | (9) |
| Various | | Amiton, $(EtO)_2PO \cdot S \cdot CH_2 \cdot CH_2 \cdot NEt_2$ | 20 | (9) |
| | | TIPA, $(Pr^iNH)_4P_2O_3$ | 300 | (11) |
| | | Paraoxon-$Pr^i$, $(Pr^iO)_2PO \cdot O \cdot C_6H_4$-4-$NO_2$ | 15 | (11) |

*Phosphates*. Most of the compounds causing paralytic effects are shown in Table 16.3, and some which fail to do so are shown in Table 16.4. The following types of compounds also do not cause paralysis: tetratolyl pyrophosphates, ditolyl phosphorochloridates, ditolyl hydrogen phosphates, dialkyl tolyl phosphates and tri-tolyl phosphites and phosphine oxides.[12, 13] Triphenyl phosphate causes paralysis in the cat but not in the chicken.[13]

At present only the following generalizations hold. All active compounds are tri-aryl phosphates. The most toxic are tri-tolyl phosphates with one or two 2-tolyl groups. The 2-Me group confers some toxicity on xylenyl phosphates, but not all xylenyl phosphates with 2-Me groups are toxic.

Generalizations are uncertain because it is difficult to prepare many of these compounds pure. The reactions used for the preparation of mixed esters normally involve stepwise reaction with $POCl_3$:

$$POCl_3 + ArOH \rightarrow ArO \cdot POCl_2 + HCl,$$

and so on. To prepare mixed tri-aryl phosphates one has, therefore, to start with pure toluols and xylenols, and to purify each intermediate carefully. Isomeric toluols and xylenols do not differ much physically, and are therefore not easily purified, and purification of the phosphorus compounds presents

TABLE 16.3. TRI-TOLYL PHOSPHATES AND HOMOLOGUES CAUSING ATAXIA IN CHICKENS

The compounds are regarded as derived from triphenyl phosphate by substitution of alkyl groups for hydrogens on the rings. Thus compound 1 is tri-2-tolyl phosphate. The compounds asterisked were given in equal daily doses until signs appeared. The total doses given are recorded under 'effective dose'. As ataxia normally takes 9 days to appear, these doses are not directly comparable with those used in the other experiments, where only single doses were given. Doses were given subcutaneously or orally. The effective doses are usually rather greater orally, but rarely twice as great. For compound No. 6 only one dosage, that shown, was tested.

| No. | Compound | | | Effective dose (mg/kg) | Refs |
|---|---|---|---|---|---|
| | $O \cdot C_6H_4$— | $O \cdot C_6H_4$— | $O \cdot C_6H_4$— | | |
| 1 | 2-Me | 2-Me | 2-Me | 200 | (12) |
| 2 | 2-Me | 2-Me | 4-Me | 25 | (12) |
| 3 | 2-Me | 4-Me | 4-Me | 50 | (12) |
| 4 | 2-Me | 3-Me | 4-Me | 20 | (12) |
| 5 | 2-Me | 3-Me | 3-Me | 20 | (14) |
| 6 | 2-Me | H | H | 500 | (12) |
| 7* | 2-Me | 3:5-$Me_2$ | 3:5-$Me_2$ | 1000 | (15) |
| 8* | 2:4-$Me_2$ | 3:5-$Me_2$ | 3:5-$Me_2$ | 10 000 | (15) |
| 9* | 2:6-$Me_2$ | 3:5-$Me_2$ | 3:5-$Me_2$ | 24 000 | (15) |

TABLE 16.4. TRI-TOLYL PHOSPHATES AND ANALOGUES WHICH DO NOT CAUSE ATAXIA IN CHICKENS

Conventions as in Table 16.3.

| No. | Compound | | | Max dose (mg/kg) | Refs |
|---|---|---|---|---|---|
| | $O \cdot C_6H_4$— | $O \cdot C_6H_4$— | $O \cdot C_6H_4$— | | |
| 1* | H | H | H | 30 000 | (13) |
| 2 | 3-Me | 3-Me | 3-Me | 5000 | (12) |
| 3 | 4-Me | 4-Me | 4-Me | 5000 | (12) |
| 4* | 2:6-$Me_2$ | 2:6-$Me_2$ | 3:5-$Me_2$ | 40 000 | (15) |
| 5* | 2:4-$Me_2$ | 2:5-$Me_2$ | 2:5-$Me_2$ | 32 000 | (15) |
| 6* | 2-Et | 2-Et | 2-Et | 2500 | (12) |

many of the usual problems. None of the work published so far contains fully adequate evidence of purity.

The syndromes produced vary somewhat both with the tri-tolyl phosphate given and with the severity of poisoning.[1] In man mild poisoning seems to produce paralysis of the flaccid type only, which disappears within 2 years or less. In more severe cases signs of spastic paralysis appear as the flaccid paralysis recedes, and may persist for many years. The severity of the spastic-type signs depends on the compound given. Tri-2-tolyl phosphate only causes spastic paralysis in very severe cases, whereas poisoning by 2-tolyl di-4-tolyl phosphate is usually complicated by spastic paralysis. In the hen and cat spastic paralysis predominates even in mild cases of poisoning by 2-tolyl di-4-tolyl phosphate, but is not observed after poisoning by tri-2-tolyl phosphate. Flaccid paralysis can be accounted for by degeneration of peripheral nerves, while spastic paralysis implies degeneration of the CNS. Histological examination of the nerves of cats and hens given the two compounds supports the distinction. Mipafox in man apparently causes paralysis of the flaccid type.

*Theories of Action*

A successful theory of action must explain both how of two compounds as similar as 2-tolyl di-4-tolyl phosphate and tri-4-tolyl phosphate one can be a powerful de-myelinating agent and the other inert, and also how two compounds as different physically and in chemical reactivity as tri-2-tolyl phosphate and Mipafox can both be active. No useful theory has yet been proposed. The theories now described are either untenable or unproven. A summary of them serves, however, as a guide to the literature, and gives some properties of the compounds.

1. *The anti-ΨChE theory. In vitro* Mipafox and DFP inhibit chicken ΨChE more rapidly than they inhibit AChE, and tri-2-tolyl phosphate inhibits neither ΨChE nor AChE. Tri-2-tolyl phosphate is, however, converted *in vitro* by rat liver and *in vivo* in several species to a compound which inhibits ΨChE faster than AChE.[16, 17] Given to chicken, all three compounds inhibit ΨChE in the CNS preferentially.[11, 18] As ΨChE is found in the myelinated tracts, it was therefore suggested that ΨChE is involved in maintaining the myelin sheath, and that de-myelination results when ΨChE is inhibited.[19] This theory is untenable. Firstly, ΨChE in the CNS of the hen can be inhibited preferentially by several compounds, such as TIPA, which do not cause paralysis;[11, 20] and secondly the ΨChE levels in hens given DFP or Mipafox have usually recovered to normal by the onset of paralysis.[11, 20] Except for tri-2-tolyl phosphate, the compounds which cause paralysis do not inhibit the ΨChE for an unusually long time, and the rates of recovery of ΨChE activity are the same after paralysing agents as after some other compounds.[11, 20]

Although the anti-ΨChE theory is untenable, the finding that tri-2-tolyl phosphate is metabolized to a compound which reacts with ΨChE is significant. The metabolite is an anti-esterase, so that the great difference between the re-activity of its parent compound and of Mipafox may be irrelevant. Tri-2-tolyl phosphate may be an indirect paralysing agent, just as Schradan is an indirect inhibitor of AChE. There is some confirmation of this view. Tri-3-tolyl phosphate and tri-4-tolyl phosphate are not converted to inhibitors *in vivo*,[16, 22] and are therefore unlikely to be indirect anti-esterases.

2. *Anti-oxidation theory.* The pathological findings resemble those found in thiamine deficiency, which probably causes nerve lesions by interfering with pyruvate oxidation. In fact both de-myelinating and other phosphorus compounds inhibit glucose and pyruvate oxidation at high concentrations, but these concentrations are about 500 times greater than those likely to be present *in vivo*,[23] and both glucose and pyruvate oxidation are normal in brains taken from hens given tri-2-tolyl phosphate.[24]

3. *Anti-lipid synthesis theory.* There is some evidence for this theory.

The rate of synthesis of lipid can be measured on excized sciatic nerves by the rate of incorporation of $^{14}C$ from $^{14}C$-acetate. The rate of synthesis by nerves from hens given DFP in paralytic doses fell shortly after dosing to about half of normal, and was still below normal 17 days later. DFP only inhibits lipid synthesis *in vitro* at about $10^{-3}$ M. Lipid synthesis may, therefore, not be inhibited by the direct action of DFP on the enzymes responsible for synthesis, but by the inhibition of some coupled reaction. This complication does not, of course, invalidate the theory.[25]

Phospholipid synthesis appears to be normal in paralysed hens. Incorporation of $^{32}P$ given as phosphate into phospholipids takes place at the same rate in normal and paralysed hens, and the $^{32}P$ is distributed in the same way between different phospholipid fractions.[26]

## NARCOTIC EFFECTS

Several compounds of low toxicity have been tested by intravenous injection into female rats. Without exception all produced some narcotic effects, ranging from inco-ordination and weakness to deep anaesthesia.[27, 28] The compounds are listed in Table 16.5. Some, such as the trialkyl phosphates, are not of a type to cause cholinergic symptoms. Others are usually regarded as anti-AChE's, and cause predominantly cholinergic symptoms when given by mouth. Thus Malathion, the quinolyl compound and Demeton-S-Me are all typical anti-AChE's (all three are indirect inhibitors) except when given intravenously. Then the first two at high doses kill before cholinergic symptoms appear, and Demeton-S-Me causes a transient anaesthesia, which gives way to cholinergic symptoms. Demeton-O-Me shows the importance of the route of administration particularly clearly. By i-v injection the $LD_{50}$ is 216 mg/kg, and the rats show no cholinergic symptoms, and die in a coma of respiratory failure. By mouth the $LD_{50}$ is 680 mg/kg, and the rats die with typical cholinergic symptoms. Probably after oral injection the compound lingers in the stomach long enough for an appreciable fraction to isomerize to Demeton-S-Me (Chapter IV), to which the oral toxicity should be attributed. The fat–water partition coefficient of Demeton-O-Me is high, so that after intravenous injection the compound is probably removed rapidly from the blood, and can no longer be converted.

TABLE 16.5. COMPOUNDS CAUSING NARCOTIC EFFECTS IN FEMALE RATS AT SUBLETHAL DOSES[27]

| Compound | mg/kg | Narcotic signs | Cholinergic signs |
|---|---|---|---|
| Demeton-S-Me | 50 | weakness | marked |
| Tributyl phosphate | 80 | anaesthesia | none |
| Triethyl phosphorothionate | 150 | anaesthesia | none |
| Demeton-O-Me | 200 | weakness | none |
| 8-quinolyl tetramethylphosphorodiamidate | 250 | anaesthesia | marked |
| Malathion | 300 | weakness | marked |
| Triethyl phosphate | 500 | anaesthesia | none |
| Trimethyl phosphate | 1800 | weakness | none |

The mechanism of action is uncertain, but a reasonable hypothesis can be put forward. The effects are marked only after injection into the blood stream, which suggests that a high concentration is necessary. Assuming rapid distribution *in vivo*, the concentrations causing narcosis are about $10^{-3}$ M. As already shown in Chapter XII such concentrations of several compounds block nerve conduction and motor end-plates. The combination of anaesthetic effects and paralysis of the respiratory musculature observed accord with this hypothesis. Narcotic block of conduction is attributed to a block in sodium ion transport across the axon membrane. Narcotics also block the passage of sodium ions through the membranes surrounding red blood cells. DFP is a typical narcotic in this respect. At $9{\cdot}1 \times 10^{-3}$ M it blocks the flow of sodium ions from cat red cells.[29]

Neither water solubility nor fat–water partition coefficients can be important. Some compounds are miscible in all proportions with water, while others are very sparingly soluble. The quinolyl compound and Demeton-O-Me cause narcotic effects of similar intensity at similar doses, although the first is miscible with water, and therefore likely to have a very low fat–water partition coefficient, and the second has a very high partition coefficient.

## REFERENCES


1. D. Henschler. *Klin. Wochensch.* **36,** 663 (1958)
2. P. L. Bidstrup, J. A. Bonnell and A. G. Beckett. *Brit. med. J.* **1,** 1068 (1953)
3. M. I. Smith, E. Elvoove and W. H. Frazier. *Publ. Hlth. Rep.* **45,** 1703 (1930)
4. J. M. Barnes and F. A. Denz. *J. Path. Bact.* **65,** 597 (1953)
5. J. C. B. Fenton. *J. Path. Bact.* **69,** 181 (1955)
6. M. I. Smith and R. D. Lillie. *Arch. Neurol. Psychiat.* **26,** 976 (1931)
7. J. B. Cavanagh. *J. Neurol. Neurosurg. Psychiat.* **17,** 163 (1954)
8. M. C. Lancaster. *Brit. J. Pharmacol.* **15,** 279 (1960)
9. D. R. Davies, P. Holland and M. J. Rumens. *Brit. J. Pharmacol.* **15,** 271 (1960)
10. J. M. Barnes, personal communication.
11. A. N. Davison. *Brit. J. Pharmacol.* **8,** 212 (1953)
12. C. H. Hine, M. K. Dunlap, E. G. Rice, M. M. Coursey, R. M. Gross and H. H. Anderson. *J. Pharmacol.* **116,** 227 (1956)
13. D. Henschler and H-H. Bayer. *Arch. exp. Path. Pharmak.* **233,** 512 (1958)
14. D. Henschler. *Arch. exp. Path. Pharmak.* **232,** 223 (1957)
15. D. Henschler. *Arch. exp. Path. Pharmak.* **237,** 459 (1959)
16. W. N. Aldridge. *Biochem. J.* **56,** 185 (1954)
17. D. K. Myers, J. B. J. Rebel, C. Veeger, A. Kemp and E. G. L. Simons. *Nature, Lond.* **176,** 259 (1955)
18. C. J. Earl and R. H. S. Thompson. *Brit. J. Pharmacol.* **7,** 685 (1952)
19. M. G. Ord and R. H. S. Thompson. *Biochem. J.* **51,** 245 (1952)
20. L. Austin and D. R. Davies. *Brit. J. Pharmacol.* **9,** 145 (1954)
21. B. Mendel and D. K. Myers. *Nature, Lond.* **170,** 928 (1952)
22. M. M. Coursey, M. K. Dunlap and C. H. Hine. *Proc. Soc. exp. Biol. N.Y.* **96,** 673 (1957)
23. K. P. Strickland, R. H. S. Thompson and G. R. Webster. *Biochem. J.* **62,** 512 (1956)
24. C. J. Earl, R. H. S. Thompson and G. R. Webster. *Brit. J. Pharmacol.* **8,** 110 (1953)
25. L. Austin. *Brit. J. Pharmacol.* **12,** 356 (1957)
26. G. R. Webster. *Biochem. J.* **57,** 153 (1954)
27. M. Vandekar. *Nature, Lond.* **179,** 154 (1957)
28. J. C. Sabine and F. N. Hayes. *Arch. industr. Hlth.* **6,** 174 (1952)
29. H. Davson and P. A. Matchett. *J. cell. comp. Physiol.* **37,** 501 (1951)

PART IV

# PHARMACOLOGY IN INSECTS

## CHAPTER XVII

# Pharmacology in Insects

INSECTS are one of the greatest competitors with man for available food supplies, and by far the most important use of phosphorus compounds is to kill them.* In spite of this almost every aspect of the action of phosphorus compounds on insects is debatable. In part this is because many of the techniques applied to the study of mammals are not applicable to the study of insects, but it is mainly because few workers have tried to tackle the problems presented. Because of the shortage of data it has to be assumed at present that phosphorus compounds act in similar ways on insects of different species. It will be surprising if this assumption is not proved to be an over-simplification as more work is done.

## NERVE FUNCTION

Most is known of the larger insects, especially of the American cockroach (*Periplaneta americana*) and of the migratory locust (*Locusta migratoria migratorioides*). These species will be called 'cockroach' and 'locust' respectively throughout.

Some features of the nervous system of the locust are shown diagrammatically in Fig. 17.1A and B. Only those parts on which anti-AChE's have been studied are shown. The reader is referred to the references for detailed diagrams. Certain microscopic features are of particular importance, and are now described, and compared with similar features in the cockroach, and other insects.

*The cercal sensory system.* The cockroach has marked cerci, which contain many sensory nerve endings, responsive to heat or touch.[3] The sensory neurones synapse in the last (sixth) abdominal ganglion. The locust has no obvious cerci, but has similar, though fewer, cercal sensory neurones, which also synapse in the last (fifth) abdominal ganglion.

*Giant axons.* These, like the cerci, are more conspicuous in the cockroach. The connectives between the abdominal ganglia contain many axons of

* Phosphorus compounds are, of course, used to control other arthropods besides insects. It would be very hard to carry out the sort of fundamental work described here on most of them, and in fact little has been done. I have, therefore, dealt only with insects.

different sizes, but about ten are very much larger in diameter than the rest, and are called giant axons.[3, 4] In the locust they have been traced as far as the suboesophageal ganglion.[5] In the cockroach they have not been traced into the thorax, but it is probable that they are distributed as in the locust.

*The myoneural system in the hind leg.* In the locust the extensor tibialis muscle in the hind leg is exceptionally powerful, and provides the motive force for jumping. It is served by one axon from the crural nerve, and by the two axons in nerve 3b. The extensor tibialis in the cockroach is only served by one nerve, which resembles nerve 3b in the locust. The flexor tibialis is served by a few axons in a branch of the crural nerve.[2]

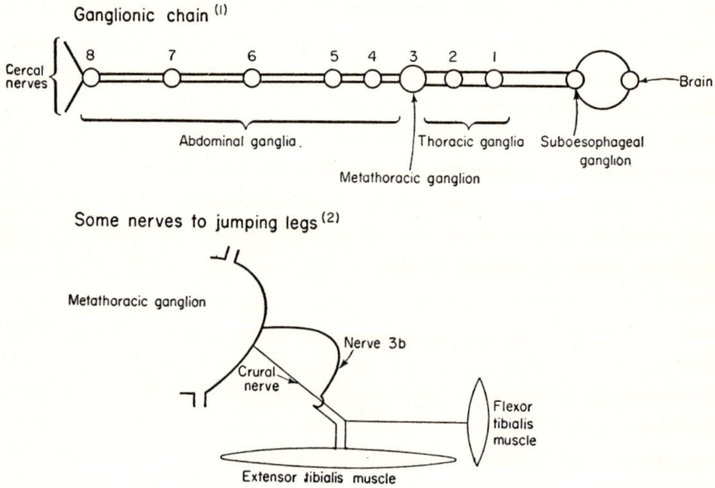


FIG. 17.1. Some parts of the nervous system in the locust. The diagrams are in no sense to scale. Accurate but more complex drawings are to be found in the references given.

A. Ganglionic chain.[1]

B. Some nerves to the jumping legs.[2]

The innervation of insect muscles is very different in detail from the innervation of mammalian voluntary muscle.[6] In insects a whole muscle is often innervated by only 1 or 2 axons, and rarely by more than 5. These branch in such a way that each muscle fibre is studded along its whole length with closely-spaced end-plates. In mammals one end-plate per fibre is common, but in insects only very short muscles are innervated in this way. The end-plates in insects are also morphologically different from those in mammals. Various forms have been described.[7, 8] Their common characteristic appears to be that they leave no length of muscle fibre greater than about 100 $\mu$ without a nerve ending. The motor axons in insects are myelinated, but much less than in mammals, and the myelin sheaths are wrapped round the axons in a different way.[8]

*Membranes.* The whole insect nervous system appears to be covered by a tough sheath. The sheath is not easily penetrated by micro-electrodes,[9] so that intracellular recording has been carried out mainly on de-sheathed preparations. The sheath was said to be impermeable to ions,[9–12] but this view no longer appears tenable. The synaptic regions do, however, appear to be surrounded by ion-impermeable membranes. The subject has been discussed on pp. 208f. The impermeability of these membranes rules out many experiments. It is, for example, impossible to perfuse an insect ganglion with ACh.

## ELECTROPHYSIOLOGY

The characteristics and mechanism of conduction in axons are the same in the giant axon of the cockroach and in mammals.[11, 13, 14]

The synaptic transmission of impulses from the cercal sensory neurones to the giant axons in the cockroach has some interesting features. Both types of neurone can carry about 500 stimuli/sec, but rates of stimulation of the sensory neurones greater than 50/sec cause partial block in synaptic transmission; and cessation of high rates of stimulation is followed by after-discharges in the giant axons.[3] In other ways the synapses are arranged as in a mammalian ganglion: many sensory neurones serve each internuncial neurone, and each internuncial neurone must be excited simultaneously by several sensory neurones to propagate an action potential.[3, 15]

The myoneural system functions differently in insects from in mammals.[6, 16, 17] An action potential in the crural nerve of the locust causes an e.p.p. to be developed nearly simultaneously beneath every end-plate along the extensor muscle. The e.p.p. is about 60 mV. It sets off an 'action potential' of about 25 mV, so that the sarcoplasm becomes transiently positive. Neither the e.p.p. *nor the action potential* are propagated, and if the action potential is suppressed by changing the ionic medium surrounding the muscle, the muscle still contracts, although not so violently. Thus the contraction depends on the e.p.p. as well as on the action potential. The close-spacing of end-plates is obviously necessary on a muscle which responds by this mechanism. The fibre in the crural nerve is 'fast', i.e. a single stimulus causes a strong twitch, and a few rapid stimuli build up a short, violent tetanus. The system fatigues rapidly. Action potentials in nerve 3b produce smaller e.p.p.'s with a slower time course. These slow fibres do not cause contraction unless the e.p.p.'s sum, and this only takes place when the frequency of stimulation is fairly fast—about 5/sec or higher. In the cockroach there is no 'fast' fibre. This corresponds to the difference in behaviour between the locust and cockroach. A stimulus makes the locust jump, but only makes the cockroach run.

As action potentials are not propagated, it is probable that an insect muscle cannot respond to direct stimulation.[6] This is difficult to prove, as the muscle is covered with nerve endings, so that stimulation of the muscle is likely to stimulate the nerve, and thereby cause a contraction.

The ganglionic connectives are not normally electrically quiet. Recordings show a continuous low level of activity.

## THE ACh THEORY OF SYNAPTIC TRANMISSION

The evidence that synaptic and myoneural transmission of nerve impulses is mediated in insects by ACh is much weaker than the evidence for such transmission in mammals. As the interpretation of the effects of phosphorus compounds depends upon the status accorded to this theory, the evidence for and against it is reviewed briefly here.

In mammals the presence in a group of neurones of ACh, AChE and choline-acetylase is taken to show that the neurones are cholinergic. Such evidence gains strength from the background of other results which are only consistent with the ACh theory of synaptic transmission—the well-known effects on ganglia and muscles of ACh, the release from ganglia and muscles of ACh during stimulation, the effects of injection of ACh into synaptic and myoneural gaps, the effects of drugs such as D-tubocurarine and atropine which antagonize ACh, and so on. In insects many of the analogous effects cannot be demonstrated, and those which can be are more difficult to interpret.

Enzymes which hydrolyse ACh have been found in every insect studied (Chapter VIII). They are distributed throughout the nervous system, and have also been found in high concentrations in several insect muscles.[18] In cockroach muscles they are mainly confined to the end-plate regions.[19] They often differ in substrate specificity from mammalian AChE's, and it has been suggested that ACh is not their natural substrate.

Compounds with some of the pharmacological properties of ACh have been separated from the nervous systems of several insect species. The compound separated from the heads of houseflies (*Musca domestica*) has been identified as ACh by a wide variety of physico-chemical, chemical and pharmacological tests.[20] There is no reason to believe the compound separated from the cockroach[21, 22] and blowfly (*Calliphora erythrocephala*)[23] is not ACh also. The separation of ACh from insect tissues can only be carried out effectively under carefully controlled conditions.[24, 25]

Choline-acetylase has been demonstrated in acetone-dried powders from the heads of blowflies (*Lucilia sericata*) and houseflies.[26, 27] The synthesis of ACh has also been demonstrated in houseflies *in vivo* by injecting $^{14}C$-acetate and separating labelled ACh chromatographically.[28]

AChE,[19] ACh[22] and choline-acetylase[29] are all distributed throughout the nervous system of the cockroach. Neither ACh nor choline-acetylase are found in the muscles.[29] The distribution is therefore like that in mammals in some ways.

None of the evidence is completely unambiguous. Thus the presence of an enzyme which hydrolyses ACh does not prove that ACh is its natural substrate; the fact that ACh is found does not prove that ACh has a physiological function; and the synthesis of ACh by an enzyme system presented with acetate, choline and various additives does not prove that the enzyme system synthesizes ACh *in vivo*. The combination of all three types of evidence makes a good case, however, for the theory that some synapses are mediated by ACh in insects.

Attempts to carry out confirmatory tests give results which cannot be interpreted with certainty.

Two preparations have been used to study the effects of drugs on synaptic transmission: the cercal nerve–nerve cord preparation of the cockroach and a reflex preparation consisting of the locust hind leg and the metathoracic ganglion. Both ganglia are very resistant to ACh, the least concentration causing block being about $10^{-2}$ M. They are more sensitive when de-sheathed, and still more so when de-sheathed and eserinized. After this treatment $10^{-4}$ M ACh causes discharges in the giant axons of the cockroach preparation, followed by block of synaptic transmission;[1, 9] and $10^{-5}$ M ACh first facilitates and then blocks the reflex in the locust preparation.[45] Even under these conditions, however, neither ganglion is as sensitive as eserinized mammalian ganglia, which are blocked by $10^{-6}$ M ACh.

Insect ganglia are also very resistant to drugs other than nicotine which affect mammalian ganglia. Thus D-tubocurarine, atropine and decamethonium are without effect except at concentrations above $10^{-2}$ M;[4, 30] and such compounds are of very low toxicity to the locust ($LD_{50}$'s mainly above 1 g/kg).[31]

Such results can either be held to show that ACh is not the mediator, or that many of the synapses are protected by additional membranes besides the enveloping sheath. There is some evidence for the second theory.

(1) When methylene blue or aniline blue was injected into the thoracic ganglion of the locust neither penetrated far into the substance of the ganglion except along the line of entry of the needle.[30] Presumably only a few synapses would be affected by any compound distributed in this way.

(2) A micro-injection of $10^{-5}$ M ACh into the last abdominal ganglion of the cockroach through a pipette 2 $\mu$ in diameter caused a burst of discharges in the giant axons. The axons then, however, became permanently unresponsive to cercal stimulation, suggesting that they had been damaged by the pipette, and that the damage may have caused the discharges.[22]

(3) The eserinized ventral cord of the cockroach did not secrete ACh into the fluid with which it was surrounded ('perfusion' fluid) when the cord was stimulated. Likewise no ACh was released on de-sheathing an untreated ganglion, but ACh was released on de-sheathing an eserinized ganglion ($10^{-4}$ M eserine) after 1 hr stimulation. As the amount obtained from the

de-sheathed ganglion was less than the increase in the ACh content of the ganglion which took place during the experiment,[22] it was probably much less than the amount released in the ganglion. (The argument is as follows. The increase represents ACh synthesized during the experiment. But neurones contain stocks of ACh, so that the release of ACh from the nerve endings during stimulation is normally faster than synthesis.) The discrepancy is most easily explained if de-sheathing is inadequate to release into the perfusion fluid all the ACh present, implying that there are other barriers to the diffusion of ACh. The idea that individual synapses may be protected is not implausible. Such protection may be necessary in insects, in which the blood can vary greatly in composition and amount.[12, 32]

If the ACh theory is accepted, the ineffectiveness of drugs such as D-tubocurarine and atropine can be explained either by the presence of protective membranes, or by differences in the properties of the receptors. In mammals different receptors respond differently to the same drug, although they are stimulated in the same way by ACh. Atropine, for example, blocks peripheral parasympathetic transmission when given at doses of about 0·01 mg/kg, but doses greater than 1 mg/kg are needed to block ganglia, and doses greater than 20 mg/kg to block myoneural transmission in the voluntary system (Chapter XI).

To conclude, it is probable that some synapses in the cockroach, locust and three species of fly are mediated by ACh, and that the ACh is held in the neighbourhood of the synapses by impermeable membranes, but the evidence is not wholly convincing.

The neurones likely to be cholinergic differ both in distribution and in their ACh content from cholinergic neurones in mammals. As the giant axons in the cockroach are stimulated by ACh, and this stimulation is increased by eserine, it is probable that the neurones which synapse with them are cholinergic. These are the sensory cercal neurones, i.e. in insects some sensory neurones are cholinergic. The flexor tibialis muscle in the locust does not respond to $10^{-1}$ M ACh. The extensor tibialis muscle sometimes responds to concentrations as low as $10^{-5}$ M ACh by contracting violently, and continuous perfusion causes a brief tetanic contraction; but the action of ACh on neither muscle is increased by eserine.[30] It therefore appears unlikely that the motor nerves are cholinergic. The crural nerve contains much lower concentrations of ACh and choline-acetylase than the ganglionic chain, and it is possible that the small amounts found are in sensory nerves, which share the same nerve trunk. On the other hand, the myoneural junctions contain AChE. In brief, some sensory nerves in insects are probably cholinergic, and some motor nerves are probably not.

The concentrations of ACh in insect nervous systems are very much higher than in the brains of mammals. Thus the cockroach ganglionic chain contains 60–140 $\mu$g/g (different concentrations in different parts),[22] and the brains of three species of fly about 30 $\mu$g/g,[24, 25] whereas rabbit brains contain about 0·8 $\mu$g/g,[33] and rat brains about 2·3 $\mu$g/g.[34] As in mammals the ACh appears to be stored in compartments in the neurones (vesicles).[35] The AChE and choline-acetylase concentrations are also often higher than in mammals. Why these high concentrations are present is unknown. Possibly cholinergic synapses in insects are less sensitive to ACh than the synapses in mammals.

## THE ACTION OF PHOSPHORUS COMPOUNDS

The effects of phosphorus compounds on synaptic transmission have been studied on the sixth abdominal ganglion of the cockroach and on the metathoracic ganglion of the locust.

Application of TEPP or eserine* to the last abdominal ganglion of the cockroach greatly increased the voltage and the duration of the e.p.p.'s set up in the internuncial neurones by a single stimulus in the cercal nerve. A brief volley of stimuli caused the e.p.p.'s to sum in the treated ganglia, so that the post-ganglionic nerves became depolarized.[15] Studies of the action potentials in the connectives are consistent with these results. The ganglion was treated with TEPP or DFP, and the cercal nerve stimulated at 30/sec.[36] Some 2–10 min after the application of $5 \times 10^{-6}$ M DFP after-discharges appeared in the minor fibres, and 10–20 min after application in the giant axons. The appearance of after-discharges in the giant axons was followed rapidly by a burst of discharges, lasting 15–30 sec, which was in no way synchronized with the cercal stimuli and persisted when cercal stimulation was stopped. The asynchronous burst gave way to block of synaptic transmission. On the anti-AChE theory such results can be explained as follows. Inhibition of AChE increases the effect of the ACh released by each stimulation of the crural nerve. Consequently the e.p.p.'s fall more slowly, and set off more than one action potential, i.e. after-discharges appear. As inhibition progresses the depolarization becomes greater, so that the after-discharges merge into a continuous, asynchronous discharge. Finally the post-ganglionic neurones become depolarized, and so fail to respond. The last two stages are particularly likely to take place in insects because the ACh cannot escape from the synaptic region.

More complete experiments show that four stages in the action of phosphorus compounds on the last abdominal ganglion can be distinguished, depending upon the concentration applied.[4] Tabun, Sarin and several homologues of Sarin were used. In the order of increasing concentrations the stages were:

* In this and other papers indirect inhibitors such as Parathion are said to act in the same way as direct inhibitors. Probably the indirect inhibitors were either impure or were metabolized to direct inhibitors under the conditions of the experiment.

(1) Prolonged after-discharges followed by synaptic block. The burst of discharges sometimes appeared without cercal stimulation. If progressive inhibition was stopped by washing the ganglion thoroughly, a fairly normal response to stimulation returned, but was accompanied by asynchronous discharges and periods of block lasting a few minutes. This state could be maintained for many hours, and was the state to which all ganglia treated with higher concentrations returned on washing.

(2) Synchronized synaptic responses were obtained. These were easily fatigued, and were sometimes superposed on asynchronous discharges. DFP, TEPP and eserine did not produce these effects.[37] The state is reminiscent of that found in a mammalian nerve-muscle preparation, where it is due to the adaptation of the post-synaptic membrane to high ACh concentrations (Chapter XII).

(3) Prolonged synaptic block was caused by considerably higher concentrations, and was maintained until the compound was washed out. The block may be caused by a blocking action of the phosphorus compounds. The only objection to this theory is the very low concentration of one compound, isoamyl methylphosphonofluoridate, which was stated to have caused block on one preparation at $5{\cdot}7 \times 10^{-10}$ M.

(4) Axonic block was caused by some compounds at about $10^{-3}$ M or above.

When the tarsus on the hind-leg of the locust is stimulated by heat or pinching the leg retracts. This is a monosynaptic reflex, initiated by the tarsal sensory nerves, synapsing in the metathoracic ganglion, and mediated via the crural nerve. The effects of phosphorus compounds on this reflex were studied.[30] The phosphorus compounds were either applied in agar jelly or paraffin wax to the metathoracic ganglion, or injected into it. Paraoxon and TEPP applied at $10^{-5}$ M caused extra kicks very rapidly. The reflex was then blocked. The block sometimes gave way to bursts of sporadic movement, and, after Paraoxon, there were periods of normal response to stimulation. Very pure Parathion caused similar effects, but at $10^{-5}$ M the first effects only appeared 2–4 hr after treatment.[45] Presumably this time is required for the metabolic conversion of Parathion to Paraoxon. By analogy with mammals the results could be explained by ACh accumulation in the end-plate region, followed by partial adaptation to ACh of the motor neurones. Violent movements followed by paralysis have also been observed in *Dysticus marginalis*, when ganglia were treated with Parathion (impure?) and the movements of the abdomen recorded.[38]

Effects on nerve-muscle preparations were studied on the desert locust (*Schistocerca gregaria*).[30] The responses of the extensor and flexor tibialis muscles were recorded separately, and the compounds were perfused continuously through the femur as dilute solutions. The crural nerve and nerve 3b were stimulated individually. The responses to indirect stimulation were wholly unaffected by the perfusion of $10^{-3}$ M TEPP or Paraoxon for 1 hr. Thus neither the fast (crural nerve-muscle) nor the slow (nerve 3b-muscle) systems are affected by anti-AChE's.

None of the effects described so far are altered by atropine or D-tubocurarine.[4, 30, 31, 36, 38] They are substantially those which might be expected from the properties of the insect nervous system described earlier in the Chapter. Neurones synapsing in the central nervous system are probably cholinergic, and synaptic events are therefore, as in mammals, first facilitated and then blocked by anti-AChE's Motor nerves are not cholinergic, and are therefore unaffected. As in mammals the general similarity in action between eserine and phosphorus compounds is evidence that the anti-AChE theory is correct.

The heart of the cockroach can be studied in isolation. The effects of phosphorus compounds on it are precisely the reverse of the effects of the same compounds on mammalian heart.[39, 40] Low concentrations of TEPP or DFP accelerate it, and potentiate the *accelerating* effect of ACh. These effects are antagonized by atropine. Higher concentrations of phosphorus compounds have no effect, and concentrations around $10^{-4}$ M stop it, i.e. stimulation gives way to block as the concentration is raised. The effects at low concentrations are wholly consistent with the anti-AChE theory.

Phosphorus inhibitors change the ACh concentration in insects. There is normally an increase, especially in the nervous system,[22, 23, 28, 41] often followed by a decrease to normal levels.[28, 42] Some free ACh is found in the blood.[41] As ACh is not released from ganglia this probably originates outside the nervous system. Similar changes in concentration have been observed in mammals (Fig. 14.1). The increase in concentration does not, of course, measure the increase in the synaptic gaps. As the general concentration in insects is very high, a small increase in overall concentration may represent a very great increase in the sensitive regions. The fall in concentration which often takes place may well be due to the re-activation of some AChE, which is known to take place rapidly in some insects (p. 206).

Symptoms of poisoning are much as might be expected from the results on isolated organs already described.[41, 43, 44] The following sequence has been observed in the cockroach treated topically with TEPP.[41] After 15–30 min the insect became hyperactive, and its movements became unco-ordinated. A period of convulsions followed, after which the insect lay prostrate, twitching from time to time. After about 7 hr the insect became motionless. Death did not take place for several days.

The early signs can, obviously, be explained by the anti-AChE theory of poisoning, but it is not clear why the insect dies. An insect breathes by diffusion, and its survival does not depend upon an unimpaired circulation, so that two causes of death in mammals cannot be causes in insects. Several theories have been put forward.

(1) The insect dies from prolonged paralysis of the nervous system due to ACh accumulation. This is the anti-AChE theory. As pointed out in Chapter VIII the AChE levels have often risen to normal at the time of death; and the ACh levels are only higher than normal in some insects.

(2) The phosphorus compounds do not kill by their action on the nervous system, but by their action elsewhere. As ionic compounds are very much less toxic than closely related non-ionic ones (Chapter VIII), and only the nervous system appears to be fully protected by membranes impermeable to ionic compounds, this theory can be rejected.

(3) The phosphorus compounds inhibit enzymes other than AChE in the nervous system, and this inhibition is less reversible than inhibition of AChE's. Until such enzymes are found, and their normal functions described, this theory cannot be discussed profitably.

(4) The abnormal nervous activity induced by phosphorus compounds damages the nervous system irreparably. This theory is consistent with the few facts available. The activity in the cockroach giant axons is highly abnormal—rapid bursts of action potentials lasting for 15–30 sec. Under normal conditions cercal stimulation causes the insect to move rapidly from the source of the stimulus. The same is true of other sensory reflexes, and it is possible that the insect nervous system is not designed to withstand continuous violent sensory stimulation for long periods. On this theory it is not necessary to assume that overstimulation results from inhibition of AChE, although in the cockroach and locust it is probable that it does. DDT apparently stimulates the sensory system directly,[30] and may have the same final effect. It must, however, be emphasized that there is no direct evidence that continuous nerve stimulation can kill an insect. Probably no conclusion can be reached until more is known of normal nerve function in insects, and of what lesions cause death.

## REFERENCES

1. F. O. Albrecht. *The Anatomy of the Desert Locust*, London Univ., Athlone Press (1953)
2. G. Hoyle. *Proc. roy. Soc.* **143B,** 281 (1955)
3. K. D. Roeder, N. K. Kennedy and E. A. Sampson. *J. Neurophysiol.* **10,** 1 (1947)
4. K. D. Roeder and N. K. Kennedy. *J. Pharmacol.* **114,** 211 (1955)
5. P. T. Haskell, personal communication.
6. G. Hoyle. *Invertebrate Physiology* p. 73. Univ. of Oregon (1957)
7. C. A. G. Wiersma. *Invertebrate Physiology* p. 143. Univ. of Oregon (1957)
8. G. A. Edwards, H. Ruska and E. de Harven. *J. biophys. biochem. Cytol.* **4,** 107 (1958)
9. B. M. Twarog and K. D. Roeder. *Biol. Bull.* **111,** 278 (1956)
10. R. D. O'Brien. *Ann. entomol. Soc. America* **50,** 223 (1957): *J. econ. Ent.* **52,** 812 (1959)
11. G. Hoyle. *J. exp. Biol.* **30,** 121 (1953)
12. K. Mellanby. *Biol. Rev.* **14,** 243 (1939)

13. T. Yamasaki and T. Narahashi. *Nature, Lond.* **182,** 1805 (1958)
14. T. Yamasaki and T. Narahashi. *J. Insect Physiol.* **3,** 146, 230 (1959)
15. T. Yamasaki and T. Narahashi. *Nature, Lond.* **182,** 1805 (1958)
16. G. Hoyle. *Proc. roy. Soc.* **143B,** 343 (1955)
17. J. del Castillo, G. Hoyle and X. Machne. *J. Physiol.* **121,** 539 (1953)
18. J. E. Casida. *Biochem. J.* **60,** 487 (1955)
19. K. Iyatomi and K. Kanehisa. *Chem. Abstr.* **52,** 14013 (1958)
20. W. Chefurka and B. N. Smallman. *Canad. J. Biochem.* **34,** 731 (1956)
21. S. J. Mikalouis and R. H. Brown. *J. cell. comp. Physiol.* **18,** 401 (1941)
22. E. H. Colhoun. *J. Insect Physiol.* **2,** 108 (1958)
23. S. E. Lewis and K. S. Fowler. *Nature, Lond.* **178,** 919 (1956)
24. S. E. Lewis and B. N. Smallman. *J. Physiol.* **134,** 241 (1956)
25. K. S. Fowler and S. E. Lewis. *J. Physiol.* **142,** 165 (1958)
26. B. N. Smallman. *J. Physiol.* **132,** 343 (1956)
27. N. Frontali. *J. Insect Physiol.* **1,** 319 (1957)
28. F. P. W. Winteringham and A. Harrison. *Nature, Lond.* **178,** 81 (1956)
29. E. H. Colhoun. *Nature, Lond.* **182,** 1378 (1958)
30. P. A. Harlow. *Ann. appl. Biol.* **46,** 55 (1958)
31. H. S. Hopf. *Ann. appl. Biol.* **39,** 193 (1952)
32. J. H. Welsh. *Neurohormones. The Hormones, III,* **97,** Academic Press, N.Y. (1955)
33. M. Michaelis, J. E. Finesinger, F. de Balbain Verster and R. W. Erickson. *J. Pharmacol.* **111,** 169 (1954)
34. J. Crossland, H. M. Pappius and K. A. C. Elliott. *Amer. J. Physiol.* **183,** 27 (1955)
35. E. H. Colhoun. *Nature, Lond.* **181,** 490 (1958)
36. B. M. Twarog and K. D. Roeder. *Ann. entomol. Soc. America* **50,** 231 (1957)
37. K. D. Roeder. *Bull. Johns Hopkins Hosp.* **83,** 587 (1948)
38. H. Fritsch and H. Krupp. *Arch. exp. Path. Pharmak.* **214,** 227 (1952)
39. B. J. Krijgsman, D. Dresden and N. E. Berger. *Bull. entomol. Res.* **41,** 141 (1950)
40. M. B. Naidu. *Bull. entomol. Res.* **46,** 205 (1954)
41. E. H. Colhoun. *Canad. J. Biochem.* **37,** 259 (1959)
42. B. N. Smallman and R. W. Fisher. *Canad. J. Biochem.* **36,** 575 (1958)
43. D. R. Wiesmann. *Mitt. biol. Reichanstalt f. Land-u. Forstwirtschaft.* **70,** 107 (1951)
44. L. E. Chadwick and D. L. Hill. *J. Neurophysiol.* **10,** 235 (1947)
45. P. A. Harlow, personal communication.

# APPENDIX

APPENDIX

# Theory of Reaction Rates in Solution

1. *Order of reaction.* The rate at which a reaction proceeds depends upon the concentrations of the reactants. The order of a reaction is a measure of this dependence under the actual experimental conditions used.

Consider the reaction between DFP and water:

$$(Pr^iO)_2PO \cdot F + H_2O \rightarrow (Pr^iO)_2PO \cdot OH + HF.$$

The rate of disappearance of DFP is proportional to the concentration of DFP. In dilute solution the concentration of water is unchanged during the reaction, so that under these conditions the rate does not depend upon the concentration of water. The rate is therefore given by:

$$d[DFP]/dt = -k[DFP].$$

(Concentrations are denoted by square brackets.) The overall rate is described as first order because it depends upon 1 concentration. It can also be described as first order with respect to DFP and zero order with respect to water. Integration of the equation gives:

$$\log_{10} [DFP]_0/[DFP]_t = kt/2{\cdot}303,$$

where $[DFP]_0$ and $[DFP]_t$ are the concentrations at the start of the reaction ($t = 0$) and at time $t$. The constant 2·303 is $\log_e 10$, and merely converts the equation to ordinary decadic logarithms. The constant $k$ is the rate constant. From the equation it can be seen that a plot of

$$\log [DFP]_0/[DFP]_t$$

against $t$ gives a straight line, of gradient $k/2{\cdot}303$. The straightness of such a plot is the usual test that a reaction is first order, and $k$ is calculated from the gradient. The constant $k$ is always given in the units: $min^{-1}$ in this book.

Now consider the reaction between DFP and hydroxyl ions. The rate depends upon the concentration of both reagents, so that:

$$d[DFP]/dt = -k_{OH}[DFP][OH^{\ominus}].$$

The reaction is second order, as it depends upon 2 concentrations. The rate constant is in the units: $litres.moles^{-1}.min^{-1}$, when, as always in this book, concentrations are given in moles/litre. Integration is simple, but a different treatment is used here. Suppose that $[OH^\ominus]$ is very much greater than [DFP]—in practice it is enough for $[OH^\ominus]$ to be more than 10 times [DFP]. Then $[OH^\ominus]$ is nearly constant throughout the reaction, so the equation becomes:

$$d[DFP]/dt = -k_{OH}[DFP][OH^\ominus] = -k[DFP],$$

where $k = k_{OH}[OH^\ominus]$. This is a pseudo first order reaction. For any given $[OH^\ominus]$ the log vs time plot is a straight line, but the gradient is proportional to $[OH^\ominus]$. These two characteristics show that a reaction is second order.

In a complex reaction with several successive steps the rate is controlled by the slowest step, and the order found is the order of this step.

2. *Molecularity of reaction.* The molecularity is the number of molecules which must come together for reaction to take place. It can never be less than the order of the reaction, but it may be more. Thus as the hydrolysis of DFP by hydroxyl ions is second order the reaction must be at least bimolecular. The hydrolysis of DFP by water must also be bimolecular, as the products show that a molecule of water and a molecule of DFP must come into contact, but the reaction is only first order because the concentration of water remains constant throughout the reaction.

Usually kinetic measurements indicate the minimum molecularity of the slowest stage. In some cases the kinetics are too complex to fit the simple schemes described, and must be analysed in terms of successive and/or simultaneous reactions. The kinetics then yield more information about the reactions taking place. This is the case with enzyme reactions.

3. *Energies of activation.* Rate constants vary with temperature, usually according to the equation:

$$k = \text{const.}\ e^{-E/RT}.$$

In logarithmic form this is: $\log_{10} k = A - E/2{\cdot}303RT$, where $T$ is the absolute temperature, $R$ is the gas constant, $A$ is an integration constant and $E$ is the energy of activation. $E$ is the energy required on the average to raise the molecules from their normal state into one active enough for reaction to take place. It is expressed in kcal/mole. The equation shows that the rate of reaction increases by a constant multiple for a constant temperature change. (There are small theoretical and observable deviations from this equation which are not considered here.) If $k$ is doubled by a 10° rise in temperature at about room temperature $E$ is about 13 kcal/mole.

4. *Probability of reaction.* The energy for a reaction comes from collisions in most cases. If every collision with enough energy led to reaction taking place the rate constant would be given by: $k = Z \cdot e^{-E/RT}$, in which $Z$ is the number of collisions per minute. In fact most collisions are often abortive, as the energy has to be concentrated in the bond broken during the reaction, and in a complex molecule there are many ways in which it can be dissipated. A probability factor may be introduced, to give:

$k = P \,.\, Z \,.\, e^{-E/RT}$. It is, however, better to express probabilities in terms of entropy, and this leads to the analogous equation:

$$k = \frac{\mathbf{k}T}{h} \,.\, e^{\Delta S/R} \,.\, e^{-E/RT},$$

where $\mathbf{k}$ is the Boltzmann constant, $h$ is Planck's constant, and $\Delta S$ is the entropy of activation. High negative $\Delta S$ indicates low probability.

The characteristics of a simple reaction are completely specified when $E$ and $\Delta S$ are known. $E$ and $\Delta S$ are, however, hard to determine, as rate constants must be found to an unusual degree of accuracy at at least three temperatures. Thus, if $\Delta S$ is the same for two reactions a difference in $E$ of 1 kcal/mole represents a fivefold difference between the rates; but this difference in $E$ can arise from errors of about 5% in the rate constants, as $E$ is calculated from the ratios of rate constants. Rate constants are therefore often compared directly, but conclusions drawn from such comparisons must be treated with caution.

5. *Acids and bases.* An acid is any group or molecule which can donate protons (hydrogen ions), and a base is any which can accept them. Thus every acid, A, has a conjugate base, $B^{\ominus}$:

$$A \rightleftharpoons B^{\ominus} + H^{\oplus}.$$

Some conjugate acid-base pairs are: $H_3\overset{\oplus}{O}$ and $H_2O$; $H_2O$ and $OH^{\ominus}$; and $\overset{\oplus}{N}H_4$ and $NH_3$. A solvent is needed to accept the protons, so that equilibrium involves two acid-base pairs. In water, for example, the dissociation of an acid can be represented by the equation:

$$A + H_2O \rightleftharpoons H_3\overset{\oplus}{O} + B^{\ominus},$$

giving the equilibrium constant: $K' = [H_3\overset{\oplus}{O}][B^{\ominus}]/[A][H_2O]$. As $[H_2O]$ is a constant, a simpler dissociation constant can be formed:

$$K_a = [H_3\overset{\oplus}{O}][B^{\ominus}]/[A].$$

$K_a$ is the acid constant and measures the tendency of the acid to donate protons, i.e. it measures the strength of the acid. The base constant is defined similarly:

$$K_b = [OH^{\ominus}][A]/[B^{\ominus}].$$

In water $[H_3\overset{\oplus}{O}][OH^{\ominus}] = K_w$, whence $K_b = K_w/K_a$. Thus the acid strength of the acid, A, and the base strength of its conjugate base, $B^{\ominus}$, are related—a strong acid has a weak conjugate base, and vice versa.

Strong acids, such as HCl, HBr, HI and $HClO_4$, are those in which the reaction with water goes to completion. Catalysis by such acids is catalysis by the oxonium ions produced. This is usually called hydrogen ion catalysis.

6. *General acid-base catalysis.*[1] From the above discussion it appears that there is no qualitative difference between oxonium ions and other acids, or between hydroxyl ions and other bases. A reaction catalysed by any one acid should be catalysed by all the rest, and similarly for bases. Assuming that acid catalysis depends upon the catalyst donating a proton, the catalytic rate constant should vary with the acid strength according to the Bronsted relation:

$$\log (k_A/p) = \text{const.} + \alpha \log (qK_a/p),$$

where $k_A$ is the catalytic constant, and the acid, A, has $p$ atoms from which it can lose a proton and its conjugate base has $q$ atoms on which it can accept them. The constant $\alpha$ must have a value between 0 and 1. General acid catalysis can only be observed readily when $\alpha$ is in the range 0·2–0·8.

If base catalysis involves the removal of a proton, then a similar equation can be derived for it. In many examples described in Chapter IV the Bronsted relation holds although the phosphorus compound attacked has no hydrogen atom which could be removed by a base. This is so, for example, with DFP, $(Pr^iO)_2PO \cdot F$, in which all the hydrogen atoms are firmly bonded to carbons. There are two ways of explaining the paradox. Firstly, the compound may be attacked by water to form an intermediate, and the base catalyses the removal of a proton from the intermediate:

$$H_2O + P \cdot X \rightleftharpoons H_2O \cdots P \cdot X \xrightarrow{+B^{\ominus}} HO \cdot P + X^{\ominus} + BH.$$

Secondly, both the base strengths and the catalytic effects of bases may have a common cause. Thus, if the mechanism of hydrolysis is:

$$B^{\ominus} + P \cdot X \rightarrow \overset{\delta\ominus}{B} \cdots P \cdots \overset{\delta\ominus}{X} \rightarrow B \cdot P + X^{\ominus},$$

$$B \cdot P + H_2O \rightarrow HO \cdot P + BH,$$

and the rate of the reaction is controlled by the first stage, the rate will vary with the ease with which the base gives up its negative charge. But this readiness to give up negative charge is also what makes $B^{\ominus}$ a base. The Bronsted relation should, therefore, still hold for a group of base-catalysed reactions which take place by the same mechanism. The problem is discussed in Chapter IV.

Although hydrolysis by water, by hydroxyl ions and by oxonium ions are only special cases of acid-base catalysis, it is convenient to regard the overall rate constant as made up of five separate constants:

$$k = k_W + k_{OH} + k_H + k_A + k_B.$$

The general theory of acid-base catalysis is discussed by Bell,[1] whose book should be consulted for more details.

7. *Enzyme reactions.* Only esterases are considered here. In the simplest case the hydrolysis of a substrate by an enzyme takes place in two stages. A

complex is formed between the enzyme and the substrate, which then breaks down to products with the release of the enzyme. The first stage is reversible, so the reactions can be written:

$$E + S \underset{k_2}{\overset{k_1}{\rightleftharpoons}} ES \xrightarrow{k_3} E + P.$$

The second reaction may be bimolecular, involving water molecules, but is first order. If the substrate is in such excess that its concentration does not change during the reaction, the first stage is pseudo first order. It is assumed that the complex goes back to enzyme and substrate faster than it breaks down to enzyme and products, i.e. that $k_2 \gg k_3$. Then, after a few seconds, once the reaction is under way, the only concentration which changes is that of the product, so that:

$$d[E]/dt = 0 = k_1[E][S] - (k_2 + k_3)[ES] \tag{A}$$

As $k_3 \ll k_2$, $k_3[ES]$ can be ignored, whence:

$$[E][S]/[ES] = k_2/k_1 = K_m \tag{B}$$

$K_m$ is the concentration of S which complexes half the enzyme. (If the initial enzyme concentration is $E_0$, as $[ES] = E_0 - [E]$, when $[ES] = [E] = E_0/2$, $K_m = [S]$.) Thus the stronger the tendency to form a complex the smaller $K_m$.

For the rate of formation of products:

$$d[P]/dt = k_3[ES] = k_3[E][S]/K_m \tag{C}$$

When $[S] \ll K_m$, $[E] = E_0$ (from B), so the rate is proportional to [S], i.e. the rate is first order in [S]. When $[S] \gg K_m$, $[ES] = E_0$, so the rate is independent of [S], i.e. it is zero order in [S]. Under these conditions the enzyme is fully complexed, so increasing the concentration more cannot further increase the rate. The maximum rate, $V$, is given by: $V = k_3E_0$.

The general relationship between the rate, $v$, $(d[P]/dt)$, and the substrate concentration is given by the Michaelis–Menton equation:

$$v = V[S]/([S] + K_m) \tag{D}$$

The rate is, of course, always proportional to the concentration of the enzyme—the above equation shows how the rate varies with [S] when the concentration of enzyme is held constant.

The basic assumption made in this treatment is that $k_2 \gg k_3$. This has to be proved, and it is often difficult to do so. It is likely to be true when the reaction is fast, and has a high temperature coefficient. When the temperature coefficient is high, the energy of activation is high, so one rate constant must be low. As $E_0$ and [S] are low, $k_1$ cannot be low, or the overall reaction would be slow. Therefore $k_3$ is low. Unless equilibrium is reached slowly (which is rare in enzyme reactions), $k_2$ must be high. The condition is thus satisfied. There is, however, some doubt as to whether the Michaelis–Menton equation holds for AChE, which hydrolyses ACh at a rate which is nearly independent of temperature.

The effect of complex formation is to increase enormously the effective concentrations of the reactants, by holding them together until reaction takes place.

In reversible inhibition an inhibitor competes with substrate for the enzyme, so that the two reaction sequences:

$$E + S \rightleftharpoons ES \rightarrow E + P$$
$$E + I \rightleftharpoons EI$$

take place simultaneously. The equilibrium between enzyme and inhibitor is analogous to that between enzyme and substrate, so there is a constant for the formation of the inhibitor–enzyme complex, $K_I$, where $K_I = [E][I]/[EI]$. It can be shown that the ratio of the rates of formation of products from the substrate with and without inhibitor present is given by the equation:

$$v'/v = ([S] + K_m)/([S] + K_m + K_m[I]/K_I),$$

when the determinations of rate are made at the same substrate concentration. The degree of inhibition observed, i.e. $(v - v')/v$, thus depends on both the inhibitor and the substrate concentrations, increasing with the former, and decreasing with the latter.

Suppose that the enzyme ionizes, and that only one form is reactive. For example, suppose that there are two protons which can be lost:

$$EH_2 \rightleftharpoons \overset{\oplus}{H} + \boxed{EH \rightleftharpoons E + \overset{\oplus}{H}},$$

and that only the EH form can react with substrate. The acid dissociation constants of EH and $EH_2$ are:

$$K_1 = [E][\overset{\oplus}{H}]/[EH], \quad \text{and} \quad K_2 = [EH][\overset{\oplus}{H}]/[EH_2].$$

The rate is proportional to the fraction of enzyme in the form EH, which is given by:

$$F = 1/(1 + K_1/[\overset{\oplus}{H}] + [\overset{\oplus}{H}]/K_2). \qquad \text{(E)}$$

A trial calculation shows that if $K_1$ and $K_2$ differ by more than a factor of 100, i.e. $pK_1 - pK_2 > 2$, a plot of rate against pH is a bell-shaped curve, with a maximum at the pH midway between $pK_1$ and $pK_2$, such that the rate is half its maximum when the pH is the same as either $pK_1$ or $pK_2$. (This is slightly in error quantitatively, but is good enough for comparisons between theory and experiment.) In a simpler case the enzyme exists in two forms, protonated and unprotonated, and only the unprotonated form catalyses. Then:

$$F = 1/(1 + [\overset{\oplus}{\mathrm{H}}]/K_1), \qquad \text{(F)}$$

where $K_1$ is the dissociation constant of the acid, EH. In this case, as the pH is raised the rate increases to a maximum, which does not fall at still higher pH's; and the rate is half the maximum when $\mathrm{pH} = \mathrm{p}K_1$.

Some other cases are considered in the text.

## REFERENCE

1. R. P. Bell. *Acid-Base Catalysis*, Clarendon Press, Oxford, 1941.

# GLOSSARY

# Glossary

THE Glossary is in four sections.

Section I is devoted to those compounds for which common names have been used in the book. For each compound the common name is given in alphabetical order, and is followed by the full scientific name on the Anglo-American system and the structural formula. One reference (occasionally two) is given to a satisfactory method of preparation. Many references have already been given in Chapter II. These are entered as, e.g. (II-7), denoting reference (7) in Chapter II. Other references are numbered in the usual way and are listed at the end of Section I. For two compounds I could find no method. The entry: as (2), means that the method given in reference (2) is likely to work well.

In Section II some other compounds often referred to in the literature are listed as in Section I, except that no methods of preparation are referred to.

In Section III all the scientific names appearing in Sections I and II are listed alphabetically, and the common names are given. Which Section the common name appears in is indicated by 'I' or 'II'.

In Section IV other names used in the literature for some compounds in Sections I and II are listed and cross-indexed.

## SECTION I
# Common Names Used in the Book

| COMMON NAME | SYSTEMATIC NAME | STRUCTURAL FORMULA | METHOD |
|---|---|---|---|
| Acethion | S-carbethoxymethyl O,O-diethyl phosphorodithioate | $(EtO)_2PS \cdot S \cdot CH_2 \cdot CO \cdot OEt$ | (1) |
| Amiton | 2-(diethoxyphosphinylthio)-ethyldiethylammonium hydrogen oxalate | $(EtO)_2PO \cdot S \cdot CH_2 \cdot CH_2 \cdot NEt_2$ (oxalate salt) | (II-67) |
| Amiton-ethiodide | 2-(diethoxyphosphinylthio)-ethyltriethylammonium iodide | $(EtO)_2PO \cdot S \cdot CH_2 \cdot CH_2 \cdot \overset{\oplus}{N}Et_3\ I^{\ominus}$ | as (22) |
| Chlorthion | 3-chloro-4-nitrophenyl dimethyl phosphorothionate | $(MeO)_2PS \cdot O \cdot C_6H_3$-3-Cl-4-$NO_2$ | (2) |
| Co-ral | 3-chloro-4-methyl-7-coumarinyl diethyl phosphorothionate | 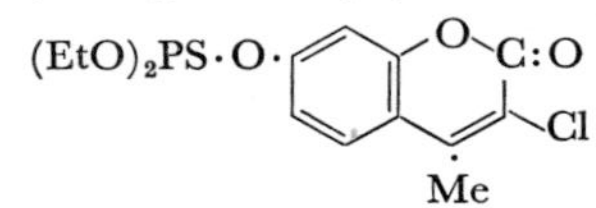 | (3) |
| DDVP | 2:2-dichlorovinyl dimethyl phosphate | $(MeO)_2PO \cdot O \cdot CH{:}CCl_2$ | (II-77) |
| Delnav | 2:3-bis(diethoxyphosphinothioylthio)-1:4-dioxane | 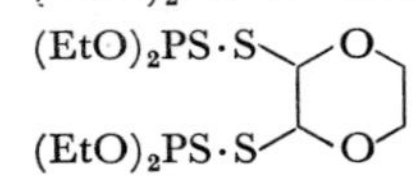 | (II-60) |
| Demeton-O | diethyl 2-ethylthioethyl phosphorothionate | $(EtO)_2PS \cdot O \cdot CH_2 \cdot CH_2 \cdot SEt$ | (II-109, 4) |
| Demeton-O-Me | dimethyl 2-ethylthioethyl phosphorothionate | $(MeO)_2PS \cdot O \cdot CH_2 \cdot CH_2 \cdot SEt$ | (4) |
| Demeton-S | O,O-diethyl S-2-ethylthioethyl phosphorothiolate | $(EtO)_2PO \cdot S \cdot CH_2 \cdot CH_2 \cdot SEt$ | (II-23) |
| Demeton-S-Me | O,O-dimethyl S-2-ethylthioethyl phosphorothiolate | $(MeO)_2PO \cdot S \cdot CH_2 \cdot CH_2 \cdot SEt$ | (II-23) |
| DFP | di-isopropyl phosphorofluoridate | $(Pr^iO)_2PO \cdot F$ | (5, 6) |
| Diazinon | diethyl 4-(2-isopropyl-6-methylpyrimidinyl) phosphorothionate | 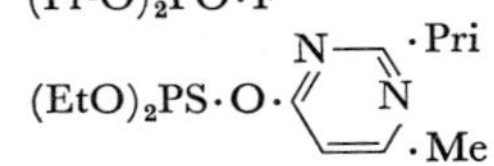 | (7) |
| Dicapthon | 2-chloro-4-nitrophenyl dimethyl phosphorothionate | $(MeO)_2PS \cdot O \cdot C_6H_3$-2-Cl-4-$NO_2$ | (8) |
| Dimefox | tetramethylphosphorodiamidic fluoride | $(Me_2N)_2PO \cdot F$ | (II-11, II-12) |
| Dimethoate | O,O-dimethyl S-(N-methylcarbamoylmethyl) phosphorodithioate | $(MeO)_2PS \cdot S \cdot CH_2 \cdot CO \cdot NHMe$ | (9) |
| Dipterex | dimethyl 1-hydroxy-2:2:2-trichloroethylphosphonate | $(MeO)_2PO \cdot CH(OH) \cdot CCl_3$ | (II-76) |
| Disyston | O,O-diethyl S-2-ethylthioethyl phosphorodithioate | $(EtO)_2PS \cdot S \cdot CH_2 \cdot CH_2 \cdot SEt$ | (10) |
| EPN | ethyl 4-nitrophenyl phenylphosphonothionate | $(EtO)(Ph)PS \cdot O \cdot C_6H_4$-4-$NO_2$ | (11) |
| Guthion | O,O-dimethyl S-(4-oxo-1:2:3-benzotriazinyl-3-methyl) phosphorodithioate | 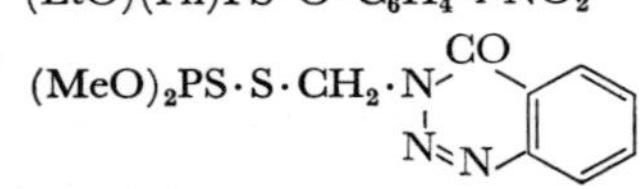 | (12) |

## Section I—*continued*

| Common name | Systematic name | Structural formula | Method |
|---|---|---|---|
| Gutoxon | O,O-dimethyl S-(4-oxo-1:2:3-benzotriazinyl-3-methyl) phosphorothiolate | 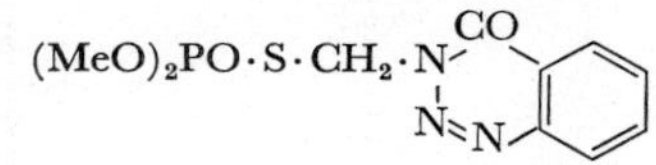 | (13) |
| HETP | mixture of ethyl polyphosphates containing 20% TEPP | | |
| Malaoxon | S-(1:2-dicarbethoxyethyl) O,O-dimethyl phosphorothiolate | $(MeO)_2PO \cdot S \cdot CH(CO \cdot OEt) \cdot CH_2 \cdot CO \cdot OEt$ | (II-112, 14) |
| Malathion | S-(1:2-dicarbethoxyethyl) O,O-dimethyl phosphorodithioate | $(MeO)_2PS \cdot S \cdot CH(CO \cdot OEt) \cdot CH_2 \cdot CO \cdot OEt$ | (II-47) |
| Mazidox | tetramethylphosphorodiamidic azide | $(Me_2N)_2PO \cdot N_3$ | (II-13) |
| Mipafox | N,N′-di-isopropylphosphorodiamidic fluoride | $(Pr^iNH)_2PO \cdot F$ | (15) |
| Paraoxon | diethyl 4-nitrophenyl phosphate | $(EtO)_2PO \cdot O \cdot C_6H_4\text{-}4\text{-}NO_2$ | (II-2) |
| Paraoxon-Me | dimethyl 4-nitrophenyl phosphate | $(MeO)_2PO \cdot O \cdot C_6H_4\text{-}4\text{-}NO_2$ | (II-2) |
| Paraoxon-$Pr^i$ | di-isopropyl 4-nitrophenyl phosphate | $(Pr^iO)_2PO \cdot O \cdot C_6H_4\text{-}4\text{-}NO_2$ | (II-2) |
| Parathion | diethyl 4-nitrophenyl phosphorothionate | $(EtO)_2PS \cdot O \cdot C_6H_4\text{-}4\text{-}NO_2$ | (II-9) |
| Parathion-Me | dimethyl 4-nitrophenyl phosphorothionate | $(MeO)_2PS \cdot O \cdot C_6H_4\text{-}4\text{-}NO_2$ | (16) |
| Parathion-$Pr^i$ | di-isopropyl 4-nitrophenyl phosphorothionate | $(Pr^iO)_2PS \cdot O \cdot C_6H_4\text{-}4\text{-}NO_2$ | as (II-9) |
| Phosdrin | 1-carbomethoxy-1-propen-2-yl dimethyl phosphate | $(MeO)_2PO \cdot O \cdot CMe{:}CH \cdot CO \cdot OMe$ | (II-87) |
| Phosphinylcholines | P,P-(fluoro)(methyl)phosphinylcholine and analogues | $(Me)(F)PO \cdot O \cdot CH_2 \cdot CH_2 \cdot \overset{\oplus}{N}Me_3$ | (II-18) |
| Potasan | diethyl 4-methylumbelliferonyl phosphorothionate | 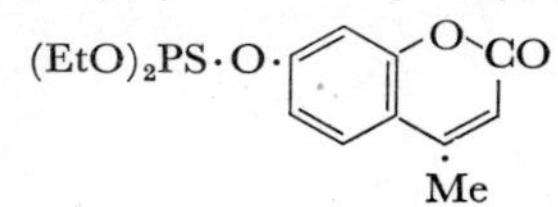 | (17) |
| Ro3-0412 | 3-(dimethoxyphosphinyloxy)-N-trimethylanilinium methosulphate | $(MeO)_2PO \cdot O \cdot C_6H_4\text{-}3\text{-}\overset{\oplus}{N}Me_3\ MeSO_4^{\ominus}$ | (II-99) |
| Ro3-0422 | 3-(diethoxyphosphinyloxy)-N-methylquinolinium methosulphate | 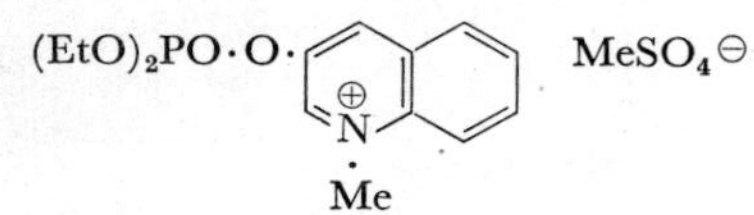 | (II-99) |
| Sarin | isopropyl methylphosphonofluoridate | $(Pr^iO)(Me)PO \cdot F$ | (23) |
| Schradan | octamethylpyrophosphoramide | $(Me_2N)_2PO \cdot O \cdot PO(NMe_2)_2$ | (II-7) |
| Sulfotepp | bis(diethyl phosphorothionic) anhydride | $(EtO)_2PS \cdot O \cdot PS(OEt)_2$ | (II-71, II-72) |
| Systox | mixture of Demeton-O and Demeton-S | | |

SECTION I—*continued*

| COMMON NAME | SYSTEMATIC NAME | STRUCTURAL FORMULA | METHOD |
|---|---|---|---|
| Tabun | ethyl N-dimethylphosphoramidocyanidate | $(EtO)(Me_2N)PO\cdot CN$ | (II-15) |
| TEPP | tetraethyl pyrophosphate | $(EtO)_2PO\cdot O\cdot PO(OEt)_2$ | (II-6) |
| Thimet | O,O-diethyl S-ethylthiomethyl phosphorodithioate | $(EtO)_2PS\cdot S\cdot CH_2\cdot SEt$ | (II-48) |
| Thionotepp | (diethyl phosphoric) (diethyl phosphorothionic) anhydride | $(EtO)_2PS\cdot O\cdot PO(OEt)_2$ | (II-39, II-40) |
| TIPA | N,N′,N″,N‴-tetraisopropylpyrophosphoramide | $(Pr^iNH)_2PO\cdot O\cdot PO(NHPr^i)_2$ | (18) |
| TOCP | tri-2-tolyl phosphate | $PO(O\cdot C_6H_4\text{-2-Me})_3$ | |
| Trithion | S-(4-chlorophenylthiomethyl) O,O-diethyl phosphorodithioate | $(EtO)_2PS\cdot S\cdot CH_2\cdot S\cdot C_6H_4\text{-4-Cl}$ | (19) |
| Trolene | dimethyl 2:4:5-trichlorophenyl phosphorothionate | $(MeO)_2PS\cdot O\cdot C_6H_2\text{-2:4:5-}Cl_3$ | (20) |
| 217-AO | 2-(diethoxyphosphinylthio)-ethyldimethylammonium hydrogen oxalate | $(EtO)_2PO\cdot S\cdot CH_2\cdot CH_2\cdot \overset{\oplus}{N}Me_2$ oxalate | (21) |
| 217-MI | 2-(diethoxyphosphinyl)-thiocholine iodide | $(EtO)_2PO\cdot S\cdot CH_2\cdot CH_2\cdot NMe_3\ I^{\ominus}$ | (21) |

## REFERENCES


1. R. D. O'Brien, C. D. Thorn and R. W. Fisher. *J. econ. Ent.* **51,** 714 (1958)
2. *U.S.* 2,701,259
3. H. R. Kreuger, J. E. Casida and R. P. Niedermeier. *J. Ag. and Fd. Chem.* **7,** 182 (1959)
4. *U.S.* 2,571,989
5. B. C. Saunders and G. J. Stacey. *J. chem. Soc.* 695 (1948)
6. G. Olah, A. Pavlath and G. Hosszang. *Acta chim. Acad. Sci. Hungary* **8,** 41 (1955)
7. S. J. Lonloudes, J. N. Kaplanis and C. C. Roan. *J. org. Chem.* **21,** 685 (1956)
8. *Ger.* 921,870
9. E. J. Honigberg and J. T. Cassaday. *J. Amer. chem. Soc.* **73,** 557 (1951)
10. *Ger.* 917,668
11. *U.S.* 2,666,727
12. *U.S.* 2,758,115
13. *U.S.* 2,843,588
14. *Ger.* 942,988
15. *Brit.* 688,787
16. S. A. Hall. *J. Amer. chem. Soc.* **72,** 2768 (1950)
17. *Brit.* 713,142
18. *Brit.* 688,766
19. *U.S.* 2,793,224
20. *U.S.* 2,599,375
21. L-E. Tammelin. *Acta chem. scand.* **11,** 1340 (1957)
22. *Brit.* 819,735
23. P. J. R. Bryant and A. H. Ford-Moore. *J. chem. Soc.* 1553 (1960)

## Section II
## Other Common Names

| Common name | Systematic name | Structural formula |
|---|---|---|
| Am. Cyanamid 12008 | O,O-diethyl S-isopropylthiomethyl phosphorodithioate | $(EtO)_2PS\cdot S\cdot CH_2\cdot SPr^i$ |
| Am. Cyanamid 12009 | O,O-diethyl S-propylthiomethyl phosphorodithioate | $(EtO)_2PS\cdot S\cdot CH_2\cdot SPr$ |
| Am. Cyanamid 12013 | O,O-di-isopropyl S-isopropylthiomethyl phosphorodithioate | $(Pr^iO)_2PS\cdot S\cdot CH_2\cdot SPr^i$ |
| Am. Cyanamid 12415 | diethyl 6-[3(2H)-pyridazinonyl] phosphorothionate | 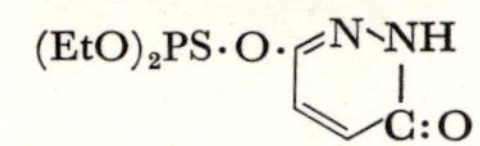 |
| Am. Cyanamid 12433 | dimethyl homologue of above | |
| Am. Cyanamid 12503 | 2-phenyl homologue of Am. Cyanamid 12433 | |
| Am. Cyanamid 18682 | O,O-diethyl S-(N-isopropylcarbamoylmethyl) phosphorodithioate | $(EtO)_2PS\cdot S\cdot CH_2\cdot CO\cdot NHPr^i$ |
| Bayer 28/63 | 2-(diethoxyphosphinylthio)-ethylethylmethylsulphonium methosulphate | $(EtO)_2PO\cdot S\cdot CH_2\cdot CH_2\cdot \overset{\oplus}{S}EtMe\ MeSO_4$ |
| Co-ral-Me | dimethyl homologue of Co-ral | |
| Dow ET-15 | methyl 2:4:5-trichlorophenyl phosphoramidothionate | $(MeO)(H_2N)PS\cdot O\cdot C_6H_2\text{-}2:4:5\text{-}Cl_3$ |
| Ethion | bis(diethoxyphosphinothioylthio)-methane | $(EtO)_2PS\cdot S\cdot CH_2\cdot S\cdot PS(OEt)_2$ |
| Metasystox | mixture of Demeton-O-Me and Demeton-S-Me | |
| Phosphamidon | 1-chloro-1-diethylcarbamoyl-1-propen-2-yl dimethyl phosphate | $(MeO)_2PO\cdot O\cdot CMe:C(Cl)\cdot CO\cdot NEt_2$ |
| Pirazinon | diethyl 4-(2-propyl-6-methyl-pyrimidinyl) phosphorothionate | 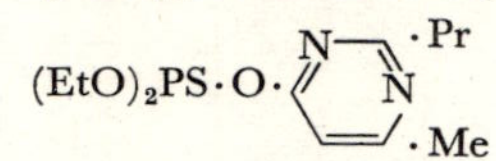 |
| Shell OS 1836 | 2-chlorovinyl diethyl phosphate | $(EtO)_2PO\cdot CH:CHCl$ |
| Soman | 1-methyl-2:2-dimethylpropyl methylphosphonofluoridate | $(Me_3C\cdot CHMe\cdot O)(Me)PO\cdot F$ |

## Section III

## Cross Index of Systematic and Common Names

The index each common name appears in is denoted by I or II.

| Systematic name | Common name |
|---|---|
| 2:3-bis(diethoxyphosphinothioylthio)-1:4-dioxane | Delnav (I) |
| bis(diethoxyphosphinothioylthio)-methane | Ethion (II) |
| bis(diethyl phosphorothionic) anhydride | Sulfotepp (I) |
| bis(N,N′-di-isopropylphosphorodiamidic) anhydride | TIPA (I) |
| bis(tetramethylphosphorodiamidic) anhydride | Schradan (I) |
| S-carbethoxymethyl O,O-diethyl phosphorodithioate | Acethion (I) |
| 1-carbomethoxy-1-propen-2-yl dimethyl phosphate | Phosdrin (I) |
| 1-chloro-1-diethylcarbamoyl-1-propen-2-yl dimethyl phosphate | Phosphamidon (II) |
| 3-chloro-4-methyl-7-coumarinyl diethyl phosphorothionate | Co-ral (I) |
| 3-chloro-4-methyl-7-coumarinyl dimethyl phosphorothionate | Co-ral-Me (II) |
| 2-chloro-4-nitrophenyl dimethyl phosphorothionate | Dicapthon (I) |
| 3-chloro-4-nitrophenyl dimethyl phosphorothionate | Chlorthion (I) |
| S-(4-chlorophenylthiomethyl) O,O-diethyl phosphorodithioate | Trithion (I) |
| 2-chlorovinyl diethyl phosphate | Shell OS 1836 (II) |
| S-(1:2-dicarbethoxyethyl) O,O-dimethyl phosphorodithioate | Malathion (I) |
| S-(1:2-dicarbethoxyethyl) O,O-dimethyl phosphorothiolate | Malaoxon (I) |
| 2:2-dichlorovinyl dimethyl phosphate | DDVP (I) |
| 3-(diethoxyphosphinyloxy)-N-methylquinolinium methosulphate | Ro3-0422 (I) |
| 2-(diethoxyphosphinylthio)-ethyldiethylammonium hydrogen oxalate | Amiton (I) |
| 2-(diethoxyphosphinylthio)-ethyldimethylammonium hydrogen oxalate | 217-AO |
| 2-(diethoxyphosphinylthio)-ethylethylmethylsulphonium methosulphate | Bayer 28/63 (II) |
| 2-(diethoxyphosphinylthio)-ethyl-triethylammonium iodide | Amiton-ethiodide (I) |
| 2-(diethoxyphosphinyl)-thiocholine iodide | 217-MI (I) |
| O,O-diethyl S-2-ethylthioethyl phosphorodithioate | Disyston (I) |
| O,O-diethyl S-2-ethylthioethyl phosphorothiolate | Demeton-S (I) |
| diethyl 2-ethylthioethyl phosphorothionate | Demeton-O (I) |
| O,O-diethyl S-ethylthiomethyl phosphorodithioate | Thimet (I) |
| O,O-diethyl S-(N-isopropylcarbamoylmethyl) phosphorodithioate | Am. Cyanamid 18682 (II) |
| diethyl 4-(2-isopropyl-6-methylpyrimidinyl) phosphorothionate | Diazinon (I) |
| O,O-diethyl S-isopropylthiomethyl phosphorodithioate | Am. Cyanamid 12008 (II) |

## Section III—*continued*

| Systematic name | Common name |
|---|---|
| diethyl 4-methylumbelliferonyl phosphorothionate | Potasan (I) |
| diethyl 4-nitrophenyl phosphate | Paraoxon (I) |
| diethyl 4-nitrophenyl phosphorothionate | Parathion (I) |
| (diethyl phosphoric) (diethyl phosphorothionic) anhydride | Thionotepp (I) |
| diethyl 4-(2-propyl-6-methylpyrimidinyl) phosphorothionate | Pirazinon (II) |
| O,O-diethyl S-propylthiomethyl phosphorodithioate | Am. Cyanamid 12009 (II) |
| diethyl 6-[3(2H)-pyridazinonyl] phosphorothionate | Am. Cyanamid 12415 (II) |
| O,O-di-isopropyl S-isopropylthiomethyl phosphorodithioate | Am. Cyanamid 12013 (II) |
| di-isopropyl 4-nitrophenyl phosphate | Paraoxon-$Pr^i$ (I) |
| di-isopropyl 4-nitrophenyl phosphorothionate | Parathion-$Pr^i$ (I) |
| N,N′-di-isopropylphosphorodiamidic fluoride | Mipafox (I) |
| di-isopropyl phosphorofluoridate | DFP (I) |
| 3-(dimethoxyphosphinyloxy)-N-trimethylanilinium methosulphate | Ro3-0412 (I) |
| dimethyl 1-hydroxy-2:2:2-trichloroethylphosphonate | Dipterex (I) |
| O,O-dimethyl S-(N-methylcarbamoylmethyl) phosphorodithioate | Dimethoate (I) |
| dimethyl 4-nitrophenyl phosphate | Paraoxon-Me (I) |
| dimethyl 4-nitrophenyl phosphorothionate | Parathion-Me (I) |
| O,O-dimethyl S-(4-oxo-1:2:3-benzotriazinyl-3-methyl) phosphorodithioate | Guthion (I) |
| O,O-dimethyl S-(4-oxo-1:2:3-benzotriazinyl-3-methyl) phosphorothiolate | Gutoxon (I) |
| dimethyl 6-[3(2-phenyl)-pyridazinonyl] phosphorothionate | Am. Cyanamid 12503 (II) |
| dimethyl 6-[3(2H)-pyridazinonyl] phosphorothionate | Am. Cyanamid 12433 (II) |
| dimethyl 2:4:5-trichlorophenyl phosphorothionate | Trolene (I) |
| ethyl N,N-dimethylphosphoramidocyanidate | Tabun (I) |
| ethyl 4-nitrophenyl phenylphosphonothionate | EPN (I) |
| S-2-ethylthioethyl O,O-dimethyl phosphorothiolate | Demeton-S-Me (I) |
| 2-ethylthioethyl dimethyl phosphorothionate | Demeton-O-Me (I) |
| P,P-(fluoro)(methyl)phosphinylcholine | phosphinylcholine (I) |
| 1-methyl-2:2-dimethylpropyl methylphosphonofluoridate | Soman (II) |
| methyl 2:4:5-trichlorophenyl phosphoramidothionate | Dow ET-15 (II) |
| octamethylpyrophosphoramide | Schradan (I) |
| tetraethyl pyrophosphate | TEPP (I) |
| N,N′,N″,N‴-tetraisopropylpyrophosphoramide | TIPA (I) |
| tetramethylphosphorodiamidic azide | Mazidox (I) |
| tetramethylphosphorodiamidic fluoride | Dimefox (I) |
| 1:2:2-trimethylpropyl methylphosphonofluoridate | Soman (II) |
| tri-2-tolyl phosphate | TOCP (I) |

## Section IV

## Some Other Names in the Literature

The other name is given first, and then the preferred name in the first two indexes. The Roman numeral in brackets shows in which subsidiary index the preferred name is to be found.

| Names sometimes used | Preferred names | |
|---|---|---|
| Am. Cyanimid 3911 | Thimet | (I) |
| Am. Cyanimid 4124 | Dicapthon | (I) |
| Bayer L13/59 | Dipterex | (I) |
| Bayer 21/199 | Co-ral | (I) |
| Bayer 21/200 | Co-ral-Me | (II) |
| Bayer 22/190 | Chlorthion | (I) |
| Bayer 16259 | Guthion-Et | (II) |
| Bayer 17147 | Guthion | (I) |
| Bayer 19639 | Disyston | (I) |
| DBD | Diazinon | (I) |
| Dow ET/57 | Trolene | (I) |
| DPF | DFP | (I) |
| Dyflos | DFP | (I) |
| E600 | Paraoxon | (I) |
| E605 | Parathion | (I) |
| E838 | Potasan | (I) |
| G-24480 | Diazinon | (I) |
| G-24622 | Pirazinon | (II) |
| GB | Sarin | (I) |
| Gusathion | Guthion | (I) |
| Hanane | Dimefox | (I) |
| Hercules AC-528 | Delnav | (I) |
| Isochlorthion | Dicapthon | (I) |
| Iso-OMPA | TIPA | (I) |
| Isosystox | Demeton-S | (I) |
| M-74 | Disyston | (I) |
| MeE605 | Parathion-Me | (I) |
| Niagara 1240 | Ethion | (II) |
| Nialate | Ethion | (II) |
| OMPA | Schradan | (I) |
| Pestox 3 | Schradan (commercial) | (I) |
| Pestox 14 | Dimefox | (I) |
| Pestox 15 | Mipafox | (I) |
| Phosphostigmines | Ro3-0412 and analogues | (I) |
| Rogor | Dimethoate | (I) |
| Ronnel | Trolene | (I) |
| Shell OS-2046 | Phosdrin | (I) |
| Stauffer R-1303 | Trithion | (I) |
| Tetram | Amiton | (I) |
| Thiometon | Disyston-Me | (II) |

# Subject Index



For trivial names, see Glossary

For trivial names, see Glossary

For trivial names, see Glossary

For trivial names, see Glossary

For trivial names, see Glossary

For trivial names, see Glossary

For trivial names, see Glossary

For trivial names, see Glossary

# Formula Index

Only those compounds which are referred to at some point by formula alone are listed. The formula for each is given in the same way as in the text, except for a few obvious simplifications. For most compounds the systematic name is given also, and, where different, the abbreviation under which it is referred to in the main index. The abbreviation is italicized. For those compounds with trivial names listed in Part I of the Glossary, trivial names only are given.

—$CH_2 \cdot CH_2 \cdot \overset{\oplus}{N}Me_3$ ($I^{\ominus}$) 217-MI
—$CH_2 \cdot CH_2 \cdot SEt$ Demeton-S
—$CH_2 \cdot CH_2 \cdot \overset{\oplus}{S}Et_2$ 2-(diethoxyphosphinylthio)-ethyldiethylsulphonium ion *sulphonium cpds*
—$CH_2 \cdot CH_2 \cdot \overset{\oplus}{S}Et \cdot CH_2 \cdot CH_2 \cdot SEt$ 2-(diethoxyphosphinylthio)-ethyl(2-ethylthioethyl)ethylsulphonium ion, *sulphonium cpds*
—$CH_2 \cdot CH_2 \cdot \overset{\oplus}{S}EtMe$ 2-(diethoxyphosphinylthio)-ethylethylmethylsulphonium ion *sulphonium cpds*
—$CH_2 \cdot CH_2 \cdot SO \cdot Et$ diethyl S-2-ethylsulphinylethyl phosphorothiolate
—$CH_2 \cdot CH_2 \cdot SO_2 \cdot Et$ diethyl S-2-ethylsulphonylethyl phosphorothiolate
—$C_6H_4$-4-$NO_2$ diethyl S-4-nitrophenyl phosphorothiolate
—$(CH_2)_3 \cdot SEt$ diethyl S-3-ethylthiopropyl phosphorothiolate, *sulphides, side-chain*
—$(CH_2)_3 \cdot \overset{\oplus}{S}EtMe$ 3-(diethoxyphosphinylthio)-propylethylmethylsulphonium ion, *sulphonium cpds*
—$C_6H_4$-4-SMe diethyl S-4-methylthiophenyl phosphorothiolate, *sulphides, side-chain*
—$CH_2 \cdot SO \cdot Et$ diethyl S-ethylsulphinylmethyl phosphorothiolate, *sulphoxides, side-chain*
—$CH_2 \cdot SO_2 \cdot Et$ diethyl S-ethylsulphonylmethyl phosphorothiolate, *sulphones, side-chain*
—$C_6H_4$-4-$SO \cdot Me$ diethyl S-4-methylsulphinylphenyl phosphorothiolate, *sulphoxides, side-chain*

$EtO \cdot PS(O \cdot C_6H_4$-4-$NO_2)_2$ di-4-nitrophenyl ethyl phosphorothionate

$(EtO)_2PS \cdot Cl$ diethyl phosphorochloridothionate, *chloridates*

$[(EtO)_2PS]_2O$ Sulfotepp

$(EtO)_2PS \cdot O$—$CH_2 \cdot CH_2 \cdot NEt_2$ 2-(diethoxyphosphinothioyloxy)-ethyldiethylamine
—$CH_2 \cdot CH_2 \cdot SO \cdot Et$ diethyl 2-ethylsulphinylethyl phosphorothionate, *Demeton-O analogues*
—$CH_2 \cdot CH_2 \cdot SO_2 \cdot Et$ diethyl 2-ethylsulphonylethyl phosphorothionate, *Demeton-O analogues*
—$C_6H_4$-4-$NO_2$ Parathion
—$C_6H_4$-4-SMe diethyl 4-methylthiophenyl phosphorothionate
—$C_6H_4$-4-$SO \cdot Me$ diethyl 4-methylsulphinylphenyl phosphorothionate
—$PO(OEt)_2$ Thionotepp

$(EtO)_2PS \cdot S$—$CH_2 \cdot CH_2 \cdot CO \cdot NH_2$ S-2-carbamoylethyl diethyl phosphorodithioate, *Acethion analogues*
—$CH_2 \cdot CH_2 \cdot CO \cdot OEt$ S-2-carbethoxyethyl diethyl phosphorodithioate, *Aethionc analogues*
—$CH_2 \cdot CHMe \cdot CO \cdot OEt$ S-1-(2-carbethoxy)propyl diethyl phosphorodithioate, *Acethion analogues*
—$CH_2 \cdot CH_2 \cdot SO \cdot Et$ diethyl S-2-ethylsulphinylethyl phosphorodithioate, *Disyston analogues*
—$CH_2 \cdot CH_2 \cdot SO_2 \cdot Et$ diethyl S-2-ethylsulphonylethyl phosphorodithioate, *Disyston analogues*
—$CH_2 \cdot CO \cdot OH$ (diethoxyphosphinothioylthio)acetic acid, *Acethion analogues*
—$CH_2 \cdot CO \cdot Me$ (diethoxyphosphinothioylthio)acetone, *Acethion analogues*

—NHMe pentamethylphosphoramide, *phosphoramides*
—NHPr$^i$ isopropyltetramethylphosphoramide, *phosphoramides*
—$NMe_2$ hexamethylphosphoramide
—OEt ethyl tetramethylphosphorodiamidate, *phosphoramides*
—OH hydrogen tetramethylphosphorodiamidate
—O·$PO(NMe_2)_2$ Schradan

$(Me_2N)(Pr^iO)PO\cdot CN$ isopropyl dimethylphosphoramidocyanidate, *cyanidates*

(MeO)(Me)PO·F methyl methylphosphonofluoridate, *fluoridates*

(MeO)(MeS)PO·S·$CH(CH_2\cdot CO\cdot OEt)\cdot CO\cdot OEt$ S-(1:2-dicarbethoxyethyl) O,S-dimethyl phosphorodithioate, *Malathion, isomerization.*

$(MeO)_2PO$—AChE (—E or —ΨChE) dimethylphosphorylated AChE, etc., *phosphorylated ChE's*
—$CH(O\cdot CO\cdot Pr^i)\cdot CCl_3$ dimethyl 1-butyroxy-2:2:2-trichloroethylphosphonate, *butyrylated Dipterex*
—Cl dimethyl phosphorochloridate, *chloridates*
—F dimethyl phosphorofluoridate, *fluoridates*

$(MeO)_2PO\cdot O$—$CH:CCl_2$ DDVP
—$C_6H_2$-2:4:5-$Cl_3$ dimethyl 2:4:5-trichlorophenyl phosphate, *Trolene, phosphate analogue*
—$C_6H_4$-4-$NO_2$ Paraoxon-Me
—$C(OMe):CCl_2$ 2:2-dichloro-1-methoxyvinyl dimethyl phosphate, *vinyl cpds*
—$PO(OMe)_2$ tetramethyl pyrophosphate

MeO·$PO(O\cdot C_6H_4$-2-$NO_2)_2$ di-4-nitrophenyl methyl phosphate

$(MeO)_2PO\cdot S$—$CH_2\cdot CH_2\cdot \overset{\oplus}{S}Et\cdot CH_2\cdot CH_2\cdot SEt$ 2-(dimethoxyphosphinylthio)-ethyl-(2-ethylthioethyl)ethylsulphonium ion, *sulphonium cpds*
—$CH_2\cdot CH_2\cdot \overset{\oplus}{S}EtMe$ 2-(dimethoxyphosphinylthio)-ethylethylmethlysulphonium ion, *sulphonium cpds*
—$CH_2\cdot CH_2\cdot SO\cdot Et$ dimethyl S-2-ethylsulphinylethyl phosphorothiolate
—$CH_2\cdot CH_2\cdot SO_2\cdot Et$ dimethyl S-2-ethylsulphonylethyl phosphorothiolate
—$CH_2\cdot CO\cdot NHMe$ dimethyl S-(N-methylcarbamoyl)methyl phosphorothiolate, *Dimethoate, oxygen analogue*
—$C_6H_4$-4-$NO_2$ dimethyl S-4-nitrophenyl phosphorothiolate

$(MeO)_2PS\cdot O\cdot CH_2\cdot CH_2\cdot SEt$ Demeton-O-Me

$(MeO)_2PS\cdot O\cdot C_6H_4$-4-$NO_2$ Parathion-Me

MeO·$PS(O\cdot C_6H_4$-2-$NO_2)_2$ di-2-nitrophenyl methyl phosphorothionate

$Me_2PO\cdot F$ dimethylphosphinic fluoride, *fluoridates*

MePO(F)·O—$CH_2\cdot CH_2\cdot \overset{\oplus}{N}Me_3$ 2-P,P-(fluoro)(methyl)phosphinylcholine
—$CHMe\cdot CH_2\cdot \overset{\oplus}{N}Me_3$ 2-P,P-(fluoro)(methyl)phosphinyl-2-methylcholine, *as above*
—$(CH_2)_3\cdot \overset{\oplus}{N}Me_3$ 3-P,P-(fluoro)(methyl)phosphinylhomocholine, *as above*

MeS·$PO(O\cdot C_6H_4$-2-$NO_2)_2$ di-2-nitrophenyl S-methyl phosphorothiolate